Professional Writing and Rhetoric

Professional Writing and Rhetoric

READINGS FROM THE FIELD

• • • • •

Tim Peeples
Elon University

Longman

New York San Francisco Boston
London Toronto Sydney Tokyo Singapore Madrid
Mexico City Munich Paris Cape Town Hong Kong Montreal

Senior Vice President and Publisher: Joseph Opiela
Acquisitions Editor: Lynn M. Huddon
Marketing Manager: Christopher Bennem
Production Manager: Douglas Bell
Project Coordination, Text Design, and Electronic Page Makeup:
Elm Street Publishing Services, Inc.
Senior Cover Designer/Manager: Nancy Danahy
Manufacturing Buyer: Lucy Hebard
Printer and Binder: The Maple-Vail Book Manufacturing Group
Cover Printer: Phoenix Color Corporation

Library of Congress Cataloging-in-Publication Data

Peeples, Tim.
Professional writing and rhetoric: readings from the field / Tim Peeples.
p. cm.
Includes index.
ISBN 0-321-09975-3
1. College readers. 2. Business writing—Problems, exercises, etc. 3. Technical writing—Problems, exercises, etc. 4. English language—Rhetoric—Problems, exercises, etc. 5. English language—Business English—Problems, exercises, etc. 6. English language—Technical English—Problems, exercises, etc. 7. Readers—Rhetoric. I. Title.
PE1479.B87 P44 2003
808'.0666—dc21 2002031279

Please visit our website at http://www.ablongman.com

ISBN 0-321-09975-3

3 4 5 6 7 8 9 10—MA—05 04 03

CONTENTS

PART 4: BECOMING A PROFESSIONAL WRITER 385

Preface

Professional Writing and Rhetoric: Readings from the Field functions as a disciplinary reader for undergraduate and entry-level master's students who are interested in pursuing studies and/or careers in professional writing. It introduces this student audience to professional writing by inviting them into conversations *about* the field by people *in* the field and, thus, fills an important gap in the books currently available within professional writing studies.

Most contemporary professional writing texts fall into one of four categories:

- how-to-oriented rhetorics generally used in service-oriented business, technical, and advanced writing-in-the-majors courses
- edited collections presenting new research and scholarship and written for an audience of advanced graduate students and professors
- advanced topic texts advancing the field's cutting edge and targeted, again, at advanced graduate students, professors, and experienced professional writers
- pedagogically and/or curricularly oriented texts discussing issues in course and program design and directed primarily toward professors

An edited reader, *Professional Writing and Rhetoric* addresses a growing need as the field expands "up" from service-oriented courses and "down" from advanced graduate programs: it offers a text that makes the field's theoretical discussions accessible to undergraduates and entry-level master's students who are majoring, minoring, or getting certificates in professional writing studies.

Professional Writing and Rhetoric guides students into the discussions that continue to form this relatively young field by (1) organizing readings rhetorically, (2) including several readings that are regularly cited in the field's literatures, (3) selecting readings that are accessible to the target population, and (4) offering pedagogical devices that aid comprehension and encourage critical reflection. The aim is not to present a "greatest hits of the field," nor to direct students' thinking and practice toward the hottest new theories, nor to challenge the thinking of those already comfortably in the field. Instead, older and newer selections are intermixed within a rhetorical framework in order to establish disciplinary background, encourage students to make connections across readings, promote reflective rhetorical practice, stimulate discussion, and encourage students to become co-inquirers within the discipline. The aim is to guide new students of professional writing and rhetoric into the disciplinary conversations with which professors and advanced graduate students are typically quite comfortable, but that once challenged and excited all of us and will still challenge and excite those new to the field.

In the past fifteen years or so, the field of professional writing has taken a strong rhetorical turn. Sometimes the characterization of that turn is reduced to "heightened audience awareness." Though certainly a central feature of the rhetorical turn, being aware of one's audience is only one feature of an integrated field of professional writing and rhetoric. Understanding the relationships between rhetoric and professional writing requires much more rigorous study than the simplified "be aware of your audience" reflects. *Professional Writing and Rhetoric* actively integrates rhetoric and professional writing in a way that encourages students to critically engage with what it means to study and practice professional writing and rhetoric.

ACKNOWLEDGMENTS

Despite the single name on the cover, this book has been the work of many hands. I would first like to thank Lynn Huddon. Without Lynn there would be no book. Through her awareness of and commitment to the field, interest in the potential of such a project, perseverance, encouragement, and ongoing expert guidance, an idea grew into a book. Kristi Olson, who played an important role in keeping the project moving along smoothly, also deserves a good deal of thanks. More locally, Lynn Melchor deserves a great deal of thanks in the assistance she offered locating the sources I needed to research and produce this book. Along the way, a wide variety of anonymous reviewers helped guide significant revisions in terms of individual selections, whole chapters, and pedagogical supports, always with a careful eye to disciplinary, course, programmatic, and student needs. Thanks to all, who, together, gave voice to the field: Stephen A. Bernhardt, University of Delaware; Thomas Deans, Kansas State University; Paul Dombrowski, University of Central Florida; Kay Harley, Saginaw Valley State University; William Hart-Davisdon, Rensselaer Polytechnic Institute; Carl G. Herndl, Iowa State University; Margaret N. Hundleby, Auburn University; Kathy Hurley, Minnesota State University–Mankato; Jeff Jablonski, University of Nevada, Las Vegas; Barry Maid, Arizona State University East; Celia Patterson, Pittsburg State University; Rachel Spilka, University of Wisconsin–Milwaukee; Dale Sullivan, University of Minnesota, Twin Cities; Barry Thatcher, New Mexico State University; and Patricia Wojahn, New Mexico State University.

A special group of colleagues, who have been part of this book with varying degrees of direct contact and interaction, also deserve my thanks: Jeff Jablonski and Graham Smart, friends and colleagues who offered helpful advice and encouraging words early in the project; Jim Porter, Pat Sullivan, Johndan Johnson-Eilola, Bud Weiser, Janice Lauer, Victor Villanueva, and Sharon Crowley, professors who played an integral part in guiding me into the conversations of the field as a co-inquirer and modeled what I hope to be for my students; and all my students in the Professional Writing and Rhetoric concentration and Professional Writing Studies minor at Elon University, who continually convince me that the work I and many of us are doing to reclaim and reshape rhetorical studies at the undergraduate level is valuable. Among those colleagues offering helpful guidance along the way, I want particularly to thank Bill Hart-Davidson, longtime friend and collaborator. This book represents but a moment—though an important one—in an enthusiastic inquiry into and commitment to rhetoric and writing that he and I have engaged, and will continue to engage, for a long time.

Most of all, though, I give my unending thanks to Margaret and Leah, dearest co-authors with whom I daily write the most important texts of all.

Tim Peeples

Professional Writing and Rhetoric

Introduction

No matter why you have chosen to pick up this book, *Professional Writing and Rhetoric* will help you approach professional writing with greater rhetorical awareness, sensitivity, and effectiveness. It does so through the voices of professionals in the field. There are any number of books to choose from if you are looking for a summary of issues in the field, or if you are looking for an introductory handbook to professional writing. These books are excellent choices if you are looking for summaries and handbooks. But if you are interested in becoming part of the disciplinary and professional conversations of the field, *Professional Writing and Rhetoric* is a good choice. The readings in this collection are written *by* professional writers *for* professional writers. Together, they form a framework for making sense of professional writing as a field, and thus they help you become part of the field's conversations.

You might worry that such a selection of readings will go over your head. Though I am confident you will find most of the readings challenging and engaging, I am also confident you will not find yourself lost. Overall, the readings have been carefully chosen and reviewed by other professionals in the field to give you an introduction that offers appropriate introductory breadth and depth. This does not mean you will find the readings easy, nor that the readings will provide you clear, uncomplicated answers to questions and issues professional writers face. On the contrary, the aim of the readings, individually and collectively, is to invite you into the ongoing conversations that make up the field of professional writing. Such conversations are necessarily complex and difficult, as is the work of professional writers. Still, the readings have been chosen, in part, because they make up an appropriate introduction for someone relatively new to the field.

In addition to reading, you are invited into this professional conversation through a variety of writing activities. Before each reading, you will find a list of terms and concepts. You are encouraged to define these terms and concepts as you read, as a way to become more comfortable with some of the jargon that defines the discipline. At the end of each reading, you will find a list of questions designed to help you make sense of the reading you have just completed. Some of the questions will help you increase your comprehension of the articles, but many others direct you to produce documents and presentations that a professional writer might be asked to produce. Through both kinds of writing activities, you become more conversant with the field. Finally, at the end of each chapter, you will find projects that help you make connections across the readings of each chapter. Most of these projects give you a chance to produce the kinds of oral and/or written documents a professional writer might produce, and a number of them will also require you to apply what you have learned from the readings. In all of these ways, *Professional Writing and Rhetoric* helps you enter the conversation of the field no matter what your past experience has been.

GETTING ORIENTED: *PROFESSIONAL WRITING AND RHETORIC* RESPONDS TO BROAD ISSUES IN THE FIELD

Professional Writing and Rhetoric does not pretend to be an objective summary of the field. Like the readings that make up its contents, this book is a part of the field's ongoing conversations. Though the book attempts to be as representative as possible of the issues that define the field, it is still a response to other conversations. In general, *Professional Writing and Rhetoric* responds to three dominant binaries or dichotomies with which not only professional writing but also higher education in general wrestles: practice vs. theory, production vs. practice, and school vs. work.

Practice vs. Theory

Professional writing courses, often carrying titles like Technical Writing, Writing for the Professions, and Business Communication, have traditionally emphasized practice over theory. Perhaps this is so because of their historical growth, developing largely out of requests from engineering and business schools to improve the communication skills of their students. Instruction in these courses focused almost entirely on *how-to* knowledge often taught as acontextual rules or generic forms. Students would practice writing memos, reports, instructions, etc., and even if these assignments were contextualized within cases, assignments routinely asked students to respond in formulaic ways (e.g., "Based on X case, write an effective bad news letter"). The aim of the course work was to give students *practice* in writing workplace documents. In such a course, it made little sense to introduce students to theoretical discussions surrounding effective communication because formalized practice requires no theorizing from the writer. By its nature, formalized practice simply requires that you master a variety of generalized response structures and skills and then practice adapting them to particular communicative situations.

A growing number of professional writing courses, though, have begun introducing students to theories of writing and rhetoric. This change has occurred for a variety of reasons more numerous and complex than can be fully explored here. For instance, the change has been driven by pedagogy as course work in professional writing classes has become more contextualized in cases and especially client-based projects. Placed within specific rhetorical situations, students (and the instructors facilitating students' learning) discover that formulaic responses are rarely effective. Writers find themselves asking questions that require them to theorize on the spot about what defines effective communication in the particular scenarios within which they find themselves. In order to engage in this context-specific action—the theorizing required of effective writers—students must have some familiarity with theoretical conversations in writing and rhetoric.

The increased need to introduce students to theories of writing and rhetoric has also been driven by contextual changes in workplace writing. For instance, along with the rapid growth in technology, came an increased demand for writers and people capable of examining issues from rhetorical/communicative perspectives. More and more writers were needed to develop the documents that would help people and machines work together. That is, as technological development exploded, there arose a great demand for communicators who could help people use these technologies and also help product developers design technologies that best met user needs. These writers needed to do more than simply adapt basic rhetorical principles and generic forms to new situations. They needed theories of writing and rhetoric to help them work through the communicative issues they faced in a host of new contexts.

At the same time that the technological explosion was creating these needs for writers, it was also creating an explosion in communication media. The old job title of "technical writer" has exploded right along with new media, creating a variety of new job titles, like Web author, document designer, information engineer, electronic publications manager, and human-centered designer. The expanded role of the professional writer caused, to a great extent, by technological growth required the writer to be able to theorize a whole new set of issues. These include issues like how are the visual and textual interrelated, how do dynamic texts like Web pages and databases affect how we write, and how is the collaborative writing process common in the workplace best managed when it is done largely across computer networks. These, again, were not issues where practice alone could prepare writers. The context encouraged the inclusion of more theory into professional writing courses.

But theory alone is not enough. *Professional Writing and Rhetoric* assumes that theory and practice should not be separated from one another: good practice requires theoretical knowledge, and good theorizing is not only itself a practice but it also requires an awareness of and responsiveness to practice. Many of the readings in this collection are theoretical, but the assignments and projects accompanying them direct you to make connections across theory and practice. *Professional Writing and Rhetoric* assumes that making such connections—from theories to practices and from practices to theories—greatly defines professional writing expertise. It is the ability to *theorize* effectively within particular rhetorical situations that makes a professional writer truly expert.

Production vs. Practice

The dominant perception of writing in both lay populations as well as scholarly ones considers writing an art of production, or a way of making texts. This perception becomes obvious in the ways that writers are defined. When someone asks a writer, "What do you do?" writers often respond with statements like, "I write grants," or "computer instructions," or "scientific articles." And at times when people meet a writer, they often react with such questions as "Oh, so do you write novels?" These examples illustrate that writers are often defined by the products they produce.

What gets lost in this definition is all of the *activity* that surrounds the production of texts, the very *social activity* that writing requires. Being a professional writer requires much more than simply sitting alone at a desk crafting neat sentences. But because of the ways writing is over-defined by its products—documents—it is often misunderstood as this solitary act of textual production. Some kinds of writing lend themselves to more solitary activity, for sure, but even in such cases, there is a great deal of (inter)action. The writer interacts with language, which is socially invented. The writer's text interacts with previous, current, and future texts both written and spoken, a relationship often called intertextuality. The writer is even interacting with readers, though they may be referred to as invoked or imagined. So, even in what looks like the most solitary situation, writers are constantly interacting. Especially in workplaces, interactions are much more visible and physical. Writers make calls, talk with others, observe people interacting with texts and products, gather to meet in conference rooms both physical and virtual, test the effectiveness of their documents, manage others involved with document production, and interact with a variety of communication technologies. Perceived this way, the work of writing clearly extends beyond its textual products, correcting the misperception that writing is a solitary act of textual production.

An exclusive focus on the products of writing not only hides the social interaction that is integral to writing, but it also clouds the nature of texts or documents as forms of social action or means by which we mediate social interaction. When we write, we are choosing one medium of action. We could stand up and shout, or we could speak, or we could move, or we could act in a variety of other ways. But we choose to act through writing. To understand this nature of texts, we could consider a document like a summons. Is a summons just an isolated product that takes on no action in the world? Obviously not. It is a text that takes up the action of summoning another party to act in a particular fashion. The document is a form of social action itself, apart from the writer. Also, no matter how beautifully a summons might be written—that is, no matter how wonderful the "product"—if it fails to summon, to act, then it certainly cannot be considered a successful document. Though product features are important to defining good writing, they cannot fully define what it means to write and write well.

Professional Writing and Rhetoric takes the position that writing is both practice and production. There is no doubt that when we write, we produce texts; and there is a great deal that writers must know about the arts of production, or *technē*. This book addresses production in Part 3, focusing on issues of production that are both particular to professional writing and often overlooked as products of writers' work. But, *Professional Writing and Rhetoric* also assumes that writing is a form of social action. As such, it happens within social contexts and has ethical consequences. Part 2 focuses on professional writing as a form of social practice, examining the contextual and ethical issues particular to the field.

School vs. Work

Like all forms of professional education, the field of professional writing often faces questions about the role course work plays in the development of the writer, as well as the role course work plays in the definition of what professional writing is. If we were to create two poles that define this binary, on one side would be the statement that school should train writers for the workplace and on the other side would be the statement that the workplace ought to reflect what is taught in school.

Professional Writing and Rhetoric takes the position that neither side of this binary is possible nor preferable. School cannot reproduce workplace contexts and, thus, "train" students for workplace writing. At the same time, the workplace reflects a very different context than school and, thus, carries with it a different set of values and purposes. As a result, the workplace cannot simply reflect school. Still, these differences do not mean that school and work are completely unrelated.

A majority of the readings illustrate how rhetorical reasoning interacts with and becomes a part of organizational contexts and scenarios. In no instance will you find one of the following happening: (1) the workplace has a ready-made answer to which the writer adapts or (2) school offers a ready-made answer that the writer simply imports into the workplace. What you will find is that workplace cultures and contexts both exert influence on and are influenced by the rhetorical knowledge writers bring with them from other contexts, including school. This interaction is exciting. It means that what you learn in school can help you shape the way writing and work get done in the workplace. It also means that what you and others learn and experience in the workplace, through internships, for instance, have a significant impact on formal education.

UNDERSTANDING PROFESSIONAL WRITING AS ORGANIZATIONALLY SITUATED AUTHORSHIP

As you can see from even the brief discussion that preceded, professional writing is much more than simply transferring writing skills from school to the workplace. Professional writing is a complex rhetorical act that, if done with expertise, requires a writer to theorize within a wide variety of rhetorical situations. The readings collected in *Professional Writing and Rhetoric*, along with their accompanying assignments and projects, come together to formulate a definition of professional writing as what might be called "organizationally situated authorship," a definition that is meant both to encourage a view of professional writing as rhetorical and to capture the breadth of the professional writer's role.

The three terms that make up the phrase "organizationally situated authorship" are chosen with care and complexly interrelated. However, they require careful "unpacking." For instance, in what ways are the professional writer and the texts produced by the writer organizationally situated? Is there a single, concrete situation that defines the context of the professional writer's work, or might there be multiple situations that are concrete, imaginative, and/or virtual? What is the difference between "authorship" and "situated authorship"? Why use "authorship" at all? In what ways are professional writers understood and treated differently when they are referred to as (and refer to themselves as) authors instead of writers? If professional writers are authors, what kinds of things do they author? These are just a few of the questions the phrase "organizationally situated authorship" provokes.

As stated earlier, *Professional Writing and Rhetoric* is itself a part of the ongoing conversations that are creating the field of professional writing. As part of these ongoing conversations, the book indirectly interjects the following question: "Does the concept of 'organizationally situated authorship,' if understood broadly, effectively capture the scope of the professional writer's work?" In addition to exploring the wide variety of questions and projects you will find throughout this book, return repeatedly to this question that the whole book poses. Throughout the readings, introductions, assignments, and projects, *Professional Writing and Rhetoric* invites you to take an active role in exploring and shaping the field. By repeatedly returning your thoughts and class discussion to this one overarching question, you are perhaps exploring one of the most crucial questions for yourself and the field: Who are you as a professional writer?

PART 1

Defining the Field

Chapters 1 and 2 have been designed to answer some of the primary questions you might have as you begin studying professional writing and rhetoric. The later chapters examine finer issues related to the field, but to prepare you to engage in these finer issues, Part 1 introduces some of the main issues that set the framework for studying the field of professional writing and rhetoric.

The first questions you might have are "What do professional writers do?" and "Where do they work?" The range of contexts in which professional writers might work is extensive. Any list would be incomplete, but the contexts include business, engineering, computer industries, environmental sciences, medicine and health care, government, social service, nonprofit organizations, advertising, marketing, publishing, and graphic design. Even this list is woefully incomplete. The task of introducing you to what professional writers do and where they work, though, exceeds the focus and length of this book. If you want to explore these questions further, you can find them in several fine books, such as *The Practice of Technical and Scientific Communication: Writing in Professional Contexts* (Jean A. Lutz and C. Gilbert Storms), *Writing a Professional Life: Stories of Technical Communicators On and Off the Job* (Gerald J. Savage and Dale L. Sullivan), and *Careers for Writers & Others Who Have a Way with Words* (Robert W. Bly). You may find one or more of these books a helpful supplement to this text.

Once you have a good sense of where professional writers work and what they do, which many of you probably already have, you will probably wonder, "What is rhetoric?" Chapter 1 serves as an introduction to rhetoric. For those who do not have a background in this ancient discipline, the chapter gives you a brief glimpse. It is assumed, though, that you have already had a course or several courses in rhetoric or that your course instructor will supplement what is presented in Chapter 1 with further instructions and possibly readings. In the first case, Chapter 1 serves as a brief warm-up to get you reflecting and talking. In the second case, Chapter 1 sets a foundation upon which your instructor can build. In either case, Chapter 1 introduces you to some of the key rhetorical issues that are discussed throughout this book.

Chapter 2 takes up the question, "What is the relationship between professional writing and rhetoric?" Implicitly, this is where *Professional Writing and Rhetoric* constructs a framework for understanding the work of professional writers as "organizationally situated authorship." Rhetoric has long studied and strategized speaking and writing within specific contexts or rhetorical situations. When professional writing and rhetoric are brought together, rhetoric's specialized knowledge of authoring within specific rhetorical situations is carried, more generally, into the organizational contexts in which professional writers work. Chapter 2 examines how this joining of rhetoric and professional writing affects the ways we understand professional writing (and even rhetoric).

CHAPTER 1

What Is Rhetoric?

INTRODUCTION

Having picked up a book on professional writing *and rhetoric*, you are most likely wondering, "What is rhetoric?" Most people have heard the term used in public discourse, but from what we hear on the news and in the newspapers, rhetoric typically carries negative connotations. When we recall the term being used, we recall politicians whose "mere rhetoric" is just a bunch of smoke and mirrors to misdirect us. Is this the extent of what rhetoric is?

Together, the readings in Chapter 1 focus on the following questions, helping you make better sense of rhetoric:

- How is rhetoric defined?
- What fields is it related to?
- What is its history?
- What is the purpose or function of rhetoric?
- What is its scope, or what kinds of issues does it address?

It's no doubt that a small selection of readings cannot adequately introduce to you a discipline like rhetoric, a discipline with such a long and dynamic history. But these readings can help you get a taste of its breadth, scope, history, and function.

You may still be wondering, though, "Why should I be concerned about rhetoric at all?" In order for you to understand how professional writing and rhetoric are related, you must first understand what rhetoric is on its own terms, which are quite extensive and complicated.

One issue you should pay attention to as you read the selections in Chapter 1 is the relationship between rhetoric, philosophy, ethics, and politics. Is rhetoric completely separate from these other disciplines? If so, is it possible and appropriate for a rhetor—one who practices rhetoric, either through speaking or writing—to employ rhetoric without any concern for ethics? Or, is it possible and effective for a rhetor to employ rhetoric without considering the political context? Looking at it in a different way, if rhetoric is a separate discipline entirely, does the definition, scope, and practice of rhetoric remain constant despite ethical, political, and philosophical changes between cultures and across histories, or is rhetoric rhetoric?

These issues may seem "merely academic" to you as they are presented here, but in professional writing contexts, they become "very real." For instance, let us imagine that you are a professional writer for a cigarette company and you have been asked to be the lead writer on a report arguing that studies on cigarette smoking show no direct relationship between smoking and heart disease. In one of the first meetings you have with your supervisors, you ask about some of the studies you have seen in- and out-of-house that seem to contradict the message you are being asked to forward. Your supervisors off-handedly discount those sources and rather abruptly suggest that your job is to "write an effective report," not to initiate your own research into the matter. In this scenario do you see any overlap between rhetoric/writing and ethics? Or do you see these as two separate disciplines? Can you "write an effective report" without addressing ethical issues within the scenario? By examining the nature of rhetoric, Chapter 1 helps you to begin formulating responses to these and many other challenging questions that arise in professional writing and rhetoric.

Rhetoric is not a discipline that has stood still over time. Formalized in ancient Greece, rhetoric has had a dynamic history. Rhetoricians—those who study and teach rhetoric—have long argued what rhetoric is, what knowledge it requires, and how one gains such knowledge. The first reading by Foss, Foss, and Trapp introduces you to some of the history of this dynamic discipline.

The second reading is the opening to Aristotle's *Rhetoric*, arguably one of the most influential classical theories of rhetoric. In his typical way of analyzing and categorizing objects of study, Aristotle defines quite clearly the scope and function of rhetoric. At the same time that his categorizing clearly marks a field of study and practice, it also raises questions about the effectiveness of such boundary setting. The third reading, also from Aristotle but from his *Nicomachean Ethics*, gives you a foundation for exploring the distinction Aristotle makes between science, art, and practical wisdom. It has been vigorously debated whether rhetoric is a science, art, or practice. Some of the fuel for this debate is found in the first line of Aristotle's *Rhetoric*, the second reading in this chapter, and briefly summarized in the introductory pages coming before George Kennedy's translation of Aristotle's text. As a student of professional writing and rhetoric, you need to explore where you stand on the categorization of rhetoric, for where you categorize rhetoric has a great impact on the scope, function, and practice of writing and rhetoric.

The final reading in the chapter comes from Roman times and is often attributed to Cicero, though this connection is not certain and even argued by many to be unfounded. This piece is interesting and important for a variety of reasons. For one, it is interesting to see how Aristotle's categorizing of rhetoric changes over time. Do you see the boundaries getting more or less sharp, for instance? It is also interesting because reading both Aristotle's *Rhetoric* and *Rhetorica Ad Herennium* gives us some comparative historical perspective on what rhetoric *is* or *becomes* in different cultures and historical contexts. This is an issue raised in the first reading in the chapter by Foss, Foss, and Trapp. This reading also introduces a clear discussion of the five canons of rhetoric: invention, arrangement, style, memory, and delivery. These are just a few of the key rhetorical issues and conversations that you must be aware of and wrestle with in order to effectively enter the conservations in the field.

FOCUSING ON KEY TERMS AND CONCEPTS

Focus on the following terms and concepts while you read through this selection. Understanding these will not only increase your understanding of the selection that follows, but you will find that, because most of these terms or concepts are commonly used in professional writing and rhetoric, understanding them helps you get a better sense of the field itself.

1. sophists
2. second sophistic
3. rhetorical canons
4. epistemological rhetoric
5. belletristic rhetoric
6. elocutionary rhetoric

Perspectives on the Study of Rhetoric

SONJA K. FOSS, KAREN A. FOSS,
ROBERT TRAPP

When we hear the word "rhetoric" used today, the meaning frequently is pejorative. More often than not, it refers to talk without action, empty words with no substance, or flowery, ornamental speech. A typical use of the term occurred at one point during the Iranian hostage crisis. When Iranian authorities asserted that the hostages might have been released from the embassy had the deposed shah of Iran remained in Panama to face extradition proceedings, a senior White House aide responded to these assertions by saying, "that sort of promise is little more than rhetoric from people who have made commitments in the past and who have been unwilling or unable to keep those commitments."[1]

Rhetoric should not engender, however, only negative connotations for us. In the Western tradition, rhetoric has a long and distinguished history as an art dating back to classical Greece and Rome. Although our focus in this book is on contemporary treatments of rhetoric, we will begin with a general overview of the rhetorical tradition. We hope this brief review will dispel the disparaging meanings associated with the term "rhetoric" and provide a foundation for understanding the contemporary perspectives explored in later chapters.

A BRIEF HISTORY OF RHETORICAL THOUGHT

The art of rhetoric is said to have originated in the fifth century B.C. with Corax of Syracuse. A revolution on Syracuse, a Greek colony on the island of Sicily, in about 465 B.C., was the catalyst for the formal study of rhetoric. When the tyrannical dictators

Source: Reprinted by permission of Waveland Press, Inc. from Sonja K. Foss, Karen A. Foss, and Robert Trapp, "Perspectives on the Study of Rhetoric," *Contemporary Perspectives on Rhetoric,* 1985, pp. 1–10. (Prospect Heights, IL: Waveland Press, Inc., 2002).

[1] Terence Smith, "U.S. Aides Discount Teheran Rhetoric," *New York Times*, March 25, 1980, p. 9.

on the island were overthrown and a democracy was established, the courts were deluged with conflicting property claims: was the rightful owner of a piece of land its original owner or the one who had been given the land during the dictator's reign? The Greek legal system required that citizens represent themselves in court—they could not hire attorneys to speak on their behalf as we can today. The burden, then, was on the claimants in these land disputes to make the best possible case and to present it persuasively to the jury.

Corax realized the need for systematic instruction in the art of speaking in the law courts and wrote a treatise called the "Art of Rhetoric." Although no copies of this work survive, we know from later writers that the notion of probability was central to his rhetorical system. He believed that a speaker must argue from general probabilities or establish probable conclusions when matters of fact cannot be established with absolute certainty. He also showed that probability can be used regardless of the side argued. For instance, to argue that someone convicted of driving under the influence of alcohol probably is guilty if arrested for a second time on the same charge is an argument from probability. But so is the opposing argument—that the person convicted once will be especially cautious and probably will not get into that same situation again. In addition to the principle of probability, Corax contributed the first formal treatment of the organization of speeches. He argued that speeches consist of three major parts—an introduction, an argument or proof, and a conclusion—an arrangement that was elaborated on by later writers about rhetoric.[2]

Corax's pupil, Tisias, is credited with introducing Corax's rhetorical system to mainland Greece. With the coming of rhetorical instruction to Athens and the emerging belief that eloquence was an art that could be taught, the rise of a class of teachers of rhetoric, called sophists, was only natural. The word *sophos* means knowledge or wisdom, so a sophist was essentially a teacher of wisdom. Sophistry, not unlike rhetoric, has a tarnished reputation, so that today we associate the sophists with fallacious or devious reasoning.

The Greeks' distrust of the sophists was due to several factors. First, the sophists were itinerant professors and often foreigners to Athens, and some distrust existed simply because of their foreign status. They also professed to teach wisdom or excellence, a virtue that traditionally the Greeks believed could not be taught. In addition, the sophists charged for their services, a practice not only at odds with tradition, but one that made sophistic education a luxury that could not be afforded by all. This in itself may have generated some ill feelings. In large part, however, the continuing condemnation accorded the sophists can be attributed to an accident of history—the survival of Plato's dialogues. Plato, to whom we will return shortly, stood in adamant opposition to the sophists, and several of his dialogues make the sophists look silly indeed.[3] While Plato's views now are considered unjustified in

---

[2] George Kennedy, *The Art of Persuasion in Greece* (Princeton, N.J.: Princeton University Press, 1963), pp. 58–61; and Bromley Smith, "Corax and Probability," *Quarterly Journal of Speech*, 7 (February 1921), 13–42.

[3] Kennedy, *The Art of Persuasion in Greece*, pp. 13–15; and Lester Thonssen and A. Craig Baird, *Speech Criticism: The Development of Standards for Rhetorical Appraisal* (New York: Ronald, 1948), pp. 36–37.

large part, an anti-sophistic sentiment nevertheless was perpetuated in his dialogues that has continued to the present day.[4]

Protagoras of Abdera (c. 480–411 B.C.) is called the initiator of the sophistic movement. He is remembered for the statement, "Man is the measure of all things," which indicates the interest the sophists as a group placed on the study of humanity as the perspective from which to approach the world. This phrase also suggests the relative position many of the sophists accorded to truth: absolute truth was unknowable and perhaps nonexistent and had to be established in each individual case.[5] A second sophist deserving of mention is Gorgias, who was the subject of one of Plato's disparaging dialogues on the sophists and their brand of rhetoric. Originally from Sicily, Gorgias established a school of rhetoric in Athens and became known for his emphasis on the poetic dimensions of language. He also is called the father of impromptu speaking because this was a favored technique at his school.[6]

Another sophist whose work is significant in the history of rhetorical thought is Isocrates (436–338 B.C.). He began his career as a speechwriter for those involved in state affairs because he lacked the voice and nerve to speak in public. In 392 B.C., he established a school of rhetoric in Athens and advocated as an ideal the orator active in public life. He believed that politics and rhetoric could not be separated; both disciplines were needed for participation in the life of the state. In addition, unlike many other teachers of his day, Isocrates encouraged his students to learn from other teachers—to take instruction with those best qualified to teach them.[7]

The sophists' emphasis on technique suggests that rhetoric had not yet achieved formal status as an area of study. The work of the Greek philosopher, Plato (427–347 B.C.), provided the foundation for such developments, although paradoxically, he also is remembered as one of the great opponents of rhetoric. Plato was a wealthy Athenian who rejected the ideal of political involvement in favor of philosophy after the death of his teacher and mentor, Socrates. At his school, the Academy, he espoused a belief in philosophical thought and knowledge, or dialectic, and rejected any form of relative knowledge or opinions as unreal. Thus, he opposed the practical and relative nature of rhetoric advocated by the sophists.

The two dialogues in which Plato's views on rhetoric emerge most clearly are the *Gorgias* and the *Phaedrus*. In the *Gorgias*, Plato set Gorgias and others against Socrates in order to distinguish true from false rhetoric, or the rhetoric as practiced by the sophists from an ideal rhetoric grounded in philosophy. Plato faulted rhetoric for ignoring true knowledge; for failing to work toward the good, which for Plato was the end toward which all human pursuits should be directed; and because it was a tech-

---

[4] That Plato's negative view of the sophists was unjustified has been asserted by numerous scholars. His views in the *Gorgias*, in particular, have come under frequent re-examination. See, for example, Bruce E. Gronbeck, "Gorgias on Rhetoric and Poetic: A Rehabilitation," *Southern Speech Communication Journal*, 38 (Fall 1972), 27–38; and Richard Leo Enos, "The Epistemology of Gorgias' Rhetoric: A Re-examination," *Southern Speech Communication Journal*, 42 (Fall 1976), 35–51.

[5] Kennedy, *The Art of Persuasion in Greece*, p. 13; and Philip Wheelwright, *The Presocratics* (Indianapolis: Odyssey, 1966), pp. 238–40.

[6] Thonssen and Baird, p. 38

[7] Russel H . Wagner, "The Rhetorical Theory of Isocrates," *Quarterly Journal of Speech*, 8 (November 1922), 323–37; and William L. Benoit, "Isocrates on Rhetorical Education," *Communication Education*, 33 (April 1984), 109–20.

nique or knack rather than an art: "[R]hetoric seems not to be an artistic pursuit at all, but that of a shrewd, courageous spirit which is naturally clever at dealing with men; and I call the chief part of it flattery. It seems to me to have many branches and one of them is cookery, which is thought to be an art, but according to my notion is no art at all, but a knack and a routine."[8]

In Plato's later dialogue, the *Phaedrus*, he used three speeches on love as analogies for his ideas about rhetoric. The first two speeches illustrate the faults of rhetoric as practiced in contemporary Athens: either it fails to move listeners at all or it appeals to evil or base motives. With the third speech, however, which Plato had Socrates deliver, he articulated an ideal rhetoric. It is based first and foremost on knowing the truth and the nature of the human soul: "any man who does not know the truth, but has only gone about chasing after opinions, will produce an art of speech which will seem not only ridiculous, but no art at all."[9] In addition to his concern for content, Plato also commented on organization, style, and delivery in the *Phaedrus*, thus paving the way for a comprehensive treatment of all areas of rhetoric.

Plato's student, Aristotle (384–322 B.C.), was responsible for first systematizing rhetoric into a unified body of thought. In fact, his *Rhetoric* often is considered the foundation of the discipline of speech communication. While Aristotle could not avoid the influence of Plato's ideas, he diverged significantly from his teacher in his treatise on rhetoric.

Aristotle was a scientist trained in classification, and this orientation emerges in the *Rhetoric*. Rather than attempting a moral treatise on the subject, as did Plato, Aristotle sought to categorize objectively the various facets of rhetoric, which he defined as "the faculty of discovering in the particular case what are the available means of persuasion."[10] The result was a philosophic and pragmatic treatise that drew upon Plato's ideas as well as on the sophistic tradition.

Aristotle devoted a large portion of the *Rhetoric* to invention, or the finding of materials and modes of proof to use in presenting those materials to an audience. He dealt as well, however, with style, organization, and delivery, or the pragmatic processes of presentation. Thus, he incorporated what now are considered to be the major canons of rhetoric that have formed the parameters of its study for centuries. The canons consist of invention, or the discovery of ideas and arguments; organization, or the arrangement of the ideas discovered by means of invention; elocution or style, which involves the linguistic choices a speaker must make; and delivery, or the presentation of the speech. Memory is the fifth canon, although Aristotle made no mention of it.

No major rhetorical treatises survived in the two hundred years after Aristotle's *Rhetoric*. This was a time of increasing Roman power in the Mediterranean, and not surprising, the next extant work on rhetoric was a Latin text, the *Ad Herennium*, written about 100 B.C. The Romans were borrowers and, as with most other aspects of Greek culture, they adopted the basic principles of rhetoric developed by the Greeks.

---

8 Plato, *Gorgias* 463.

9 Plato, *Phaedrus* 262.

10 Aristotle, *Rhetoric* 1.2 1355b. For a comparison of the rhetorics of Aristotle and Plato, see Everett Lee Hunt, "Plato and Aristotle on Rhetoric and Rhetoricians," in *Studies in Rhetoric and Public Speaking* (New York: Russell & Russell, 1962), pp. 3–60.

The Romans were practical people, however, and the more pragmatic aspects of rhetoric were the ones that appealed most to them. They added little that was new to the study of rhetoric but rather organized and refined it as a practical art.

The *Ad Herennium* appears to be a representative Roman text in that it is essentially Greek in content and Roman in form. A discussion of the five canons constitute the essence of this schoolboys' manual, but the practical aspects, not their theoretical underpinnings, are featured. The systematization and categorization that characterized the *Ad Herennium's* approach to rhetoric were typical of the Roman treatises that followed.[11]

Cicero (106–43 B.C.) represents the epitome of Roman rhetoric, since in addition to writing on the art of rhetoric, he was himself a great orator. His earliest treatise on the subject was *De Inventione* (87 B.C.), which he wrote when only twenty years old. Although he considered it an immature piece in comparison to his later thinking on the subject, it offers another model of the highly prescriptive nature of most Roman rhetorical treatises.

Cicero's major work on rhetoric was *De Oratore* (55 B.C.), in which he attempted to restore the union of rhetoric and philosophy by advocating that rhetoric be taught as the single art useful for dealing with all practical affairs. He drew heavily on Isocrates' ideas in advocating an integration of natural ability, comprehensive knowledge of all the liberal arts, and extensive practice in writing. As a practicing orator, Cicero developed the notion of style more fully than did his predecessors and devoted virtually an entire treatise, *Orator* (46 B.C.), to distinguishing three types of style—the plain, the moderate, and the grand.[12]

A final Roman rhetorician deserving of mention is the Roman lawyer and educator, M. Fabius Quintilian (35–95 A.D.). In his *Institutes of Oratory* (93 A.D.), Quintilian described the ideal training of the citizen-orator from birth through retirement. He defined the orator as "the good man speaking well," and his approach was not rule bound as were many Roman rhetorics.[13] He was eclectic and flexible, drawing from Plato, Aristotle, Isocrates, and Cicero and also integrating his own teaching experiences into traditional theory. His work was so systematic that it not only serves as an excellent synthesis of Greek and Roman rhetorical thought, but it was an important source of ideas on education throughout the Middle Ages.

With the decline of democracy in Rome, rhetoric entered an era when it essentially was divorced from civic affairs. A series of emperors were in power, and anyone who spoke publicly in opposition to them was likely to be punished. Rhetoric, then, was relegated to a back seat and became an art concerned with style and delivery rather than with content. This period, from about 150 to 400 A.D., often is referred to as the Second Sophistic because of the excesses of delivery and style similar to those for which the early sophists were criticized.

The Middle Ages (400–1400 A.D.) followed the Second Sophistic, and during this period, rhetoric became aligned with preaching, letter writing, and education. The

---

[11] George Kennedy, *The Art of Rhetoric in the Roman World* (Princeton, N.J.: Princeton University Press, 1972), pp. 106–08.

[12] For a summary of Cicero's style, see Thomas R. King, "The Perfect Orator in *Brutus*," *Southern Speech Journal*, 33 (Winter 1967), 124–28.

[13] Thonssen and Baird, p. 92.

concern with preaching as an oratorical form might be said to have begun with St. Augustine (354–430 A.D.). Many call Augustine a bridge between the classical and medieval periods; nevertheless, he is the only major thinker on rhetoric associated with the Middle Ages. As Christianity became increasingly powerful, rhetoric was condemned as a pagan art; many Christians believed that the rhetorical ideas formulated by the pagans of classical Greece and Rome should not be studied and that possession of Christian truth was accompanied by an automatic ability to communicate that truth effectively. St. Augustine, however, had been a teacher of rhetoric before converting to Christianity in 386. Thus, in his *On Christian Doctrine* (426), he argued that preachers need to be able to teach, to delight, and to move—Cicero's notion of the duties of the orator—and that to accomplish the aims of Christianity, attention to the rules of effective expression was necessary.[14] Because St. Augustine believed such rules were to be used only in the expression of truth, he revitalized the philosophic basis of rhetoric that largely had been ignored since Quintilian.

Letter writing was another form in which rhetoric found expression during the Middle Ages. Many political decisions were made privately through letters and decrees; in addition, letter writing became a method of record keeping for both secular and religious organizations as they increased in size and complexity. Letter writing, too, was necessary in order to bridge the distances of the medieval world, which no longer consisted of a single center of culture and power as was the case with the classical period.[15] Thus, principles of letter writing, including the conscious adaptation of salutation, language, and format to a particular addressee, were studied as rhetoric.

Finally, rhetoric played a role in education in the Middle Ages as one of the three great liberal arts. Along with logic and grammar, rhetoric was considered part of the *trivium* of learning, much as our three Rs of reading, writing, and arithmetic function today.[16] While the emphasis shifted among these arts from time to time, each was treated in a highly practical rather than a theoretic manner.

The Renaissance, from 1400 to 1600 A.D., signaled the end of the Middle Ages but did little to alter substantially the course of rhetorical thought. Few innovations were introduced; instead, the classical writers were emphasized and many of the Greek and Latin treatises that had been presumed lost were discovered in monasteries. The concern with style and expression that characterized the Middle Ages continued with perhaps even more excess, prompting it to be labeled an age of "social ingratiation."[17]

Peter Ramus (1515–1572) was a well-known French scholar of the Renaissance who typified the position accorded to rhetoric during this period. Essentially, he made rhetoric subordinate to logic by placing invention and organization under the rubric of logic and leaving rhetoric with only style and delivery.[18] This dichotomizing and

---

[14] James J. Murphy, "Saint Augustine and the Debate About a Christian Rhetoric," *Quarterly Journal of Speech*, 56 (December 1960), 400–10; and Saint Augustine *On Christian Doctrine* xvii, 34.

[15] Nancy L. Harper, *Human Communication Theory: The History of a Paradigm* (Rochelle Park, N.J.: Hayden, 1979), p. 71; and James Murphy, *Rhetoric in the Middle Ages: A History of Rhetorical Theory from Saint Augustine to the Renaissance* (Berkeley: University of California Press, 1974).

[16] Donald Lemen Clark, *Rhetoric in Greco-Roman Education* (Westport, Conn.: Greenwood, 1957), p. 12.

[17] Douglas Ehninger, "On Rhetoric and Rhetorics," *Western Speech*, 31 (Fall 1967), 244.

[18] Wilbur Samuel Howell, *Logic and Rhetoric in England, 1500–1700* (New York: Russell & Russell, 1956), p. 148.

departmentalizing of knowledge made for easy teaching, and Ramus' taxonomy was perpetuated for generations through the educational system.

The period from 1600 to 1900 is known as the age of modern rhetoric. Francis Bacon (1561–1626) is a figure who bridges the rhetoric of the Renaissance and that of modern rhetoric. He was concerned with the lack of scholarly progress during the Middle Ages and sought to promote a revival of secular knowledge through an empirical examination of the world. He introduced ideas about the nature of sensory perception, arguing that our sensory interpretations are highly inaccurate and should be subjected to reasoned, empirical investigation. His definition of rhetoric contained this notion of rationality: "the duty of Rhetoric is to apply Reason to Imagination for the better moving of the will."[19] Bacon, then, anticipated the decline in the church's influence, the renewed interest in rhetoric, and the focus on psychological and cognitive processes that would become important to the study of rhetoric in the next centuries.

Three trends in rhetoric characterized the modern period—epistemological, belletristic, and elocutionist. Epistemology is the study of the origin, nature, methods, and limits of human knowledge. Epistemological thinkers sought to recast classical approaches in terms of modern developments in psychology. They attempted to understand rhetoric in relation to underlying mental processes and contributed to the development of a rhetoric firmly grounded in a study of human nature.

George Campbell (1719–1796) and Richard Whately (1758–1859) exemplify the best of the epistemological tradition. Campbell was a Scottish minister, teacher, and author of *The Philosophy of Rhetoric* (1776). He drew on Aristotle, Cicero, and Quintilian as well as the faculty psychology and empiricism of his times. Faculty psychology attempted to explain human behavior in terms of five powers or faculties of the mind—understanding, memory, imagination, passion, and will—and Campbell's definition of rhetoric was directed to these faculties: "to enlighten the understanding, to please the imagination, to move the passions, or to influence the will."[20] Campbell's approach to evidence suggests his ties to the rational, empirical approach to knowledge gaining prominence in his day. He distinguished three types of evidence—mathematical axioms, derived through reasoning; consciousness, or the result of sensory stimulation; and common sense, an intuitive sense share by virtually all humans.

Richard Whately, like Campbell, was a preacher, and his *Elements of Rhetoric*, published in 1828, often is considered the logical culmination of Campbell's thought.[21] His view of rhetoric was similar to Campbell's in its dependence on faculty psychology, but he deviated in making argumentation the focus of the art of rhetoric: "The *finding* of suitable arguments to prove a given point, and the skilful [sic] *arrangement* of them, may be considered as the immediate and proper province of Rhetoric, and of that alone."[22] He also is remembered for his analysis of presumption

---

[19] Hugh C. Dick, ed., *Selected Writings of Francis Bacon* (New York: Modern Library, 1955), p. x; and Harper, pp. 100, 109.

[20] George Campbell, *The Philosophy of Rhetoric*, ed. Lloyd F. Bitzer (1776; rpt. Carbondale: Southern Illinois University Press, 1963), p. 1.

[21] Douglas Ehninger, "Introduction," in Richard Whately, *Elements of Rhetoric*, ed. Douglas Ehninger (1828; rpt. Carbondale: Southern Illinois University Press. 1963), p. xv.

[22] Whately, p. 39.

and burden of proof, which paved the way for modern argumentation and debate practices. The epistemologists, then, combined their knowledge of classical rhetoric and contemporary psychology to create rhetorics based on an understanding of human nature. In this, they offered audience-centered approaches to rhetoric and paved the way for contemporary concerns with audience analysis.

The second direction rhetoric took in the modern period is known as the belles lettres movement; the term, in French, literally means "fine or beautiful letters." It referred to literature valued primarily for its aesthetic qualities more than for its informative value. Belletristic rhetorics were distinguished by their breadth—rhetoric was considered to consist not only of spoken discourse but of writing and criticism as well. In addition, the scholars of this school believed that all the fine arts, including rhetoric, poetry, drama, music, and even gardening and architecture, could be subjected to the same critical standards.[23] Thus, the critical component to rhetoric gained an importance not seen in earlier approaches.

Hugh Blair (1718–1800) stands as a representative figure of the belletristic period. In his *Lectures on Rhetoric and Belles Lettres*, based on a series of lectures he delivered at the University of Edinburgh, he presented an overview of the relationship among rhetoric, literature, and criticism. One of his most innovative contributions was his discussion of taste, or the faculty that is capable of deriving pleasure from contact with the beautiful. Taste, according to Blair, is perfected when a sensory pleasure is coupled with reason—when reason can explain the source of that pleasure.[24] Blair's ideas on rhetoric proved extremely popular and laid the foundations for contemporary literary and rhetorical criticism.

The elocutionary movement, the third rhetorical trend of the modern period, reached its height in the mid-eighteenth century. It developed in response to the poor delivery styles of contemporary preachers, lawyers, and other public figures and because the canon of delivery had been neglected, for the most part, since classical times. Like the epistemologists, the elocutionists were concerned about contributing to a more scientific understanding of the human being and believed that their observations on voice and gesture—characteristics unique to humans—constituted one such contribution.[25] The elocutionists also sought to determine the effects of delivery on the various faculties of the mind, thus continuing the link with modern psychology. Despite a stated concern for invention, however, many elocutionary treatises were not much more than prescriptive and often highly mechanical techniques for the management of voice and gestures.

Gilbert Austin's guidelines are representative of the highly stylized approach of the elocutionists. He offered this advice to the speaker, for instance, about eye contact and volume: "He should not stare about, but cast down his eyes, and compose his countenance: nor should he at once discharge the whole volume of his voice,

---

[23] James L. Golden, Goodwin F. Berquist, and William E. Coleman, *The Rhetoric of Western Thought*, 3rd ed. (1976; rpt. Dubuque, Iowa: Kendall/Hunt, 1983), pp. 107–108.

[24] Hugh Blair, *Lectures on Rhetoric and Belles Lettres* (London: William Baynes and Son, 1825), p. 24.

[25] Golden, Berquist, and Coleman, pp. 175–76.

but begin almost at the lowest pitch, and issue the smallest quantity; if he desire to silence every murmur, and to arrest all attention."[26] As another example, James Burgh believed each emotion could be linked with a specific, external expression; he categorized seventy-one emotions and their particular manifestations. Thomas Sheridan (1719–1788), who wrote *A Course of Lectures on Elocution* in 1762, was perhaps the most famous elocutionist. Sheridan not only was in the forefront in terms of criticizing the speakers of his day, but he sought to establish a universal standard of pronunciation for the English language in addition to offering the usual techniques for delivery.[27]

The elocutionists have been criticized for their excesses in terms of style and delivery and for the inflexibility of their techniques. Their efforts to derive an empirical science of delivery based on observation, however, foreshadowed the use of the scientific method to study all aspects of human communication, and their theories had a tremendous effect on how speech was taught in American classrooms in the nineteenth century.

The twentieth century has seen a renewed interest in the study of rhetoric, and this era has become known as the contemporary period. While the elocutionists had narrowed the focus of rhetoric to delivery, contemporary rhetorical scholars have revitalized rhetoric as an art that includes the canons of invention, organization, and elocution, as well as delivery. Contemporary scholars also tend to be eclectic, drawing not only on the rhetorical treatises of classical Greece and Rome and other periods but on a variety of contemporary disciplines such as psychology, sociology, literary criticism, English, and philosophy as well. Currently, then, rhetoric has regained some of its earlier importance as a broad liberal art that is more than simply the expression of ideas or considerations of style apart from substance or action.

DEVELOPING YOUR UNDERSTANDING

1. **Explain what impact the following sophistic maxim has on the definition, scope, and function of rhetoric: "Man is the measure of all things."**
2. **Based on the abbreviated rhetorical history presented by Foss et al, identify several different definitions of rhetoric. Analyze and discuss their similarities and differences. Then, present your own definition of rhetoric and explain your rationale.**
3. **Referring to the rhetoricians discussed in Foss et al, summarize the different ways rhetoric, philosophy, politics, and ethics have been related (or separated) in definitions of rhetoric.**
4. **Develop your own definition of rhetoric (if you have not done so already for question 2). In your definition of rhetoric, describe the relationships between rhetoric, philosophy, politics, and ethics. Then, assess how your position is supported and/or challenged by the history of rhetoric as presented by Foss et al.**

---

[26] Gilbert Austin, *Chironomia or a Treatise on Rhetorical Delivery*, ed. Mary Margaret Robb and Lester Thonssen (1806; rpt. Carbondale: Southern Illinois University Press, 1966), p. 94.

[27] Thomas Sheridan, *A Course of Lectures on Elocution* (London: W. Strahan, 1762).

FOCUSING ON KEY TERMS AND CONCEPTS

Focus on the following terms and concepts while you read through this selection. Understanding these will not only increase your understanding of the selection that follows, but you will find that, because most of these terms or concepts are commonly used in professional writing and rhetoric, understanding them helps you get a better sense of the field itself.

1. dialectic
2. art
3. artistic pisteis
4. nonartistic pisteis
5. invent
6. ethos
7. pathos
8. logos
9. topos/topoi
10. deliberative rhetoric
11. judicial rhetoric
12. demonstrative rhetoric

Excerpts from Book I, Rhetoric

ARISTOTLE translated by GEORGE A. KENNEDY

Books 1–2 discuss the means of persuasion available to a public speaker from logical argument, the presentation of the speaker's character, and moving the emotions of the audience. Although this part of rhetoric has come to be known as "invention," Aristotle himself offers no general term for it until the transition section at the end of book 2, where he refers to it as *dianoia*, "thought." Throughout books 1 and 2, understanding the available means of persuasion is treated as constituting the whole of rhetoric, properly understood; and until the last sentence of 2.26 there is no anticipation of discussion of style and arrangement in book 3. Books 1–2 are a unit and probably made up the whole of the *Rhetoric* as it once existed.

CHAPTERS 1–3: INTRODUCTION

Chapter 1: Introduction to Rhetoric for Students of Dialectic

The *Rhetoric* shows signs of being addressed to different audiences, probably reflecting differing contexts in which Aristotle lectured on rhetoric at different times in his career. Though much of the work provides practical instruction on how to compose a speech, useful to any citizen, some parts seem to be addressed primarily to students of philosophy. What is now regarded as the first chapter of book 1 was apparently originally addressed to students who had completed a study of dialectic (such as is found in the *Topics*) and who had little knowledge of rhetoric, though they may have been aware of the existence of handbooks on the subject. For them Aristotle explains the similarities between dialectic as they know it and rhetoric as he understands it but does not comment on the differences. The chapter as a whole is very Platonic and contains echoes of several of Plato's dialogues.

Dialectic, as understood by Aristotle, was the art of philosophical disputation. Practice in it was regularly provided in his philosophical school, and his treatise known as *Topics* is a textbook of dialectic. The opening chapters of the *Topics* may be found in Appendix I.C. The procedure in dialectic was for one student to state a thesis (e.g., "Pleasure is the only good") and for a second student to try to refute the thesis by asking a series of questions that could be answered by *yes* or *no*. If successful, the interlocutor led the respondent into a contradiction or logically undefensible position by means of definition and division of the question or by drawing analogies; however, the respondent might be able to defend his position and win the argument. Dialectic proceeds by question and answer, not, as rhetoric does, by continuous exposition. A dialectical argument does not contain the parts of a public address; there is no introduction, narration, or epilogue, as in a speech—only proof. In dialectic only logical argument is acceptable, whereas in rhetoric (as Aristotle will explain in chapter 2), the impression of character conveyed by the speaker and the emotions awakened in the audience contribute to persuasion. While both dialectic and rhetoric build their arguments on commonly held opinions (*endoxa*) and deal only with the probable (not with scientific certainty), dialectic examines general issues (such as the nature of justice) whereas rhetoric usually seeks a specific judgment (e.g., whether or not some specific action was just or whether or not some specific policy will be beneficial). Epideictic is a partial exception to this. Platonic dialogues make extensive use of dialectic as Socrates seeks to refute the position of an opponent—for example, Gorgias, Polus, and Callicles in the *Gorgias*. Platonic dialogues also contain rhetorical passages expressive of Socrates' character and appeals to the emotions of the hearer, as in his second speech in the *Phaedrus*.

After discussing the similarities between dialectic and rhetoric, Aristotle criticizes (sections 3–11) the *arts*, or handbooks, of previous writers, which he finds unsatisfactory in several ways. These handbooks are now lost; and the only surviving treatise on rhetoric from the classical period other than Aristotle's is a slightly later work known as the *Rhetoric to Alexander*. Into this discussion are inserted parenthetical remarks (sections 7–9) on the specificity desirable in framing good laws, a subject of interest to students of political philosophy but of limited relevance to rhetorical theory. The chapter concludes (sections 12–14) with a discussion of why rhetoric is useful—remarks that can be thought of as addressed primarily to students of philosophy who, under the influence of Plato, may regard the subject of rhetoric as trivial. A general Greek audience would probably have assumed that rhetoric was useful and been more dubious about dialectic, which could easily seem pedantic hairsplitting, as it did to Isocrates (see, e.g., *Against the Sophists* and the prooemion to the *Encomium of Helen*).

Chapter 1 creates acute problems for the unity of the treatise. Aristotle here seems firmly to reject using the emotions, identifies rhetoric with logical argument, and gives no hint that style and arrangement may be important in rhetoric (as will emerge in book 3). In section 6 he even seems to say that the importance of the justice of a case are not appropriate issues for a speaker to discuss; they should be left for the audience to judge. But the justice of a speaker's case, its importance, and its amplification subsequently will be given extended treatment. Some interpreters seek to force the point of view of chapter 1 into conformity with what follows by making very careful distinctions about what Aristotle is saying. This involves claiming, for example, that *pisteis*, "proofs," in section 3 already includes the use of character and emotion as means of persuasion, that *verbal attack*, *pity*, and *anger* in section 4 refer to *expressions* of emotion rather than to the reasoned use of an understanding of psychology and motivation. Section 6 can be made consistent with later parts of the work if Aristotle is regarded as saying that the speaker's interpretation of what is just or important should not be allowed to color the audience's judgment. It can be stressed that a speaker needs to understand tricks that may be used by an opponent but should not employ them himself. Despite other possible interpretations, it is probably better to acknowledge frankly that chapter 1 is inconsistent with what follows, that it is far more austere in tone than Aristotle's general view of rhetoric, and that the difference results from addressing different audiences and from the attempt to link the study of dialectic with that of rhetoric. Aristotle either failed to revise the chapter or has let stand a deliberately provocative critique of the teaching of rhetoric in his own time as a way of emphasizing the needs for greater attention to logic, thus justifying the writing of a rhetoric handbook by a philosopher. The chapter might even be compared to Socrates' provocative description in the *Gorgias* of contemporary rhetoric as a form of flattery, a view that Socrates, too, subsequently modifies. The result is to encourage a dialogue between the reader and the text of the *Rhetoric* about the moral purpose and valid uses of rhetoric.

The first chapter is one of the earliest examples of an introduction to the study of a discipline (the beginning of the *Topics* is another) and is thus an antecedent of the Greek *prolegomenon* or Latin *accessus* commonly found at the beginning of technical works in later antiquity and the Middle Ages.

[1354a] 1. Rhetoric[1] is an *antistrophos*[2] to dialectic; for both are concerned with such things as are, to a certain extent, within the knowledge of all people and belong to no separately defined science.[3] A result is that all people, in some way, share in both; for all, to some extent, try both to test and maintain an argument [as in dialec-

---

[1] *Hē rhētorikē* (the rhetorical), a feminine singular adjective used as an abstract noun; cf. *dialektikē*, *poiētikē*. Neither dialectic nor rhetoric assume knowledge of any technical subject, and both build a case on the basis of what any reasonable person would believe. Aristotle takes the term *rhetoric* from Plato; others usually spoke of the "art of speech"; see Schiappa 1990.

[2] *Antistrophos* is commonly translated "counterpart." Other possibilities include "correlative" and "coordinate." The word can mean "converse." In Greek choral lyric, the metrical pattern of a *strophē*, or stanza, is repeated with different words in the *antistrophē*. Aristotle is, however, probably thinking of, and rejecting, the analogy of the true and false arts elaborated by Socrates in the *Gorgias*, where justice is said to be an *antistrophos* to medicine (464b8) and rhetoric, the false form of justice, is compared to cookery, the false form of medicine (465c1–3). Isocrates (*Antidosis* 182) speaks of the arts of the south (called philosophy, but essentially political rhetoric) and the arts of the body (gymnastic) as *antistrophoi*. This view is equally unacceptable to Aristotle, for whom rhetoric is a tool, like dialectic, though its subject matter is derived from some other discipline, such as ethics or politics; see *Rhetoric* 1.2.7. Aristotle thus avoids the fallacy of Plato's *Gorgias* where Socrates is obsessed with finding some kind of knowledge specific to rhetoric. On later interpretations of *antistrophos* se Green 1990.

[3] The first sentence of the treatise, with its proposition and supporting reason, is an example of what Aristotle will call an enthymeme. The reader should become sensitive to the constant use of enthymemes throughout the text, often introduced by the particular *gar* (for).

tic] and to defend themselves and attack [others, as in rhetoric]. 2. Now among the general public, some do these things randomly and others through an ability acquired by habit,[4] but since both ways are possible, it is clear that it would also be possible to do the same by [following] a path; for it is possible to observe[5] the cause why some succeed by habit and others accidentally,[6] and all would at once agree that such observation is the activity of an art [*tekhnē*].[7]

3. As things are now,[8] those who have composed *Arts of Speech* have worked on a small part of the subject; for only *pisteis*[9] are artistic (other things are supplementary), and these writers say nothing about enthymemes, which is the "body" of persuasion,[10] while they give most of their attention to matters external to the subject; 4. for verbal attack and pity and anger and such emotions of the soul do not relate to fact but are appeals to the juryman.[11] As a result, if all trials were conducted as they are in some present-day states and especially in those well governed, [the handbook writers] would have nothing to say; 5. for everyone thinks the laws ought to require this, and some even adopt the practice and forbid speaking outside the subject, as in the Areopagus too,[12] rightly so providing; for it is wrong to warp the jury by leading them into anger or envy or pity: that is the same as if someone made a straightedge rule crooked before using it. 6. And further, it is clear that the opponents have no function except to show that something is or is not true or has happened or has not happened;[13] whether it is important or trivial or just or unjust, in so far as the lawmaker has not provided a definition, the juryman should somehow decide himself and not learn from the opponents.[14]

---

[4] The former hardly know what they are doing; but the latter, by trial and error, have gained a practical sense of what is effective.

[5] *Theorein*, lit. "see" but with the implication "theorize." This is an instance of the visual imagery common in the *Rhetoric*.

[6] Here, as often, Aristotle reverses the order of reference: *accidentally* refers back to *randomly*. Such *chiasmus* is a common feature of Greek.

[7] In contrast to Socrates in the *Gorgias*, Aristotle has no doubt that rhetoric is an art. Awareness of the cause of success allows technique to be conceptualized and taught systematically. On Aristotle's understanding of an "art," see the passage from *Nicomachean Ethics* 6.4 in Appendix I.B.

[8] In 1.2.4 Aristotle again criticizes contemporary technical writers. He thus appears to be thinking primarily of the handbooks of the mid–fourth century, such as those by Pamphilus and Callippus cited in 2.23.21. Aristotle collected the doctrines of some handbooks in a lost work, *Synagōgē tekhnōn*; see Appendix I.D. Plato provides a brief summary of the earlier ones in *Phaedrus* 266d–67d.

[9] *Pistis* (pl. *pisteis*) has a number of different meanings in different contexts: "proof, means of persuasion, belief," etc. In 1.2.2–3 Aristotle distinguishes between artistic and nonartistic *pisteis*, and divides the former into three means of persuasion based on character, logical argument, and arousing emotion. Here in chap. 1 readers familiar with dialectic have no knowledge yet of persuasion by character or emotion and will assume that *pistis* means "logical proof." In 3.17.15 *pistis* means "logical argument" in contrast to character presentation.

[10] *Body* is here contrasted with "matters external" in the next clause. Though Aristotle does not say so, one might speculate that the soul, or life, of persuasion comes from ethical and emotional qualities.

[11] The handbooks offered examples of argument from probability, but they did not recognize its logical structure. The concept of the logical syllogism and its rhetorical counterpart, the enthymeme (to be discussed in chap. 2), are Aristotelian contributions. The handbooks probably treated the emotions in discussing the prooemium and epilogue (on which see Aristotle's account in 3.13,19) and in separate collections or discussions such as the *Eleoi* of Thrasymachus (see *Rhetoric* 3.1.7).

[12] In Aristotle's time the jurisdiction of the Athenian court of the Areopagus was chiefly limited to homicide cases. That its rules of relevance were strict is also attested in Lycurgus' speech *Against Leocrites* 12.

[13] On the possible implications of this statement for Aristotle's view of a "general rhetoric," see Wieland 1968; but there is no other passage in Aristotle expressly supporting the view Wieland advances.

[14] On the problems created by this statement, see the introductory comment to this chapter.

The following passage on framing laws resembles some of what Plato says in *Laws* 9.875–76[15] and is apparently a parenthetical remark of Aristotle to students of political philosophy; he may well have said something of this sort to young Alexander. Section 9 will take up where section 6 leaves off.

(7. It is highly appropriate for well-enacted laws to define everything as exactly as possible and for as little as possible to be left to the judges:[16] first because it is easier to find one or a few than [to find] many who are prudent and capable of framing laws and judging; **[1354b]** second, legislation results from consideration over much time, while judgments are made at the moment [of a trial or debate], so it is difficult for the judges to determine justice and benefits fairly; but most important of all, because the judgment of a lawmaker is not about a particular case but about what lies in the future and in general, while the assemblyman and the juryman are actually judging present and specific cases. For them, friendliness and hostility and individual self-interest are often involved, with the result that they are no longer able to see the truth adequately, but their private pleasure or grief casts a shadow on their judgment. 8. In other matters, then, as we have been saying, the judge should have authority to determine as little as possible; but it is necessary to leave to the judges the question of whether something has happened or has not happened, will or will not be, is or is not the case; for the lawmaker cannot foresee these things.)

9. If this is so, it is clear that matters external to the subject are described as an art by those who define other things; for example, what the introduction [*prooimion*] or the narration [*diēgēsis*][17] should contain, and each of the other parts; for [in treating these matters] they concern themselves only with how they may put the judge in a certain frame of mind,[18] while they explain nothing about artistic proofs; and that is the question of how one may become *enthymematic*.[19] 10. It is for this reason that although the method of deliberative and judicial speaking is the same and though deliberative subjects are finer and more important to the state than private transactions, [the handbook writers] have nothing to say about the former, and all try to describe the art of speaking in a lawcourt, because it is less serviceable to speak things outside the subject in deliberative situations;[20] for there the judge judges about matters that affect himself, so that nothing is needed except to show that circumstances are as the speaker says.[21] But in judicial speeches this is not enough; rather, it is serviceable to

---

[15] A suggestion made to the translator by Eckhardt Schütrumpf.

[16] This "philosophical" position is somewhat modified in 1.13.13, when Aristotle considers the practical problems involved.

[17] The *Arts*, or handbooks of rhetoric, were organized around discussion of what should be said in each of the separate parts usually found in a judicial speech. These included *prooimion* (introduction), *diēgēsis* (narration), *pistis* (proof), and *epilogos* (conclusion) and sometimes additional parts. See 3.13–19.

[18] This was regarded as a major function of the prooemium (cf. 3.14.9–11) and epilogue (3.19.1).

[19] The meaning of this term will be explained in the next paragraph.

[20] The *Arts* of rhetoric to which Aristotle refers were certainly largely concerned with techniques useful in the law courts; but speeches like Demosthenes' *On the Crown* show that these could be as fine and as politically significant as speeches in the democratic assembly and were by no means limited to "private transactions," or contracts, as Aristotle insinuates. In the manuscripts the sentence continues, "and deliberative oratory is less mischievous than judicial, but of more general interest." This is probably an addition by a later writer.

[21] In deliberative rhetoric the "judges" are members of a council or assembly making decisions about public matters that affect themselves.

gain over the hearer; for the judgment is about other people's business and the judges, considering the matter in relation to their own affairs and listening with partiality, lend themselves to [the needs of] the litigants but do not judge [objectively]. **[1355a]** Thus, as we said earlier, in many places the law prohibits speaking outside the subject [in court cases]; in deliberative assemblies the judges themselves adequately guard against this.

11. Since it is evident that artistic method is concerned with *pisteis* and since *pistis* is a sort of demonstration [*apodeixis*][22] (for we most believe when we suppose something to have been demonstrated) and since rhetorical *apodeixis* is enthymeme (and this is, generally speaking, the strongest of the *pisteis*) and the enthymeme is a sort of syllogism [or reasoning] (and it is a function of dialectic, either as a whole or one of its parts, to see about every syllogism equally), it is clear that he who is best able to see from what materials, and how, a syllogism arises would also be most enthymematic—if he grasps also what sort of things an enthymeme is concerned with and what differences it has from a logical syllogism; for it belongs to the same capacity both to see the true and [to see] what resembles the true, and at the same time humans have a natural disposition for the true and to a large extent hit on the truth; thus an ability to aim at commonly held opinions [*endoxa*] is a characteristic of one who also has a similar ability to regard to the truth.[23]

The Usefulness of Rhetoric

That other writers describe as an art things outside the subject [of a speech] and that they have rather too much inclined toward judicial oratory is clear; 12. but rhetoric is useful [first] because the true and the just are by nature[24] stronger than their opposites, so that if judgments are not made in the right way [the true and the just] are

---

[22] *Apodeixis* = "demonstration," usu. with logical validity (as in scientific reasoning) but occasionally more generally, including probable argument (as here).

[23] On *endoxa* see *Topics* 1.1 in Appendix I.C. The student is assumed already to understand, from earlier study of logic and dialectic, the concepts of *pistis*, *apodeixis*, and *enthymema*. Enthymeme literally means "something in the mind" and had been used by Alcidamas and Isocrates to mean "idea" expressed in a speech. In *Prior Analytics* 2.27 an enthymeme is defined as "a syllogism from probabilities or signs." Aristotle sometimes uses *syllogismos* loosely to mean "reasoning," *enthymema* to mean a consideration in whatever form it is put. A valid syllogism in the technical sense is a logical certainty, "true," and most perfectly seen only when expressed symbolically, e.g., "If all A is B, and some A is C, then all C is B." The traditional example in post-Aristotelian logic is, "If all men are mortal, and Socrates is a man, then Socrates is a mortal." In 1.2.14 Aristotle says that "few" of the premises of enthymemes are necessarily true, thus slightly modifying the definition in the *Analytics*. In 1.2.13 and 2.22.3 he says that an enthymeme need not express all its premises. The Aristotelian distinction between syllogism and an enthymeme thus seems largely one of context—tightly reasoned philosophical discourse in the case of syllogism versus popular speech or writing with resulting informality in the expression of the argument in an enthymeme. In public address an argument may be a worthwhile consideration even if it is not absolutely valid. An example of a typical enthymeme might be "Socrates is virtuous; for he is wise" or "Since/If Socrates is wise, he is virtuous." Here the premises are only probable and a universal major premise, "All the wise are virtuous" is assumed. For Aristotle's own examples of enthymemes, see 2.21.2 and the end of 3.17.17.

[24] Aristotle believed that truth was grounded in nature (*physis*) and capable of apprehension by reason. In this he differs both from Plato (for whom truth is grounded in the divine origin of the soul) and from the sophists (for whom judgments were based on *nomos* [convention], which in turn results from the ambivalent nature of language as the basis of human society).

necessarily defeated [by their opposites]. And this is worthy of censure.[25] Further, even if we were to have the most exact knowledge, it would not be very easy for us in speaking to use it to persuade some audiences. Speech based on knowledge is teaching, but teaching is impossible [with some audiences]; rather, it is necessary for *pisteis* and speeches [as a whole] to be formed on the basis of common [beliefs], as we said in the *Topics*[26] about communication with a crowd. Further, one should be able to argue persuasively on either side of a question, just as in the use of syllogisms, not that we may actually do both (for one should not persuade what is debased)[27] but in order that it may not escape our notice what the real state of the case is and that we ourselves may be able to refute if another person uses speech unjustly. None of the other arts reasons in opposite directions; dialectic and rhetoric alone do this, for both are equally concerned with opposites.[28] Of course the underlying facts are not equally good in each case; but true and better ones are by nature always more productive of good syllogisms and, in a word, more persuasive. In addition, it would be strange if an inability to defend oneself by means of the body is shameful, while there is no shame in an inability to use speech; **[1355a]** the latter is more characteristic of humans than is use of the body. 13. And if it is argued that great harm can be done by unjustly using such power of words, this objection applies to all good things except for virtue, and most of all to the most useful things, like strength, health, wealth, and military strategy; for by using these justly one would do the greatest good and unjustly, the greatest harm.[29]

14. That rhetoric, therefore, does not belong to a single defined genus of subject but is like dialectic and that it is useful is clear—and that its function is not to persuade but to see the available means of persuasion in each case, as is true also in all the other arts; for neither is it the function of medicine to create health but to promote this as much as possible; for it is nevertheless possible to treat well those who cannot recover health. In addition, [it is clear] that it is a function of one and the same art to see the persuasive and [to see] the apparently persuasive, just as [it is] in dialectic [to recognize] a syllogism and [to recognize] an apparent syllogism;[30] for sophistry is not a matter of ability but of deliberate choice [*proairesis*] [of specious arguments].[31] In the case of rhetoric, however, there is the difference that one person

---

[25] On the text and interpretation of this sentence, see Grimaldi, 1980–88, 1:25–28. Judgments will not be made in the right way if the facts and reasons are not brought out persuasively. To do this, the speaker needs a knowledge of rhetoric.

[26] *Topics* 1.1.2; see Appendix I.C.

[27] *What is debased (ta phaula)* refers to whatever is bad, cheap, or morally and socially useless. This principle, important as a response to the criticisms of Plato, appears only in a parenthetical remark and is not repeated in the prescriptive parts of the treatise.

[28] There is, however, the difference that in dialectic, opposite trains of argument are actually expressed in the dialectical situation, whereas in rhetoric the speaker has usually tried to think out the opposing arguments before speaking to be able to answer them if need arises. But occasionally, an orator will both express and refute an opposing argument in the course of a speech or even be seen debating with himself about what is right.

[29] Another possible echo of instruction to Alexander.

[30] Rhetoric uses both logically valid arguments and probabilities. The jump to sophistry in the next sentence perhaps implies a recognition that "the apparently persuasive" and "an apparent syllogism" include fallacious arguments that initially sound valid in an oral situation but will not hold up under scrutiny. Both the orator and the dialectician need to be able to recognize these.

[31] In modern linguistic terminology, *sophist* is the "marked" member of the pair *dialectician/sophist* in that the

will be [called] *rhētōr*[32] on the basis of his knowledge and another on the basis of his deliberate choice, while in dialectic *sophist* refers to deliberate choice [of specious arguments], *dialectician* not to deliberate choice, but to ability [at argument generally]. Let us now try to discuss the method itself: how and from what sources we may reach our objectives.[33] Starting again, therefore, as it were from the beginning, after defining what rhetoric is, let us say all that remains [to be said about the whole subject].

Chapter 2: Definition of Rhetoric; *Pisteis,* or the Means of Persuasion in Public Address; Paradigms, Enthymemes, and Their Sources; Common Topics; *Eidē* and *Idia*

1. Let rhetoric be [defined as] an ability, in each [particular] case, to see the available means of persuasion.[34] This is the function of no other art;[35] for each of the others[36] is instructive and persuasive about its own subject: for example, medicine about health and disease and geometry about the properties of magnitudes and arithmetic about numbers and similarly in the case of the other arts and sciences. But rhetoric seems to be able to observe the persuasive about "the given," so to speak. That, too,

---

first includes the second; but *rhētōr* is "unmarked" and may be interpreted either as any effective speaker or as a speaker who uses tricky arguments.

[32] In classical Greek, *rhētōr* means any public speaker, though often referring to a person who plays a leadership role in public debate or is active in the law courts. In the Roman period, *rhētōr* frequently means "rhetorician, "teacher of rhetoric." Latin *orator* (orig. "envoy") and thus English "orator," are translations of *rhētōr* but take on an implication of eloquence not necessarily present in the Greek word.

[33] For some speculations on Aristotle's objectives, see Lord 1981. Aristotle's own objective is clearly an understanding of the nature, materials, and uses of rhetoric; but he has pointed out that the art is useful, and as the treatise unrolls it will often take on the tone of a prescriptive handbook on how to compose a persuasive speech.

[34] Aristotle uses the phrase *estō dē*, "Let X be. . ." commonly of a working hypothesis rather than a final definition and occasionally to resume a definition made earlier. The definition here is anticipated in 1.1.14 on the *ergon* of rhetoric. He identifies the genus to which rhetoric belongs as *dynamis*: "ability, capacity, faculty." In his philosophical writing *dynamis* is the regular word for "potentially" in matter or form that is "actualized" by an efficient cause. The actuality produced by the potentiality of rhetoric is not the written or oral text of a speech, or even persuasion but the art of "seeing" how persuasion may be effected. In *Nicomachean Ethics* 6.4 (see Appendix I.B) he defines all art as a reasoned capacity to make something and says that it is concerned with the coming-into-being of something that is capable of either being or not being. Art is thus for him not the product of artistic skill, but the skill itself. Later rhetoricians often amplify Aristotle's definition by adding *through speech*; the root of the word *rhetoric*, *rhē–*, refers specifically to speech. Though he uses *poetics* to refer to arts other than poetry (dance, painting, sculpture), he never uses *rhetoric* to refer to any art except that of speech. As is clear from chap. 3, Aristotle primarily thinks of rhetoric as manifested in the civic context of public address; but he often draws examples of rhetoric from poetry or historical writing, and in the *Poetics* (19.1456a–b) the "thought" of a speaker in tragedy is said to be a matter of rhetoric. *In each case (peri hekaston)* refers to the fact that rhetoric deals with specific circumstances (particular individuals and their actions). *To see* translates *theorēsai*, "to be an observer of and to grasp the meaning or utility of." English *theory* comes from the related noun *theoria*. *The available means of persuasion* renders *to enkekhomenon pithanon*, "what is inherently and potentially persuasive" in the facts, circumstances, character of the speaker, attitude of the audience, etc. *Endekhomenon* often means "possible."

[35] Dialectic comes closest but deals with general questions, not specific cases; and for dialectic the final term, *means of persuasion (pithanon)*, would presumably become *means of reasoning (syllogismos)*; see *Topics* 1.1 in Appendix I.C.

[36] Except, of course, dialectic.

is why we say it does not include technical knowledge of any particular, defined genus [of subjects].

2. Of the *pisteis*, some are atechnic ["nonartistic"], some entechnic ["embodied in art, artistic"].[37] I call atechnic those that are not provided by "us" [i.e., the potential speaker] but are preexisting: for example, witnesses, testimony of slaves taken under torture,[38] contracts, and such like; and artistic whatever can be prepared by method and by "us"; thus, one must *use* the former and *invent*[39] the latter. **[1356a]** 3. Of the *pisteis* provided through speech there are three species: for some are in the character [*ēthos*] of the speaker, and some in disposing the listener in some way, and some in the argument [*logos*] itself, by showing or seeming to show something.[40]

4. [There is persuasion] through character whenever the speech is spoken[41] in such a way as to make the speaker worthy of credence; for we believe fair-minded people to a greater extent and more quickly [than we do others] on all subjects in general and completely so in cases where there is not exact knowledge but room for doubt.[42] And this should result from speech, not from a previous opinion that the speaker is a certain kind of person;[43] for it is not the case, as some of the technical writers propose in their treatment of the art, that fair-mindedness [*epieikeia*] on the part of the speaker makes no contribution to persuasiveness;[44] rather, character is almost, so to speak, the controlling factor in persuasion.

5. [There is persuasion] through the hearers when they are led to feel emotion [*pathos*] by the speech; for we do not give the same judgment when grieved and rejoicing or when being friendly and hostile. To this and only this we said contempo-

---

[37] Later writers sometimes call these *extrinsic* and *intrinsic*, respectively. Aristotle discusses atechnic proof in 1.15. In 3.16.1 he also refers to the "facts" in a epideictic speech as atechnic.

[38] In Greek law, the evidence of slaves was only admissible in court if taken under torture. There was much debate about its reliability; see 1.15.26.

[39] *Heurein*, "to find out"; *heuresis* becomes the regular word for rhetorical invention.

[40] *Ēthos* in Aristotle means "character," esp. "moral character," and except in 2.21.16 is regarded as an attribute of a person, not of a speech. Aristotle does not use the term in the technical sense of "rhetorical ethos," the technique or effect of the presentation of character in a discourse. "Disposing the listener in some way" is defined in sec. 5 below as leading the hearers to feel emotion (*pathos*). Again, *pathos* is an attribute of persons, not of a speech. The shorthand ethos–pathos–logos to describe the modes of persuasion is a convenience but does not represent Aristotle's own usage.

[41] Aristotle is not thinking of style and delivery but of the thought and contents.

[42] Here and in 1.9.1 and 2.1.5–7 the role of character in a speech is regarded as making the speaker seem trustworthy. The extended discussion of types of character in 2.12–17 relates to the somewhat different matter of the adaptation of the character of a speaker to the character of an audience. Aristotle's later treatment of character in rhetoric is in fact somewhat wider than in this initial definition.

[43] Aristotle thus does not include in rhetorical ethos the authority that a speaker may possess due to his position in government or society, previous actions, reputation for wisdom, or anything except what is actually contained in the speech and the character it reveals. Presumably, he would regard all other factors, sometimes highly important in the success of rhetoric, as inartistic; but he never says so. One practical reason for stressing character as revealed within the speech was that Greek law required defendants to speak on their own behalf, and they were often lacking in external authority. They could commission a speech from a professional speechwriter (lagographer) and then memorize it for delivery in court. Lysias, in particular, had great success in conveying a favorable impression or moral character (*ēthopoiia*) in the many speeches he wrote for defendants.

[44] Some handbook writers perhaps rejected an appearance of fair-mindedness as too mild and favored an uncompromising attitude. Aristotle's point is that an appearance of fair-mindedness gives the speaker an initial advantage.

rary technical writers try to give their attention. The details on this subject will be made clear when we speak about the emotions.[45]

6. Persuasion occurs through the arguments [*logoi*] when we show the truth or the apparent truth from whatever is persuasive in each case.

7. Since *pisteis* come about through these [three means], it is clear that to grasp an understanding of them is the function of one who can form syllogisms and be observant about characters and virtues and, third, about emotions (what each of the emotions is and what are its qualities and from what it comes to be and how). The result is that rhetoric is a certain kind of offshoot [*paraphues*] or dialectic and of ethical studies (which it is just to call politics).[46] (Thus, too, rhetoric dresses itself up[47] in the form of politics, as do those who pretend to a knowledge of it,[48] sometimes through lack of education, sometimes through boastfulness and other human causes.) Rhetoric is partly [*morion ti*] dialectic, and resembles it, as we said at the outset; for neither of them is identifiable with knowledge of any specific subject, but they are distinct abilities of supplying words. Concerning their potentiality and how they relate to each other, almost enough has been said.

8. In the case of persuasion through proving or seeming to prove something, just as in dialectic **[1356b]** there is on the one hand induction [*epagēgē*] and on the other the syllogism and the apparent syllogism, so the situation is similar in rhetoric; for the *paradeigma* ["example"] is an induction, the *enthymema* a syllogism. I call a rhetorical syllogism an enthymeme, a rhetorical induction a paradigm.[49] And all [speakers] produce logical persuasion by means of paradigms or enthymemes and by nothing other than these. As a result, since it is always necessary to show something either by syllogizing or by inducing (and this is clear to us from the *Analytics*),[50] it is necessary that each of these be the same as each of the others.[51] 9. What the difference is between a

---

45 In 2.2–11. Aristotle's inclusion of emotion as a mode of persuasion, despite his objections to the handbooks, is a recognition that among human beings judgment is not entirely a rational act. There are morally valid emotions in every situation, and it is part of the orator's duty to clarify these in the minds of the audience. On this question in general, see Johnstone 1980; 1–24.

46 In calling rhetoric an *antistrophos* of dialectic in 1.1.1, and an offshoot of dialectic and ethical studies here, and "partly dialectic" and like it in the next sentence, Aristotle avoids use of the formal categories of genus and species. He cannot very well call rhetoric a species of dialectic, since it contains elements—the persuasive effect of character and emotion in particular—that are not proper to dialectic; but at the same time he stresses the logical side of rhetoric and thus its relationship to dialectic. He does not entertain the possibility that dialectic should be regarded as a species of rhetoric, perhaps because dialectic deals with universals, rhetoric with specifics; dialectic is logically prior. Also, to make rhetoric the more general term would lead to the celebration of it as the most characteristic and worthwhile human activity, as Isocrates regarded it. For Aristotle, that honor belongs to philosophy—hence his attempt to find metaphors to describe rhetoric as a mixture of logical, political, and ethical elements. In *Nicomachean Ethics* 1.2.4–6 he says that politics is an "architectonic" subject, of which generalship, economics, and rhetoric are parts.

47 *Hypoduetai*, an echo of Plato, *Gorgias* 464c.

48 Gorgias, Polus, Isacrates, and their followers.

49 Aristotle will discuss the paradigm at greater length in 2.20 and the enthymeme in 2.22. The first three sentences of this paragraph, found in all manuscripts, are double-bracketed by Rudolf Kassel in his Berlin 1976 edition of the Greek text, which is Kassel's way of indicating passages that he regarded as later additions by Aristotle to the otherwise completed treatise. These are interesting suggestions, but essentially subjective in each case.

50 *Prior Analytics* 2.23; *Posterior Analytics* 1.1.

51 Not identical, in which case there would be no need for two sets of terms, but *essentially* the same in their

paradigm and an enthymeme is clear from the *Topics* (for an account was given there earlier of syllogism and induction):[52] to show on the basis of many similar instances that something is so is in dialectic induction, in rhetoric paradigm; but to show that if some premises are true, something else [the conclusion] beyond them results from these because they are true, either universally or for the most part, in dialectic is called syllogism and in rhetoric enthymeme. 10. And it is also apparent that either species of rhetoric[53] has merit (what has also been said in the *Methodics*[54] is true in these cases too); for some rhetorical utterances are paradigmatic, some enthymematic; and similarly, some orators are paradigmatic, some enthymematic. Speeches using paradigms are not less persuasive, but those with enthymemes excite more favorable audience reaction. 11. The cause—and how each should be used—we shall explain later;[55] now we shall explain these things themselves more clearly.

Since the persuasive is persuasive to someone (and is either immediately plausible and believable in itself or seems to be shown by statements that are so) and since no art examines the particular—for example, the art of medicine does not specify what is healthful for Socrates or for Callias but for persons of a certain sort (this is artistic, while particulars are limitless and not knowable)—neither does rhetoric theorize about each opinion—what may seem so to Socrates or Hippias—but about what seems true to people of a certain sort, as is also true with dialectic.[56] For the latter does not form syllogisms from things at random (some things seem true even to madmen) but from that [which seems true] to people in need of argument, and rhetoric [forms enthymemes] from things [that seem true] to people already accustomed to deliberate among themselves.[57] **[1357a]** 12. Its function [*ergon*] is concerned with the sort of things we debate and for which we do not have [other] arts and among such listeners as are not able to see many things all together or to reason from a distant starting point. And we debate about things that seem to be capable for admitting two possibilities; for no one debates things incapable of being different either in past or future or present, at least not if they suppose that to be the case; for there is nothing more [to say]. 13. It is possible to form syllogisms and draw inductive conclusions either from previous syllogisms or from statements that are not reasoned out but require a syllogism [if they are to be accepted] because they are not commonly believed [*endoxa*]; but the former of these [i.e., a chain of syllogism] is necessarily not easy to follow because of the length [of the argument] (the judge is assumed to be a simple

---

underlying structure. In formal logic an induction consists of a series of particular observations from which a general conclusion is drawn; in rhetoric it takes the form of a particular statement supported by one or more parallels, with the universal conclusion left unstated. Similarly, an enthymeme rarely takes the full syllogistic form of major premise, minor premise, and conclusion; more often a conclusion is offered and supported by a reason, as in the first sentence of the *Rhetoric*. On the logic of this passage see Schröder 1985. Schröder does not agree with Kassel's view that it is a later addition.

[52] There is some discussion of syllogism in *Topics* 1.1, and 1.12 offers a definition of induction with an example: "If the skilled pilot is best, and [similarly] the charioteer, then in general the skilled is the best in each thing."

[53] The species using example or that using enthymeme.

[54] A lost logical work by Aristotle of which the extant *On Interpretation* may have been a part; see Rist 1989, 84.

[55] In 2.20–24.

[56] Dialectic builds its proof on the opinions of all, the majority, or the wise; cf. *Topics* 1.1 in Appendix I.C.

[57] Translating the text as conjectured by Kassel.

person),[58] and the latter is not persuasive because the premises are not agreed to or commonly believed. Thus, it is necessary for an enthymeme and a paradigm to be concerned with things that are for the most part capable of being other than they are—the paradigm inductively, the enthymeme syllogistically—and drawn from few premises and often less than those of the primary syllogism;[59] for if one of these is known, it does not have to be stated, since the hearer supplies it: for example, [to show] that Dorieus has won a contest with a crown it is enough to have said that he has won the Olympic games, and there is no need to add that the Olympic games have a crown as the prize; for everybody knows that.[60]

14. Since few of the premises from which rhetorical syllogisms are formed are necessarily true (most of the matters with which judgment and examination are concerned can be other than they are; for people deliberate and examine what they are doing, and [human] actions are all of this kind, and none of them [are], so to speak, necessary) and since things that happen for the most part and are possible can only be reasoned on the basis of other such things, and necessary actions [only] from necessities (and this is clear to us also from the *Analytics*),[61] it is evident that [the premises] from which enthymemes are spoken are sometimes necessarily true but mostly true [only] for the most part. Moreover, enthymemes are derived from probabilities [*eikota*] and signs [*sēmeia*], so it is necessary that each of these be the same as each [of the truth values mentioned];[62] 15. for a probability [*eikos*] is what happens for the most part, not in a simple sense, as some define it, but whatever, among things that can be other than they are, is so related to that in regard to which it is probable as a universal is related to a particular.[63] **[1357b]** 16. In the case of signs [*sēmeia*], some are related as the particular to the universal, some as the universal to the particular. Of these, a necessary sign is a *tēkmerion*, and that which is not necessary has no distinguishing name. 17. Now I call necessary for those from which a [logically valid] syllogism can be formed; thus, I call this kind of sign a *tēkmerion*; for when people think it is not possible to refute a statement, they think they are offering a *tekmērion*, as though the matter were shown and concluded [*peparasmenon*]. (*Tekmar* and *peras*

---

58 By *judge (kritēs)* Aristotle means a member of the assembly or of a jury. In Athenian legal procedures there were no professional judges in the modern sense. The democratic juries of the Athenian courts ranged in size from 201 to 5,001, drawn by lot from the male citizen body.

59 The fully expressed syllogism that is logically inherent in the enthymeme.

60 Later writers (see Appendix I.F) often regard an enthymeme as an abbreviated syllogism in which one premise, usually the major, is not expressed but is assumed, e.g., "Socrates is mortal, for he is a man," assuming "all men are mortal." Aristotle notes that this is often the case, but it is not a necessary feature of the enthymeme. The real determinant of an enthymeme in contrast to a syllogism is what a popular audience will understand. Aristotle regards rhetoric, and thus the enthymeme, as addressed to an audience that cannot be assumed to follow intricate logical argument or will be impatient with premises that seem unnecessary steps in the argument. The underlying logical structure should, however be present.

61 *Prior Analytics* 1.8, 1.12–14, 1.27; *Posterior Analytics* 1.6, 1.30, 2.12.

62 I.e., probabilities correspond to things true for the most part, signs to things necessarily true. But Aristotle will modify this in what follows: some signs are necessary, others only probable. Both probabilities and signs are statements about human actions, though they may be based on physical manifestations, as the following examples show.

63 Grimaldi (1980–88, 1:62) instances "Children love their parents": it is a "probability" because a general observation—universal in form, probably, but not necessarily true in particular instances. "Some" may refer to handbook writers who discussed argument from probability.

["limit, conclusion"] have the same meaning in the ancient form of [our] language.) 18. An example of signs [*sēmeia*] related as the particular to the universal is if someone were to state that since Socrates was wise and just, it is a sign that the wise are just. This is indeed a sign, but refutable, even if true in this case; for it is not syllogistically valid. But if someone were to state that there is a sign that someone is sick, for he has a fever, or that a woman has given birth, for she has milk, that is a necessary sign. Among signs, this is only true of a *tekmērion*; for only it, if true, is irrefutable. It is an example of the relation of the universal to the particular if someone said that it is a sign of fever that someone breathes rapidly. This, too, is refutable, even if true [in some case]; for it is possible to breathe rapidly and not be feverish. Thus, what probability is and what sign and *tekmērion* are and how they differ has now been explained. In the *Analytics*[64] they are defined more clearly, and the cause explained why some are not syllogistic and others are.

19. It has been explained that a paradigm is an induction and with what kinds of things it is concerned. It is reasoning neither from part to whole nor from whole to part but from part to part, like to like, when two things fall under the same genus but one is better known than the other.[65] For example, [when someone claims] that Dionysius is plotting tyranny because he is seeking a bodyguard; for Peisistratus also, when plotting earlier, sought a guard and after receiving it made himself tyrant, and Theagenes [did the same] in Megara, and others, whom the audience knows of, all become examples of Dionysius, of whom they do not yet know whether he makes his demand for this reason. All these actions fall under the same [genus]: that one plotting tyranny seeks a guard.[66]

[1358a] The sources of *pisteis* that seem demonstrative [*apodeiktikai*] have now been explained. 20. But in the case of enthymemes, a very big difference—and one overlooked by almost everybody—is one that is also found in the case of syllogisms in dialectical method; for some [enthymemes] are formed in accord with the method of rhetoric, just as also some syllogisms are formed in accord with the method of dialectic, while others accord with [the content of] other arts and capabilities, either those in existence or those not yet understood.[67] Hence, [the differences] escape notice of the listeners; and the more [speakers] fasten upon [the subject matter] in its proper sense, [the more] they depart from rhetoric or dialectic.[68] This statement will be clearer if explained in more detail.

---

[64] *Prior Analystics* 2.27.

[65] There is an "unmeditated inference," or unspoken recognition of the universal proposition. See Hauser 1985, 171–79.

[66] It could be argued that seeking a bodyguard is a "sign" of intent to establish a tyranny, and certainly paradigms and signs have some similarity; but Aristotle seems to think of a paradigm as useful in indicating motivation or the probable course of events that the audience might not otherwise anticipate, whereas a sign is usually an existing fact or condition that anyone might recognize. More important to him, however, is the logical difference that the paradigm moves from the particular premises to a particular conclusion, with the universal link not necessarily expressed (just as the universal major premise of an enthymeme need not be expressed), whereas the sign moves either from universal to particular or particular to universal.

[67] It is characteristic of Aristotle to feel that there were other subjects not yet systematically studied.

[68] This passage is regarded as textually corrupt by the editors. Kassel indicates that something has been lost after *listeners*; Ross rejects *the more*. The basic thought is that people do not realize that rhetoric and dialectic, though they have a method, lack content or facts and must borrow these from other disciplines, such as politics or ethics. Enthymemes are rhetorical strategies but also usually substantive arguments; and the more the argu-

The "Topics" of Syllogisms and Enthymemes

Topos literally means "place," metaphorically that location or space in an art where a speaker can look for "available means of persuasion." Rhetoric itself can be said to operate in civic space. Although the word accords with Aristotle's fondness for visual imagery, he did not originate its use in the sense of "topic"; Isocrates, early in the fourth century, had so used it, and probably others did before him. In Isocrates' *Encomium of Helen* (section 4) *topos* refers to forms of eristical argument, such as fact or possibility—what Aristotle will call *koina.* In the same speech (section 38) *topos* refers to the use of an ancient witness, Theseus' opinion of Helen—what Aristotle regards as "nonartistic" *pistis.* The word may also already have been used in mnemonic theory of the physical setting against which an object or idea could be remembered. Neither in *Topics* nor in *Rhetoric* does Aristotle give a definition of *topos*, another sign that he assumed the word would be easily understood; he does, however, give his own special twist to its meaning, usually distinguishing it from *koina* and *idia* and using it primarily of strategies of argument as discussed in 2.23. See Sprute 1982, 172–82.

21. I am saying that dialectical and rhetorical syllogisms are those in which we state *topoi*, and these are applicable in common [*koinēi*] to questions of justice and physics and politics and many different species [of knowledge]; for example, the *topos* of the more and the less;[69] for to form syllogisms or speak enthymemes from this about justice will be just as possible as about physics or anything else, although these subjects differ in species.[70] But there are "specifics"[71] that come from the premises of each species and genus [of knowledge]; for example, in physics there are premises from which there is neither an enthymeme nor a syllogism applicable to ethics; and in ethics [there are] others not useful in physics. It is the same in all cases. The former [the common *topoi*] will not make one understand any genus; for they are not concerned with any underlying subject. As to the latter [the specifics], to the degree that someone makes better choice of the premises, he will have created knowledge different from dialectic and rhetoric without its being recognized; for if he succeeds in hitting on first principles [*arkhai*], the knowledge will no longer be dialectic or rhetoric but the science of which [the speaker] grasps the first principles.[72] 22. Most en-

---

ment comes from the premises of politics, ethics, or other subjects, the more the enthymeme becomes an argument of that discipline and the less it is purely rhetorical. In practice, the limits are never reached; any argument has some strategy (what Aristotle will call "topics" in 2.23) and some content (what he will call *idia* and discuss in 1.4–14 and 2.1–17). Some possible implications of this passage are discussed by Garver 1988, but he twists the meaning of some of Aristotle's words (*metabainō*, *tynkhanō*, etc.) to create problems that perhaps do not exist.

[69] To be discussed in 2.23.4 (the chapter on topics).

[70] The *topos* does not tell one anything about these subjects but can be applied to each; for example, "If it is just to punish offenses, it is more just to punish great offenses," "If a small force will move a body, a larger force will move it as well" and "If public revenues will support a large army, they will support a smaller army."

[71] *Idia* (n. pl. of the adj. from *eidos*), "specificities, specific or particular things." The word is chosen to denote things characteristic of the species. Aristotle here does not call these specifics topics, but he does so refer to them in 1.15.19; and in sec. 22, as well as in 1.61, he speaks of them as *stoikheia*, which he says later (2.22.13, 2.26.1) are the "same" as topics. Thus, many rhetoricians have found it convenient to speak of "special, specific, particular, material" topics belonging to the separate disciplines, in contrast to "common" or "formal" topics, which are rhetorical or dialectical strategies of argument.

[72] For the concept of "first principles" see note on 1.7.12. Part or all of a discourse may be thought of as falling in a spectrum, varying from the most general and popular to the most technical. A speech in a law court, for example, will become less "rhetorical" and more "jurisprudential" as it undertakes detailed discussion of the

thymemes are derived from these species that are particular and specific, fewer from the common [topics].[73] Just as in the case of *topoi*, so also in the case of enthymemes, a distinction should be made between the species and the *topoi* from which they are to be taken. By "species" I mean the premises specific to each genus [of knowledge], and by *topoi* those common to all. But let us take up first the genera [*gene*] of rhetoric so that having defined how many there are, we may separately take up their elements[74] and premises.[75]

Chapter 3: The Three Species of Rhetoric: Deliberative, Judicial, and Epideictic

1. The species [*eidē*] of rhetoric are three in number; for such is the number [of classes] to which the hearers of speeches belong. A speech [situation] consists of three things: a speaker and a subject on which he speaks and someone addressed,[76] **[1358b]** and the objective [*telos*] of the speech relates to the last (I mean the hearer). 2. Now it is necessary for the hearer to be either a spectator [*theoros*] or a judge [*kritēs*], and [in the latter case] a judge of either past or future happenings. A member of a democratic assembly is an example of one judging about future happenings, a juryman an example of one judging the past. A spectator is concerned with the ability [of the speaker].[77] 3. Thus, there would necessarily be three genera of rhetorics;[78] *symbouleutikon* ["deliberative"], *dikanikon* ["judicial"], *epideiktikon* ["demonstrative"]. Deliberative advice is ei-

---

law. In terms of valid proof it is desirable to do this, but too technical a speech will not be comprehensible to the judges.

[73] This is because of the need for "content": rhetoric constantly employs the special knowledge of other arts, such as politics or ethics.

[74] Elements (*stoikheia*) are the same as topics; see 2.22.13, 2.26.1.

[75] Aristotle's use of *genos*, *eidos*, and *idia* in this passage may make it somewhat difficult to follow; but he is probably not seeking to make a logical statement about the relationship of genus and species. In a general way he can be said to view knowledge as a genus of which particular forms, (e.g., physics, politics, and ethics) are species (*eidē*). The premises of the *eidē* are their *idia*. In the concluding sentence he also calls the kinds of rhetoric *genē* (genera), but in the first sentence of the next chapter will call them *eidē* (species) and in 3.3 reverts to *genē*. See n. 78.

[76] Eighteenth-century rhetoricians add *the occasion* to Aristotle's three factors in the speech situation, and modern authorities have suggested other approaches, e.g., "addresser, message, addressee, context, common code, and contact" (Roman Jakobson).

[77] This sentence is rejected by Kassel as an insertion into the text by a later reader, perhaps rightly. The audience in epideictic is not called upon to take a specific action, in the way that an assemblyman or juryman is called upon to vote; but epideictic may be viewed as an oratorical contest, either with other speakers or previous speakers (cf., e.g., Isocrates, *Panegyricus* 1), and in 2.18.1 Aristotle notes that the spectator also is in this sense a judge. The definition of epideictic has remained a problem in rhetorical theory, since it becomes the category for all forms of discourse that are not specifically deliberative or judicial; later ancient rhetoricians regarded it as including poetry and prose literature, and since Renaissance times it has sometimes included other arts like painting, sculpture, and music as well. Aristotle, however, thinks of epideictic only as a species of oratory as he knew its forms in Greece, including funeral orations like that by Pericles in Thucydides' *History of the Peloponnesian War* (2.35–46) and the *Encomia* of Helen by Gorgias and Isocrates. In such speeches, praise corrects, modifies, or strengthens an audience's belief about civic virtue or the reputation of an individual.

[78] The appearance here of "rhetorics" in the plural is very unusual in Greek and probably results from the use of *genē* in the plural. Aristotle may use *genē* here of the kinds of rhetorics earlier called *eidē* because in the next sentence he is going to divide them further into species.

ther protreptic ["exhortation"] or apotreptic ["dissuasion"]; for both those advising in private and those speaking in public always do one or the other of these. In the law court there is either accusation [*katēgoria*] or defense [*apologia*]; for it is necessary for the disputants to offer one or the other of these. In epideictic, there is either praise [*epainos*] or blame [*psogos*]. 4. Each of these has its own "time": for the deliberative speaker, the future (for whether exhorting or dissuading he advises about future events); for the speaker in court, the past (for he always prosecutes or defends concerning what has been done); in epideictic the present is the most important; for all speakers praise or blame in regard to existing qualities, but they often also make use of other things, both reminding [the audience] of the past and projecting the course of the future.[79] 5. The "end"[80] of each of these is different, and there are three ends for three [species]: for the deliberative speaker [the end] is the advantageous [*sympheron*][81] and the harmful (for someone urging something advises it as the better course and one dissuading dissuades on the ground that it is worse), and he includes other factors as incidental: whether it is just or unjust, or honorable or disgraceful; for those speaking in the law courts [the end] is the just [*dikaion*] and the unjust, and they make other considerations incidental to these; for those praising and blaming [the end] is the honorable [*kalon*] and the shameful, and these speakers bring up other considerations in reference to these qualities. 6. Here is a sign that the end of each [species of rhetoric] is what has been said: sometimes one would not dispute other factors; for example, a judicial speaker [might not deny] that he has done something or done harm, but he would never agree that he has [intentionally] done wrong; for [if he admitted that,] there would be no need of a trial. Similarly, deliberative speakers often grant other factors, but they would never admit that they are advising things that are not advantageous [to the audience] or that they are dissuading [the audience] from what is beneficial; and often they do not insist that it is not unjust to enslave neighbors or those who have done no wrong. And similarly, those who praise or blame do not consider whether someone has done actions that are advantageous or harmful [to himself] **[1359a]** but often they include it even as a source of praise that he did what was honorable without regard to the cost to himself; for example, they praise Achilles because he went to the aid of his companion Patroclus knowing that he himself must die, though he could have lived. To him, such a death was more honorable; but life was advantageous.

---

[79] In practice, as in funeral orations, speakers usually praise past actions but with the intent of celebrating timeless virtues and inculcating them as models for the future.

[80] *Telos*, the final objective of the speaker and his art, which is actualized in the persuasion of an audience. Later rhetoricians sometimes called these "final headings." Each *telos* often becomes a specific topic in a speech; see, for example, the discussions of expedience and justice in the speeches of Cleon and Diodotus in the Mytilenian debate in Thucydides 3.37–48.

[81] *Sympheron* is often translated "expedient"; literally, it means whatever "brings with it" advantage (Lat. *utilitas*). Later rhetoricians were troubled by the moral implication and sought to modify what they saw as Aristotle's focus on expediency in political discourse; see esp. Quintilian 3.8.1–3. Since Aristotle has said in 1.1.12 that we must not persuade what is bad, he would presumably recommend that a speaker seek to identify the enlightened, long-term advantage to the audience. "Advantageous" or "beneficial" seems the best translation. In sec. 6 Aristotle recognizes that in practice deliberative speakers are often indifferent to the question of the injustice to others of some action.

Propositions Common to All Species of Rhetoric

No technical term appears in this chapter to denote the four subjects of propositions described here, but in 2.18.2 they are called *koina*, "common things," "commonalties," in contrast to *idia*, "specifics." They are discussed in greater detail in 2.19. Since the *koinon* "greater and smaller" discussed in section 9 seems similar to the topic of "the more and the less" mentioned in 1.2.21, these *koina* have often been called "topics" or "common topics." Grimaldi (1980–88, 1:85–86) objects to this, with some reason, though in 3.19.2 Aristotle speaks of "topics" of amplification and seems to be referring to 2.19. Generally, however, Aristotle keeps them distinct: the topic of "the more and the less," discussed separately in 2.23.4, is a strategy of argument, always involving some contrast, whereas "greater and smaller," discussed in 1.7, 14 and 2.19.26–27, are arguments about the degree of *magnitude* (that term occurs in 2.18.4) or importance of something and are analogous to such questions as whether something is possible or has actually been done. Whether something is possible, actually true, or important are fundamental issues in any speech; and thus Aristotle mentions them immediately after identifying the basic issues of the advantageous, the just, and the honorable.

7. It is evident from what has been said that it is first of all necessary [for a speaker] to have propositions [*protaseis*] on these matters.[82] (*Tekmēria* and probabilities and signs are rhetorical propositions. A syllogism is wholly from propositions, and the enthymeme is a syllogism consisting of propositions expressed.)[83] 8. And since impossibilities cannot be done nor have been done, but possibilities [alone can be done or have been done], it is necessary for the deliberative, judicial, and epideictic speaker to have propositions about the possible and the impossible and [about] whether something has happened or not and [about] whether it will or will not come to be. 9. Further, since all speakers, praising and blaming and urging and dissuading and prosecuting and defending, not only try to show what has been mentioned but that the good or the evil or the honorable or the shameful or the just or the unjust is great or small, either speaking of things in themselves or in comparison to each other, it is clear that it would be necessary also to have propositions about the great and the small and the greater and the lesser, both generally and specifically; for example, [about] what is the greater or lesser good or injustice or justice, and similarly about other qualities.[84] The subjects about which it is necessary to frame propositions have [now] been stated. Next we must distinguish between each in specific terms; that is, what deliberation, and what epideictic speeches, and thirdly, what lawsuits, are concerned with.

DEVELOPING YOUR UNDERSTANDING

1. **Aristotle argues that the function of rhetoric is not to persuade but to discover the available means of persuasion. Explain how "persuasion" and "discovering the means of persuasion" are distinct. Then, explain how the distinction affects the scope and purpose of the rhetor's/professional writer's work?**

---

[82] The advantageous, the just, the honorable, and their opposites.

[83] The propositions inherent in the underlying syllogism are not necessarily all expressed in the related enthymeme; some may be assumed.

[84] The subjects of propositions common to all species of rhetoric are thus the possible and impossible, past fact (or its nonexistence), future fact (or its nonexistence), and degree of magnitude or importance. These are discussed further in 2.19.

2. According to Aristotle's characterization, rhetoric is a generalizable art that, when applied to particular situations, helps guide speaking/writing. Briefly explain this characterization of rhetoric. Then, discuss what professional writers gain from an art of rhetoric, as well as the limitations such an art poses for professional writers.
3. Aristotle argues that "rhetoric is a certain kind of offshoot [*paraphues*] of dialectic and of ethical studies (which it is just to call politics)." Referring to this passage, found in Chapter 2, section 7, as well as others, summarize the relationship that Aristotle builds between rhetoric and other arts, sciences, and practices.
4. Identify and explain the functions of rhetoric as argued by Aristotle. Refer to specific passages.

FOCUSING ON KEY TERMS AND CONCEPTS

Focus on the following terms and concepts while you read through this selection. Understanding these will not only increase your understanding of the selection that follows, but you will find that, because most of these terms or concepts are commonly used in professional writing and rhetoric, understanding them helps you get a better sense of the field itself.

1. science
2. induction
3. syllogism
4. art
5. the variable
6. the invariable
7. practical wisdom
8. deliberation
9. virtue

Excerpts from "Book VI, Intellectual Virtue" in Nicomachean Ethics: The Chief Intellectual Virtues

ARISTOTLE translated by DAVID ROSS

SCIENCE—DEMONSTRATIVE KNOWLEDGE OF THE NECESSARY AND ETERNAL

3. Let us begin, then, from the beginning, and discuss these states once more. Let it be assumed that the states by virtue of which the soul possesses truth by way of affirmation or denial are five in number, i.e. art, scientific knowledge, practical wisdom, philosophic wisdom, intuitive reason; we do not include judgment and opinion because in these we may be mistaken.

Now what scientific knowledge is, if we are to speak exactly and not follow mere similarities, is plain from what follows. We all suppose that what we know is not even

---

Source: From *The Nicomachean Ethics* by Aristotle, translated by David Ross (World's Classics, 1980), pp. 140–143. Reprinted by permission of Oxford University Press.

capable of being otherwise; of things capable of being otherwise we do not know, when they have passed outside our observation, whether they exist or not. Therefore the object of scientific knowledge is of necessity. Therefore it is eternal; for things that are of necessity in the unqualified sense are all eternal; and things that are eternal are ungenerated and imperishable. Again, every science is thought to be capable of being taught, and its object of being learned. And all teaching starts from what is already known, as we maintain in the Analytics also; for it proceeds sometimes through induction and sometimes by syllogism. Now induction is the starting-point which knowledge even of the universal presupposes, while syllogism proceeds from universals. There are therefore starting-points from which syllogism proceeds, which are not reached by syllogism; it is therefore by induction that they are acquired. Scientific knowledge is, then, a state of capacity to demonstrate, and has the other limiting characteristics which we specify in the Analytics, for it is when a man believes in a certain way and the starting-points are known to him that he has scientific knowledge, since if they are not better known to him than the conclusion, he will have his knowledge only incidentally.

Let this, then, be taken as our account of scientific knowledge.

ART—KNOWLEDGE OF HOW TO MAKE THINGS

4. In the variable are included both things made and things done; making and acting are different (for their nature we treat even the discussions outside our school as reliable); so that the reasoned state of capacity to act is different from the reasoned state of capacity to make. Hence too they are not included one in the other; for neither is acting making nor is making acting. Now since architecture is an art and is essentially a reasoned state of capacity to make, and there is neither any art that is not such a state nor any such state that is not an art, art is identical with a state of capacity to make, involving a true course of reasoning. All art is concerned with coming into being, i.e. with contriving and considering how something may come into being which is capable of either being or not being, and whose origin is in the maker and not in the thing made; for art is concerned neither with things that are, or come into being, by necessity, nor with things that do so in accordance with nature (since these have their origin in themselves). Making and acting being different, art must be a matter of making, not of acting. And in a sense chance and art are concerned with the same objects; as Agathon says, "art loves chance and chance loves art." Art, then, as has been is a state concerned with making, involving a true course of reasoning, and lack of art on the contrary is a state concerned with making, involving a false course of reasoning; both are concerned with the variable.

PRACTICAL WISDOM—KNOWLEDGE OF HOW TO SECURE THE ENDS OF HUMAN LIFE

5. Regarding practical wisdom we shall get at the truth by considering who are the persons we credit with it. Now it is thought to be the mark of a man of practical wisdom to be able to deliberate well about what is good and expedient for himself, not in some particular respect, e.g. about what sorts of things conduce to health or to strength, but about what sorts of thing conduce to the good life in general. This

is shown by the fact that we credit men with practical wisdom in some particular respect when they have calculated well with a view to some good end which is one of those that are not the object of any art. It follows that in the general sense also the man who is capable of deliberating has practical wisdom. Now no one deliberates about things that are invariable, nor about things that it is impossible for him to do. Therefore, since scientific knowledge involves demonstration, but there is no demonstration of things whose first principles are variable (for all such things might actually be otherwise), and since it is impossible to deliberate about things that are of necessity, practical wisdom cannot be scientific knowledge nor art; not science because that which can be done is capable of being otherwise, not art because action and making are different kinds of things. The remaining alternative, then, is that it is a true and reasoned state of capacity to act with regard to the things that are good or bad for man. For while making has an end other than itself, action cannot; for good action itself is its end. It is for this reason that we think Pericles and men like him have practical wisdom, viz. because they can see what is good for themselves and what is good for men in general; we consider that those can do this who are good at managing households or states. (This is why we call temperance (sophrosune) by this name; we imply that it preserves one's practical wisdom (sozousa tan phronsin). Now what it preserves is a judgment of the kind we have described. For it is not any and every judgment that pleasant and painful objects destroy and pervert, e.g. the judgment that the triangle has or has not its angles equal to two right angles, but only judgments about what is to be done. For the originating causes of the things that are done consist in the end at which they are aimed; but the man who has been ruined by pleasure or pain forthwith fails to see any such originating cause—to see that for the sake of this or because of this he ought to choose and do whatever he chooses and does; for vice is destructive of the originating cause of action.)

Practical wisdom, then, must be a reasoned and true state of capacity to act with regard to human goods. But further, while there is such a thing as excellence in art, there is no such thing as excellence in practical wisdom; and in art he who errs willingly is preferable, but in practical wisdom, as in the virtues, he is the reverse. Plainly, then, practical wisdom is a virtue and not an art. There being two parts of the soul that can follow a course of reasoning, it must be the virtue of one of the two, i.e. of that part which forms opinions; for opinion is about the variable and so is practical wisdom. But yet it is not only a reasoned state; this is shown by the fact that a state of that sort may forgotten but practical wisdom cannot.

DEVELOPING YOUR UNDERSTANDING

1. Prepare a table or matrix that captures the distinctions between science, art, and practical wisdom as outlined by Aristotle, and be prepared to discuss your table/matrix in class.
2. Characterize what rhetoric would be if it were categorized as a science, an art, and a practice/action. Then, compare and contrast the three depictions, drawing out the key similarities and differences.
3. Aristotle used the variable and invariable as one set of features to distinguish the virtues. Referring to other readings in this chapter, and perhaps sources outside this book with which you are familiar, argue whether you consider rhetoric to be concerned with the variable or the invariable, and explain the impact you see such a distinction making on what rhetoric is and is not.

FOCUSING ON KEY TERMS AND CONCEPTS

Focus on the following terms and concepts while you read through this selection. Understanding these will not only increase your understanding of the selection that follows, but you will find that, because most of these terms or concepts are commonly used in professional writing and rhetoric, understanding them helps you get a better sense of the field itself.

1. epideictic
2. deliberative
3. judicial
4. invention
5. arrangement
6. style
7. memory
8. delivery
9. six parts of discourse

Excerpts from Book I, Rhetorica Ad Herennium

MARCUS TULLIUS CICERO translated by HARRY CAPLAN

BOOK I

1 I. My private affairs keep me so busy that I can hardly find enough leisure to devote to study, and the little that is vouchsafed to me I have usually preferred to spend on philosophy. Yet your desire, Gaius Herennius, has spurred me to compose a work on the Theory of Public Speaking, lest you should suppose that in a matter which concerns you I either lacked the will or shirked the labour. And I have undertaken this project the more gladly because I knew that you had good grounds in wishing to learn rhetoric, for it is true that copiousness and facility in expression bear abundant fruit, if controlled by proper knowledge and a strict discipline of the mind.

That is why I have omitted to treat those topics which, for the sake of futile self-assertion, Greek writers[a] have adopted. For they, from fear of appearing to know too little, have gone in quest of notions irrelevant to the art, in order that the art might seem more difficult to understand. I, on the other hand, have treated those topics which seemed pertinent to the theory of public speaking. I have not been moved by hope of gain[b] or desire for glory, as the rest have been, in undertaking to write, but

Source: Reprinted by permission of the publishers and Trustees of the Loeb Classical Library from *Cicero: Volume I ~ Rhetorica and Herenium,* Loeb Classical Library Volume L 403, translated by Harry Caplan, Cambridge, Mass: Harvard University Press, 1954, pp. 3–11. The Loeb Classical Library ® is a registered trademark of the President and Fellows of Harvard College.

[a] The beginning of Book 4 further sets forth the author's attitude to the Greek writers on rhetoric (who these are specifically is uncertain); *cf.* also 3. xxiii. 38. For his attitude to philosophical studies see the end of Book 4.

[b] Apparently textbooks on public speaking sold well; see Theodor Birt, *Rhein. Mus.* 72 (1917/18). 311–16.

have done so in order that, by my painstaking work, I may gratify your wish. To avoid prolixity, I shall now begin my discussion of the subject, as soon as I have given you this one injunction: Theory without continuous practice in speaking is of little avail; from this you may understand that the precepts of theory here offered ought to be applied in practice.

2 II. The task of the public speaker is to discuss capably those matters which law and custom have fixed for the uses of citizenship, and to secure as far as possible the agreement of his hearers.[c] There are three kinds[d] of causes which the speaker must treat: Epideictic, Deliberative, and Judicial.[e] The epideictic kind is devoted to the praise or censure of some particular person. The deliberative consists in the discussion of policy and embraces persuasion and dissuasion.[f] The judicial is based on legal controversy, and comprises criminal prosecution or civil suit, and defence.[g]

Now I shall explain what faculties the speaker should possess, and then show the proper means of treating these causes.[h]

3 The speaker, then, should possess the faculties of Invention, Arrangement, Style, Memory, and Delivery.[i] Invention is the devising of matter, true or plausible, that would make the case convincing.[j] Arrangement is the ordering and distribution of the matter, making clear the place to which each thing is to be assigned. Style is the adaptation of suitable words and sentences to the matter devised. Memory is the

---

[c] The definition is that of Hermagoras, to whom the function (εργον) of the perfect orator is τὸ τεθὲν πολιτικὸν ζητήμα διατιθεσθαι κατὰ τὸ ἐνδεχόμενον πειοτικῶς . See Sextus Empiricus, *Adv. Rhet.* 62, ed. Fabricius, 2. 150. *Cf.* Cecero, *De Inv.* 1. v. 6.

[d] γένη.

[e] ἐπιδεικτικόν, ουμβουλευτικόν, δικανικόν. The scheme is Aristotelian (*rhet.* 1. 3, 1358b) but in essence older. The author's emphasis in the first two books, on the judicial kind, is characteristically Hellenistic (*e.g.*, Hermagorean). The better tradition indicates that originally rhetoric was concerned with the judicial kind, and was later extended to the other two fields. For a study of the *three genera* see D. A. G. Hinks, *Class. Quarterly* 30 (1936). 170–6. *Cf.* Cicero, *De Inv.* 1. v. 7.

[f] προτροπή ανδ ἀποτροπή.

[g] κατηγορία, δική, απολογία.

[h] 2. ii. 2 below.

[i] εὕρεσις, τάξις or οἰκονομία, λέξις or ἑρμηνεία or φράσιζ, μνήμη, ὑποκρισιζ. The pre-Aristotelian rhetoric, represented by the *Rhet. Ad Alexandrum*, treated the first three (without classifying them); Aristotle would add Delivery (*Rhet.* 3. 1, 1403b), and his pupil Theophrastus did so (see note on 3. xi. 19 below). When precisely in the Hellenistic period Memory was added as a fifth division by the Rhodian or the Pergamene school, we do not know. These faculties (*res*; see also 1. ii. 3) are referred to in 2. i. 1 below (*cf.* 1.iii.4) as the speaker's *functions* (*officia* = εργα του ῥήτορος). Quintilian, 3. 3. 11ff., considers them as departments or constituent elements of the art (*paries rhetorices*) rather than as *opera* (= *officia*); so also here at 3. i. 1, 3. viii. 15, 3. xvi. 28, and Cicero, *De Invi.* 1. vii. 9. εργον is an Aristotelian concept (*cf.* the definition of rhetoric in *Rhet.* 1. 1–2, 1355b), and Aristotle was the first to classify the (major) functions. Our author here gives the usual order of the divisions; so also Cicero, *De Oratore* 1. 31. 142. Diogenes Laertius, 7. 43, presents the Stoic scheme: Invention, Style (φράσις), Arrangement, and Delivery. A goodly number of rhetorical systems were actually based on these εργα (*e.g.*, in most part Cicero's and Quintilian's); others were based on the divisions of the discourse (μόρια λόγου). See K. Barwick, *hermes* 57 (1922). 1 ff.; Friedrich Solmsen, *Amer. Journ. Philo.* 62 (1941). 35–50, 169–90. Our author conflates the two schemes he has inherited; see especially 1. ii. 3–iii. 4, 2. i. 1–ii. 2, and the Introduction to the present volume, p. xviii.

[j] The concept goes back at least as far as Plato (*e.g.*, *Phaedrus* 236 A); see Aristotle, *Rhet.* 1. 2 (1355b), on finding artistic proofs.

firm retention in the mind of the matter, words, and arrangement. Delivery is the graceful regulation of voice, countenance, and gesture.

All these faculties we can acquire by three means: Theory, Imitation, and Practice.[k] By theory is meant a set of rules that provide a definite method and system of speaking. Imitation stimulates us to attain, in accordance with a studied method, the effectiveness of certain models in speaking. Practice is assiduous exercise and experience in speaking.

Since, then, I have shown what causes the speaker should treat and what kinds of competence he should possess, it seems that I now need to indicate how the speech can be adapted to the theory of the speaker's function.

4 III. Invention is used for the six parts of a discourse: the Introduction, Statement of Facts, Division, Proof, Refutation, and Conclusion.[l] The Introduction is the beginning of the discourse, and by it the hearer's mind is prepared[m] for attention. The Narration or Statement of Facts sets forth the events that have occurred or might have occurred.[n] By means of the Division we make clear what matters are agreed upon and what are contested, and announce what points we intend to take up. Proof is the presentation of our arguments, together with their corroboration.[o]

---

[k] τέχνη (also παίδεία, ἐπιστήμη, μάθησις, *scientia, doctrina*), μίμησις, γυμνασία (also ἄσκησις, μελέτη, εμπειρία, συνηθεια, *declamatio*). The usual triad, Nature (φυσις, *nature, ingenium, facultas*), Theory, and Practice, can be traced back to Protagoras, Plato (*Phaedrus* 269 D), and Isocrates (*e.g.*, *Antid.* 187; *Adv. Soph.* 14–18, where Imitation is also included). *Cf.* also Aristotle in Diogenes Laertius 5. 18; Cicero, *De Inv.* 1. i. 2. *De Oratore* 1. 4. 14; Dionysius Halic. in Syrianus, *Scholia Hermog.*, ed. Rabe, 1. 4–5; Tacitus, *Dialog. de Orator.*, ch. 33; Plutarch, *De liberis educ.* 4 (2 A); and see Paul Shorey, *Trans. Am. Pilol. Assn.* 40 (1909). 185–201. Imitation is presumed to have been emphasized in the Pergamene school of rhetors under Stoic influence. Quintilian, 3. 5. 1, tells us that it was classed by some writers as a fourth element, which he yet subordinates to Theory. On Imitation *cf.* Antonius in Cicero, *de Oratore* 2. 21. 89 ff.; Dionysius halic., *De Imitat.* (*Opuscula* 2. 197–217, ed. Usener-Rader-macher); Quintilian, 10. 1. 20 ff.; Eduard Stemplinger *Das Plagiat in der Griech. Lit.*, Leipzig and Berlin, 1912, pp. 81 ff.; Kroll, "Rhetorik", coll. 1113 ff.; Paulus Otto, *Quaestiones selectae ad libellum qui est περι υψους* spectantes, diss. Kiel, 1906, pp. 6–19; G. C. Fiske, *Lucilius and Horace*, Madison, 1920, ch. 1; J. F. D'Alton, *Roman Literary Theory and Criticism*, Longon, New York, and Toronto, 1931, pp. 426 ff. Richard McKeon, "Literary Criticism and the Concept of Imitation in Antiquity," *Mod. Phiol.* 34, 1 (1936). 1–35, and esp. pp. 26 ff.; D. L. Clark, "Imitation : Theory and Practice in roman Rhetoric," *Quart. Journ. Speech* 37, 1 (1951). 11–22. "Exercise" refers to the *progymnasmata*, of which our treatise and Cicero's *De Inv.* show the first traces in Latin rhetoric, and to the "suasoriae" (*deliberations*) and "controversiae" (*causae*) in which the treatise abounds. See also 4. xliv. 59 (Refining). The divorce between *praeexercitamenta* and *exercitations* belongs to the Augustan period.

[l] The author's treatment of the parts of a discourse differs from that of Aristotle, who, in *Rhet.* 3. 13 (1414a) ff., discusses them—Proem, Statement of Facts, Proof, and Conclusion—with all three kinds of oratory in view, not only the judicial, under Arrangement. Note that Invention is applied concretely to the parts of the discourse; in 1. xi. 18 ff. below the Issues are subjoined to Proof and Refutation. *Cf.* Cicero, *De Inv.* 1. xiv. 19. The Stoic scheme included Proem, Statement of Facts, Replies to Opponents, and Conclusion (Diogenes Laertius 7. 43).

[m] παρασκευάζεται.The concept if Isocratean. *Cf. Rhet. Ad Alex.*, ch. 29 (1436a); Dionysius Halic., *De Lys.* 17; Anon. Seg. 5 and 9 (Spengel-Hammer 1 [2]. 353–4); Rufus 4 (Spengel-Hammer 1 [2]. 399); Anon., in Rabe, *Proleg. Sylloge*, p. 62.

[n] This definition is translated directly from a Greek original; see Hermogenes, *Progymn.* 2 (ed. rabe, p. 4), Syrianus, *Scholia Hermog.* (ed. Rabe, p. 4), Syrianus, *Scholia Hermog.* (ed. Rabe 2. 170), Theon 4 (Spengel 2. 78). *Cf.* Cicero, *De Inv.* 1. xix. 27.

[o] *Cf.* Cicero, *De Inv.* 1. xxiv. 34.

[p] *Cf.* Cicero, *De Inv.* 1. xlii. 78 (*reprehensio*).

Refutation is the destruction of our adversaries' arguments.[p] The Conclusion is the end of the discourse, formed in accordance with the principles of the art.

Along with the speaker's functions, in order to make the subject easier to understand, I have been led also to discuss the parts of a discourse, and to adapt these to the theory of Invention. It seems, then, that I must at this juncture first discuss the Introduction.[q]

DEVELOPING YOUR UNDERSTANDING

1. Referring to the pages you have from Book I of *Rhetorica Ad Herennium*, summarize the function and scope of rhetoric. Compare and contrast what *Rhetorica Ad Herennium* puts forward with what Aristotle's *Rhetoric* puts forward.
2. Using the little bit of *Rhetorica Ad Herennium* you have available here and any other sources from this chapter, support, contend with, or modify how Foss, Foss, and Trapp (in the first reading from the chapter) represent Roman rhetoric.
3. Summarize the way *Rhetorica Ad Herennium* states rhetoric can be learned. Locate if and how other sources from this chapter identify how rhetoric can be learned, comparing and contrasting these sources with *Rhetorica Ad Herennium*. Finally, explain how what you discover about the education of the rhetor can be applied to your own education.

---

[q] προλόγος, probably.

CHAPTER 1

Projects

1. Chapter 6 of *Professional Writing and Rhetoric* argues that professional writing is defined, in great part, by its focus on readers or users; that is, professional writing and rhetoric is user- or reader-centered. Using the readings from Chapter 1, as well as other primary and secondary rhetorical sources you find through independent research, examine (meaning support, contend with, and/or modify) the following claim: Rhetoric is an audience-, reader-, or user-centered art.

2. Despite the fact that rhetoric is one of the oldest disciplines and was traditionally one of the core liberal arts, you will probably find yourself in many situations where you are asked to explain what rhetoric is. Your family will probably ask you, as they wonder what it is you are studying in this course. Interviewers will probably ask you, as they wonder what rhetoric adds to you as a potential employee. And if you happen to mention "rhetoric" to people at work, they will probably ask you what it is, also.

   Construct a scenario in which you might be asked to define rhetoric for someone. For this scenario, develop a text that defines the discipline. Since several scenarios that might call for a definition of rhetoric would not also typically call for a written document (i.e., they would call for a verbal response), you should approach this project creatively. Consider a variety of "texts" (maybe a movie-short, a children's book, a cartoon, or a taped dramatic dialogue) you could produce.

3. Obviously, four selected readings have not covered the entire discipline of rhetoric. There remain numerous issues/topics, rhetoricians, theoretical approaches, and histories that have not been represented, or represented well enough by the readings in this chapter.

   As a class, and with the help of your instructor, develop a list of issues/topics, rhetoricians, theoretical approaches, and/or histories that you could research further. Individually or in small groups, choose a research area, develop a research question into which you will inquire, and write a research plan to help guide your inquiry. Your aim should be to develop, as a class, a set of oral and written reports that you can bring back to the class in order to develop a more robust sense of rhetoric. The written reports should be published (in either print or online form) and incorporated into the class as additional readings/resources.

CHAPTER 2

What Is the Relationship Between Professional Writing and Rhetoric?

INTRODUCTION

The title of this book, *Professional Writing and Rhetoric*, can be read in one of two ways. You might read the title and think the book is about two different fields that are related: professional writing as one field and rhetoric as a separate but related field. Or, you might read it and think it is about an interdisciplinary field: professional writing and rhetoric as integrally joined in a way that forms a new field of study and work. The readings in Chapter 2, together, examine some of the relationships between professional writing and rhetoric.

Questions you should consider as you read these selections are as follows:

- How is professional writing affected when writing is understood as socially situated or contextualized?
- How is professional writing affected when writing is understood as a practice?
- How is professional writing affected when writing is understood as a productive art?
- How is professional writing affected when the writer is understood to be a rhetor and an author?
- How is professional writing affected when it is understood as organizationally situated authorship?

Each of these questions arises when professional writing is considered a rhetorical practice and art.

One of the concerns of rhetoric is understanding the relationship between communication and specific situations, what we often refer to as the rhetorical situation. Some, for instance, have argued that rhetorical situations are external realities that rhetors walk into, realities that create moments for rhetorical action. In other words, rhetorical situations exist prior to speaking or writing. Others have argued that rhetorical situations are "invented" in rhetors' speeches or texts. According to this position, audiences and readers enter into whatever rhetorical situations the rhetor creates for them. In other words, rhetorical situations do not precede speaking or writing; they are created through speaking and writing. Lester Faigley, in "Nonacademic Writing: The Social Perspective," examines this issue as it relates to professional writing.

Another issue within rhetoric questions whether rhetoric is a productive art—a way of making texts—or if it is better understood as a practice—a way of doing or acting. In Chapter 1, you read about how Aristotle defined rhetoric in terms of productive art and practice, and though Aristotle seems to argue that rhetoric is a productive art, his works, as well as others within the rhetorical tradition, seem to suggest that rhetoric is also a form of practice. Is professional writing one or the other, or a combination of the two? Such a question has great impact on what professional writers are, do, and are responsible for. Carolyn Miller, in "What's Practical about Technical Writing?" examines some of the ways this issue affects how we understand who professional writers are, what they do, and what they should do.

A third issue that has been central to rhetoric over the ages is defining what it means to be a rhetor or a writer. When we think about rhetors and/or writers, a great range of terms can be used to define such people, each with varying degrees of authority: scribe, orator, poet(ess), stenographer, scholar, wordsmith, leader, editor, publisher, document specialist, and author are just some of the terms we might conjure. Which one of these is best suited to the professional writer? And once we choose one, in what ways does the term affect what we consider part of the writer's expertise? These are some of the issues taken up by Susan Regli in "Whose Ideas?: The Technical Writer's Expertise in Inventio" and Jennifer Slack, David Miller, and Jeffrey Doak in "The Technical Communicator as Author: Meaning, Power, Authority."

The issues and questions that have been part of the long history of rhetoric, thus, have a great deal of relevance to professional writing. These few selected readings offer only a glimpse at how the two fields are related, but when examined together, they point to a field of professional writing and rhetoric that can be defined as "organizationally situated authorship." As you read the selections in this chapter, reflect on this one broad question: "How would you define organizationally situated authorship?"

FOCUSING ON KEY TERMS AND CONCEPTS

Focus on the following terms and concepts while you read through this selection. Understanding these will not only increase your understanding of the selection that follows, but you will find that, because most of these terms or concepts are commonly used in professional writing and rhetoric, understanding them helps you get a better sense of the field itself.

1. communicative chains
2. writing as a social act
3. discourse community

NONACADEMIC WRITING: THE SOCIAL PERSPECTIVE

LESTER FAIGLEY
University of Texas at Austin

The past several years have seen a great deal of interest in the writing people do as part of their work. As other chapters in this book will indicate, this job-related writing is worthy of our interest and serious study. In exploring this sort of writing, researchers can take one or a combination of three major theoretical perspectives—the *textual* perspective, the *individual* perspective, and the *social* perspective. In this chapter I discuss the foundations of the social perspective and how it might contribute to research in nonacademic writing. Although the social perspective is least well established, I will argue that it can be a fruitful perspective from which to study nonacademic writing. To illustrate the three theoretical perspectives, I will refer to the following four examples of writing situations in nonacademic settings.

- An editor working for a major publisher in New York neglects to answer a query from an editor in another division. A few days later she writes a brief memo on the company's memo stationery, apologizing for her failure to respond. She uses the excuse that the request became buried on her desk. She follows the company's memo format, but she adds a letterlike closing that says: "Excavatingly yours."
- A supervisor of bank examiners in Colorado has the responsibility of teaching newly hired examiners how to write reports. Since examiners travel extensively and are not well paid, his staff is young and turns over rapidly. The supervisor is now revising an examiner's report on a small bank in southwestern Colorado that has made several questionable loans. In the margins he notes several problems with the report: the lack of reasons for several conclusions, the omission of important factual details, and general wordiness. At the end he explains to the young examiner why the overall tone is inappropriate for an examiner's report. He reminds the examiner that *the report will be read at a board of directors' meeting and that it will be the basis for any reform* in the bank's management. He tells him to stick to the specific regulations that were violated and to avoid derogatory remarks about the practices of rural banks.
- A nurse in Boston changes jobs and begins work at a psychiatric hospital. At his previous job at a large general hospital, the nurse's section of a patient's chart was a checklist. The psychiatric hospital, however, requires discursive notes on the chart. The nurse photocopies a few examples during his first day on the new job. He uses these examples as models when writing the chart for a schizophrenic patient. He observes that the notes are written in phrases and

---

Source: Faigley, Lester. "Nonacademic Writing: The Social Perspective." *Writing in Nonacademic Settings,* ed. Lee Odell and Dixie Goswami, 1985, pp. 231–248.

that certain abbreviations occur frequently, such as *pt.* for *patient.* He begins describing his patient's behavior:

> Very anxious and agitated, seclusive to room except when preoccupied with phone. Poor personal hygiene. With much coaxing, pt. finally took a bath but refused to wash hair. Pt. very paranoid. States "Someone is trying to burn down my trailer."

- A wildlife biologist works for an environmental engineering firm in Houston. She serves as project manager for an ecological survey of the proposed site for a liquefied natural-gas terminal on Matagorda Bay in Texas. She is part of a team that is preparing an environmental-impact statement for a major oil company, and she is composing on a computer the final report on terrestrial ecology. This report will be submitted with other reports on aquatic ecology and hydrology. Major subsections of the report include (1) wildlife habitats, (2) checklists of species, (3) endangered species, and (4) commercially important species. In writing the "checklists of species" subsection, the biologist relies on several master checklist files stored on computer diskettes. She loads the master checklist file for birds, a file that includes all species known to the Texas coast. She edits the file using her field notes on the birds she sighted while visiting the site, marking either "present," "absent," or "probable" beside each species. The biologist knows that the Environmental Protection Agency (EPA) gives special attention to endangered flora and fauna, and she includes in the report a separate subsection for habitats of endangered species. She documents her own findings with independently published sightings.

PERSPECTIVES FOR RESEARCH

Each of the writing situations just described—the editor's memo, the bank examiner's report, the nurse's notes, and the environmental-impact statement—differs substantially from typical classroom writing tasks. An overriding question for researchers of nonacademic writing is how these differences might best be understood and described. The three perspectives mentioned at the beginning of this chapter represent general lines of research that attempt to answer this question. They are, in fact, collections of approaches, collapsed and simplified here for purposes of comparison.

The Textual Perspective

The primary concerns of linguistics and literary criticism during much of the twentieth century have been the description of formal features in language and texts. Following from the assumptions of these traditions, much writing research has analyzed features in texts. This line of inquiry has long been dominant in the study of business and technical writing. One goal of this research has been to describe features that typify particular genres, such as what elements appear in the introduction of a marketing forecast. Another goal has been to produce more "readable" texts. *Readability* has been defined traditionally in terms of quantifiable linguistic features such as sentence length and word length—the basis for the popular readability formulas of Flesch and Gunning (reviewed in Selzer, 1983). Only recently have discussions of readability included factors such as the suitability of texts for potential readers (see Redish, et al., chapter 3, this volume).

If researchers who take the textual perspective were asked to examine the four situations I cited at the beginning, they would collect and analyze the texts the writers produce. They might, for example, compared the specialized vocabularies of the environmental-impact statement, the examiner's report, and the nurse's notes. They might compute T-unit length and clause length for each example. They might analyze the topics of individual sentences and determine how these sentence topics form topical progressions. They might, for example, study documents' tables of contents in order to identify conventions of organization. They might look at errors in the nurse's writing and measure "improvement." And they might comment on stylistic variations such as the closing of the editor's memo. Results of these studies would be used to make generalizations about specific kinds of texts—generalizations that are sometimes stated prescriptively as rules for style and format.

The Individual Perspective

This perspective has been strongly influenced by recent theory and research in psychology. For much of this century, linguistics and psychology in the United States were dominated by behaviorism, which declared mental strategies to be unobservable and beyond scientific investigation. During the 1950s and 1960s, however, behaviorist assumptions encountered serious objections. In linguistics, Noam Chomsky argued persuasively that behaviorist theory could not account for the complexity of human language acquisition, and thereby changed the direction of American linguistic research. In psychology, further challenges arose from several sources, two of which later became important in the study of writing. The European cognitive-developmental tradition—best known through the work of Jean Piaget—influenced American researchers studying the development of the thinking reflected in children's writing. A second tradition of cognitive psychology in the United States engaged researchers in creating general theoretical models of the reasoning that attends the writing process. Both these new lines of inquiry in psychology directed the attention of some writing researchers to strategies writers use in composing. For example, Emig (1971) tried to identify some of the strategies high school students used when they composed. Emig's work was followed by numerous other studies of the composing processes of elementary, secondary, and college students. The 1970s movement toward process-oriented inquiry into how children and young adults learn to write eventually led to studies of how nonacademic writers compose (e.g., Gould, 1980).

For researchers who take the individual perspective, a text is not so much an object as an outcome of an individual's cognitive processes. The primary attention shifts away from the text to an individual writer's emerging conception of the writing task. Researchers taking the individual perspective would likely examine how writers make certain choices during composing. They would inquire about writers' goals in composing, either by retrospective interviews or by asking writers to voice their thoughts while they composed. They would consider how an individual's formulation of a writing task directs the production of the resulting text. For example, researchers might observe how the biologist divides the task of writing the environmental impact statement into segments and what she hopes to accomplish in each section. They might observe how the editor at a publishing house goes about creating a persona as she writes the memo and how she understands that persona to fulfill a larger purpose—in this case, gaining the reader's acceptance of an oversight. They might study protocols of the nurse's composing or consider the time he devotes to each stage of writing (e.g., Does he ever revise?). They might take the bank examiner's case as an example of failure to develop an appropriate sense of audience. One goal of these studies would be to describe the processes that are effective and those that are ineffective so that effective strategies can be taught to ineffective writers.

The Social Perspective

The social perspective also focuses on the process of composing, but this perspective understands process in far broader terms. In the social perspective, writing processes do not start with "prewriting" and stop with "revising." Researchers taking a social perspective study how individual acts of communication define, organize, and maintain social groups. They view written texts not as detached objects possessing meaning on their own, but as links in communicative chains, with their meaning emerging from their relationships to previous texts and the present context. The social perspective, then, moves beyond the traditional rhetorical concern for audience, forcing researchers to consider issues such as social roles, group purposes, communal organization, ideology, and finally theories of culture.

If we consider the examples at the beginning of the chapter, we see that neither the textual perspective nor the individual perspective gives us a way to understand how a wildlife biologist learns to write an environmental-impact statement, or why the nurse's section of a patient's chart at one hospital would not require any writing, or how the supervisor's editing of the bank examiner's report affects the audit of the bank and its consequences, or even why the editor's closing is funny. These questions all involve social relations, tensions, or conflicts that go beyond the text as a physical object and the writer as an isolated strategist. To ask these questions is to assume that writing, like operating a jackhammer, arguing a lawsuit, or designing an office building, is a social act that takes place in a structure of authority, changes constantly as society changes, has consequences in the economic and political realms, and shapes the writer as much as it is shaped by the writer. Questions like these could be avoided as long as researchers studied student compositions, but they arise as soon as we leave the academic setting with which we are familiar. Consequently, a writing researcher taking the social perspective needs not only new methods of research, but also a theory that explains how we can participate daily in an all-encompassing social world and yet still see the structure of that world. Before turning to questions of research methodology from the social perspective, I first will look at how such a theory might be constructed.

FOUNDATIONS FOR A SOCIAL THEORY OF WRITING

A central tenet of the social perspective is that communication is inextricably bound up in the culture of a particular society. Consequently, a researcher of writing who takes the social perspective must have some way of defining and describing that society in terms broader than the traditional rhetorical conception of audience (see

Nystrand, 1982). For those of us who have been trained to appreciate literary texts as works of solitary artistic genius rather than expressions of a culture, the task of describing a society seems formidable—if not impossible. There is some comfort, however, in knowing that others ill equipped in theory and method have stumbled onto vast social questions concerning language and have not only survived but even changed basic notions about how we communicate. One such group of explorers—an appropriate metaphor here—were anthropological linguists who attempted to describe the languages of Africa and Asia during the years following World War II. They found that traditional definitions of language and methods of linguistic analysis were no better suited for the astonishing diversity of language in newly emerging nations than was the wool clothing earlier explorers wore to the trophics. These linguists met speakers of the "same" language living a few villages apart who could not understand each other. In many small villages they found that everyone was fluent in two languages or dialects, and that a speaker's choice of one of them often conveyed the social standing of the speaker or listener.

To cope with this diversity, linguists developed the notion of a *speech community*, which Gumperz (1971) defines as "any human aggregate characterized by regular and frequent interaction by means of a shared body of verbal signs and set off from similar aggregates by significant differences in language usage" (p. 114). This notion of a speech community became a basis for the new discipline of *sociolinguistics*. Sociolinguists employed the idea of a speech community to examine how language is used to maintain social identity. For example, Blom and Gumperz (1972) studied a small Norwegian village where all residents spoke both Bokmål, one of the two forms of standard Norwegian, and a local dialect. They found that choices between the two dialects varied among speakers within the community. In some cases, choices between the two dialects signaled certain attitudes and beliefs. A similar phenomenon occurs in my neighborhood in Austin, Texas, where most residents are bilingual in English and Spanish. My neighbors typically greet each other in Spanish, then often switch to English if they wish to engage in prolonged conversation, then signal the conclusion of the conversation by returning to Spanish. Differences in language use can establish social identity even among speakers of the same language (cf. Hymes, 1972). For example, speakers of English would likely understand the literal meaning of utterances of inner-city blacks such as "Your momma so black, she sweat chocolate," but they might not understand that such insults comprise a form of verbal play called the "dozens" (Labov, 1972).

Although the notion of a speech community offers us some insights into the social dimensions of writing, the concept of a community connected by writing must be defined by different criteria. Many of the linguistic markers of speech communities (e.g., differences in pronunciation) do not have simple parallels in written language. Further, written language is actually a collection of genres. Written language is composed in and comes to us through many forms—in shopping lists, in newspapers, in dictated letters, in scripted newscasts, in signs, in receipts. As many commentators on literacy have noted, written language can be understood outside the writer's immediate community or outside the writer's lifetime (which is also true for electronically recorded spoken language).

We need, therefore, an alternative concept to accommodate some of the special circumstances of written language—a concept we might label a *discourse community*.

In one sense, all persons literate in a language constitute a discourse community. But few, if any, texts are written for everyone who is capable of deciphering the words. Texts are almost always written for persons in restricted groups (cf. Bazerman, 1979). Persons in these groups may be connected primarily by written texts, as is the case with scholars on different continents who participate in a scholarly debate. Or they may belong to the same organization that has an in-house language and certain local discourse conventions. The key notion is that within a language community, people acquire specialized kinds of discourse competence that enable them to participate in specialized groups. Members know what is worth communicating, how it can be communicated, what other members of the community are likely to know and believe to be true about certain subjects, how other members can be persuaded, and so on.

Scholars for a long time have recognized that academic disciplines are a type of discourse community, each with its own language, subject matters, and methods of argument. In this seminal book, *The Use of Argument*, Toulmin (1958) theorizes that although arguments have basic structural similarity, they also are distinguished by fields. He offers academic disciplines as examples of fields, pointing out that patterns of arguments in fields such as physics are very different from those in disciplines such as history or law. Willard (1983) broadens Toulmin's account of a field to include instances of ordinary discourse. In addition to academic disciplines, which Willard calls *normative fields*, Willard distinguishes *encounter fields* (communication among strangers), *relation fields* (communication among associates, friends, and spouses), and *issue fields* (schools of thought that often cross disciplines such as Freudianism). Willard describes fields as rhetorical in operation. Fields sanction what knowledge is accepted, what subjects might be investigated, and what kinds of evidence and rhetorical appeals are permitted.

The academic discourse communities receiving the most study to date have been the sciences, with most attention coming from an extensive research program in the sociology of science (reviewed in Bazerman, 1983). Following Merton's (1957) observation that the growth of scientific knowledge reflects its social organization, many researchers have examined groups, subgroups, and hierarchies among scientists. Researchers have considered how scientific articles serve the social organization as both a means of communication and a means of earning rewards. Hagstrom (1965) drew the analogy of the scientific article as a form of primitive "gift giving," where the scientist offers the "gift" with the expectation of receiving some sort of later recognition from the community. Latour and Woolgar (1979) argue for a different model, where scientists publish to earn credibility, which in turn furthers their interest in the "game" of science. Another issue in this research program is the nature of scientific knowledge. The old notion of an independent and rational body of scientific knowledge has collided with many demonstrations of the human construction of scientific facts (e.g., Feyerabend, 1975; Toulmin, 1972), and "new" scientific knowledge has been shown to emerge from an agreed-on body of old knowledge (e.g., Price, 1963).

It is tempting to import wholesale the research issues raised in the sociology of science for the study of nonacademic writing. But before any such ambitious research program can begin, certain questions of definition must be addressed. One of the most crucial is how to differentiate academic and nonacademic writing. In examining nonacademic writing, we find many overlapping communities. For example, the biologist writing an environmental-impact statement abides not only by certain disciplinary conventions in biology, certain legal forms determined by the Environmental Protection Agency, and certain unstated and stated conventions particular to her company, but also by a complex set of conventions of political language (consider the use of the term *endangered species*). If the notion of discourse communities is to be illuminating, it must not be used without attending to how such communities might be identified and defined and how communities shape the form and content of specific texts. Chapter 9 by Miller and Selzer in this volume suggests how analyses of texts written in specific communities might proceed.

In the case of academic or professional discourse, it is relatively easy to see writing as a social activity. It is more difficult to see how a "private" act of writing, such as an entry in a diary, might be construed as a social act. Take an extreme example, where the writer of a diary encodes her entries in a cipher that only she knows. Theoreticians of the social perspective, such as Lev Vygotsky, would argue that such a coded diary entry would be no less a social act than the environmental-impact statement. Vygotsky (1962) contends that there is no such thing as "private" language, or even "private" thought:

> Thought development is determined by language, i.e., by the linguistic tools of thought, and by the sociocultural experiences of the child . . . The child's intellectual growth is contingent on his mastering the social means of thought, that is, language . . . Verbal thought is not an innate, natural form of behavior but is determined by a historical-cultural process [p. 51].

The historical-cultural process to which Vygotsky refers is simply that children do not learn words from a dictionary but through hearing them uttered in social situations to convey specific intentions and to achieve specific ends (see Bizzell, 1982). Words carry the contexts in which they have been used. Granted, Vygotsky does discuss "inner speech," but his conception of inner speech is not the same as private language. Although inner speech is not voiced, it consists of fragments of speech the speaker has drawn from the community in which he or she lives. More important, inner speech takes the form of a dialogue, which implies the continuous presence of an "other."

Vygotsky's contemporary, M. M. Bakhtin, applied these same notions to written texts. It is not clear whether Bakhtin and Vygotsky knew each other or influenced each other. (Bakhtin remains mysterious in other ways as well. Apparently some of his works were published under the names of his associates.) In *Marxism and the Philosophy of Language*, originally published under the name V. N. Vološinov in 1929, Bakhtin claims the textual perspective (which he calls "abstract objectivism") distorts the nature of written language by separating a text from its context. Bakhtin goes on to say that the textual perspective mistakenly assumes that meaning can be separated from a specific situation, that the textual approach inevitably emphasizes parts at the expense of the whole. He also faults approaches that center on the individual; these approaches he claims, miss the nature of language. Like Vygotsky, he insists that language is dialogic, that a text is not an isolated, monologic utterance, but "a moment in the continuous process of verbal communication" [Vološinov, 1973, p. 95]. A text is written in orientation to previous texts of the same kind and on the same subjects; it inevitably grows out of some concrete situation; and it inevitably provokes some response, even if it is simply discarded. In short, the essence of a text—any text—is inextricably tied up in chains of communication and not in the linguistic forms on the page or in the minds of individual writers.

RESEARCH ON WRITING FROM THE SOCIAL PERSPECTIVE

A broadly defined research program that explores writing from the social perspective would first examine what constitutes a discourse community. It would probe the fluid and multiple nature of discourse communities, and how communities overlap and change. Such a research program would examine how a particular discourse community is organized by its interactions and by the texts it produces (see chapter 8 by Paradis et al. in this volume). It would examine what subjects are considered appropriate in that community and how those subjects are determined. It would examine how genres evolve within a community. Finally, it would investigate how a community sanctions certain methods of inquiry.

Such a research program would integrate considerations of individual writers and particular texts into a broader view of the social functions of writing. It would explore how individual writers come to know the beliefs and expectations of other members of the community, and how individuals can alter the community's beliefs and expectations. It would consider how individuals cope with texts—how they learn to read texts and how to make meaning in texts in a particular community. It would investigate how conventions shape and are shaped by the processes of writing and reading. It would examine not only how individuals learn to represent themselves in a text, but how that representation emerges in response to a specific situation. In addition to the familiar aspects of the composing process, this research program would consider how all language is interaction, how all texts entail contexts, and how texts accomplish interactions between writers and readers rather than embodying meaning entirely by themselves. Consequently, this research program would not only examine an individual's composing processes, but would also follow the completed text, examining how it is disseminated, who has access to it, who reads it and who doesn't, what is read, what actions people take upon reading it, and how it influences subsequent texts.

Moreover, this research program would not separate the study of texts from the study of technologies used to create texts. These technologies include not only writing implements, but also symbol systems and the knowledge to interpret those systems. New technologies arise in response to needs, and members of discourse communities must know how to apply new technologies to existing functions for writing (see chapter 4 by Halpern in this volume; see also Faigley & Miller, 1982; Halpern & Liggett, 1984; Williams, 1981). For example, in writing the endangered-species subsection of the environmental-impact statement, the biologist uses computer software to form a pie chart that illustrates the percentages of wildlife habitat affected on the proposed site. The knowledge that readers use to interpret the pie graph is as critical a technology to this particular writing act as the technology that led to the development of the computer hardware and software.

The central questions for research taking the social perspective are ones that concern the contexts in which texts are written and read. These questions will be addressed in theoretical, historical, and empirical research. Theoreticians who adopt the social perspective can look to a long tradition of scholarship in rhetoric and more recent work in semiotics (e.g., Barthes, 1968); literary criticism (e.g., Fish, 1980); the philosophy of science (e.g., Popper, 1963); social psychology (e.g., Vygotsky, 1962); and cultural anthropology (e.g., Geertz, 1983). Historians can examine the functions of writing in small communities or the effects of literacy on large ones. Empirical researchers must be able to connect theoretical approaches to the mundane writing events of everyday life.

POSSIBILITIES FOR EMPIRICAL RESEARCH

In the social study of language, two major lines of empirical inquiry have emerged—one quantitative and the other qualitative. Both quantitative and qualitative approaches can be valuable in studying nonacademic writing. The quantitative approach is exemplified by work in sociolinguistics, such as Labov's (1966) findings that certain linguistic features are stratified by social class. The qualitative approach is exemplified by research in anthropology that is collectively known as *ethnography* (see chapter 14 by Doheny-Farina and Odell in this volume). Because qualitative research offers the potential for describing the complex social situation that any act of writing involves, empirical researchers are likely to use qualitative approaches with increasing frequency.

But if researchers take a qualitative approach, what do they examine? Let us consider again the situations posed at the beginning of this chapter. In the case of the editor, researchers might begin with the apparent tension between the constraints of the memo form and the tone sought by the editor—a tension that prompts innovation. Examining the causes of this tension leads to issues of the use of language by those whose business is the production of language, the use of language between two people at the same level of the corporate structure, and the use of language to personalize an apparently impersonal form. For example, researchers might collect instances of personalization (e.g., handwritten additions, capitalization or underlining, second-person address, private references) and ask writers why they chose to make personal additions.

In the case of the bank examiner, researchers might observe, over the course of a year, the supervisor's interaction with three or four trainees. In teaching the trainees how to write examiners' reports, the supervisor must also teach the trainees about the social organization of a bank. By understanding the social organization, examiners can help to correct the problems they uncover. To study how this social knowledge is transmitted, researchers would record the oral as well as the written communication between the supervisor and the trainees. They likely would interview trainees at different times to discover how social understanding evolves, and they would be sensitive to the reactions of bankers to the examiners' reports.

The case of the nurse also concerns the way writers under someone else's authority learn the conventions of a community. At one hospital, nurses are allowed only a checklist. At the other, they can—and must—write; but at the same time, they must use certain conventions associated with the practice of psychiatric medicine and with the particular hospital. Researchers should be interested in how nurses acquire and internalize these conventions. Researchers might also wish to observe how these written reports are used by physicians in diagnosing and treating patients.

In the case of the environmental-impact statement, a researcher who takes the social point of view might try to identify the sources of the set format for such documents. One might also want to consider the effects of this format on the kinds of information that can and cannot be considered. Ohmann (1976), for instance, has analyzed the conventions of the Pentagon papers and their effect on U .S. policies in the Vietnam War. Or one might consider ways in which a specific report differs from the conventional format of the environmental-impact statement. Is there a tension, traceable in the structure of the statement, between the format and the issues of the particular case? For instance, is one section much longer than usual? Is the tone of the opening different from previous statements? What is revised in the course of writing, and by whom? What cannot be revised?

All these lines of inquiry spring from three general questions:

1. What is the social relationship of writers and readers, and how does the text function in this social relationship?
2. How does this kind of text change over time?
3. How does the perspective of the observer define and limit the observation of this text?

This last question forces researchers to consider what it means to observe and what it means to interpret. Debates over these issues have occupied cultural anthropologists for the past two decades. Anthropologists have developed two broad notions of *ethnography*: an older notion concerned with observation and a newer notion concerned with interpretation. Both notions are important to the study of nonacademic writing.

The older notion is useful for its focus on how to observe. One anthropologist says that ethnography involves the attempt to "record and describe the culturally significant behaviors of a particular society" (Conklin, 1968, p. 172). He goes on to say that

> ideally, this description, an ethnography, requires a long period of intimate study and residence in a small, well-defined community, knowledge of the spoken language, and the employment of a wide range of observational techniques including prolonged face-to-face contacts with members of the local group, direct participation in some of that group's activities, and a greater emphasis on intensive work with informants than on the use of documentary or survey data [p. 172].

In a traditional conception of ethnography, an anthropologist lives (usually for a year or longer) in the culture being studied (usually a technologically primitive culture) and collects copious data by observing, interviewing, charting patterns, and collecting case studies. Although not every method might be used, the ethnographer will surely use more than one method in collecting data, and the chief data source will be the ethnographer's diary. The ethnographer tries to avoid value judgments and abandons assumptions from his or her own culture. Hymes (1980) says that ethnographic investigation is always open-ended.

The newer notion of ethnography is sometimes called *interpretative anthropology*. One of its chief practitioners is Geertz, whose essay, "Thick Description: Toward an Interpretative Theory of Culture" (in Geertz, 1973), argues that a culture can be "read" not by starting with abstract concepts but by first microscopically examining the culture's most salient activities. Geertz's famous essay on the Balinese cockfight (1973) demonstrates how a single event can provide "a metasocial commentary upon the whole matter of assorting human beings into fixed hierarchical ranks and then organizing the major part of collective existence around the assortment" (p. 448). The function of the cockfight "is interpretative; it is the Balinese reading of Balinese experience, a story they tell themselves about themselves" (p. 448). In a similar way a researcher of nonacademic writing can "read" in a manager's striking out the formal salutation "Dear Mr. Wittenburg:" and inserting by hand "Kent—" in a memo to a subordinate a great deal about how the community of the workplace is socially organized and maintained. As Geertz says, "Small facts speak to large issues" (p. 23).

The potential for qualitative research in nonacademic writing is great, but researchers should heed the warnings of anthropologists. One of the most critical is the insistence on a cross-cultural perspective. Some anthropologists question whether valid ethnographies are possible by members of the same culture. These anthropologists argue that the experience of living in another culture makes the ethnographer aware of how much a sense of belonging to a culture depends on shared knowledge and beliefs. Although very few writing researchers will attempt ethnographies of the kind done by anthropologists, the need for contrastive analysis still exists. Researchers of nonacademic writing must continually reflect on their own perspective—on what they are likely to observe and not observe, and on how their own assumptions about writing and the world affect how they interpret what they observe (see Boon, 1982; Clifford, 1983).

Researchers should also be aware of the history of writing systems. Contemporary archaeologists have found that the development of writing systems grew out of economic necessity. The purposes of writing for the first five hundred years apparently were strictly commercial and administrative (Driver, 1948). Most surviving tablets record the property and accounts of temples; religious, historical, and legal functions for writing came later. Today we are in the midst of large-scale changes in the nature and uses of writing systems—changes brought about by electronic technology and again stimulated by changing economic and social needs. Computerized information services were first established to provide immediate access to financial news and other economic information, but these data bases quickly spread to more general kinds of information and even to hobbies. Electronic mail is as old as the telegraph, but with the advent of computer and satellite technology it has become an increasingly pervasive communications system, extending rapidly beyond the workplace. The point here is that writing technologies arise from perceived needs within communities. If world trade were less complex, the need to develop electronic communication technologies would be proportionately less. Consequently, the changing nature of nonacademic writing cannot be understood without examining changes in communities that produce nonacademic writing.

Researchers who take the social perspective show us that writing in a complex society is diverse and that our definitions of literacy must necessarily be pluralistic. They show us that writing is an act not easily separated from its functions in a particular discourse community. They increase our awareness of the social importance of what we teach. In chapters that follow, Odell (chapter 7); Paradis, Dobrin, and Miller (chapter 8); and Miller and Selzer (chapter 9) explore some of the complex relationships between writing and the social, organizational, and professional contexts in which that writing is done.

ACKNOWLEDGMENTS

I am grateful for the responses of Phyllis Artiss, Charles Bazerman, Patricia Gambrell, Greg Myers, Martin Nystrand, Walter Reed, and Beverly Stoeltje to earlier drafts of this chapter.

References

Barthes, R. (1968). *Elements of semiology*. (A. Lavers & C. Smith, Trans.). New York: Hill and Wang.

Bazerman, C. (1979). Written language communities: Writing in the context of reading. (ERIC Document ED 232 159.)

Bazerman, C. (1983). Scientific writing as a social act: A review of the literature of the sociology of science. In P. V. Anderson, R. J. Brockmann, & C. Miller (Eds.), *New essays in technical writing and communication: Research, theory and practice*. Farmingdale, NY: Baywood.

Bizzell, P. (1982). Cognition, convention, and certainty: What we need to know about writing. Pre/Text, 3, 213–243.

Blom, J. P., & Gumperz, J. J. (1972). Social meaning in linguistic structures. In J. J. Gumperz & D. Hymes (Eds.), *Directions of sociolinguistics*. New York: Holt, Rinehart and Winston.

Boon, J. A. (1982). *Other tribes, other scribes*. Cambridge: Cambridge University Press.

Clifford, J. (1983). On ethnographic authority. *Representations*, *1*, 118–146.

Conklin, H. C. (1968). Ethnography. In D. Sills (Ed.), *International encyclopedia of the social sciences*. London: Macmillian.

Driver, G. R. (1948). *Semitic writing from pictograph to alphabet*. London: Oxford University Press.

Emig, J. A. (1971). *The composing processes of twelfth graders* (NCTE Research Report No. 13). Urbana, IL: National Council of Teachers of English.

Faigley, L., & Miller, T. (1982). What we learn from writing on the job. *College English*, *44*, 557–569.

Feyerabend, P. (1975). *Against method: An outline of an anarchistic theory of knowledge*. Atlantic Highlands, NJ: Humanities Press.

Fish, S. (1980). *Is there a text in this class?* Cambridge, MA: Harvard University Press.

Geertz, C. (1973). The interpretation of cultures. New York: Basic Books.

Geertz, C. (1983). *Local knowledge: Further essays in interpretative anthropology*. New York: Basic Books.

Gould, J. D. (1980). Experiments on composing letters: Some facts, some myths, some observations. In L. W. Gregg & E. R. Steinberg (Eds.), *Cognitive processes in writing*. Hillsdale, NJ: Lawrence Erlbaum.

Gumperz, J. J. (1971). *Language in social groups*. Stanford, CA: Stanford University Press.

Hagstrom, W. O. (1965). *The scientific community*. New York: Basic Books.

Halpern, J., & Liggett, S. (1984). *Computers and composing: How the new technologies are changing writing*. Carbondale, IL: Southern Illinois University Press.

Hymes, D. (1972). Introduction: Toward ethnographies of communication. In P. Giglioli (Ed.), *Language and social context*. Baltimore: Penguin.

Hymes, D. (1980). What is ethnography? In D. Hymes (Ed.), *Language in education: Ethnolinguistic essays*. Arlington, VA: Center for Applied Linguistics.

Labov, W. (1966). *The social stratification of English in New York City*. Arlington, VA: Center for Applied Linguistics.

Labov, W. (1972). *Language in the inner city: Studies in the Black English Vernacular*. Philadelphia: University of Pennsylvania Press.

Latour, B., & Woolgar, S. (1979). *Laboratory life: The social construction of scientific facts*. Beverly Hills, CA: SAGE Publications.

Merton, R. K. (1957). *Social theory and social structure*. New York: Free Press.

Nystrand, M. (1982). Rhetoric's "audience" and linguistics' "speech community": Implications for understanding writing, reading, and text. In M. Nystrand (Ed.), *What writers know: The language, process, and structure of written discourse*. New York: Academic Press.

Ohmann, R. (1976). *English in America: A radical view of the profession*. New York: Oxford University Press.

Popper, K. (1963). *Conjectures and refutations: The growth of scientific knowledge*. New York: Harper & Row.

Price, D. J. (1963). *Little science, big science*. New York: Columbia University Press.

Selzer, J. (1983) What constitutes a "readable" technical style? In P. V. Anderson, R. J. Brockmann, & C. Miller (Eds.), *New essays in technical writing and communication: Research, theory, and practice*. Farmingdale, NY: Baywood.

Toulmin, S. (1958). *The uses of argument*. Cambridge: Cambridge University Press.

Toulmin, S. (1972). *Human understanding* (Vol. 1). *The collective evolution of scientific concepts*. Princeton, NJ: Princeton University Press.

Vološinov, V. N. (1973). *Marxism and the philosophy of language*. (L. Matejka & I. R. Titunik, Trans.). New York: Seminar Press. (Original work published 1929)

Vygotsky, L. S. (1962). *Thought and language*. (E. Hanfmann & G. Vakar, Trans.). Cambridge, MA: MIT Press (Original work published 1934)

Willard, C. A. (1983). *Argumentation and the social grounds of knowledge*. University, AL: University of Alabama Press.

Williams, R. (1981). Communications technologies and social institutions. In *Contact: Human communication and its history*. New York: Thames and Hudson.

DEVELOPING YOUR UNDERSTANDING

1. **Explain the following: "The social perspective [of writing], then, moves beyond the traditional rhetorical concern for audience, forcing researchers [and professional writers] to consider issues such as social roles, group purposes, communal organization, ideology, and finally theories of culture." Refer to the readings in Chapter 1, or to knowledge you have of rhetoric from past course work, to support your explanation.**
2. **Faigley argues that questions about the impact of social context on writing "could be avoided as long as researchers studied student compositions, but they arise as soon as we**

leave the academic setting with which we are familiar." In fact, questions of social contexts' impacts on writing can be—and frequently are—ignored in the workplace. Referring to the scenarios with which Faigley opens, analyze how ignoring questions of social context might negatively affect you and your work if you were placed in the contexts of those writers.

3. Faigley argues that professional writers taking "the social perspective must have some way of defining and describing . . . society in terms broader than the traditional rhetorical conception of audience" and that "discourse community" has arisen as a more effective concept. Argue whether Faigley's argument and the replacement of discourse community for audience renders rhetoric obsolete or inadequate for contemporary workplace writing, or whether his argument extends and/or revises a rhetorical perspective of professional writing.
4. Assuming that a document in the workplace is "'a moment in the continuous process of . . . communication,'" explain what the professional writer needs to know and be able to do in order to participate effectively in this ongoing process.
5. Write an exploratory essay, using the following as an introduction:

    Under the "Research on Writing from the Social Perspective" and "Possibilities for Empirical Research" sections, Faigley identifies numerous research issues/questions that are important for fleshing out a better understanding of a social perspective on nonacademic writing. For instance, Faigley suggests that writing researchers examine "what subjects are considered appropriate in [particular] communities and how those subjects are determined." Though these issues/questions are significant and interesting for the purpose of research, they can also be reformulated as suggested practices or rhetorical maxims for professional writers adopting a social perspective. Rather than posing the previous example as a research question, that is, it can be turned into a suggested practice: "professional writers should pay careful attention to the subjects that are appropriate in particular communities and take time to explore how the communities within which they are writing have determined what are appropriate subjects."

    After reviewing the research issues/questions posed by Faigley in these two sections, I have determined that the following three issues/questions can be reformulated as the most significant suggestions/maxims for the professional writer driven by a social perspective: 1, 2, and 3. The essay that follows reformulates these issues/questions, argues why these are the most significant to a social perspective on professional writing, and explains how the suggested practices would affect the professional writer's work.

FOCUSING ON KEY TERMS AND CONCEPTS

Focus on the following terms and concepts while you read through this selection. Understanding these will not only increase your understanding of the selection that follows, but you will find that, because most of these terms or concepts are commonly used in professional writing and rhetoric, understanding them helps you get a better sense of the field itself.

1. practical (low sense)
2. practical (high sense)
3. theoretical knowledge
4. practical knowledge
5. productive knowledge
6. technē
7. poiesis
8. praxis
9. phronesis

What's Practical About Technical Writing?

CAROLYN R. MILLER

Courses and programs in technical writing are both praised and damned for being "practical." Other writing courses are practical, to be sure: in general, practical rhetoric emphasizes that discourse is a means for pursuing a goal. Thus, freshman composition aims to help students be more effective as students, technical writing aims to help them be more effective as engineers or accountants or systems analysts, and the writing instruction that accompanies many literature courses aims to help them to be more effective as reader-critics. But since technical writing is singled out for being practical, it is worth considering what makes it so.

THE MEANING OF "PRACTICAL"

Most immediately, the practical seems to be concerned with getting things done, with efficient and effective action. Furthermore, efficiency and effectiveness seem more important for some types of action than for others; that is, some actions themselves have practical aims (rather than aesthetic or ritual ones), actions concerned with the material necessities of making a living or managing a household. One can thus *be* practical (or impractical) *about* practical action. *Being* practical suggests a certain attitude or mode of learning, an efficiency (or goal directedness) that relies on rules proved through use rather than on theory, history, experience, or general appreciation. Practical rhetoric therefore seems to concern the instrumental aspect of discourse—its potential for getting things done—and at the same time to invite a how-to, or handbook, method of instruction. Technical writing partakes of both these dimensions of practical rhetoric.

The rhetoric of the early Greeks also involved both dimensions. They emphasized that rhetoric was an art (or techne). This meant (to Aristotle, at least) that rhetoric was conceptualized and teachable (not a knack, as Plato had feared) but neither certain nor absolute (not a science, as Plato had hoped). Greek rhetoric thus initiated both a handbook tradition of instruction and a counterposed theoretical appreciation for the multiplicity of relations between means and ends.

Richard Bernstein has suggested that there are both "low" and "high" senses of "practical," two senses that parallel the handbook and theoretical traditions of rhetoric. It is the low sense, Bernstein says, that calls to mind "some mundane and bread-and-butter activity or character. The practical man is one who is not concerned with theory (even anti-theoretical or anti-intellectual), who knows how to get along in the rough and tumble of the world" (x). The high sense, which derives from the Aristotelian concept of praxis and underlies modern philosophical pragma-

---

Source: Reprinted by permission of the Modern Language Association of America from Carolyn Miller, "What's Practical About Technical Writing?" *Technical Writing: Theory and Practice*. Ed. Bertie E. Fearing and W. Keats Sparrow, 1989, pp. 14–24.

tism, concerns human conduct in those activities that maintain the life of the community. One of the many reasons for the discrepancy between these two senses of the practical highlights the dilemma of technical writing, which is usually called practical in the low sense (by both its friends and its enemies, incidentally). This reason has to do with the social structure of the Greek city-state, which permitted the free citizen to be concerned with the good of the polis without being much concerned with bread-and-butter activities. The reason, of course, is the institution of slavery. Manual labor and most commercial activity were performed by noncitizens—slaves, foreigners, women. These activities were "preconditions" to the fulfillment of human potential in self-government, according to Nicholas Lobkowicz: "One would almost be tempted to say that the Greeks considered all 'prepolitical' activities prehuman and that only in the political life were they able to see a way of life which transcended the animal realm" (22). Technical writing, the rhetoric of "the world of work," of commerce and production, is thus associated with what were low forms of practice from the beginning. In a world in which it is more dishonorable to own slaves than it is to work for a living, we might question whether this association should prevail.

A CONCEPTUAL CONTRADICTION

Before trying to suggest what it might mean to apply the higher sense of practical to technical writing, I want to indicate some difficulties in accepting the low sense uncritically, as many technical writing teachers have. These difficulties are revealed by a contradiction within the self-justifying discourse of technical writing pedagogy: the attempt to hold both that nonacademic rhetorical practices are inadequate (and therefore need improvement through instruction) and that they serve as authoritative models (and therefore define goals for instruction). We seem, that is, uncertain about where to locate norms, about whether the definition of "good writing" is to be derived from academic knowledge or from nonacademic practices. Most teachers will recognize the contradiction in the familiar dilemma of having to admit to students the discrepancy between practices that are supposed to be effective and those that are actually preferred and accepted.

The first side of the contradiction is the familiar justification for teaching technical writing. We teach it because when students graduate and begin writing on the job, they do not do very well. In the technical writing textbook I use, the first chapter, "Why Study Technical Communication?" documents the "inadequate communication skills of many technical professionals" (Olsen and Huckin 7). For example, it quotes a survey about recently graduated civil engineers showing that writing and speaking are the areas of competence most important to civil-engineering practice but that about two-thirds of recent graduates are judged "inferior" in these areas; results for mechanical and electrical engineers are similar. Complaints about technical writing from senior officials in science and industry include "foggy language," failures of emphasis and coherence, illogical reasoning, poor organization—a familiar litany. Most technical writing textbooks begin with the same rationale, that nonacademic rhetorical practices are wanting. The justification for academic instruction is that academics know something that can help improve professional practices.

The second side of the contradiction derives from the research that interested faculty members have begun to do on rhetorical practices in business, industry, and science. This research is justified not only by the academic assumption that knowledge is a good thing but also (and often primarily) by the belief that knowledge of nonacademic practices is necessary to define goals for teaching practical rhetoric. As Paul Anderson puts it, "We [educators] must first understand the profession, then design our curricula accordingly. Only if we understand intimately the job we intend to prepare our students to perform can we create effective professional programs" ("What Technical" 161).

One of the favorite research projects is the survey, which can show what kinds of work-related writing the population surveyed does, how important it seems to be, what its common problems are, and what qualities and features are valued. In reviewing selected surveys, Elizabeth Tebeaux notes discrepancies between instructional assumptions and industrial practices and concludes that "several curricular changes are clearly mandated" in order to "meet the communication needs of writers in industry" (422). Anderson reviewed fifty surveys, because they can provide "teachers with important insights they can use as they design courses in business, technical and other forms of career-related writing" ("What Survey" 4). Many surveys, such as those by Marcus Green and Timothy Nolan and by Bill Coggin, have been proffered as authoritative sources of information abut what a curriculum should accomplish for its graduates. Ethnographic research has also been justified in instructional terms: according to Stephen Doheny-Farina, for example, "By learning more about nonacademic contexts for writing, we are learning more about the kinds of rhetorical demands faced by many of our college graduates," and this knowledge "can inform the teaching of writing" (159).

Major national grants have gone to researchers engaged in work justified in these same ways, a clue to the institutionalization of this line of reasoning, as well as to its extension from technical writing to composition in general. The Fund for Improvement of Post-Secondary Education (FIPSE) sponsored a project on writing-program evaluation at the University of Texas; the project produced a report saying that "before any college writing program can be judged effective or ineffective, we must know first if what it teaches has value to its graduates in later life. Like any educational program, the overall effectiveness of writing programs must be judged according to the needs of the population they serve" (Faigley et al. 1–2). Another FIPSE grant went to Wayne State for a university-industry collaborative effort on research and curriculum development in professional writing. The researchers present cooperation between academics and practitioners as the way to "ensure that students are prepared for the diverse communication tasks outside the university" (Couture et al. 392–93). FIPSE has also sponsored research on collaborate writing in the workplace by Lisa Ede and Andrea Lunsford, who cite as a major problem "the dichotomy between current models and methods of teaching writing . . . and the actual writing situations students will face upon graduation"; this dichotomy results, in part, from "our lack of detailed understanding about on-the-job writing" ("Research" 69). The National Institute of Education earlier sponsored work by Lee Odell and Dixie Goswami on writing in nonacademic settings; their study also suggests that our ability to teach writing will be "enhanced" by more complete understanding of how people come to write successfully on the job ("Writing" 257).

PRACTICE AS DESCRIPTIVE OR PRESCRIPTIVE

In its eagerness to be useful—to students and their future employers—technical writing has sought a basis in practice, a basis that is problematic. I do not mean to suggest that academics should keep themselves ignorant of nonacademic practices; indeed, much of the research I cited above has been extremely illuminating. But technical writing teachers and curriculum planners should take seriously the problem of how to think about practice. The problem leads one to the complex relation between description and prescription. Odell warns against mistaking one for the other: "we must be careful not to confuse *what is* with *what ought to be* . . . We have scarcely begun to understand how organizational context relates to writing, and we have almost no information about which aspects of that relationship are helpful to writers and which are harmful" (278). Anderson also warns us about this mistake: in presenting a model of the technical writing profession for use in designating curricula, he cautions that the model "represents an ideal. It is built around the *best* practices of the profession, not around *common* practice—or malpractice" ("What Technical" 165). He gives as examples usability testing (not common but good) and readability formulas (common but bad). Neither Odell nor Anderson, however, gives us much help in understanding what is helpful and what is harmful, what is good practice and what is malpractice. Even David Dobrin's discussion of the contradictions involved in teaching to the standards of employers, although it recommends both curricular and corporate reform, relies finally on accepting practices of the workplace on their own terms; teachers should "make people at work better able to deal with others" ("What's the Purpose" 159).

At this point, it is worth recalling an earlier (unfounded) study of writing in nonacademic settings, "Writing, Out in the World," a chapter of Richard Ohmann's *English in America*. Ohmann avoids the contradiction of taking practice as both imperfect and authoritative by positing a wider perspective from which to make such judgments; he requires, as Odell and Anderson and Dobrin do not, a basis for evaluating a practice other than that of the practice itself. The nonacademic writing Ohmann examined is that of futurists and forecasters, of foreign-policy analysts, and of the government officials who wrote the memorandums we call "The Pentagon Papers." Ohmann sought to establish, not that academic writing is different from writing in the workplace, but that they are dangerously similar; he concludes that academic instruction in writing "has helped, willy nilly, to teach the rhetoric of the bureaucrats and technicians" (205). He claims that the

> writing of the powerful and influential shares some characteristics with the required writing of their college-age sons and daughters; that these characteristics are fairly important to the style of thinking and planning that guides the most powerful country in the world; and that this style has some systematically dangerous features when it operates not in the classroom but on the stages of history. (173)

A similar and more direct charge has been made recently by Susan Wells, who claims that "the ideology of technical writing explicitly assents to its instrumental subordination to capital; the aim of the discipline as a whole is to become a more responsive tool" (247). Being useful is not necessarily good, according to these Marxist critics, but little in the discourse of technical writing allows for this conclusion or explores its consequences. Because the Marxist critique features practical activity as a central concept, it raises questions that are particularly germane to technical writing, questions about whose interests a practice serves and how we decide whose interests should be served.

PRACTICE AND HIGHER EDUCATION

The uneasy relation between nonacademic practice and academic instruction has been part of academic discussions about technical writing from their beginnings in the late nineteenth century, as Robert Connor's historical work has shown. Connors documents recurrent debates over whether practical or humanistic goals should prevail in technical writing courses (or, as they were commonly called, "engineering English"), whether, that is, such study should prepare technical students for work or for leisure. Moreover, these debates reflect a larger debate in American higher education, about the appropriate relation between vocational preparation and cultural awareness. In mid-nineteenth century, this debate transformed the American college curriculum, according to the educational historian Frederick Rudolph, who points specifically to the Morrill Act of 1862 and the founding of Cornell in 1866. The first president of Cornell, Andrew White, "confronted all the choices that had been troubling college authorities: practical or classical studies, old professions or new vocations, pure or applied science, training for culture and character or for jobs" (117). White opted for pluralism, for providing many courses of study in preparation for many kinds of lives: "the Cornell curriculum . . . multiplied truth into truths, a limited few professions into an endless number of new self-respecting ways of moving into the middle class" (119). In a similar vein, Laurence Veysey's study of the emergence of the American university in the nineteenth century traces the development of "utility" as a basis for education. During this period, according to Veysey, "America was a scene of vocational ambition," both in terms of individual aspirations and in terms of the desire for public service. At the same time, the notion of public service broadened to include practical and technical occupations, not just the gentlemanly occupations for which earlier education had been preparatory. "Vocational training," says Versey, "directly affected the undergraduate curriculum of the new university" (66).

Other commentators have emphasized that the relation between instruction and practice is part of a more general condition, the subsistence of higher education in a socioeconomic matrix. Clark Kerr, in *The Uses of the University*, says that "the life of the universities for a thousand years has been tied into the recognized professions in the surrounding society, and the universities will continue to respond as new professions arise" (111). (This view, of course, implies that the classical curriculum served as preparation not for leisure but for the upper-class vocations of law, politics, and the ministry.) John Kenneth Galbraith has noted that "it is the vanity of educators that they shape the educational system to their preferred image. They may not be without influence, but the decisive force is the economic system" (236). More specifically, in his critique of nonacademic writing, Ohmann comments that

> the constraints upon English from the rest of the university and especially from outside it are strong. . . .[T]he writers of the textbooks and the planners of courses . . . can hardly

> ignore what passes for intellectual currency in that part of the world where vital decisions are made or what kind of composition succeeds in the terms of that part of the world. (206)

Current enthusiasm for "industry-university collaboration" in applied research and development is perhaps the most recent manifestation of this general and necessary relation. But there is also a repertoire of accepted mechanisms for channeling the relation—internships, advisory councils, certification of graduates, and procedures for justifying and accrediting programs. These mechanisms are used in educational programs for the established professions, like law, medicine, engineering, and teaching, as well as in several areas of practical rhetoric with relatively long curricular histories, like journalism and public relations. For the most part, the channels these mechanisms create are one-way: influence flows primarily from nonacademic practices to the academy. The gradient is reflected in the language at the industry-university interface, which includes, on the one hand, "demand," "need," "value" and, on the other, "response," "service," "utility." My own university, a land-grant institution, provides a case in point. Its "Mission Statement" declares that the university "has responsibility for the academic, research, and public service programs in areas of primary importance to the State's economy." University policies concerning proposals for new degree programs require statements concerning the proposed program's relation to the institutional mission, to student demand, and to "manpower" needs in the state.

Teachers of technical writing have advocated applying the mechanisms of nonacademic influence to their new programs, using the same kinds of language. Internship programs should be adopted in technical communication programs, according to a recent review of literature, because they encourage students to relate their study of theory to practice, permit faculty members to "keep in touch with" current practices, and enable employers "to influence college programs" (Gloe 18–19). Advisory councils are advocated because they "integrate the endeavors of the two worlds [academic and business-industrial] directly and in a[n] . . . effective manner" (Brockmann 137). (Certification has been discussed within the Society for Technical Communication, but there is insufficient consensus in the profession to arrive at standards ["Certification" 6]; accreditation is now being investigated by the society [*Strategic Plan*].)

Such language echoes the discourse of other professional programs, programs that have provided precedents for technical communication.

> **Library Science**
> It is widely believed and reported that a chasm of mutual ignorance and indifference separates librarians and library educators from one another. . . . All sectors of practice regularly and strongly express a desire for more influence over the content and character of professional education. (Clough and Galvin 2)

> **Public relations**
> Practitioners and educators must act in concert to guide public relations in the direction of professionalism. (Commission on Graduate Studies in Public Relations 5)

> **Information science**
> Lack of communication between the employers of information professionals and the institutions that educate and train them is one reason that educational institutions are not meeting needs and demands of the changing environment and new technologies. (Griffiths, abstract)
>
> **Business**
> MBA curricula must be reevaluated and, perhaps, restructured if they are to meet business expectations, and—from the point of view of business—if they are to better prepare students for the real world in which they will build their careers. (Jenkins and Reizenstein 24)
>
> **Journalism**
> What training and preparation do radio and television journalists consider important for a career in their field? Answers . . . should contain valuable insights for the broadcast journalism educator. (Fisher 140)
>
> **Training and Development**
> Training activities involve a wide variety of skills, abilities, knowledge, and information. . . . An interdisciplinary approach to T&D preparation is important, given the range of competencies required. (Reed 11)

This discourse is infected by the assumptions that what is common practice is useful and what is useful is good. The good that is sought is the good of an existing industry or profession, with existing structures and functions. For the most part, these are tied to private interests, and to the extent that educational programs are based on existing nonacademic practices, they perpetuate and strengthen those private interests—they do indeed make their faculties and their students "more responsive tools." As the minutes of one meeting of the advisory council to the School of Engineering at my university indicate, regular contact between the university and industry "makes students more valuable to industry."

PRAXIS AND TECHNE

My discussion so far has relied on a set of related oppositions that pervade the discourse of higher education:

theory versus practice
academy versus industry
ivory tower versus marketplace
idle speculation versus vocationalism
inquiry versus action
gentleman-scholar versus technician-dupe
contemplation versus application
general versus particular
knowing-that versus knowing-how
science versus knack

In this form the oppositions are probably unresolvable, and the best we can hope for is Anderson's notion that they should form a "creative tension" (Introd. 6).

Another approach is to suspect the worst: that a dichotomy so widespread must be (at least partly) false. And in fact, Aristotle's characterization of rhetoric as an art, rather than a science or a knack, cuts through these oppositions with a middle term—techne. As he defines it in the *Nicomachean Ethics*, "a productive state that is truly reasoned" (VI, iv), techne requires both particular and general knowledge, both knowing-how and knowing-that; techne is both applicable and conceptualized. Donald Schön's recent critique of professional education relies on the same middle term: it is "art," he says, that professionals display in practice, and it is art that unifies theory and application in a process he calls "reflection-in-action." Aristotle's *techne rhetorike*, or treatise or rhetorical art, joins theory and practice by deriving knowing how from knowing that, prescription from description. Although positivist philosophy claims that this derivation is fallacious ("you can't get 'ought' from 'is' "), one of the major insights of Marx, according to Bernstein, is to deny the positivist fallacy. Marx (as well as Aristotle) is able to derive from description of existing social practices the shape of human need and potential—which provide the basis for prescription.

But to understand Aristotle's *Rhetoric* only as a techne is to miss what Aristotle himself has to say about practice. Understood as techne, Aristotle's treatise would fall within the handbook tradition, as a set of instructions that helps one produce texts. Such a treatise would concern productive knowledge, or *episteme poietike*, one of three kinds of knowledge in Aristotle's system: theoretical (concerned with knowing for its own sake), practical (concerned with doing), and productive (concerned with making). According to George Kennedy, Aristotle does not make the connection between rhetoric and productive knowledge (as he does for poetics) but treats rhetoric as theoretical knowledge concerned with "discovering" the available means of persuasion (63).

The remaining alternative—that Aristotelian rhetoric is practical, rather than theoretical or productive—has been argued by Richard McKeon, and its implications have been explored by Eugene Garver. To see rhetoric as practical, in Aristotle's system, is to emphasize action over knowledge or production; rhetoric becomes a form of conduct, like the related practical realms of ethics and politics, which are constant background presences in the *Rhetoric*. Aristotle distinguishes carefully in the *Nicomachean Ethics* between production and practice, poiesis and praxis: as distinct from "science," or theoretical knowledge, both concern the variable, or that which can be other than it is; but they differ in that production "aims at an end other than itself," the product, and practice aims at its own performance, at "doing well." The reasoning appropriate to production takes the form of techne, art or technique, and the reasoning appropriate to performance, or conduct, takes the form of *phronesis*, prudence; for Aristotle there can be no art, or technical knowledge, of conduct. Prudence is the reasoning that makes one "capable of action in the sphere of human goods" (*NE* 6: v). Like techne, prudential reasoning is situated to undermine the oppositions that plague discussions of professional education, for it necessarily concerns both universals and particulars: it applies knowledge of human goods to particular circumstances (*NE* 6: vii; Garver 645). Unlike techne, however, which is concerned with the useful (that is, with the quality of a product given a set of expectations for it), prudence is concerned with the good (that is, with the quality of the expectations themselves).

Aristotle's concept of praxis has also informed some recent thinking about human action. As the central concept in Marx, praxis highlights the way in which the human person "is the result of his [or her] own work" (Bernstein 39; see also Lobkowicz 418–20). Human belief structures and social relations are understood to be used in practical relations between human beings and objects. Schön's account of professional practice emphasizes the "knowing inherent in intelligent action" (50). Moreover, practices, as Alasdair MacIntyre has insisted, create not only knowledge but their own goods, and because practices are necessarily social, these goods require "subordinating ourselves within the practice in our relationship to other practitioners" (191). The insights for the academic are that practice creates both knowledge and value and that the value created comprehends the good of the community in which the practice has a history.

Understanding practical rhetoric as a matter of *conduct* rather than of production, as a matter of arguing in a prudent way toward the good of the community rather than of constructing texts, should provide some new perspectives for teachers of technical writing and developers of courses and programs in technical communication. For example, it provides a reasonable basis for the necessary combination of academic and nonacademic contributions to curriculum. If praxis creates knowledge, academics should indeed know about nonacademic practices. But the academy does not have to be just a receptacle for practices and knowledge created elsewhere. The academy itself is also a set of practices, including those of observation, conceptualization, and instruction—practices that create their own kind of knowledge. Such knowledge allows the academy to provide a standpoint for inquiry into and criticism of nonacademic practices. We ought not, in other words, simply design our courses and curricula to replicate existing practices, taking them for granted and seeking to make them more efficient on their own terms, making our students "more valuable to industry"; we ought instead to question those practices and encourage our students to do so too. Wells's "pedagogy for technical writing" suggests that we should aim "to work within the structures of technical discourse so that students can negotiate their demands but also be aware of the limited but real possibility of moving beyond them" (264). My own earlier sketch of a new pedagogy similarly suggested the need to promote both competence and critical awareness of the implications of competence ("Humanistic" 617). I might now supplement critical awareness with prudential judgment, the ability (and willingness) to take socially responsible action, including symbolic action.

An understanding of practical rhetoric as conduct provides what a techne cannot: a locus for questioning, for criticism, for distinguishing good practice from bad. That locus is not the individual or any particular set of private interests but the human community that is created through conduct; this community is the basis for practice in Bernstein's "high" sense. While the good that praxis in this higher sense creates may include the interests of individuals and industry, it is larger and more complex; the relevant community is not the working group or the corporation but the larger community within which the corporation sells its products, pays taxes, hires employees, lobbies, issues stock, files lawsuits, and is itself held accountable to the law.

Through praxis we make ourselves and each other in interaction: Aristotle emphasizes the political dimension of this interaction, Marx the economic. But whether our everyday activities are primarily those of governing a community or those of making a living, they have both political and economic dimensions. If technical writing is the rhetoric of "the world of work," it is the rhetoric of contemporary praxis. In teaching such rhetoric, then, we acquire a measure of responsibility for political and economic conduct.

DEVELOPING YOUR UNDERSTANDING

1. Summarize what is sometimes referred to as the "the is vs. the ought" controversy, or nonacademic professional writing practice vs. academic instruction. Identify where you stand on the controversy. In your response, examine the strengths and weaknesses of both sides.
2. Summarize the distinctions between rhetoric/writing as a productive vs. a practical art. In your summary, refer to the aims of each and the kinds of knowledge required of each. Also, compare and contrast how each perspective would affect the professional rhetor's/writer's work.
3. If we assume that professional writing should be understood as "a matter of *conduct* rather than of production, as a matter of arguing in a prudent way toward the good of the community rather than of constructing texts," explain what you believe professional writers would do at work. Refer to the following scenario to contextualize your explanation:

   You work as the communications officer for a regional environmental watchdog organization in Northern Michigan. You are the only full-time writer on the payroll, though you have one part-time staff writer and an intern from a nearby university. The organization also has a director, a financial officer, two full-time environmental scientists, and a host of volunteers.

   In the past three years, sport fishers in the region—a major tourist industry that is the pet of a state senator—have noticed a significant decrease in the size and population of various fish species. Your staff scientists' field tests have identified increased water pollutants that could be traced to several different industries upstream, but their resources and data make it hard to confirm any source.

   The organization for which you work decides that they need to initiate a communications campaign to address the problem. You begin your work by . . .

FOCUSING ON KEY TERMS AND CONCEPTS

Focus on the following terms and concepts while you read through this selection. Understanding these will not only increase your understanding of the selection that follows, but you will find that, because most of these terms or concepts are commonly used in professional writing and rhetoric, understanding them helps you get a better sense of the field itself.

1. invention
2. knowledge (as a verb)
3. rhetoric of persuasion
4. rhetoric of interaction

Whose Ideas? The Technical Writer's Expertise in Inventio

SUSAN HARKNESS REGLI
Carnegie Mellon University, Pittsburgh, Pennsylvania

ABSTRACT

Compelling arguments from researchers studying the rhetoric of science have convinced both scientists and humanists that technical writing involves invention, or discovery of the available means of argument. If we agree that *inventio* is crucial to technical writing, however, we encounter a problem: namely, that the rhetor engaged in invention as part of a technical writing process does not necessarily have expertise in the subject matter of the composition. What, then, is the expertise that the technical writer contributes to the invention process? Working from the notion that knowledge is an activity rather than a commodity [1], I argue that a technical writer's expertise in invention lies in an ability to adapt rhetorical heuristics to situations of interdisciplinary collaboration. This focus expands our understanding of how invention works when the goal of communication is producing knowledge across disciplinary boundaries, rather than winning an argument with persuasive techniques.[1]

In an article titled "Ciceronian Rhetoric and the Rise of Science: Plain Style Reconsidered," Michael Halloran and Merrill Whitburn discuss the need for students to be equipped with a broad range of rhetorical devices. They lament that in many technical writing textbooks, "Students are expected to muddle through by stylistic instinct, an approach that reinforces the myth of the born writer" [2, p. 68]. In this article, I will argue that if we do not work to articulate rich techniques for invention in the education of technical writers, we inadvertently reinforce the myth of the technical writer as a born scribe—a fortuitously gifted communicator who by instinct knows how to "clean up" the products of the real "inventors" of technical information. More specifically, I will suggest that to articulate those rich techniques, rhetoricians must recognize and examine the expertise they have in interdisciplinary collaboration, a realm where the eristic techniques of the law court simply will not suffice to effectively facilitate communication.

An excerpt from a review of changes brought about by the high-tech workplace, written by Elizabeth Tebeaux, highlights where the emphasis often lies in discussions of technical writing in the "Information Age":

Source: Regli, Susan Harkness. "Whose Ideas? The Technical Writer's Expertise in *Inventio.*" *Journal of Technical Writing and Communication,* Volume 29, Number 1, pp. 31–40. Baywood Publishing Company, Inc., Amityville, NY (1999). Reprinted with permission.

[1] A previous version of this article was presented at the 1998 *Conference on College Composition and Communication*, April 1–4, Chicago, Illinois.

> Communication skills will be redefined as the ability to handle vast amounts of information and to adapt it for a wide range of users: the foremost problem that will confront employees in the information society emanates from the quantity of information being continuously generated . . . the central question becomes how to handle it—access it, manage it, use it, and help others use it [3, p. 138].

In considering the task of invention in technical writing, the critical question I find myself wanting to ask about this statement is "By whom or what is the information generated?" Later in the piece, Tebeaux recognizes changes that need to be made to technical writing curricula based on the growing need to interact with professionals with varied academic backgrounds, to study audiences, and to communicate with other cultures. In each of these instances, however, the focus remains on managing, analyzing, and presenting data that seems to have been generated elsewhere. In one paragraph, the need for "group problem solving" skills is acknowledged, as Tebeaux recommends that classrooms incorporate laboratory settings where students from diverse backgrounds can "practice network problem solving" [3, p. 142]. The majority of the argument, however, positions the technical writer in the role of information manager rather than information generator.

Even when invention in scientific and technical discourse is discussed specifically, the role of the technical writer in the process is unclear. In an important bibliographic article, Carolyn Miller discusses works that take seriously the idea that scientific writing involves discovery of the available means of argument and therefore scientific writing involves invention [4]. The majority of the article discusses the invention process in science and technology. The discussion of the technical writer's contribution to this process is limited to a final section of the article that mentions those few technical writing and pedagogical resources that do include some treatment of invention. The works discussed, however, primarily seem to address either invention based on scientific expertise or invention in generic writing processes, but not a cross-section of the two areas of expertise. A short article by J. W. Allen, Jr. ("Introducing Invention to Technical Students") is the only work Miller mentions that ascribes the difficulties in discussing invention within technical writing classes to "the problem of distinguishing technical and scientific inquiry from rhetorical invention." Other than in Allen's article, it seems that the question of how the information that the technical writer will manage is originally produced is not addressed.

Which highlights the central difficulty that I wish to address in this article. Since we value technical writing as a legitimate academic discipline and profession, we obviously do not hold onto the belief that scientists and technical specialists are solely responsible for discovering the arguments in a technical document; in this model the technical writer is too easily relegated to being a glorified typesetter, a font-chooser, or a human grammar check. If we accept the notion that invention is crucial to technical writing, however, we encounter a problem: namely, that the rhetor engaged in invention as part of a technical writing process does not necessarily have expertise in the subject matter of the composition. If we believe the technical writer is a professional rhetor and not simply a document technician, we have to ask ourselves where, if it exists at all, is the expertise that the technical writer contributes to the invention process?

DO WE "KNOW" OR DO WE "KNOWLEDGE"?

One way to begin thinking about this question is to examine what we mean when we say that a writer works with generated information or someone else's knowledge to produce a document. Viewpoints from two scholars have expanded my own thinking about this question.

First, John Gage has proposed a way of understanding knowledge as an activity in which people can be engaged, rather than as a static entity that can be passed from one person to another:

> . . . knowledge can be considered as something that people *do* together, rather than as something which any one person, outside of discourse, *has*. . . . Rhetoric can be viewed as dialectical, then, when knowledge is seen as an *activity*, carried out in relation to the intentions and reasons of others and necessarily relative to the capacities and limits of human discourse, rather than a *commodity* which is contained in one mind and transferred to another [1, p. 207].

This concept is difficult to fully comprehend because it runs counter to our usual ways of thinking about knowledge transmission and the communication process; knowledge has typically been a noun, not a verb. But thinking about knowledge as a verb sheds some light on two models of technical writing, models that differ in their assumptions about where information originates. Take, for example, the role of a technical writer in a software engineering environment. Using the most limited model of technical writing, one might say that the software engineer creates a "piece" of knowledge as a commodity and gives it to a technical writer, who then "packages" it in words to be shipped along with the software. In a richer model, the writer is a rhetor who treats knowledge as an activity: the rhetor's expertise lies in knowing how to "perform" knowledge in a communal, dialectical context—how to orchestrate the conversation of a team of specialists working to invent, develop, produce, and test software products.

Writing more recently, Ronald Schleifer discusses the limitations of models of thinking and reading that "create the impression that these isolated activities are natural, individual, and context-free" [5, p. 440]. As an alternative, he looks to the collaborative work of scientists as described by Bruno Latour and Steve Woolgar in *Laboratory Life*. Exchanges among scientists are broken up by Latour and Woolgar into four categories: empirical facts, practical procedures, theory (possibilities of knowledge), and gossip. But Schleifer notes that:

> What is striking about the categories . . . is the ways in which the "subjects" of these knowledges constantly shift . . . The roles of master and apprentice are repeatedly transformed. This is precisely what the Cartesian model of the subject thinking—focused solely on "the reformation of my own opinions" but not on the *position* of the thinker—cannot accommodate; it is what the model of the isolated reader of a novel . . . does not allow [5, pp. 447–448].

Schleifer is advocating, in the scientific context of interdisciplinary collaboration, not a view of language that focuses on the position of the individual thinker, but a view of language that "focuses on the community out of which knowledge arises *at a particular time*" [5, p. 439]. Meaning making arises out of repeated exchanges at particular times among members of a group who are at once both isolated and undeniably communal.

The ideas articulated by Gage and Schleifer, although arising out of different theoretical and practical considerations, articulate a very different view of meaning making than that which would underlie the notion that technical writers simply obtain information and data from unnamed sources and package it to create communication between the makers of a product and the users. The view of knowledge as a collaborative activity of meaning making across disciplinary boundaries seems much more accurate to describe a situation in which technical writers work as professionals with expertise in invention—because invention, in this context, is a matter of elaborate communication among a variety of specialists, and technical writers should be specialists in elaborate communication. In order for this expertise in communication to be articulated and advanced, however, technical writers—and rhetoricians in general—need to explore how to create new techniques and adapt existing heuristics to support this process of interdisciplinary collaboration.

MODERN RHETORICAL EXPERTISE: PERSUASION . . . OR COLLABORATION?

In "Writing in the Content Areas: Some Theoretical Complexities," David Kaufer and Richard Young narrate a story that makes very real the areas of difficulty that scholars can encounter when working across disciplinary boundaries. The article addresses the problems that arose when two rhetoricians began working with a biologist to incorporate writing into a biology curriculum, ultimately to help students learn how to write as professional biologists. The participants in the pilot project found repeatedly that neither set of experts could succeed in the project by calling solely on their own expertise: "Velez and Young [the rhetoricians] were learning more about relevant features of biology education, and Kauffman [the biologist] was learning more about writing. Despite their mutual efforts to pinpoint the locus of expertise, the more they interacted the less well defined the center of expertise seemed to become" [6, p. 89]. The interaction that ensued "produced changes on both sides" [6, p. 101], and the effort necessary to engage in and explore this interaction seems to be a large part of the prohibitive "commitment of time and dollars" that made the pilot project ill-suited as a university-wide model without further research and refinement [6, p. 88].

The critical problem that makes the interaction between rhetoricians and the biologists take so much time and call on so many resources seems to be that we do not know enough about how to facilitate productive yet efficient collaboration. Recent research on collaboration in industry and academia has certainly increased our understanding of collaborative situations [e.g., 7, 8], but we have not extensively examined how to apply what we know about communication to these situations. As rhetoricians, we inherit a tradition that knows a great deal about how to persuade an audience, about how to win an argument, and about how to perform in a competitive situation like a law court where one party will be the winner and the other will lose. By contrast, we know much less about how to exchange knowledge profitably, about how to clarify differences of definition and associations while recognizing ambiguities, and about how to identify and solve problems while incorporating many varied perspectives. We know a great deal and have extensive codification schemes for the rhetoric of persuasion. In contrast, we know very little about the rhetoric of interaction.

One way to understand what we don't know is by looking at the writing process of an individual versus the writing process of a group. Kaufer and Young mention the interaction between problem spaces that Bereiter and Scardamalia identified in the workings of individuals as they write: "Our contention is that this interaction between the two problem spaces [content space and rhetorical space] constitutes the essence of reflection in writing" [9, p. 302]. When experts interact to plan or to write together, however, the problem spaces are numerous, each mind has its own content spaces, and the expertise is inconveniently divided into these separate content spaces. The reflective rhetor in this communication situation cannot simply move back and forth between the content space and the rhetorical space in his or her own mind; rather, the task of communication is to enable the difference spaces in the different minds to interact productively with one another, even as the locus of expertise is shifting constantly.

We need to develop more extensive heuristics for how to navigate between these problems spaces—for how to locate sources of misunderstanding, discover the relevant loci of expertise among multidisciplinary specialists, and synthesize the expertise of many into a document or plan of action. Some of what we already know will be useful to us; some of the skills and theory will have to be created out of our experience and our ingenuity. The important guideline to recognize, however, is that while persuasion is operative in any communication situation, it is not the foremost goal of the productive collaborative situation. Douglas Ehninger called the rhetoric of the current period (from the 1930s to the present) a "sociological" rhetoric: he observed that while there are many specialized strands of interest in rhetorical studies, at bottom they all "view rhetoric as an instrument for understanding and improving human relations" [10, p. 333]. What this may mean in practice is that the law court is no longer the paradigm case for rhetorical studies; instead, the collaborative, often ad-hoc group working together to make improvements is the situation toward which we need to direct our attention.

A PRELIMINARY PLAN

In this section, I will attempt to sketch what we already know about the expertise a rhetorician can bring to a collaborative situation, and what we might need to add or elaborate to create a plan for the action of a technical writer or the curriculum of a technical writing program. I will continue to use the model of the technical writer on a software development team to focus the description of this plan.

Strategies for Collaborative Problem Identification and Inquiry

Perhaps the most critical task for the technical writer on a team of specialists is to keep the team from proceeding too quickly, before the real problem has been articulated. I would recommend training in tagmemic problem solving strategies to fend off hasty conclusions and to aid the team in identifying exactly which needs the software is being designed to fulfill. In addition, these strategies could be extended to be used as a group problem solving heuristic: if each member of the team articulates what he or she sees as the crucial problem(s) to be solved, and team members read, reflect on, and discuss one another's statements, the group can better clarify where the points of disagreement or misunderstanding lie in terms of what goals the team is striving to meet.

Situation Analysis

Working hand-in-hand with tagmemic analysis, the process of having the team (again both separately and together) identify how they understand the elements of the situation for which the software is being developed will help clarify what each member sees as the appropriate audience, the exigence that is being addressed, and the constraints that need to be satisfied [11, pp. 6–8]. While Bitzer did not originally intend for his notion of the rhetorical situation to be implemented as an analytical heuristic, I believe that it could be profitably used to find places of misunderstanding, and especially to allow each member of the team to articulate the constraints under which they believe they are operating. The lack of knowledge about one another's constraints is often one of the richest sources of miscommunication in a collaborative situation; having each member articulate his or her view of the rhetorical situation could alleviate some of this confusion before the development process proceeds down dead-end paths.

Audience Analysis

As part of the analysis of situation, and also in recognition that the audience (or consumer) for the product is part of the multidisciplinary community that needs to exchange information, the technical writer should be equipped both to communicate with the potential audience for the product and to create a thorough representation of audience expectations. We have made great strides in encouraging technical writers to speak to the audience directly to find out what their needs and expectations are, to uncover the processes they use to do their work, and sometimes to redesign these processes so that poor procedures are not simply re-instantiated in the software program. Understanding the audience as part of the multidisciplinary team by which the software product is created extends the justification and meaningfulness of these procedures.

Analysis of the Structure of Information

Often, in the process of deciphering information from other fields of expertise (especially if the expert is not available for direct consultation), a technical writer will need to figure out the meaning of unfamiliar words or concepts. A deep understanding of how information is typically structured according to the principles of grammar, usage, and genre conventions is often useful in verifying the sense of sentences when the vocabulary words or underlying concepts are from unfamiliar domains. In addition, for those writing tasks where the "knowledge" has indeed been supplied by an existing database or set of printed manuals, an understanding of the structure of information is crucial to organizing large bodies of data for presentation to audiences. This task is particularly important in new online environments where the amount of available data is virtually unlimited and presentation is constrained only by how much an audience can absorb.

Production Skills

At some point in the process, documentation does need to be produced, including both verbal and graphic representations of knowledge. Current traditional rhetoric's emphasis on production and some handbooks of technical writing tend to deal solely with this aspect of the technical writer's job. But in the new paradigm, the technical writer may not be the sole person generating and revising the documentation. Drafts typically need to be passed from one member of the team to another so that changes can be suggested, questions can be asked, and revisions can be incorporated with the agreement of all members of the team (and each person will often focus on a different set of primary concerns for the documentation). This process of collaborative revision is an area in which more research and development is needed, both in terms of process and tools to support the interaction.

EXTENDING THE FIELD

Merrill Whitburn argued convincingly that the field of technical writing needs the expertise found in English departments, and that English departments need the focus of practical concerns (and marketability) that technical writing can offer [12]. I have tried in this article to extend his argument for the value of the field and I would claim that by looking at the constraints and possibilities available to a professional who specializes in technical writing, we can better understand where our expertise as modern rhetors lies, and where we ought to direct our inquiry to increase it. The plan I have outlined is an attempt to understand this expertise by examining the environment in which a technical writer operates. Unintentionally, it is also a fair articulation of what anyone receiving training in rhetoric ought to encounter at some point in his or her education. The reason for this parallel is that in the field of rhetoric as a whole, I believe we need to turn our attention away from simply aiming at persuasion, and from simply addressing discipline-specific academic environments. Inquiry as we know it is becoming increasingly multidisciplinary, collaborative, multisector (as industry, academia, and government work together more frequently), and interactive. The more we can come to understand our expertise in this relatively new environment, the better we are positioned as effective professionals. Examination of the field of technical writing—specifically of the role of invention in that field—can help us to increase this understanding.

References

1. J. T. Gage, An Adequate Epistemology for Composition: Classical and Modern Perspectives, in *Landmark Essays on Rhetorical*, R. Young and Y. Liu (eds.), Hermagoras Press, California, pp. 203–219, 1994.

2. S. M. Halloran and M. D. Whitburn, Ciceronian Rhetoric and the Rise of Science: The Plain Style Reconsidered, in *The Rhetorical Tradition and Modern Writing*, J. J. Murphy (ed.), The Modern Language Association of America, New York, pp. 58–72, 1982.
3. E. Tebeaux, The High-Tech Workplace: Implications for Technical Communication Instruction, in *Technical Writing: Theory and Practice*, B. E. Fearing and W. K. Sparrow (eds.), The Modern Language Association of America, New York, pp. 136–144, 1989.
4. C. Miller, Invention in Technical and Scientific Discourse: A Prospective Survey, in *Research in Technical Communication*, M. Moran and D. Journet (eds.), Greenwood Press, Connecticut, pp. 117–162, 1985.
5. R. Schleifer, Disciplinary and Collaboration in the Sciences and Humanities, *College English*, 59, pp. 438–452, 1997.
6. D. Kaufer and R. Young, Writing in the Content Areas: Some Theoretical Complexities, in *Theory and Practice in the Teaching of Writing: Rethinking the Discipline*, L. Odell (ed.), Southern Illinois University Press, Carbondale, Illinois, pp. 71–103, 1993.
7. M. B. Debs, Recent Research on Collaborative Writing in Industry, *Technical Communication*, 38, pp. 476–484, 1991.
8. L. Jorn, A Selected Annotated Bibliography on Collaboration in Technical Writing, *Technical Communication Quarterly*, *2:1*, pp. 105–115, 1993.
9. C. Bereiter and M. Scardamalia, *The Psychology of Written Composition*, Lawrence Erlbaum Associates, Hillsdale, New Jersey, 1987.
10. D. Ehninger, On Systems of Rhetoric, in *Professing the New Rhetorics: A Sourcebook*, T. Enos and S. C. Brown (eds.), Prentice Hall, Englewood Cliffs, New Jersey, pp. 327–339, 1994.
11. L.F. Bitzer, The Rhetorical Situation, *Philosophy and Rhetoric*, *1*, pp. 1–14, 1968.
12. M. D. Whitburn, The Ideal Orator and Literary Critic as Technical Communicators: An Emerging Revolution in English Departments, in *Essays on Classical Rhetoric and Modern Discourse*, R. J. Connors, L. S. Ede, and A. A. Lunsford (eds.), Southern Illinois University Press, Carbondale, Illinois, pp. 226–247, 1984.

DEVELOPING YOUR UNDERSTANDING

1. Regli argues that we should understand the professional writer as a "professional rhetor" rather than a "document technician." Compare and contrast these two role-types, detailing what each knows, what each does, and what each values.
2. Entrenched workplace structures and work processes often more powerfully define what a professional writer is, does, and can be than anything else. Describe how the following brief workplace scenario defines what the professional writer is, does, and can be. Then, assess what would have to change in terms of workplace structures and work processes for the professional writer to be and do what Regli advocates—"technical writers should be specialists in elaborate communication."

   Dominique is a technical writer in the Publications Department of a mid-sized software development company. She has just been given the task of writing print and online documentation for part of a financial planning software package for a major bank that will use the software to calculate and explain bank loans to clients on-the-spot.

Her department has a director, who also functions as the chief editor. The director, and a few senior writers, become part of product development teams about midway through the product development process. As part of these teams, the director and senior writers learn the software and its purposes, and on occasion they become involved in user interface design. Once a product is nearing the end of its development cycle, the director distributes to staff writers, like Dominique, various documentation writing tasks. The team of writers have a few early meetings, where the director and a small team of developers explain the software. Then the director lays out each technical writer's task. If staff writers have questions, they have ready access to the director and lead developers. The director regularly checks on staff writers' individual progress throughout their writing process. The team of staff writers typically meet twice more, once for a draft review and again for an editorial review.

3. Too often we define writers by their abilities to produce what might be called "target documents"—those final, finely polished documents that are meant to communicate some sort of coherent message to an audience. Target document production emphasizes the writer's abilities in arrangement, style, and delivery. Regli argues, however, for a writer whose most valued ability is invention. Describe the kinds of writing/documents that Regli's model professional writer might produce to aid the inventional work that is her expertise.

FOCUSING ON KEY TERMS AND CONCEPTS

Focus on the following terms and concepts while you read through this selection. Understanding these will not only increase your understanding of the selection that follows, but you will find that, because most of these terms or concepts are commonly used in professional writing and rhetoric, understanding them helps you get a better sense of the field itself.

1. participatory communication
2. transmission view
3. translation view
4. articulation view
5. encode
6. decode
7. sender
8. receiver
9. mediator

The authors explore the parallels to be found by comparing descriptions of the technical communicator with differing views of the communication process—the transmission, translation, and articulation views of communication. In each of these views, the place of the technical communicator and of technical discourse shifts with respect to the production of meaning and relations of power. The authors argue from the standpoint of the articulation view for a new conception of the technical communicator as author and of technical communication as a discourse that produces an author.

The Technical Communicator as Author: Meaning, Power, Authority

JENNIFER DARYL SLACK, DAVID JAMES MILLER, JEFFREY DOAK
Michigan Technological University

In his essay, "What Is an Author," Michel Foucault observes that in our culture, the name of an author is a variable that accompanies only certain texts to the exclusion of others: a private letter may have a signatory, but it does not have an author; a contract can have an underwriter, but it does not have an author; and, similarly, an anonymous poster attached to a wall may have a writer, but he cannot be an author. (124)

From this, Foucault concludes that "the function of an author is to characterize the existence, circulation, and operation of certain discourses within a society" (124). At its most mundane, this is simply to note the fact that certain discourses are granted the privilege of authorship while others are denied this privilege. It is more remarkable to notice, with Foucault, that this very fact suggests an inversion of the way in which we typically understand the relation between an author and a discourse: Rather than authors producing certain discourses, certain discourses are understood to produce authors. To grant authorship to a discourse is to grant that discourse a certain authority. In a peculiar turn of events, this authority comes to reside in the author, the author produced by the discourse itself. Thus it becomes evident that authorship is a manner of valorizing certain discourses over against others. As such, authorship empowers certain individuals while at the same time renders transparent the contributions of others.

The discourses created by technical communicators have not been considered authored discourses; the technical communicator may be a transmitter of messages or a translator of meanings, but he or she is not—or at least not until now—considered to be an author. We have come to see that technical communicators, as well as other professional communicators, are engaged in the process of what Marilyn Cooper has called *participatory communication*. In "Model(s) for Educating Professional Communicators," Cooper writes:

> I am defining communication as participatory communication and the role of . . . communicators as one of . . . working together to create common interests, to construct the ideals of our society, [and in light of these ideals] to examine the ends of [our] action. Professionals who communicate should be involved in this endeavor too. . . . It is [at least part] of the function of professional communicators—whether they know it or not. (12)

Source: Slack, J. et al. "The Technical Communicator as Author: Meaning, Power, Authority." *Journal of Business and Technology Communication* 7 (1993), pp. 12–36.

THE RELEVANCE OF COMMUNICATION THEORY

There are striking parallels to be found by comparing descriptions of the technical communicator (descriptions and redescriptions of the role, task, and ethos of that communicator) with the progressive development of our theoretical understanding of the communication process itself. The most remarkable of these parallels may well lie in the emerging evidence of a symmetry between disparate images of the technical communicator and distinct—although ultimately interrelated—models of communication. What we propose is that, by comparing different images of the technical communicator with parallel developments in the study of communication, a new theoretical and practical image of that communicator—the technical communicator as author—can begin to be established. Reflecting on the historical development of communication theory over the course of the past 10 years, scholars in communication have come to acknowledge that, at least with respect to the study of mass communication, two basic models of communication have gained ascendancy and, although this is less widely acknowledged, that a third is now gaining ground (see, for example, Fiske, *Introduction*; Carey). For our purposes, it is more useful to speak of these models not as models per se but as distinct views of communication. This is the case because, at bottom, each of these models seeks to express the morphology common to a collection of theories that otherwise appear more or less disparate. In this regard, the term *model* is misleading. It appears to set one theory of communication over against other such theories rather than gathering a number of specific theories together in a general conceptual classification. We have no interest in a general conceptual classification. We have no interest here in pitting one theory of communication against another. We are concerned with what these views, together, can teach us about the place of the technical communicator.

The first of these views—what we will refer to as the transmission view of communication—can be delimited in terms of a concern, for the most part, with the possibilities and problems involved in message transmission, that is, in conveying meaning from one point to another. The second—what we will call the translation view of communication—can be understood in terms of a primary concern with the constitution of meaning in the interpretation and reinterpretation of messages. The third—what we will call the articulation view of communication—can be grasped as a concern principally with the ongoing struggle to articulate and rearticulate meaning. With respect to each of these views of communication, the place of the technical communicator is located differently. In the first, the transmission view of communication, the technical communicator is a purveyor of meanings; in the second, the translation view of communication, the technical communicator is a mediator of meanings; in the third, the articulation view of communication, the technical communicator is an author who among others participates in articulating and rearticulating meanings.

Corresponding to variations in the place of the technical communicator as purveyor, mediator, or articulator of meanings, the place of the technical communicator—and of technical discourse itself—shifts in different relations of power. In the transmission view, the technical communicator remains the neutral vehicle facilitating the exercise of power. In the translation view, the technical communicator works to create symmetry within the negotiation of differential relations of power between sender and receiver. In the articulation view, the technical communicator is complicit in an ongoing articulation and rearticulation of relations of power. Ultimately, looking through the lens of articulation—as we do in this article—the different locations of the technical communicator implicate one another. That is, the technical communicator and technical discourse purvey, mediate, and articulate meaning. Likewise, the technical communicator and technical discourse facilitate, sustain, generate, and disrupt relations of power. But only by looking through the lens of articulation can we rearticulate the technical communicator and technical discourse as participating fully in the articulation of meaning and thereby fully empower the discourse as authorial.

CHANGING CONCEPTIONS OF MEANING AND POWER: TRANSMISSION

Of the three views of communication, the transmission view has been the most clearly delineated. It has been extensively critiqued and often maligned such that it is nearly requisite to begin any introductory text on communication theory with an explanation and rejection of it. For the most part, contemporary communication theories are proposed in contradistinction to it. There are, consequently, many different versions of the position and ongoing disagreements about its precise historical and theoretical contours (see, for example, "Ferment in the Field"). In general, however, the transmission view combines three defining characteristics:

1. the conception of communication as the transportation of messages
2. the conception of the message—the meaning encoded by a sender and decoded by a receiver—as a measurable entity transmitted from one point to another by means of a clearly delineated channel.
3. the conception of power as the power of the sender to effect, by means of this message, a desired mental and/or behavioral change in the receiver. This power is the power of the sender over the receiver.

The term communication has its origins in the concept of transportation (Williams; *Oxford English Dictionary [OED]*). Communications were the paths of transportation by means of which people at the centers of power could exercise control over those in the peripheries. The ability to move messages in a timely fashion across space by means of such communications was a necessary condition for political, economic, and religious domination. The emphasis in the historical development of new technologies of communication (from walkers, runners, horses, smoke signals, semaphore, print, telegraph, telephone, television, satellites, computers, fax machines, etc.) has been the transmission of knowledge and information in such a way as to *exercise control over space and people faster and farther*.

The implications for how meaning has been understood in communication theory are made clear by examining how communication as transportation gets tied to a theory of transmission. The work of Shannon and Weaver can be credited as a principal determinant in the shaping of such a view. Largely mathematical in character,

Shannon and Weaver's conception of communication is as an explicitly linear form: The sender wishes to transmit meaning, but to do so it must be encoded in the form of a message (Shannon and Weaver called this *information*). The message is sent over a channel to a receiver who then decodes it to get out of it the meaning that was encoded. The process, when perfectly executed, results in the receiver's decoding exactly the same message that the sender intended to encode.

This basic model has been amended and elaborated on extensively (see, for example, Fiske, *Introduction*; McQuail and Windahl), but its orientation to meaning remains essentially the same. Meaning is something that is "packaged up" by the sender, shipped out, and "unwrapped" by the receiver, who can then act or think accordingly. Of course, there are numerous points in the process where difficulties can render the transmission less than perfect. The sender may encode the message poorly such that the message fails to contain the intended meaning. The decoder may decode poorly, not reading the intended meaning properly. There may also be "noise" in the channel that distorts the message so that, consequently, the meaning it contains is not received in the form in which it was sent. (Noise may take many forms, from static on the telephone line to the wandering mind of the listener during the transmission.)

In the transmission view of communication, meaning is a fixed entity; it moves in space "whole cloth" from origin to destination. Communication is successful when the meaning intended by the sender is received accurately, where accuracy is measured by comparing the desired response to the message with the actual response. Communication fails when these responses diverge. In the case of failure, the communicator must locate and correct the source of the failure in the process of encoding or in the noise of the transmission. Power is simply that which is exercised when the communication is successful. The sender has power when the receiver behaves in the intended manner. Power, like meaning, is something that can be possessed and measured; its measure is to be found in the response of the receiver.

Such a view of communication appears to dominate the early stages of the theory and practice of technical communication as it emerged within the college curricula of engineering schools. Based on research done by Robert J. Connors, we would characterize this phase in technical communication as dominating the field from the late 1800s until the 1950s but persisting into the present. In this phase, technical writing and engineering writing are treated as synonymous, and the task of the technical writing course is to teach engineers or their surrogates to encode the engineers' ideas (meanings) accurately and to provide a clear channel for transmission.

Technical writing courses developed this way in response to a series of changes in the practice of engineering and the development of the engineering curriculum. As engineering and its curriculum became more technically specialized and less humanistic during the period of rapid industrialization that followed the Civil War, complaints about the unbalanced education of students in technical schools mounted. Among other deficiencies, engineers, it was claimed, "couldn't write." To correct this imbalance, courses in engineering English (later technical writing; later still, technical communication) were developed. As Connors points out, by this time the notion of the "two cultures" split was so firmly in place that, as we would put it, the kind of meanings that required encoding were sufficiently different to warrant a completely different kind of English course (331). Engineering English courses were designed, among other things, to teach students to encode the special meanings of engineers.

Education in this phase has two components: the education of engineers and the education of surrogate engineers. Both are firmly anchored in a transmission view. The earliest, but again still persistent, effort to inculcate the skill of technical writing is to teach the engineer—as sender—to be a better encoder through the use of proper language, grammar, and style. Through such training, the intent is that engineers will learn to encode messages such that they will match their intentions. Further, in teaching engineers to transmit those properly coded messages using the proper forms, the intent is to ensure that the proper channels are chosen and that the transmission is sent with minimal noise. In James Souther's review of the evolution of technical writing course content, he demonstrates that the first kinds of courses to develop were those focusing on the "effective use of language, grammar, and style" (3), later focusing on teaching the different forms, reports, and letters routinely used in the engineering profession.

Developing later, and rapidly growing alongside the engineer writer, the surrogate engineer—the technical writer—has become at least as important in the horizons of technical communication. The conjuncture of the increased demands placed on highly specialized engineers and the growing awareness of the complexity and difficulty of encoding their ideas (meanings), gives shape to the development of technical writing as a discipline in its own right (Connors places this in the 1920s). Course work and textbooks began appearing that were directed toward the technical writing student in particular rather than toward the engineer. In spite of this specialization, the technical writer is assumed to be a mere surrogate, or stand-in, for the actual (but busy) sender, the engineer.

The technical writer's job in this period dominated by the transmission view of communication is to assure that messages are accurately encoded and that they are transmitted with minimal noise over clear channels. In fact, the professional technical writer, as surrogate engineer, is rendered essentially transparent in the process, ideally *becoming* the clear channel itself. The very definition of technical writing often affirms this commitment to the transparency of the communication-as-channel. This is often explicit, as Michael Markel writes as recently as 1988:

> Technical writing is meant to fulfill a mission: to convey information to a particular audience or to affect that audience's attitudes in a particular way. To accomplish these goals, a document must be clear, accurate, complete, and easy to access. It must be economical and correct. The writer must be invisible. The only evidence of his or her hard work is a document that works—without the writer's being there to explain. (6)

It is relatively easy to understand the location of meaning and the conception of power as they operate in this phase. Meaning is posited to be in the intentions of the sender, that is, the engineers. Meaning is simply transferred over a clear channel. Technical writers are not seen as adding or contributing to meaning. In fact, if they are, they are not doing their job! After all, they are not engineers themselves; nor are they the source of the meaning to be transmitted. Nor does meaning originate in any sense in the receiver.

Because meaning resides only in the sender's intentions, and the technical writer is merely a surrogate encoder, when communication is successful (i.e., the intended response achieved), the recognition, responsibility, and power is attributed only to the sender. However, if communication fails, it is exceedingly easy to fault the encoding process, that is, the work of the technical writer. *Miscommunication*, as this failure is called, can be attributed to the weak use of language (inadequate encoding), failure to include appropriate information (inadequate encoding), or poor standards for documentation (noisy channel) (see, for example, Kostur and Hall).

Power, then, must be understood as possessed by the sender and measured by the ability of the message to achieve the desired result in the receiver. To communicate is to exercise power. The sender has no power if the receiver does not respond appropriately. Miscommunication, the principal measure of failure in this phase, occurs when there "is disparity between the message intended and the message received" (Kostur and Hall 19). Technical writers, who are rendered transparent and seen as contributing no meaning, possess no power (and therefore cannot exercise it) whenever communication is deemed successful. To be transparent is, after all, to provide a clear channel for the sender to exercise his or her power. Interestingly, however, if a message fails, technical writers can always be held responsible and called on to do a better job at encoding or transmission. They possess, then, a kind of negative power—by virtue of their potential status as "inadequate surrogates"—to manage the processes of encoding and transmission poorly and take the responsibility for miscommunication.

The persistence of thinking in these terms is evident in much of the professional and educational realities of technical communicators. The extent that their education focuses on stylistics, the proper use of forms, and skill at operating the technologies of communication—to the detriment of the kinds of knowledge and skills we introduce later—is testament to that persistence. Technical communicators are taught, for example, that the highest goal they can achieve is "clarity and brevity," which suggests a transparency that belies what they really do. On the job, the role of surrogate encoder is attested to by the extent that the communicator is treated as low in the organizational hierarchy, as working *for* the real sender, and as expert mainly in questions of style, form, editing, and media management. To transmit the sender's meaning as a perfectly executed message is the role of this communicator.

CHANGING CONCEPTIONS OF MEANING AND POWER: TRANSLATION

The second of the views specified at the outset, the translation view of communication, a view characterized by a fundamental concern over the constitution of meaning in messages in which power is negotiated between sender and receiver, has not been as clearly delineated as the transmission view. There are numerous contenders in the struggle to define the view developed in contradistinction to the transmission view of communication, and the successor has not yet been fully agreed on. There are in our reading several characteristics that the approaches to the second view seem to share

1. the conception of communication as a practice
2. the conception of meaning as produced through the interaction of sender and receiver
3. the conception of power as *negotiated*.

If you look back at our discussion of the transmission view of communication, you will note a conspicuous absence: The receiver in the process of communication is absent in any way other than as passive recipient of the communicated or miscommunicated message. Receivers add no meaning; they have no power. Reception is considered to be essentially unproblematic. If the message is encoded properly and sent over a clear channel, it should have the desired impact on the receiver. In contrast to this view, theorists of the translation view consider the activity of the receiver to be just as constitutive of the communication process as that of the sender. Communication is not a linear process that proceeds from sender to receiver, but a process of negotiation in which sender and receiver both contribute—from their different locations in the circuit of communication—to the construction of meaning. The nature of this process of negotiation can be understood by illustrating its operation in Stuart Hall's elaboration of what he has called a theory of "encoding and decoding."

Hall describes communication as a practice in which sender, message, and receiver are but "different moments" in a "complex structure of relations." Communication is "a structure produced and sustained through the articulation of linked but distinctive moments—production, circulation, distribution/consumption, reproduction" ("Encoding" 128). Each moment has its own distinctiveness and modality and contributes to the circulation that constitutes the communication. Hall describes it this way:

> The process . . . requires, at the production end, its material instruments—its "means"—as well as its own sets of social (production) relations—the organization and combination of practices within media apparatuses. But it is in the *discursive* form that the circulation of the product takes place, as well as its distribution to different audiences. Once accomplished, the discourse must then be translated—transformed, again—into social practices if the circuit is to be both completed and effective. If no "meaning" is taken, there can be no "consumption." If the meaning is not articulated in practice, it has no effect. ("Encoding" 128)

The acts of encoding and decoding are thus both active processes in the circuit of meaning production. The sender encodes meaning (meaning 1) based on the frameworks of knowledge, relations of production, and technical infrastructure within which the sender operates. A *meaningful* product is produced (a technical report, for example). But the receiver also actively decodes a meaning (meaning 2) based on potentially *different* frameworks of knowledge, relations of production, and technical infrastructure. There is no necessary correspondence (or symmetry) between meaning 1 and meaning 2, because each operates semiautonomously. It is as though the practices of encoding and decoding are practices of *translation*, from social practices to discourse and then back into social practices.

When there is symmetry between the translation processes, we can talk about equivalence between the two moments—a way of rethinking the concept of *understanding*. And when there is a lack of symmetry, we can talk about a lack of equivalence—a way of rethinking the concept of *misunderstanding*. Misunderstanding cannot be explained fully by inadequate skill at encoding or by the presence of noise in the channel. Any asymmetry can also be understood as an outcome of alternative practices of encoding and decoding (Morley).

Some translation approaches continue to use a concept such as misunderstanding because they persist in privileging the encoding process. Hall, for example, posits the encoded meaning (meaning 1) as the "dominant or preferred meaning" ("Encoding" 134). Then in comparing the symmetry between the preferred, encoded meaning and various decoded meanings, decodings are determined to be within the dominant, or preferred, code (dominant decoding); against it (oppositional decoding); wildly unrelated to it (aberrant decoding); or in a negotiated relationship to it (negotiated decoding) (Morley).

Some translation approaches have sought to dispense with the privileging of encoded meanings and render both moments as more *equally* constitutive. These approaches, such as that of John Fiske (*Television*), use conceptions of an "open text," conceptions such as polysemy and Bakhtin's heteroglossia. Heteroglossia asserts that "all utterances . . . are functions of a matrix of forces practically impossible to recoup" (qtd. in Fiske, *Television* 89). Polysemy asserts that a text is not merely a bearer of meanings. Rather, a text identifies and limits "an arena within which the meanings can be found. . . .[W]ithin those terms there is considerable space for the negotiation of meaning" (84). The more open a text, the greater the range within which receivers are free to make their own meanings.

Meanings are thus located in several places: in the practice of encoding, in the discursive product, and in the practice of decoding. In the passage of these forms, "no one moment can fully guarantee the next moment" (Hall, "Encoding" 129). Meaning is fluid and elusive, never really fixed at any moment.

Power is displaced and fluid along with meaning. There is power in the practice of making meaning. Because both encoders and decoders generate meaning, both exercise power. This is no longer simply the power of sender over receiver but the differential power of each to bring their own context to bear in the making of meaning (Fiske, *Television*).

Despite the fluidity of meaning, the translation view deals uneasily with differential relations of power. The receiver can work with the product (or text) only as it has been encoded, and that limits the openness of the text. This situation still privileges the practice of encoding. As Hall puts it,

> Polysemy must not, however, be confused with pluralism. Connotative codes are *not* equal among themselves. Any society / culture tends, with varying degrees of closure, to impose its classifications of the social and cultural and political world. These constitute a *dominant cultural order,* though it is neither univocal nor uncontested. This question of the "structure of discourses in dominance" is a crucial point. The different areas of social life appear to be mapped out into discursive domains, hierarchically organized into *dominant or preferred meanings*. ("Encoding" 134)

These dominant, or preferred, meanings must *work* to exercise power—to bring decodings into symmetry with the encodings. But decoders—always active in the decoding process—variously exercise their power to disrupt the circulation of power by decoding differently and articulating meanings differently into practice. Communication is thus *an ongoing struggle for power,* unevenly balanced toward encoding.

Currently, the field of technical communication seems to be struggling with (sometimes against) the implications for the role of the technical communicator as translator. The most obvious marker of this shift is that the technical *writer* becomes the technical *communicator* with the recognition that communicators have something to add beyond skillful encoding and clear channel. But there is much more than a name change here. To be expert in the practice of communication, to be a *communicator* in the process, signifies changes in understanding the power of the receiver as well as of the technical communicator—changes that open a virtual Pandora's box that can never again be closed.

There are a number of new things to attend to now (sometimes old things in new ways): (a) Because the process of encoding is always a process of trying to fix already slippery meanings, it is important for the communicator to understand the context of the sender. Hence familiarity with the technical field of the sender will work to ensure that in the translation process, the preferred meanings are the ones that get fixed. (b) Because the process of encoding is always an imperfect translation, it is important for the communicator to become expert at understanding and manipulating language as polysemic. Hence familiarity with the principles of rhetoric and composition and skill at using their tools will work to ensure that the communicator will know how to fix meanings. (c) Because the receivers of technical communications have the power to decode differently depending on the contexts within which they operate, the communicator must understand how those audiences decode. Hence rhetoric (as the art of persuasion), composition, audience analysis, and reader-response research will help to ensure that communicators know how to encode such that particular audiences are most likely to decode symmetrically. (d) Further, once it is recognized that there is always a struggle to fix otherwise slippery meanings, the communicator must acknowledge and work with the differential relations of power within which sender and receiver operate. Hence attention to power and ethics is essential.

These concerns all become well represented in the field of technical communication from the 1950s on, although attention to power and ethics seems least represented, for reasons discussed later. The evidence of these changing priorities can be seen in the growing recognition of the unique contribution that can be offered by technical communicators as experts rather than as surrogates. This recognition is self-reflexive, which may account for the developing professionalization of technical communication. Evidence can also be seen in the changing textbooks and instruction in technical communication (Connors; Souther). Although stylistics, grammar, editing, and the use of media still play a major part in the education of technical communicators, it has also become essential to add to their educational repertoire work in rhetoric and composition, linguistics, problem solving, audience analysis, and ethics.

There are still employers, educators, and students whose understanding of communication is linked to the thinking of the first, or transmission, view. They have difficulty understanding the role of all this theory in just getting the job done (see, for example, Vaughan 80). But what they fail to understand is that to execute the job with sophistication—to work toward the negotiation of symmetry between encoder and decoder—the theory must be brought to bear on the practice of communication. That requires attention to the complex and variable contexts within which senders and receivers produce meanings and how those contexts connect in the circuits of meaning and power.

Technical communication education is still in the process of sorting out those connections, establishing the balance between theory and practice. Becoming well established is the need to *go on theorizing,* to recognize that technical communication is not simply a skill but an academic and practical discipline that requires us to push the boundaries of theory if we are to understand what works and why.

But there is more to say about meaning, power, and ethics. The promise (for some the pestilence) released from the Pandora's box of the translation view of communication is the power of the technical communicator as translator. Given the fluidity of meaning and the polysemy of any text, a translator can never be transparent. Lawrence Grossberg describes the position of the translator in this view: "Translation involves the retrieval and reconstitution of two different traditions, of two different sets of possibilities and closures. It always involves us in compromise, not only of the text's language, but of the translator's as well" ("Language" 221). The technical communicator, by virtue of the nature of the language, then, *must* add, subtract, select, and change meaning. This ushers in the recognition that the communicator, too, exercises power, that is, the communicator—operating from within a different context—makes meaning too. That recognition requires attention to ethics grounded in an understanding of how power works.

There seems to be a subtle recognition in the field that the communicator has power, but coming to terms with the nature of that power gets lost in the demarcation of encoding and decoding, of sender and audience, as the principal sites of investigation. Most educators acknowledge that it would be a good idea for students to understand politics, power, and ethics, but there is very little explanation offered to suggest what they might do with that knowledge on the job. But one thing is certain: A technical communicator cannot be just a technical writer anymore. What, then, do technical communicators offer? We think there are some answers suggested if we look ahead through the lens of ongoing theorizing in communication.

CHANGING CONCEPTIONS OF MEANING AND POWER: ARTICULATION

The third of the views specified at the outset, the articulation view of communication, a view characterized by concerns with the struggle to articulate and rearticulate meaning and relations of power, can be delineated in contrast to both the transmission and the translation views. The transmission view acknowledges that senders do have meanings that they desire to encode and that they do often desire a particular response to that message from the receiver. However, the transmission view limits our recognition of the full fluidity of meaning. The translation view reconstitutes transmission to add an understanding of the receiver's contribution to the constitution of meaning and introduces the constitutive role of a mediator. However, translation based on the model of encoding and decoding limits our understanding of the full authorial contribution and power of the mediator.

The translation view opens the space for the attribution of authorial power (the Pandora's box) but leaves it undertheorized. The opening is evident in Grossberg's assertion (cited previously) that the language of the translator must be taken into consideration. The way through that opening is provided in the very language of encoding and decoding, specifically in thinking through Hall's suggestion that meaning is "articulated in practice" and that meaning and discourse are "transformed . . . into social practices . . . if the circuit [of meaning] is to be both completed and effective" ("Encoding" 128). The articulation view allows us to move beyond a conception of communication as the polar contributions of sender and receiver to a conception of an ongoing process of articulation constituted in (and constituting) the relations of meaning and power operating in the entire context within which messages move. That context includes not just the context of the sender and receiver (the frameworks of knowledge, relations of production, and technical infrastructure) but of the mediator(s) as well. And *mediator* here can no longer be thought of as just the technical communicator but as the channels (including media and technologies) of transmission as well.

Articulation is a concept that has been drawn from the work of Antonio Gramsci, considered by Ernesto Laclau, influenced by structuralism (especially Althusser) and postmodernism (see, for example, Deleuze and Guattari), and developed into an identifiable theoretical position by Hall ("On Postmodernism"; "Race"; "Signification"). Grossberg has elaborated on the role of power in this position ("Critical Theory"). Articulation asserts that any identity in the social formation must be understood as the nonnecessary connection between the elements that constitute it. Each identity is actually a particular connection of elements that, like a string of connotations, works to forge an identity that can and does change (Hall, "Signification"). An identity might be a subject, a social practice, an ideological position, a discursive statement, or a social group. The elements that constitute these identities are themselves identities; therefore, they too must be understood as nonnecessary, changing connections between other elements. The way in which elements connect or combine is described as an articulation. As Jennifer Daryl Slack has described, articulations, the connections between elements that forge identities, have the following characteristics:

> (a) Connections among the elements are specific, particular, and nonnecessary—they are forged and broken in particular concrete circumstances; (b) articulations vary in their tenacity; (c) articulations vary in their relative power within different social configurations; and (d) different articulations empower different possibilities and practices. (331)

Any identity might be compared to a train, which is constituted of many different types of train cars in a particular arrangement (or articulation). Each car is connected (or articulated) to another in a specific way that, taken as a whole (as a series of articulations), constitutes the identity *train*. Any specific train is thus a specific, particular set of articulations—an identifiable object with relatively clear-cut boundaries. But these specific articulations are nonnecessary; that is, there is no absolute necessity that they be connected in just that way and no guarantees that they will remain connected that way. So, for example, we could disconnect (disarticulate) and reconnect (rearticulate) cars in a different order to constitute a new identity *train*.

To say that articulations vary in their tenacity is to acknowledge that some connections are more difficult to disarticulate/rearticulate than others. Yard police, for example, may or may not let us in to change the order of the cars. Or the kinds of connections between the cars may be variously difficult to manipulate.

Some articulations are more resistant to rearticulation than others; that is, some are more *tenacious* than others. When a connection between elements is particularly resistant, the identity *train* remains intact and effective over a long period. When an articulation is effective, it is said to be powerful in that it delineates what is real and possible from what is not. Different arrangements make possible different possibilities and practices. If we disarticulate the engine, for example, the rest of the train will not move. And, in the process, we may have rearticulated the elements in such a way as to necessitate a new identity. Is a string of cars without an engine a train? Is a single engine a train? We take the answers to both to be maybe. On the other hand, a disarticulated car of the type that usually completes a train will probably not be thought of as constituting the identity *train*. The term "caboose" might have to suffice. But a train without a caboose is usually still thought to be a train.

Articulation thus points to the fact that any identity is culturally agreed on or, more accurately, struggled over in ongoing processes of disarticulation and rearticulation. For example, clearly, one element of what makes a train a train (and not, say, just a caboose) depends on our agreed-on cultural conception of *train*. To stretch this a bit, we could say that we have an ideology regarding what we empower as a train. The ideology of *train* articulates to the arrangement of the cars such that we may call a lone engine a train but not a lone caboose. But that ideology is itself an identity constituted by its articulations, one of which is the past practices of putting trains together. Given changes in those practices, say, for example, giving cabooses their own little engines to get around, we may rearticulate our ideology of *train* such that lone cabooses are more like lone engines and deserve, perhaps, the status, train. Alternatively, we may alter the identity *train* by working to rearticulate it on ideological grounds alone. We may, as teachers, for example, decide to teach people a different definition (identity) of *train* so that a lone engine or a lone caboose is rearticulated as constituting the identity. The success of our attempts at rearticulating identities, whether purposeful or not, depends on the tenacity of the various articulations that constitute it at any particular conjuncture.

To extend this now beyond more easily identifiable identities, social practices, ideological positions, discursive statements, social groups, and so on are also articulated identities whose meanings are continually and variously rearticulated. Dictionaries define the most widely accepted (or acceptable) identities, but there are frequently different, alternative articulations that are either archaic or emerging. One need only read a bit of the *OED* to begin to get a feel for how dramatically articulations can change (although the *OED* only hints at the range of connections that constitute the articulations). Raymond Williams's *Keywords* tracks changing articulations of key identities in Western thought and provides excellent cases of rearticulation.

The concepts of meaning and power are dramatically refigured in articulation theory. Meanings cannot be entities neatly wrapped up and transmitted from sender to receiver, nor can they be two separate moments (meaning 1 contributed by the sender and meaning 2 contributed by the receiver) abstractly negotiated in some sort of a circuit. Like any identity, meaning—both instances and the general concept—can be understood as an articulation that moves through ongoing processes of rearticulation. From sender through channels and receivers, each individual, each technology, each medium *contributes* in the ongoing process of articulating and rearticulating meaning. Power is no longer understood as simply the power of a sender over a receiver or as the negotiated symmetry of the sender's or receiver's meanings but as that which draws and redraws the lines of articulation. As Grossberg has put it, power "organizes the multiplicity of concrete practices and effects into predefined identities, unities, hierarchical categories, and apparently necessary relationships" ("Critical" 92). Power is thus what works to *fix meanings*, that which empowers some possibilities and disempowers others. Grossberg explains that empowerment is "the enablement of particular practices, that is, as the conditions of possibility that enable a particular practice or statement to exist in a specific social context and that enable people to live their lives in different ways" (95).

We can expand our understanding of the role of the technical communicator and of technical discourse significantly by tracking the implications of an articulation view of communication. First, by using the lens of articulation theory, we have here been able to track the changes in the theory and practice of technical communication as themselves rearticulations of elements (or identities) such as technical communicator, meaning, author, channel, sender, power, receiver, and so on. Second, however, that very lens works to rearticulate the location of the technical communicator in the process of communication, specifically in that technical communicators must now be understood as articulating or rearticulating meaning in (and variously contributing to or changing) relations of power. To gain access to those rearticulations, we will again consider the question of authorship as raised by Foucault at the beginning of this article.

It is tempting here to begin to lay out all of the elements that articulate to the notion of *author* as it moves through the stages of transmission, translation, and finally to articulation itself. These articulations would include elements such as the conception of authors as individuals, individuals as the source of meanings, the conception of meaning as a fixed entity—the practice of attributing ownership to ideas, capitalist relations of property and appropriation, a notion of the power of ideas, and a particular conception of progress ("if it's new, it's better"). However interesting that task might be, we must limit our treatment here to some very specific articulations that direct our attention to the questions of meaning and power in the theory and practice of technical communicators.

In the transmission view of technical communication, authority is articulated to scientific and technical discourse as an objective and neutral reporting of facts. Humanities types may author meaning, but scientists, engineers, and, by extension, technical writers, merely (albeit skillfully) re-present what is already objectively "out there." These are not meanings, but objective, disembodied facts. Consequently, technical communications (like the posters or contracts mentioned by Foucault) often have no authors. When technical documents are *conveyed* by named individuals, these are again not authors in the sense of originating meaning—these are simply not discourses that produce an author. Even in these cases, however, for reasons considered later, technical communicators are rarely listed among the conveyers.

Technical documents and writing in science and engineering do often name authors (what Foucault calls the writer). In this case, the author remains the sender in the transmission sense but, articulated now to the concept of conveyance of scientific fact, as an authority. Rarely, again, is authorship in these cases extended to the professional technical communicator. In part, the attribution of authorship here to the scientist or engineer at the expense of the technical communicator must be explained in terms of the tenacity of other articulated elements: the neutrality of scientific discourse, the practice of attributing ownership to ideas, a conception of invention as the expression of individual genius, capitalist relations of property and appropriation, and the persistence of the elevation of the scientific discourse over humanistic discourse (see Horkheimer and Adorno). In other words, specific relations of power articulated to a particular conception of science account for the specific identity of authorship in the sciences and the exclusion of the technical communicator from that attribution.

To evoke *author* in theory or practice from within the transmission view evokes, like a chain reaction of connotations, all these articulations, which struggle—whether purposefully or not—to hide the work that goes into fixing the identity of that work. These articulations are nonnecessary; that is, there is no necessity that they be connected in just this way and no guarantee that they will remain connected in this way. Indeed, translation works to rearticulate the question of authorship, although its challenge is incomplete.

Although the translation view suggests a more elevated role for the translator, it does not grant authority. To put it another way, the translator is seen as an expert, but only in mediating, not authoring, meanings. This is even the case in the humanities, where debates ensue over whether or not to give translators the same credit in tenure and promotion reviews as authors. In technical communication, the unique skill of individuals may be recognized as acts of mediation, but as an activity, the discourse still does not grant them authorship. Again, we suggest that this is in part due to the tenacity of some of those same relations of power discussed earlier: the practice of attributing ownership to ideas, the conception of invention as the expression of individual genius, and capitalist relations of property and appropriation.

By resting on the conception of author as articulated to the contribution of meaning, by challenging the articulation to differential relations of power between sender, translator, and receiver as being somehow evident, and to the conception of science as objective fact finding, we would advance the rearticulation of technical communicators (along with media and technology) as having authorial power. We cannot grant technical communicators status as authors merely in the scientific sense of *conveyers of fact*. That would be to deny the insight of even the translation view that asserts that the discourse of the translator (whether the translator be scientist, technology, medium, or technical communicator) must be understood as involved in the compromise. Rather, technical communicators are theoretically situated in the process of articulating meaning just as prominently as are the sender and the receiver. The process of communication is then not simply a transmission or a translation but an articulation of voices, much like what Bakhtin has characterized as the orchestration of "heteroglot, multi-voiced, multi-styled, and often multi-languaged elements" (265).

It should be obvious that different articulations empower different possibilities and disempower others. When technical communicators are not articulated to authorship, their possible contributions are severely constricted. Whether they desire it or not, technical communicators are seen as variously adding, deleting, changing, and selecting meaning. Again, whether they desire it or not, they are always implicated in relations of power. Their work is at least *complicit* in the production, reproduction, or subversion of relations of power. This is necessarily the case, even when the acceptance of the transmission or translation view may occlude the nature of the work that they do. Technical communicators *are* authors, even when they comply with the rules of discourse that deny them that recognition. When they are denied that recognition, the measure of their success can only be complete compliance with the articulations of meaning, power, and authorship from the standpoint of the transmission and translation views.

The consequences of extending authorship to technical communicators are significant. With the recognition that the communicator articulates and rearticulates meaning comes the responsibility for that rearticulation. No contribution is really transparent; it is only rendered transparent in relations of power. So, just as the power of technical communicators is recognized (as they are empowered), so too must they be held responsible.

IMPLICATIONS FOR PEDAGOGY AND PRACTICE

We heard recently of an industry recruiter who—venting some frustration over graduates knowing more theory than was good for them on the job—said, "We want robots!" This frustration has, we submit, several sources. First, and most obvious, we take this to be a plea for technical communicators to perform their transmission function well. We would not dispute the need to be able to perform skillfully using effective grammar, editing, media management, and so on.

But there is more in the recruiter's frustration. Second, then, this plea points to the fact that the field is growing rapidly in the tension between transmission, translation, and articulation. Although that tension is generative, it does not result in easily written job descriptions, clear definitions of the technical communicator's role, task, and ethos. Sometimes there is a lack of clear vision and agreement—among practitioners and their employers—about what is expected of a technical communicator and what it is he or she has to offer. In addition, however, that plea for robots suggests that there exists a particularly tenacious articulation between the conception of communication as the transparent transmission of messages, the neutrality of science and engineering, and perhaps even of the ethical neutrality of the ethics of capitalism. In fact, to behave as such a robot is to be complicit with the meanings thus articulated.

It is possible to look at some of the turmoil in the education of technical communicators and some confusions about the work of these graduates in terms of a field trying to come to terms with consequences of the technical communicator as author. The difficulties are twofold: One the one hand, the theoretical development of an articulation view has not advanced far enough to form a firm foundation for pedagogy and models of work. On the other hand, the changes that would result from this rearticulation—although theoretically and practically defensible—are not likely to come easily.

Nevertheless, because professional communicators contribute to the process of articulating meaning, whether they choose to or not, they must be able to analyze critically the ethical implications of the meanings they contribute to. Such knowledge is all the more important given the current tendency to define their work as (ethically) transparent. In a sense, technical communicators need to be shaken from the somnambulistic faith that their work is ethically neutral. Steven Katz's examination of a virtually "perfect" technical document proposing changes in a vehicle designed to asphyxiate prisoners during the Nazi holocaust ought to put an end to any assertion of ethical neutrality. It is not simply *how well* we communicate that matters. *Who* we work for and *what* we communicate matters.

The nearly ubiquitous calls for technical communicators to learn more about the technical content of their work (see, for example, Institute of Electrical and Electronics Engineers), even to participate in the early stages of project design, can be understood as easily articulated to the conception of the communicator as author. Such technical knowledge can provide the backdrop for sound, ethical decision making, as well as for competent transmission and translation.

In addition to ethics and technical knowledge, it seems equally essential that technical communicators have a superior grasp of the relationship between technology and discourse and between science and rhetoric (Horkheimer and Adorno; Miller; Wells; Sullivan; Katz). It is essential that we learn to analyze critically the articulations evoked in the language of technology and science. In a sense, technical communicators need to be shaken from the somnambulistic faith that their work is linguistically neutral.

Finally, we would add to the education of technical communicators knowledge of how organizations operate—in the form of organizational communication or organizational behavior. It is remarkable how little most of us understand the relationship between power, knowledge, and organizations. It is time that we give up the faith that the goal of communication is always clarity and brevity. In practice, the politics of organizations and organizational politics often have as their goals limiting, obscuring, or hiding information (Wells; Katz; Butenhoff). Naïveté about how organizations work articulates well to the myth of the technical communicator as engaging in an ethically and linguistically neutral activity.

To send out technical communicators with this kind of knowledge is to send them out armed.[1] It is impossible for technical communicators to take full responsibility for their work until they understand their role from an articulation view. Likewise, it is impossible to recognize the real power of technical discourse without understanding its role in the articulation and rearticulation of meaning and power. This understanding would thus empower the discourse of technical communicators by recognizing their full authorial role.

Note

1. We invite our readers to explore the consequences of this view for the role, task, and ethos of technical communicators as advocates for their constituencies: their employers, clients, and audiences. As advocates, they would be more like lawyers than their current status acknowledges. Although technical communicators have less in terms of codified law or precedent on which to draw, they could be understood as advocating for, counseling, advising, defending, or building cases. This change in status complicates the relationship to their constituents: The counsel of communicators might be accepted, rejected, or

resisted (or litigated against!). But just as a lawyer's duty is to inform employers or clients of the possible consequences of their actions, so too should it be the technical communicator's duty to inform employers or clients of the consequences of their rhetoric!

In addition, this view suggests that the expertise of technical communicators is applicable to the articulation of meaning well beyond the confines of science and engineering (or business). Instead, its scope can easily be understood as encompassing situations in which the transmission, translation, and articulation of specialized knowledge is at issue.

Finally, we do not offer this invitation with any pretense that advocacy or authorship will *simplify* the role, task, and ethos of technical communicators. We offer no apology, however, for we are advocating here changes that are already underway, even if they are not very well understood.

References

Althusser, Louis, *For Marx*. Trans. Ben Brewster. New York: Random House, 1970.

Bakhtin, M. M. *The Dialogic Imagination: Four Essays*. Trans. Caryl Emerson and Michael Holquist. Ed. Michael Holquist. Austin: University of Texas Press, 1981.

Butenhoff, Carla, "Bad Writing Can Be Good Business." *Readings in Business Communication*. Ed. Robert D. Gieselman. Champaign, IL: Stipes, 1986. 128–31.

Carey, James. *Communication as Culture: Essays on Media and Society*. Boston: Unwin Hyman, 1989.

Connors, Robert J. "The Rise of Technical Writing Instruction in America." *Journal of Technical Writing and Communication* 12 (1982): 329–52.

Cooper, Marilyn M. "Model(s) for Educating Professional Communicators." *The Council for Programs in Technical and Scientific Communication [CPTSC] Proceedings 1990*. San Diego, CA, 12 October. Ed. James P. Zappen and Susan Katz. CPTSC, 1990. 3–13.

Deleuze, Gilles, and Felix Guattari. "Rhizome." *Ideology & Consciousness* 8 (1981): 49–71. "Ferment in the Field" [Special issue]. *Journal of Communication* 33.3 (1983).

Fiske, John. *Introduction to Communication Studies*. New York: Methuen, 1982.

—. *Television Culture*. New York: Methuen, 1987.

Foucault, Michel. "What Is an Author?" *Language, Counter-Memory, Practice: Selected Essays and Interviews*. Ed. Donald F. Bouchard, Ithaca, NY: Cornell University Press, 1977. 113–38.

Gramsci, Antonio. *Selections from the Prison Notebooks*. Ed. and trans. Quentin Hoare and Geoffrey Smith. London: Lawrence & Wishart, 1971.

Grossberg, Lawrence. "Critical Theory and the Politics of Empirical Research." *Mass Communication Review Yearbook*. Ed. M. Gurevitch and M. R. Levy. London: Sage, 1987. 86–106.

—. "Language and Theorizing in the Human Sciences." *Studies in Symbolic Interaction* 2 (1979): 189–231.

Hall, Stuart. "Encoding/Decoding." *Culture, Media, Language*. Ed. Stuart Hall et al. London: Hutchinson, 1980. 128-38.

—. "On Postmodernism and Articulation: An Interview with Stuart Hall." *Journal of Communication Inquiry* 10 (1986): 45–60.

—. "Race, Articulation and Societies Structured in Dominance." *Sociological Theories: Race and Colonialism*. Ed. UNESCO. Paris: UNESCO, 1980. 305–45.

—. "Signification, Representation, Ideology: Althusser and the Post-Structuralist Debates." *Critical Studies in Mass Communication* 2 (1985): 91–114.

Horkheimer, Max, and Theodore Adorno. *Dialectic of Enlightenment*. New York: Herder & Herder, 1972.

Institute of Electrical and Electronics Engineers (IEEE). *The Engineered Communication: Designs for Continued Improvement*. Vols. 1 and 2. International Professional Communication Conference Proceedings. Orlando, FL, 30 Oct.–1 Nov. New York: IEEE, 1991.

Katz, Steven B. "The Ethic of Expediency: Classical Rhetoric, Technology, and the Holocaust," *College English* 54.3 (1992): 255–75.

Kostur, Pamela, and Kelly Hall. "Avoiding Miscommunication: How to Analyze and Edit for Meaning." *The Engineered Communication: Designs for Continued Improvement*. Vol. 1. International Professional Communication Conference Proceedings. Orlando, FL, 30 Oct.–1 Nov. New York: Institute of Electrical and Electronics Engineers, 1991. 18–25.

Laclau, Ernesto. *Politics and Ideology in Marxist Theory*. London: Verso, 1977.

Markel, Michael H. *Technical Writing Essentials*. New York: St. Martin, 1988.

McQuail, Denis, and Sven Windahl. *Communication Models for the Study of Mass Communications*. New York: Longman, 1981.

Miller, Carolyn R. "Technology as a Form of Consciousness: A Study of Contemporary Ethos." *Central States Speech Journal* 29 (1978): 228–36.

Morley, David. *The "Nationwide" Audience*. London: BFI, 1980.

Oxford English Dictionary. 1971.

Shannon, C., and W. Weaver. *The Mathematical Theory of Communication*. Champaign: University of Illinois Press, 1949.

Slack, Jennifer Daryl. "Contextualizing Technology." *Rethinking Communication*. Paradigm Exemplars 2. Ed. Brenda Dervin et al. Newbury Park, CA: Sage, 1989. 329–45.

Souther, James W. "Teaching Technical Writing: A Retrospective Appraisal." *Technical Writing Theory and Practice*. Ed. Bertie E. Fearing and W. Keats Sparrow. New York: Modern Language Association, 1989. 2–13.

Sullivan, Dale L. "Political-Ethical Implications of Defining Technical Communication as a Practice." *Journal of Advanced Composition* 10.2 (1990): 375–86.

Vaughan, David K. "The Engineer: Neglected Target of Technical Writing Instruction." *The Engineered Communication: Designs for Continued Improvement*. Vol. 1. International Professional Communication Conference Proceedings. Orlando, FL, 30 Oct.–1 Nov. New York: Institute of Electrical and Electronics Engineers, 1991. 80–83.

Wells, Susan. "Jürgen Habermas, Communicative Competence, and the Teaching of Technical Discourse." *Theory in the Classroom*. Ed. Cary Nelson. Urbana: University of Illinois Press, 1986. 245–69.

Williams, Raymond. *Keywords*. London: Fontana, 1976.

DEVELOPING YOUR UNDERSTANDING

1. Summarize a view of the technical communicator/professional writer as author, explaining what it means to be an author in an organizational setting.
2. Compare and contrast how the three models—transmission, translation, and articulation views—understand *meaning* and *power*.
3. Explain how each of the three models/views affects (a) the professional writer's assumptions about writing, (b) the professional writer's primary goals, and (c) what the professional writer does.
4. Like any profession, professional writing claims an area of expertise. However, some professions, and especially relatively young ones like professional writing, struggle to more clearly determine their area of expertise. Slack et al trace professional writing's struggle through the three models/views. According to each model, explain what the professional writer's area of expertise is.
5. Assume you have just taken a job at an organization that has a transmission view of writing, but that you have a strong *translation* or *articulation* view (choose one). Describe the conflicts you will most likely face in your work because of yours and your employer's differing views of writing. Then, list some of the strategies you might use to deal with these conflicts in a way that lets you pursue your work from, at least partially, your view.
6. Explain the following: ". . . whether they desire it or not, [professional writers] are always implicated in relations of power. Their work is at least *complicit* in the production, reproduction, or subversion of relations of power. . . .[Professional writers] *are* authors, even when they comply with the rules of discourse that deny them that recognition." Examine what this statement implies about your responsibilities as a professional writer.

CHAPTER 2
Projects

1. In many future instances, including the process of developing a professional writing portfolio, you will be required to define your field of work. Looking ahead to these future scenarios, this project asks you to develop a definition of your field of work.

   In order to write your definition, first identify some of the places in the articles from this chapter where the authors have noted, explicitly or implicitly, that rhetoric and professional writing expand upon and complement one another. Take time to summarize and illustrate these points of connection. This work will help you conceptualize and organize your report.

   Your target tasks are to write two reports, one long and academic, the other short and written for non-academic audiences. Your long report should summarize and synthesize your findings concerning the connections between rhetoric and professional writing. Your short report should define the field of professional writing and rhetoric and it should be appropriate for a professional writing portfolio section where you define your understanding of the field to potential employers. Both reports should describe how "rhetoric," "professional writing," and "professional writing and rhetoric" are similar and different.

2. A helpful way to learn about the field of professional writing and rhetoric is to observe and interview a practicing professional writer. This project invites you to study "organizationally situated authorship" by observing and interviewing a professional writer in your area.

   In pairs or groups of three, locate a professional writer who is willing to be interviewed and observed at work. Based on your interests, develop questions about organizational authorship that you would like to explore. Plan a series of interviews and observations that will help you with your inquiry. (The length and complexity of your study should reflect the time you have and your status as a lower-division undergraduate, upper-division undergraduate, masters, or doctoral student.) Develop a report of your findings that you can share with your class, your instructor, and the professional writer you studied.

   Note: You do not have to be in a big city to find professional writers or professionals taking on tasks that are primarily writing-focused. On your campus, you will find several people who are, at least in part, professional writers. Certainly, your campus publications department will have practicing professional writers. Check campus research centers, computer support services, admissions, athletics, museums, theaters, the president's office, and community outreach centers. Even at smaller colleges, someone in these or related offices will be taking on professional writing tasks.

   You should be aware, though, that many people you could study may not consider themselves to be professional writers (e.g., Web designers, public relations specialists, and even publications managers/editors); also, many people who might call themselves professional writers (e.g., poets and freelance creative writers) will not easily fit within the broad definition of professional writing and rhetoric adopted by this text.

PART 2

Professional Writing as a Social Practice

Chapters 3, 4, and 5 include readings that examine more closely how professional writing and rhetoric is affected when writing is understood as a form of practice—a way of doing or acting. In Part 1, several selections raised the issue that professional writing and rhetoric can be understood as either a productive art or a form of practice. It is not necessary to see it as one or the other; it can also be seen as both. *Professional Writing and Rhetoric* approaches it as both, dedicating Part 2 to readings focused on the practice of writing and dedicating Part 3 to readings focused on writing as a form of production.

Chapter 3 examines how writing within organizations affects the practices of professional writers. Professional writing is certainly always located or situated within contexts that are larger than organizations: civic, legal, and environmental are just a few of these larger contexts. Still, the primary context within which the practice of professional writing occurs is organizational. The readings in Chapter 3 explore not only how we define such contexts and how they affect what we can and cannot do, but also how we can be active within such contexts.

The readings in Chapter 4 explore the relationship between rhetorical and ethical action. Anytime a profession is defined as a sort of action or practice, it automatically becomes embroiled in ethical issues. That is, when a person acts, the questions "Did they act well?", "Did they act appropriately?", and "Should they have acted differently?" will always arise. Such questions are part of the realm of ethics. The readings in Chapter 4 examine such questions as they relate to professional writing.

Finally, Chapter 5 includes readings that examine how technological contexts affect the practices of writers. If you think of medicine as a practice, what images are constructed in your mind? Do you see beds, medical tools, and perhaps biomedical machinery? Perhaps you imagine musical instruments, specialized lighting of some sort, oils, and herbs? In either case, your images of the practice of medicine are embedded within certain kinds of technological contexts. The same holds true for any kind of practice, including professional writing. Just as medical practice is greatly affected by its technological context, so is professional writing. And just as medical practitioners must understand how their technological contexts have influenced, currently influence, and might later influence their work, professional writers must also address such issues. The readings in Chapter 5 enhance awareness of how professional writing is influenced by technological contexts.

CHAPTER 3

Professional Writing as Organizationally Situated Action

INTRODUCTION

Many people who are attracted to writing as a profession are drawn by the free, relatively unrestrained writing experiences they have had and fondly remember. To think, then, about being constrained by organizational contexts is not only a little distasteful to many budding writers, but it's also quite unnerving. Many students in their first professional writing classes wonder (quite reasonably), "Will I have room for any creativity if I become a professional writer?" The answer is "yes," but the creativity of a professional writer comes with knowledge of how organizational contexts can affect writing. Therefore, the readings in Chapter 3 are selected so that you can begin examining questions about organizational context, such as the following:

- What is organizational context?
- How is it related to rhetorical situation?
- How does organizational context constrain writing?
- How does it open up possibilities for writing?
- What are some of the ways writers can achieve effective organizationally contextualized action, even in the face of organizational constraints?

In classical rhetoric, place, location, context, or situation was a critical component of rhetorical education and practice. The practicing rhetor would pay very close attention to the complex relationships between time, place, speaker, and audience. Both rhetors and students of rhetoric were very aware that what may be appropriate and effective in one context may not be appropriate and effective in another. They would carefully and strategically adjust their speech or text to the situation at hand. Such awareness and application to speaking and writing defined rhetoric as an artful practice.

It would seem obvious that one would adjust her writing to the situation at hand, no? Perhaps you will find it surprising then that learning and practicing rhetoric as an art of situational appropriateness often fell, and can still fall far away from what writers learn and practice. The rise of *belles letters*—literally "beautiful letters," which emphasized style—particularly hurt the teaching and practice of writing as a contextually sensitive art. Writers began to teach and practice an art that was focused almost entirely on the production of beautifully worded texts. Good and effective writing became equated with beautifully written and crafted writing.

After world War II, teachers, researchers, and practitioners of professional writing began to pay more attention to the wide range of writing produced in the workplace. Thus, attention to classical concerns for artfully responding to context were reawakened. This reawakening happened because teachers, researchers, and professional writers were discovering that beautifully crafted writing was not always effective writing. They had rediscovered what classical rhetoric had long taught.

The readings in Chapter 3 examine the issues professional writers must understand and be able to respond to as they become authors within organizational contexts. Driskill's is the first of three readings in this chapter, and it helps you get a broad and general sense of the organizational context within which writing is practiced and produced. The second reading by Katz examines more specifically one practice of organizational writing: document review or document cycling. More specifically, Katz examines the possibilities for what she calls individuation within this rhetorical process. Finally, Spilka examines the roles both writing and orality play in the routine but complex practice of negotiating multiple audiences within organizational contexts.

FOCUSING ON KEY TERMS AND CONCEPTS

Focus on the following terms and concepts while you read through this selection. Understanding these will not only increase your understanding of the selection that follows, but you will find that, because most of these terms or concepts are commonly used in professional writing and rhetoric, understanding them helps you get a better sense of the field itself.

1. **organizational situation**
2. **rhetorical situation**
3. **structural-functionalism**
4. **Shannon-Weaver model**
5. **writing as instrumental**
6. **external sources of meaning**
7. **internal sources of meaning**

Understanding the Writing Context in Organizations

LINDA DRISKILL
Rice University

New attention has recently been given to writers' knowledge of situations and procedures in organizations. The success of many business documents seems to depend on factors outside the genre features taught in textbooks or beyond commonly investigated cognitive processes. Studying the writing decisions of analysts in a state agency, Odell found that knowledge of other departments' needs, an understanding of the agency's interests, and experience with readers' reactions to similar documents affected individuals' writing goals as well as many decisions on content, organization, and word choice.[1] The chief value of context is its usefulness in explaining the types of meanings writers attempt to express, and readers expect to interpret, in specific situations.

THE IMPORTANCE OF THE WRITING CONTEXT

The way the writing context can influence the creative and interpretive processes of writers and readers can be seen in the example of a new mutual fund's brochure headline. The new fund used a market timing approach to investing, which means that it followed technical indicators to attempt to invest in stocks only when stock prices were rising. The headline for the direct mail piece sounded full of punch to the advertising agency writer:

> When you want both safety and growth for your capital, timing is everything! And the time is right, right now.

The interplay of different meanings for *timing* and *time* were better than so much of that dry investment language, the writer thought, and he went on with another subheading: "The easy and strategic way of taking advantage of stock market trends." The headline looked attractive to the marketing people, who were eager to spread their enthusiasm for their new fund.

However, the lawyer for the industry's regulatory body, the National Association of Securities Dealers (NASD), judged the language unacceptable and did not approve the piece. The language implied that the reader stood only to benefit by investing in this fund. Further, it implied that the investor's money would be safe as the value of the investment grew. The risks of the investment were not mentioned. As a result, despite talent and creative effort, considerable expense and time were lost.

The error was both the fault of the agency writer, who lacked knowledge of the NASD's standards, and of the company, which had not hired writers with legal expertise

Source: Driskill, Linda. "Understanding the Writing Context in Organizations." *Writing in the Business Professions,* ed. Myra Kogen, 1989, pp. 125–145.

or structured its review process to assure detection of the unacceptable language.[2] Many investment brokers use only literature that has been approved by the NASD because they fear lawsuits by investors. A plaintiff would surely have a greater chance of success if unapproved literature were involved.

An awareness of the effects of specific situations, company procedures, and factors inside and outside the company has come to be known as the "business savvy" that only the experienced can apply in a writing situation. Many writing instructors, for example, would not know of the NASD and its standards for the literature of investment companies. Such awareness can be the difference between an expert writer and a novice, yet not all experienced workers are expert writers. Employees and managers, as well as teachers, consultants, and researchers, need good analytic tools and guides for writing decisions. This article presents a conceptual tool to help writers systematically tap the contextual sources of corporate savvy that affect communication success. It first discusses current theoretical models' inattention to the context for writing decisions. This article presents a conceptual tool to help writers and the meaning of documents. It then presents components of a model of the organizational context for communication and discusses how the model can systematize organizational savvy for the benefit of teachers, consultants, and writers in companies.

WHY CURRENT MODELS NEGLECT CONTEXT

Current models and theories of business communication tell little about the effects of context on writing processes. Most theoretical positions seem to have one of three orientations: One group attends to *particular aspects of communication events*, including genres (such as letters, reports, meetings, and presentations), the individual writer's processes, or communication technologies. The second emphasizes *communication systems and their abstract properties*, such as flexibility and direction of flow. The third, recently proposed by Faigley, urges interdisciplinary research into the social aspects of writing.[3]

Approaches Attending to Particular Aspects of Communication

Genres and traditional rhetorical modes (comparison, analysis, etc.) have been the bases of communication courses focusing on types of communication events or genres: the formal report, the interview, the sales letter, etc. Many textbooks are organized to serve such courses, which emphasize features of format and abstract patterns of organization, rather than (1) what is meant or understood, and (2) how these meanings matter in the context of the organizational situation.[4] These courses focus on the means for expressing meaning, not the meanings themselves.

Another narrowly focused approach has been the study of individuals' writing behavior, usually in a laboratory setting with fictional writing assignments. Courses based on this approach have emphasized individual writing strategies, especially for invention and arrangement. Studies of individuals' cognitive processes can help distinguish between experts' and novices' strategies and identify types of writing plans. Most of this research has involved fictional settings because of a desire to standardize the situation and facilitate comparison.

Recently, attention has been focused on the effects of different technologies, such as electronic mail, dictating systems, and word processing on communication. These studies tend to overlook context and to focus instead on the technology as the source of behavior. These studies sometimes are linked with investigation of individuals' processes or with surveys of workplace practices.[5] Each of these focused approaches may reveal valuable insights, but each is likely to be incomplete, to overlook some aspects of the writing context.

Systems Approaches to Organizational Communication

Although the systems approach seems to involve writing contexts, its theorists are concerned neither with meaning nor with transactions among individuals. For example, the structural-functionalist communication scholars, whose assumptions are consistent with structural-functional management theory, think of the company as a large, abstract machine:

> Structural-functionalism requires that traits or concepts that are vital to the continuance and performance of the organizational be specifically identified. Furthermore, the investigator is charged with the task of specifying the *mechanisms* within the organization that bring about the desired levels of those traits. Consequently, if degree of flexibility, directionality of message flow, message initiation, innovation / maintenance messages are the traits under scrutiny, structural-functionalism requires one to search for those key factors that lead to different levels of each trait. As more is learned abut the factors, it becomes more feasible to bring the traits under control, and so effectively "manage" communication in the organization. (emphasis added)[6]

The structural-functionalists, like many other organizational communication theorists, were heavily influenced by the communication model published by Shannon and Weaver in 1949.[7] Based on telecommunication systems, their theory is concerned with the generation of information, its flow rate along its channel, and ways to mathematically encode information to reduce "noise" in the system (Figure 1).

Shannon and Weaver were not concerned with why people needed to communicate with one another or with the content of the messages. Although Osgood subsequently criticized the Shannon-Weaver model because it did not deal with meaning, the model was irresistibly easy to grasp for people familiar with transportation systems, and it was adopted by scholars from many fields, including biochemistry, genet-

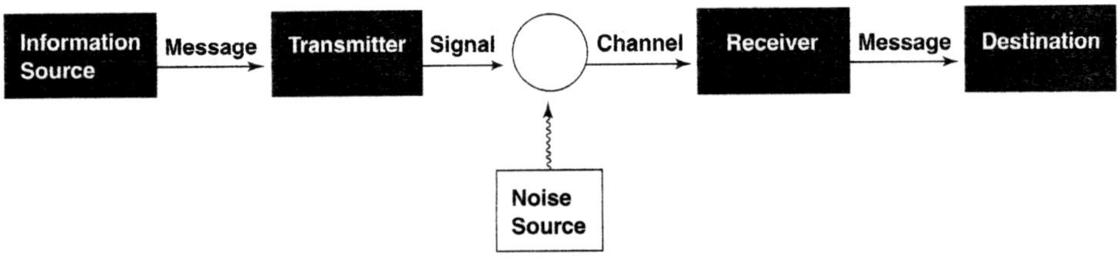

Figure 1. Shannon-Weaver model of communication. The model does not represent meaning or intentions of persons.

Source: Shannon and Weaver 1949, 7. Copyright 1948 by the Board of Trustees of the University of Illinois. Reprinted by permission of the University of Illinois Press.

ics, chemistry, and business communication.[8] This model influences the works of major communications scholars, such as Berlo, Lasswell, McCroskey, and Schramm, each of whom modified the model somewhat.[9]

Schramm revised the Shannon-Weaver model in three elaborations designed to indicate that communication takes place in an environment, involves people (not just information sources), and produces feedback. Schramm's modifications certainly offered a more complete representation of communication that the Shannon-Weaver model did, but the categories of "environment" and "feedback" are still too general to produce detailed analyses of the communication context, communications processes, or products.

In "Nonacademic Writing: The Social Perspective," Faigley reviews the development of genre and cognitive perspectives, but omits the systems approach, which has actually had great prominence in business communication. He suggests a "social perspective" in which writing is defined as an action "that takes place in a structure of authority, changes constantly as society changes, has consequences in the economic and political realms, and shapes the writer as much as it is shaped by the writer." According to Faigley, those taking the social perspective must move beyond the traditional rhetorical concern for audience to consider issues such as social roles, group purposes, communal organization, ideology, and finally, theories of culture.[10]

Faigley's intent is to create categories of research perspectives, each of which includes many specific approaches to the study of writing. He hopes to foster a new appreciation of the relevance of other disciplines' methods and premises for the study of writing by describing developments in several disciplines. Faigley uses *social* in a broad sense that does not reconcile the many specific meanings of the term used by sociology, psychology, anthropology, and other disciplines. The model of context proposed in this article is intended as an example of the approach Faigley would classify as "social." However, this discussion will avoid *social* as a theoretical term because of the multiple definitions it has in other disciplines. Context can help explain what a document means, what ideas it contains, why the writer would try to express his or her ideas in a particular way, and why readers who occupy particular roles in different parts of an organization would be likely to respond to a document in particular ways. Context has this power because it is a source of meaning for writers and readers. Experience in their particular roles in an organization context has taught them to view specific topics in particular ways, to interpret particular information according to certain formal or informal rules, and to value certain styles as preferred or appropriate.

Meaning in business communication has its primary source in the writing context because communication involves actions and goals; it is instrumental. Writers in businesses seek to create meanings that produce sales, cooperation, approval, compliance, or agreement. Meaning in business writing is not limited to subject or topic knowledge. The professional may indeed have stored in memory academic knowledge learned outside of a business or professional setting, but access to such knowledge is gained via constraints and objectives that occur in a particular situation.

Any subject or issue is framed by the perceived external environment (society, government, competitors, resources, markets) as well as the perceived internal environment of the company (size, structure, technology, culture, individuals, roles, and forms of argument or reasoning). Perceptions of the external and internal environments converge to define the situations in which workers participate. Almost all these situations have rhetorical or communication requirements, because most business functions require communication. Advertising new positions available; soliciting bids from vendors and suppliers; applying for licenses; consulting with lawyers, lenders, and advertising media; promoting and selling products and services: all involve communication.

This emphasis on the external and internal environments as sources of meaning tend to deemphasize the personality of the individual writer or reader as a source of meaning. The persona of the organizational writer is defined by a somewhat different set of features than is that of the poet, political orator, or personal friend who writes in a nonorganizational or academic setting. In most business situations, the roles of writers and readers, their powers of action and expertise as members of the organization, are more important than other aspects of their personal identity. Nevertheless, the writer or speaker does have the creative power to transform the sources of meaning and to develop original solutions to organizational problems and novel writing strategies. The training of the individual in the reasoning methods of specific professional disciplines and the range of writing plans known by the writer may strongly affect the action of the individual writer. The national or regional culture (for example, "good ole boy" cultures) may also be important.

Thus, a rhetorical situation, with its range of reader/audience roles, purposes, sets of properties, genres, individuals, and temporal and technological constrains, must be seen as embedded within a complex context that affects both writers and readers. The "subject" or "topic" is not context-free, but situated, involved in what the members of the organization must know, feel, or believe in order to accomplish their goals. Columb and Williams have proposed a descriptive technique for describing the multiple cues writers in professional situations can embed to elicit specific expectations and invoke particular domains of information.[11]

EXTERNAL SOURCES OF MEANING: MUTUAL FUNDS INDUSTRY EXAMPLE

Context as a source of meaning can be understood more easily if we separate those sources of meaning external to the firm from those within the firm. These two different types exert different kinds of influences in varying degrees and are involved at different times and in different circumstances. In most firms, external sources matter less frequently; internal sources affect virtually every document. A regulatory body can be called a source of meaning because writers consult its definitions and criteria when representing their ideas.

External sources of meaning are interpreted, not absolute, influences on writers and readers. Some management scholars assume that language and reality are isomorphic, that reality is what language declares it to be. This belief is illustrated by the way management scholars speak about an organization's environment as an independent entity, not recognizing that organizations construct their own definitions of their environments, primarily through language usage. Recent debate over the usefulness of economic indicators illustrates how "facts" of the environment, such as the "money supply" and "credit availability," are interpretations, not absolutes, of the firm's environment.

Smircich argues that instead of treating the organization's environment and the organization itself as objects or givens, managers must become aware of the language processes essential to everyday corporate life:

> The possibility of organized action hinges on the emergence and continued existence of common modes of interpretation that allow day-to-day activities to be taken for granted. In the context of group interaction, it is this routinization that we refer to as being organized. When groups encounter novel situations, new interpretations must be constructed to sustain organized activity. The process of negotiating meanings for these events may alter current understandings and thereby change the formerly taken-for-granted way of life.[12]

In the case of the mutual fund industry, mentioned earlier, several organizations, groups, and factors affect how writers in mutual funds companies interpret information and compose documents. A mutual fund is an investment company that sells shares of its investment portfolio to investors and uses the money to purchase securities, such as bonds, stocks, gold certificates, U.S. government securities, or other investment instruments. Writers in such companies may use external sources of meaning (Figure 2) to assess opportunities, obtain information, analyze audiences for company documents, and create writing plans.

The recent history (1984–86) of mutual funds that specialize in U.S. government securities illustrates the dynamic effect of the external environment as a context for

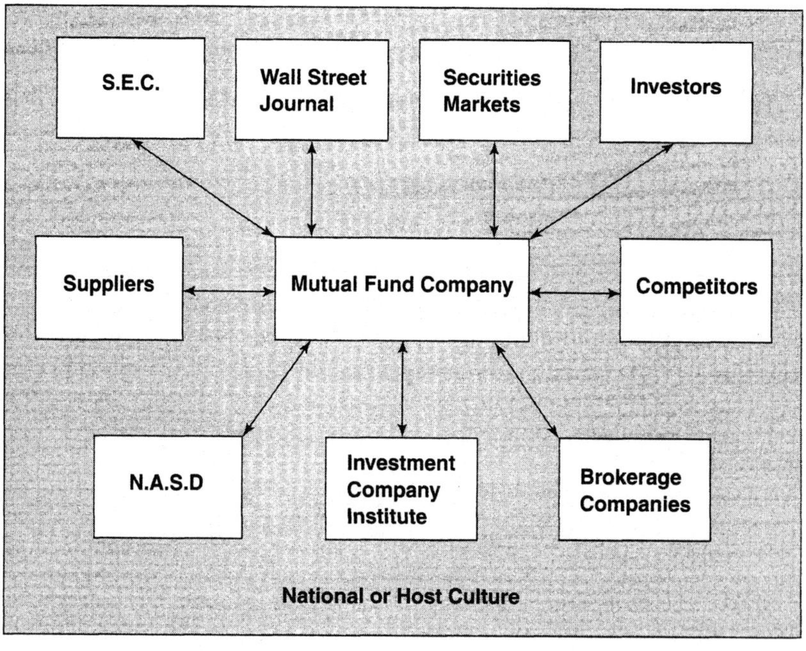

Figure 2. Model of the communication context showing external sources of meaning in the mutual funds industry.

writing. Until late in 1984, only a few funds concentrated their assets in U.S. government securities, such as treasury bonds, treasury bills, and mortgage-backed certificates such as "Ginnie Maes." These investment instruments are often traded in units of $100,000 or more, amounts that formerly had kept smaller individual investors from owning them. The attractive features of these funds were (1) the high rates of interest that were being paid and passed along to the owners of the mutual fund shares, and (2) the fact that the securities owned by the funds were backed by the U.S. government, which had never defaulted on any payment of principal or interest when due. The ads emphasized the annualized rate of interest currently paid and they usually included words such as "safety," "security," or "guaranteed," along with such patriotic symbols as the domes of capitols, flags, and eagles (Figure 3). [13]

In 1985, investment companies created many more of these funds that specialized in U.S. government securities. Advertisements began to appear that attracted billions of dollars into these new funds.[14] The Securities and Exchange Commission (SEC), however, perceived two problems with these attractive new funds. First, since nearly all the funds were new, they had no performance record, over time, on which investors might base their estimates of future performance, and the SEC was worried that investors would rely on the current high annualized rates being advertised. Second, although the government would indeed guarantee that the rate of interest would be paid, the value of the mutual fund shares was *not* guaranteed; instead, it would fluctuate according to interest rates. If interest rates on other investments rose higher than those being paid on the securities owned by the fund, the value of fund shares would decline. This risk, called interest rate risk, was believed to be poorly understood by investors.

The NASD began to send back comments on ads submitted for review and requested qualification of the language in the ads. NASD lawyers, for example, recommended that *safety* be changed to *a high degree of safety* (Figure 4). In the fall of 1985, the SEC asked the mutual fund trade association, the Investment Company Institute (ICI), to deal with the problems arising from misunderstood statements about safety, and to make uniform the widely varying practices in calculating and reporting the yield rates for these funds. Weeks went by as meetings of representatives from more than a thousand mutual fund companies met at the ICI. Concerned about the poten-

FRANKLIN

U.S. Government Securities Fund

High Yield and Safety
12.15%

Figure 3. Partial text from early Franklin U.S. Government Securities Fund advertisement emphasizing yield and safety.
Source: *The Wall Street Journal*, November 1983.

High Income
For Your IRA,
With A High
Degree of Safety

Franklin U.S. Government Securities Fund

12.38%

Figure 4. Partial text from Franklin U.S. Government Securities Fund advertisement modified to "high degree of safety."
Source: *The Wall Street Journal*, March 1985.

tial risk of lawsuits, companies began changing their advertising, even before the ICI could reach any agreement, removing the yield figures (and the explanations of how they were calculated), and changing more and more to metaphorical language to suggest indirectly the attractiveness of the product (Figure 5).

Interest rates on government securities declined in early 1986 because the yields dropped on the new certificates and bonds the funds could buy. At the same time, the marketplace was exerting an influence on one marketing point: high yields. By mid-March 1986, few government securities funds were advertising yields. Only after the ICI memorandum of agreement was completed in June 1986, did yields begin to reappear in the ads, now consistently defined and presented in uniform phrasing and letter heights. The external environment, with its complex structure of audiences, information sources, and influences, had clearly affected what mutual fund companies managing government securities funds decided to say in their publications and how writers of these ads created meaning.

INTERNAL SOURCES OF MEANING: THE CHALLENGER ACCIDENT EXAMPLE

Internal as well as external sources of meaning affect writers in companies. The structure, size, and technology of the organization will affect the roles people play and the ways rhetorical situations are defined.[15] In the 1960s the contingency theorists at Harvard showed that the volatility and complexity in a firm's environment dictate the amount, type, and frequency of information the organization processes to accomplish its mission. Since a firm's structure is a vehicle for gaining access to and commu-

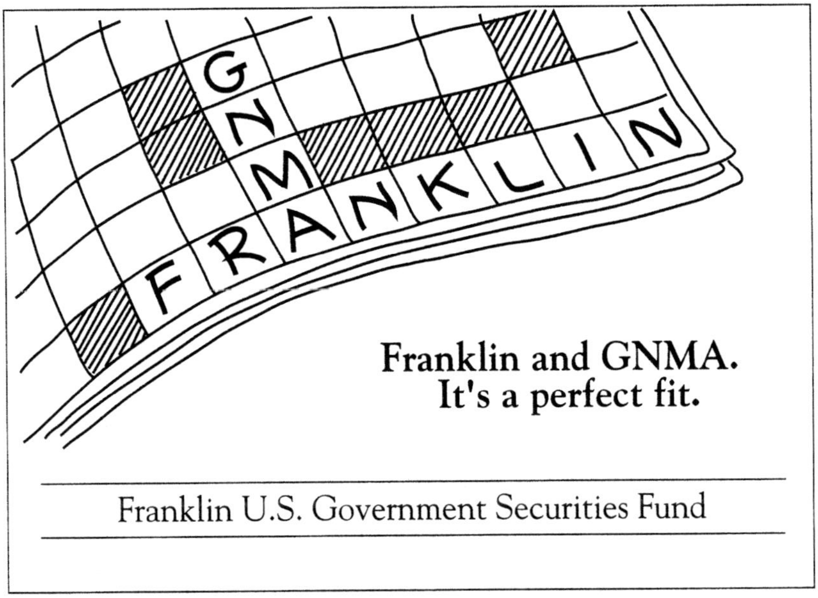

Figure 5. Partial text and illustration from Franklin U.S. Government Securities Fund advertisement using metaphor.
Source: *The Wall Street Journal*, April 28, 1986.

nicating information, organizations try to adapt their structures to secure and disseminate most efficiently the information they need from the environment.[16]

These theorists have been criticized for not paying more attention to other factors within the firm that affect communication, such as corporate culture and the individuals of the firm. Individuals are also sources of meaning and their preferences can affect writing practices. Powerful executives can also affect how writing is produced; their preferences tend to become maxims of the company culture. Space does not permit discussion of all aspects of the model proposed in Figure 6.

Corporate Culture: As management consultants and scholars interested in nonquantitative measures of corporate behavior focused attention on the distinctive practices of individual companies in the late 1970s and '80s, a picture of the power of shared values, norms, roles, rituals, and "the company way" began to emerge. Such features of a company compose what has been called its *corporate culture*. In a discussion of the variety of anthropological theories of culture whose concepts might be applied to the study of corporate culture, Allaire and Firsirotu define *corporate culture as a system of shared and meaningful symbols manifested in myths, ideologies, values, and multiple cultural artifacts*. They show that adopting a particular definition of culture commits one to specific conceptual assumptions and ways of studying culture.[17]

They argue in favor of a definition of corporate culture that separates the sociostructural system of the firm from its cultural system. For the purpose of understanding communication processes, we need to be able to separate culture and structure. If culture cannot be separated from structure, then the effects of these processes on

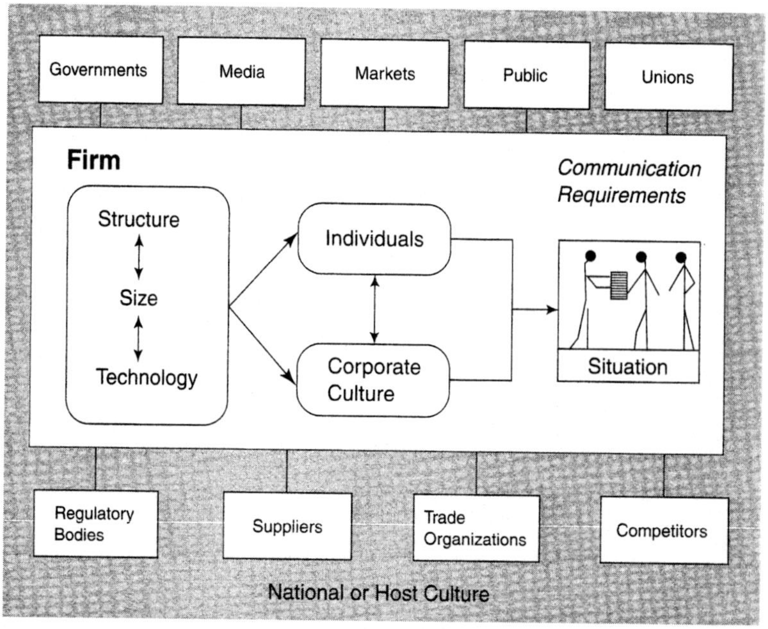

Figure 6. Model of how external environment and firm's characteristics define business situations and communication requirements.

communication cannot be separated. Yet, the structure and the shared values and beliefs of the organization may have quite different effects on writing practices. In a survey of one financial services company, two groups of employees expressed the same attitudes toward problem solving, but they differed significantly in their ability to solve problems. One group had no problem-solving unit or routine process; the other had a weekly meeting at which a special committee could discuss problems and make decisions. Not surprisingly, the second group was much more successful in dealing with problems. Structure, not culture, was the obstacle in that firm.

Like organizational structure, but different in its operation, culture is a powerful determinant of the definition of situation and of rituals and procedures: Who speaks to whom? Who listens to whom, when, and why? Corporate culture contributes many of the interpretive standards that affect writers' choices of content, persuasive approach, and word choice. In one company, I was told to delete *hope* from a draft. "We don't hope for anything around here," I was advised, "We decide what we want and then we make it happen."

Not all organizations have strong cultures—strong values, norms, and beliefs that guide action. Ouchi and others have classified cultures as ranging from those whose members are fully involved to those whose members are only slightly involved.[18] One would expect that in organizations where "anything goes so long as you get the work out," either communication processes would vary or external sources, such as the professional standards for accountants, engineers, and others, would influence communication practices.

It is important to note, however, that not all strong cultures facilitate communication; Bate reported difficulties experienced by companies whose cultures repressed communication about problems, prevented naming of individuals who were the source of trouble, and resisted cooperative problem solving and the expression of emotions. Communication and participation could be improved only by attacking the pervasive beliefs of the companies' cultures; and that is not an easy matter.[19] We need to include corporate culture in our models of communication, not because it plays a uniform role in all corporate communication, but because it accounts for a complex of interdependent factors whose configuration affects what people say, write, and read. Models of organization communication that assume uniformity in many areas of organization life cannot account for the variety of documents and events in companies. Recognizing corporate culture as a source of meaning will reduce some of the confusion and enable us to identify other influences more easily.

With a variety of techniques, organizational communication scholars could investigate how culture influences the creation of written or spoken language. Rhetorical analysis of transcribed protocols might be able to show how norms and values are transmitted, enacted, negotiated, and affirmed. Scholars also might analyze objects, such as written reports, videotapes, and marketing materials, for cultural properties and for their function in rituals. Rymer has analyzed narratives and anecdotes used by managers to identify important issues, to show relationships among events and actions, and to motivate employees in a midsized manufacturing firm.[20]

Incorporating corporate or organizational culture into models and theories of corporate communication should, therefore, enable us to describe and account for different attitudes toward communication, variations in the meanings expressed by documents, variations in preferences for modes and genres, types of analogies and anecdotes, types of arguments, and roles of writers and readers. Such an array of considerations would substantially expand the degree of organizational savvy that an experienced writer might bring to bear on a single writing task.

Definitions of Situation/Prescriptive Paradigms: Throughout this article I have described the effects of the writing context, both within and without the firm, as though writers consulted their understanding of context directly in making writing decisions. More typically, I believe that these understandings are concentrated in groups of ideas associated with particular definitions of situation. Frequently, writers will respond to questions such as "What kind of situation would you call this?" or "What does a writer do in such a situation?" with lengthy rhetorical prescriptions for audience adaptation, genre choices, production schedules, stylistic preferences, and argument strategies.

Definitions of situation reflect the values of corporate culture, the requirements of organizational structure, the influences of the firm's external environment, and ways of thinking and arguing that derive from the individual's training, education, and professional role. Writers usually define situation in terms of the work of the company or a department's routines and operations. A situation involves nonrhetorical elements: actions such as delivering goods to a particular location, manufacturing, operating machinery, or making calculations. Often associated with this definition, however, are one or more *rhetorical situations*. For each rhetorical situation there is an associated set of roles, terms, concepts, reasoning procedures, and history that serves as a guide to thinking, believing, and acting.

The definition of the rhetorical situation controls to a large extent which events or perceptions count as facts, which concepts apply to these facts, and which assumptions are used to evaluate them. The definition of situation determines which words are chosen as appropriate to the subject, which roles are available, which range of actions is appropriate, and with whom one is to communicate and how. The reasoning processes preferred by individuals seem to be heavily influenced by their education and professional training. Engineers frequently create narrative arguments, arguments that are stories explaining what happened when and under what circumstances. Managers more frequently use social science reasoning in which much of the "reasoning" is actually justification of assumptions underlying the model applied to the subject. Understanding the differences between the reasoning of different groups of professionals within a company or organization may be a primary key to anticipating the organization and use of evidence in documents produced by that group or person.

The Space Shuttle Challenger Accident Case: When a writer implements an inappropriate rhetorical situation, serious, even tragic, problems can occur, as suggested in the report of the Presidential Commission on the Space Shuttle Challenger Accident.[21] The Commission concluded that the mechanical cause of the Challenger accident was the failure of the pressure seal in the aft field joint of the right Solid Rocket Motor (Vol. 1, p. 72). The Commission also found as a contributing cause a flawed decision-making process:

> Testimony reveals failures in communication that resulted in a decision to launch . . . based on incomplete and sometimes misleading information, a conflict between engineering data and management judgments, and a NASA management structure that permitted internal flight safety problems to bypass key Shuttle managers. (Vol. 1, p. 82)

The report suggests that the critical failures occurred during two teleconferences and an intervening caucus or meeting of the Morton Thiokol engineers involved in the production of the Solid Rocket Motor. These electronically conducted meetings were part of the preflight readiness review process held in the 24-hour period before the space shuttle flight began. The NASA managers and the Thiokol engineers appear to have begun the meeting with a shared understanding of the rhetorical situation (purposes, roles, type of reasoning), but in this instance the NASA managers' model-based logic and the Thiokol engineers' analogical reasoning from a few specific instances produced a tragic conflict.

Participants from NASA, especially Lawrence B. Mulloy, the Solid Rocket Booster project manager at the Marshall Spacecraft Center who was in charge of the teleconferences, talked about the rhetorical situation as a collaborative probing of the data to determine whether the model of assumptions on which previous launch decisions had been based justified a change in that model. As a consequence of this approach, NASA officials were determined to treat a potential safety problem with the O-ring seals as a deterrent to launch only if data could be presented that invalidated the decision model used in the past. Mulloy looked at the teleconference as an encounter in which NASA and Thiokol would review the "Launch Commit

Criteria" and determine whether any of these conditions essential for launching would be violated by the predicted conditions on the morning of January 28, 1986.

The Thiokol engineers recommended that NASA should not launch at a temperature colder than the coldest previous launch (53ºF). The implication of this recommendation was that the shuttle should not be launched on the following day, when temperatures were expected to be less than 30ºF. Mulloy was very certain in his testimony about the rhetorical moves appropriate to his position in that circumstance:

> *Chairman Rogers:* Didn't you take that to be a negative recommendation?
>
> *Mr. Mulloy:* Yes sir. That was an engineering conclusion, which I found this conclusion without basis and I challenged its logic. Now, that has been interpreted by some people as applying pressure. I certainly don't consider it to be applying pressure. Any time that one of my contractors . . . come to me with a recommendation and a conclusion that is based on engineering data, I probe the basis for their conclusion to assure that it is sound and that it is logical. (Vol. 5, p. 829)
>
> We were simply looking at the engineering data and reviewing those engineering data. The concern, of course, that was being expressed was for the low ambient temperatures that were predicted for the night and the effect those low ambient temperatures would have on the propellant mean bulk temperature and on the joint particularly. (Vol. 5, p. 829)

In Mulloy's judgment, his communication tactics did not constitute pressure on the Thiokol engineers. Mulloy would not allow Thiokol to use any other reasoning process than the provision of data which showed that a launch commit criterion would be violated; but the Thiokol engineers did not have that kind of data at their disposal. Mulloy had a list of criteria that constituted a model for his decision making; the engineers had limited concrete data from a few flights and laboratory tests. The engineers who had handled the charred O-rings from the coldest previous flights were frustrated by NASA's unwillingness to consider the implications of charts showing the history of O-ring erosion on previous flights and pictures of damaged O-rings, as the testimony of Roger Boisjoly describes:

> And there was an exchange amongst the technical people on that data as to what it meant. . . . But the real exchange never really came until the conclusions and recommendations came in.
>
> At that point in time, our vice president, Mr. Bob Lund, presented those charts and he presented the charts on the conclusions and recommendations. And the bottom line was that the engineering people would not recommend a launch below 53 degrees Fahrenheit. (Vol. 1, p. 91)
>
> One of my colleagues that was in the meeting summed it up best. This was a meeting where the determination was to launch, and it was up to us to prove beyond a shadow of a doubt that it was not safe to do so. This is in total reverse to what the position usually is in a preflight conversation or a flight readiness review. It is usually exactly opposite that. . . . (Vol. 1, p. 93)

Although Mulloy maintained that customary argument structure had been followed for review of the evidence (he invited the Commission to call other witnesses who would confirm that he had handled the meeting as usual), the Thiokol people felt that the purpose of the rhetorical situation had been reversed. They were used to arguing inductively from example. Once they had had sufficient examples to provide statistically sound proof for NASA's model of launch criteria, the two reasoning processes, though different, had allowed agreement. When Thiokol had too little data, NASA managers were unwilling to look at the implications of specific examples.

After NASA Manager George Hardy, deputy director of the Marshall Space Flight Center, declared the Thiokol recommendation "appalling," and Mulloy asked whether Thiokol wanted him to wait until April to launch, Thiokol management began to feel the company's interests as sole supplier of the rocket engines were threatened and asked for a meeting among Thiokol people with the teleconference lines switched off. During this exclusive meeting of Thiokol people, a senior manager explicitly revised the rhetorical situation by asking the vice president, Lund, to change roles, "to take off his engineering hat and put on his management hat" (Vol. 1, p. 94). Chairman Rogers followed up on this testimony by asking Lund, "How do you explain the fact that you seemed to change your mind when you changed your hat?" Mr. Lund was not able to answer this question directly. Apparently, management interests differed sufficiently from engineering interests to produce a different conclusion, and Thiokol subsequently agreed that no launch criteria would be violated and the launch could proceed. Mulloy did not convey these concerns to the top two levels of the review process and the shuttle Challenger exploded shortly after the launch on January 28, 1986.

ORGANIZATIONAL SITUATIONS AND RHETORICAL SITUATIONS

Because of rapid changes in business environments and within companies, many rhetorical situations must be redefined to achieve greater congruence between organizational situations and rhetorical situations.

In the example of the mutual fund's reliance on market timing, the agency writer saw the situation as "writing a brochure for a client . . . essentially, marketing a parity product by claiming extra attention for it, making it stand out on the shelf." The writer perceived investors as breakfast cereal buyers, the sole audience for the message on the box. He needed to understand that, although he was writing a brochure for a client, the rhetorical situation involved audiences other than consumers and marketing professionals. He need to include in the rhetorical theater other powerful actors, including regulatory associations, competitors, lawyers, investment brokers, as well as investors. By using a broader model of the sources of meaning in the writing context, practitioners and teachers alike can construct more accurate definitions of organizational situations and rhetorical situations to guide their decision making.

IMPLICATIONS FOR TEACHING

Teachers can use the model described and the results of research to improve instruction. Recognizing the force of culture, technology, and situations can enrich our production and use of cases in the classroom. Brockman identifies six features of a successful case, including "fullness of the rhetorical contexts," which he associates with purpose, audience, and role.[22] A full rhetorical context should go beyond these three factors to include the relation between the organizational situation and the rhetorical situation, and the culture, values, history, and ways of thinking that determine the criteria for judging communication practice in a real organization.

Further, by studying rhetorical situations, we may identify how these provide roles for individuals trained in particular disciplines, particular ways of thinking and arguing. We can help students anticipate how the skills learned in accounting, finance, real estate, strategic planning, and other business functions will be applied in communication, and we will be able to describe more precisely the relationship between business communication and other management disciplines.

We must teach students to analyze organizational and rhetorical situations and to develop strategies for achieving greater congruence between them, given the culture, size, and technology of the organization. Finally, we should emphasize the excitement and pleasure of dealing with the demands of rhetorical situations. Creativity and personal involvement are essential for meeting the complex challenges of real organizational contexts. Too often, technical and business communication has been taught as a dry, mechanical skill devoid of personal interest. When we recognize the importance of the context for writing in organizations, we see the significance of the issues resolved through communication processes. Writing well is not merely conforming to genre conventions, as some of the genre-based approaches have implied. Communicating in organizational contexts is essential to the vitality, and even to the survival, of organizations and society in a technical era.

Notes

1. Lee Odell, "Relations between Writing and Social Context," in Lee Odell and Dixie Goswami, ed., *Writing in Nonacademic Settings* (New York: Guilford Press, 1985), 249–80.

2. Brief examples throughout the article, such as this one, are drawn from my consulting experience.

3. Lester Faigley, "Nonacademic Writing: The Social Perspective," in Lee Odell and Dixie Goswami, ed., *Writing in Nonacademic Settings*, 231–48.

4. Exceptions to these texts would be Marya Holcombe and Judith Stein's *Writing for Decision Makers* (Belmont, Calif.: Lifetime Learning, 1981); and Mathes and Stevenson's *Designing Technical Reports*, both of which emphasize the effect of the organization's structure and problem-solving activities on meaning.

5. Jeanne W. Halpern and Sarah Liggett, *Computers and Composing: How the New Technologies Are Changing Writing* (Carbondale: Southern Illinois University Press, 1984).

6. Richard V. Farace, Peter R. Monge, and Hamish M. Russell, *Communicating and Organizing* (Reading, Mass.: Addison-Wesley, 1977), 93. *Structural-functional* refers to the relation between a firm's structure and the business functions performed. Most firms attempt to group together workers with similar goals and expertise to foster cooperation and efficiency.

7. Claude Shannon and Warren Weaver, *The Mathematical Theory of Communication* (Urbana: University of Illinois Press, 1949). The image of reified language in this model

has been thoroughly analyzed by Ragnar Rommetveit, "Prospective Social Psychological Contributions to a Truly Interdisciplinary Understanding of Ordinary Language," *Language and Social Psychology* 2; 2, 3, 4 (1983), 89–104.

8. C. E. Osgood, "Psycholinguistics: A Survey of Theory and Research Problems," *Journal of Abnormal and Social Psychology* 49 (October 1954).

9. David K. Berlo, *The Process of Communication: An Introduction to Theory and Practice* (New York: Holt, Rinehart, and Winston, 1960); Harold D. Lasswell, "The Structure and Function of Communication in Society," in John Byrson, ed., *The Communication of Ideas* (New York: Harper and Row, 1948), 37–51; J. C. McCroskey, *An Introduction to Rhetorical Communication* (Englewood Cliffs, N.J.: Prentice-Hall, 1972); Wilbur Schramm, *The Process and Effects of Mass Communication* (Urbana: University of Illinois Press, 1954).

10. Lester Faigley, "Nonacademic Writing."

11. Gregory G. Colomb and Joseph M. Williams, "Perceiving Structure in Professional Prose: A Multiply Determined Experience," in *Writing in Nonacademic Settings*, 87–128.

12. Linda Smircich, "Implications for Management Theory," Linda Putnam and Michael Pacanowsky, ed., *Communication and Organizations: An Interpretive Approach* (Beverly Hills, Calif.: Sage Publications, 1983), 221.

13. Figures 3, 4, and 5 show the ads of funds managed by only one company, Franklin Funds, because of lack of space for additional figures. However, by consulting the *Wall Street Journal* for this period, the reader can see that statements are generally true about funds of this type.

14. *1986 Mutual Fund Fact Book* (Washington, D.C.: Investment Company Institute, 1986).

15. Effects of size, structure, and technology have been studied for over twenty–five years, especially by the Tavistock group in England and by the contingency theorists at Harvard. The sociotechnical models can be useful for analyzing patterns of communication, but other sources of meaning must be considered as well. T. Burns and G. M. Stalker, *The Management of Innovation* (London: Tavistock Publications, 1961); Joan Woodward, *Industrial Organizations: Theory and Practice* (London: Oxford University Press, 1965).

16. Paul R. Lawrence and Jay W. Lorsch, *Organization and Environment* (Boston: Harvard Business School, 1967). For a historical review, see Henry Mintzberg, *The Structure of Organizations: A Synthesis of the Research* (Englewood Cliffs, N.J.: Prentice Hall, 1979).

17. Yvan Allaire and Mihaela E. Firsirotu, "Theories of Organizational Culture," *Organization Studies* 5, 3 (1984): 193–226.

18. Alan L. Wilkins and William G. Ouchi, "Efficient Cultures: Exploring the Relation between Culture and Organizational Performance," *Administrative Science Quarterly* 28, 3 (1983): 468–81.

19. Paul Bate, "The Impact of Organizational Culture on Approaches to Organizational Problem Solving," *Organization Studies* 5, 1 (1984): 43–66.

20. Jone Rymer Goldstein, "Myths and Stories in Corporate Communication" (Paper presented at the Association for Business Communication Convention, Chicago, November 1985).

21. *Report of the Presidential Commission on the Space Shuttle Challenger Accident*, 5 vols. William P. Rogers, chairman (Washington, D.C.: U.S. Government Printing Office, 1986).

22. R. John Brockmann, "What Is a Case?" in R. John Brockman, ed., *The Case Method in Technical Communication: Theory and Models* ([Lubbock, Tex.]: Association of Teachers of Technical Writing, 1984), 1–16.

DEVELOPING YOUR UNDERSTANDING

1. Referring to Driskill's discussion of current models of business communication, summarize the three orientations to professional writing. Create a professional writing scenario from your own or another's experience and describe how each of the three orientations would affect how writers in such a situation would approach the task of writing.
2. Given only Figure 2 as a visual aid, prepare a ten-minute presentation for your class that defines how each of the external sources affects the writing context for the mutual fund brochure headline.
3. Referring to Figure 6 as a model, prepare a visual that summarizes Driskill's textual analysis of the Challenger accident. Your visual should identify the key organizational features of the accident and visually define the relationships between them.
4. Driskill claims that organizational environments are interpreted, not isomorphic. Explain what she means. Then, using the mutual fund brochure headline or Challenger accident situation as an example, compare and contrast the processes, contents, and purposes of viewing organizational contexts as interpreted vs. isomorphic.
5. Driskill identifies a number of ways writing is affected by corporate culture and definitions of organizational and rhetorical situations. Referring to Driskill's article and your own experience, generate a list of the ways writing can be affected by these two organizational variables. Then summarize how these two variables impacted the Challenger accident.
6. Assess the effectiveness of Figure 6. Your assessment should identify the strengths and weaknesses of the model and be supported by professional writing examples, either real or imagined.
7. Describe how organizational contexts can both constrain and motivate professional writers. Then, examine some organizational strategies professional writers might employ to (a) negotiate the constraints and (b) take advantage of the opportunities found in organizational contexts. Use specific examples to illustrate your discussion.

FOCUSING ON KEY TERMS AND CONCEPTS

Focus on the following terms and concepts while you read through this selection. Understanding these will not only increase your understanding of the selection that follows, but you will find that, because most of these terms or concepts are commonly used in professional writing and rhetoric, understanding them helps you get a better sense of the field itself.

1. socialization
2. individualization
3. organizational individuation
4. personal authority
5. social authority
6. situational authority
7. authority
8. agent for change

WRITING REVIEW AS AN OPPORTUNITY FOR INDIVIDUATION

SUSAN M. KATZ

Organizational assimilation—including socialization and individualization, as described in chapter 1—is not a temporary process that affects workers only in the early months or years of their careers; rather these processes occur repeatedly, either as the individual changes jobs or as changes occur in the organization affecting the individual's role. Thus we can say with some certainty that an understanding of these processes is important for understanding how learning and change take place in organizations. However, I would suggest that there is an additional dimension to assimilation, which I call *individuation*,[1] that occurs concurrent with socialization and individualization. While organizational communication provides some basis for proposing a concept such as individuation, there has been no research that examines how individuation might occur. That is, we don't know how newcomers become recognized for their distinctive talents and skills or how that recognition relates to their ability to gain authority, challenge the status quo, and become agents for change in organizations. This chapter will explore the concept of individuation and will demonstrate how the unique aspects of nonacademic writing review facilitate this process.

SOCIALIZATION, INDIVIDUALIZATION, AND INDIVIDUATION

As discussed previously, socialization is the process by which newcomers learn about and adapt to the goals, values, and appropriate behaviors of the organization; it is a process whereby the organization attempts to shape the individual to meet organizational needs. Individualization refers to the efforts that a newcomer makes to resist that shaping and change particular aspects of his or her role or the organization to

Source: *The Dynamics of Writing Review: Opportunities for Growth and Change in the Workplace,* Susan M. Katz, "Writing Review as an Opportunity for Individuation," 1998, pp. 73–98.

[1] The term "individuation" is used in psychology to refer to the development of distinctive personalities (Jung, 1923), and in this sense is most often used to describe child and adolescent development. Individuation has also been used to describe the willingness of individuals to publicly differentiate themselves from others in their environment (Whitney, Sagrestano, & Maslach, 1994). The newer definition includes a greater focus on voluntary individuation (such as when an individual creates a distinctive public persona represented by body piercing or tattoos), as well as the involuntary individuation that occurs in situations where the individual may be individuated by others based on physical attributes (such as when a woman or African-American is hired in a previously all-male or all-white setting).

In either definition, individuation refers to the separation of the individual from the "other." In the Jungian definition the other is the mother or, more generally, the family (Fromm, 1941; Mahler, 1968). In Whitney, Sagrestano, and Maslach's (1994) definition, the other is any social group the individual chooses to separate from or is involuntarily differentiated from.

When I speak of organizational "individuation," I am thus building on earlier concepts, but adapting them to the organizational environment.

meet personal needs. However, significantly missing from discussions of individualization is any explanation of how newcomers are able to effect such resistance and change. The concept of individuation can help explain why some newcomers are able to have a profound influence not only on their own role within the organization, but on the organization itself.

Organizational individuation is the process by which the individual develops his or her particular character within the organization. The process can be viewed as a series of behaviors:

- The individual performs assigned tasks as he or she thinks appropriate based on training and/or experience.
- More experienced workers (particularly supervisors) observe the manner in which the individual (particularly a newcomer) performs these required tasks.
- These observations and evaluations lead to a perception about the capabilities of the individual.
- These perceptions may involve the recognition of unique attributes (skills, talents, or other qualities).

Depending on the value of these attributes to the organization and on the particular circumstances prevailing within the organization at any given moment, this recognition may lead to the granting of the authority necessary to effect change.

In summary, individuation is a process that occurs as the talents, skills, or other attributes that make an individual distinct become evident through the normal interactions and productive efforts of that individual. Opportunities for individuation occur in the writing review process, where a supervisor or other experienced reviewer can recognize and evaluate the skills displayed by the writer in the text under review and in the interaction during the review process.

Just as the collaborative nature of nonacademic review facilitates socialization processes, this interactive experience is particularly well suited to individuation. Individuation can only occur when the individual is allowed to express his or her thoughts, ideas, and opinions. In writing review sessions, the reviewers in this study were anxious to hear from the newcomers so that they could produce the best possible document to meet the needs of a specific context. The reviewers all acted as collaborators who were willing to share their authority over the text under review.

In the review of a specific text, the supervisors often had a better sense of many of the factors influencing the document than the newcomers. However, the supervisors were also open to ideas and suggestions from those newcomers. In situations where suggestions are encouraged, newcomers have opportunities to demonstrate their talent and skills. The demonstration of these talents and skills can result in individuation when those attributes are recognized by the supervisor or by the group as unique to the individual. Furthermore, when these talents and skills are seen as superior to similar talents and skills of others, the newcomer may be individuated as an "expert."

Thus another way to think about individuation is to frame it in terms of expertise. When a newcomer is individuated during writing review, it is often some form of expertise that is being recognized.[2] Research in many different fields suggests that ex-

---

[2] This is not to say that expertise is the only basis for individuation. However, expertise played a significant role in the individuation of several participants in this study, as will be described later in this chapter.

pertise can lead to increased authority.[3] A brief review of the connection between expertise and authority will provide a context for a discussion of writing review as an opportunity for individuation.

The Expertise/Authority Connection

In *The Nature of Expertise*, Chi and Glaser (1988) provide us with characteristics—such as large quantities of knowledge in a particular domain and automated solutions to common problems—that define an expert. However, the many discussions of expertise in *The Nature of Expertise* (Chi, Glaser, & Farr, 1988) deal with well-defined domains, such as chess or typing. When we move to less well-defined domains, such as writing, interpersonal communication, or management, we have more difficulty creating a precise, consistent definition of expertise. Moreover, although a connection between expertise and authority seems to be generally accepted in many fields, the connection between expertise and authority differs depending on the disciplinary perspective of the scholar investigating the connection.

Personal Authority

In rhetoric and composition, Geisler (1994) has described expertise as deriving from knowledge and ability in two spaces: a domain content space and a rhetorical process space. As individuals gain content knowledge within a particular domain and the concomitant rhetorical skills to express themselves in that domain, they move toward expertise. As individuals gain this expertise, they gain authority to participate in the conversation of a discipline (Penrose & Geisler, 1994). Along similar lines, Carter (1990) describes an expert writer as an individual who has developed both general and local knowledge. That is, an expert writer is one who has learned general strategies of writing applicable to any domain as well as the specific conventions of a particular community. The source of the individual's authority to speak is a personal authority based on the knowledge that he or she possesses.

These descriptions of authority based on expertise do not exist in a vacuum, but they do not rely on any external source; thus I have given them the designation, "personal authority." Although the expert is working/writing/interacting within some context, the authority comes from the individual's personal confidence in his or her ability and right to speak within a disciplinary community. That is, individuals develop confidence as they gain experience within a discipline and recognize that "there is authority to spare" (Penrose & Geisler, 1994, p. 517). All professionals must first belong to a disciplinary community before they can belong to an organizational community, and they carry the personal authority developed within that disciplinary community into the organizational community. In the workplace, that personal authority continues to provide some level of self-confidence to the individual.

While these definitions of expertise based on personal authority do not ignore context, they do ignore the authority that comes from context. More specifically,

---

[3] There are many sources of authority in organizations in addition to expertise, but a discussion of those sources goes beyond the scope of this study. For information on other sources of authority in organizations, see Pfeffer (1992).

they ignore the authority that comes when others in the community recognize and accept the individual's expertise.

Social Authority

Like Geisler, Penrose, and Carter, cognitive psychologists Sternberg and Frensch (1992) acknowledge the importance of personal authority in their discussion of expertise, but they add a social component to their definition. Sternberg and Frensch call the personal component of expertise "cognitive competence" (p. 191). Cognitive competence, which focuses on the knowledge of the individual, encompasses the domain content knowledge/rhetorical process knowledge model of expertise as well as the general/local knowledge model. However, Sternberg and Frensch suggest that models of expertise that only look at cognitive competence "miss an important aspect of expertise, namely, its attributed aspect. In the real world—as opposed to many psychological laboratories—expertise is, in large part, an attribution. A person is an expert because she is regarded as such by others" (p. 194). When an individual is viewed as an expert by a community (e.g., disciplinary, organizational, or workgroup community), he or she gains certain benefits including greater respect, access to resources, and power and influence (p. 196). This greater power and influence that is granted to experts is what I am referring to as social authority.

Situational Authority

Situational authority is based on the immediate needs of a particular organization at a particular time. As organizational communication scholars Conrad and Ryan (1985) tell us, "a highly expert individual will gain power from this personal attribute [expertise] only if provided opportunities to publicly use that expertise to solve significant organizational problems" (p. 240). In other words, the individual's expertise has to be valued by the organization if it is likely to lead to authority. Situational authority may be temporary: It is likely that once the problem has been solved or abandoned, the expert will lose the authority granted for the duration of the problem-solving effort.

It is probable that individuals must first have confidence in their own expertise, must have personal authority, before they can gain an attribution of expertise from a community. It is difficult, but not impossible, to imagine someone who has been identified as an expert by a group who does not recognize that expertise him- or herself. However, it is impossible to imagine an individual gaining situational authority without first having the social authority from attribution of expertise. As Littlepage, Robison, and Reddington (1997) demonstrate, "accurate recognition of expertise can facilitate the utilization of expertise" (p. 145).

Authority and Newcomers

Persona, social, and situational authority are additive: Each source contributes to the amount of authority the expert is able to wield, each source builds on the other. Any one source delivers some authority; in combination they deliver greater authority. Personal authority develops throughout an individual's life; social and situational authority develop anew each time an individual joins a community. A newcomer to an organization may bring a certain level of personal authority, but the social and situational must develop over time, as colleagues come to know and respect the newcomer's expertise.

We would not expect a newcomer, particularly an entry-level newcomer, to gain social or situational authority easily or rapidly. However, the newcomers in my study did gain varying levels of authority fairly quickly, typically through their *rhetorical* expertise. That is, individuals who gained authority:

- had a well-developed sense of the importance of audience,
- consciously thought about the purpose of the documents they wrote,
- created well-supported arguments, and
- wrote clear, coherent, grammatically correct prose.

These attributes became obvious to others in their organizations during the writing review sessions that focused on the texts produced by these newcomers. The newcomers' rhetorical expertise led to authority in these instances because:

- the newcomers had confidence in their own rhetorical skills and were not hesitant to make suggestions (they had personal authority);
- the newcomers participated in writing review sessions that highlighted their rhetorical skills (giving them an opportunity for individuation);
- the supervisors recognized and valued the rhetorical expertise displayed by the newcomers (giving them social authority);
- the newcomers' skills were needed to solve problems within the organization (giving them situational authority).

Not all the newcomers in this study gained significant levels of authority, but we can look at those who did, and those who did not, and use the concepts of personal, social, and situational authority based on expertise to analyze their ability to gain authority, challenge the status quo, and become agents for change in their organizations.

GAINING AUTHORITY

In studies of organizational behavior, the concepts of authority, influence, and power are used in many different ways. Bacharach and Lawler (1980) suggest that "some authors tend to equate [authority and influence]; some tend to equate power with influence and assert that authority is a special case or power; and others see authority and influence as distinctly different dimensions" (p. 27). In addition, some scholars say that authority and influence are subsets of power; others say that power is a function of authority; yet others suggest that all power can be subsumed under the concept of influence (pp. 27–28). For the purposes of this discussion, "authority" refers to the ability of an individual to influence or make decisions affecting the work of the organization, including not only his or her own work, but also the work of others.

Darlene and Personal Authority

Darlene gained authority from all three sources, personal, social, and situational. She had personal authority based on her self-confidence as an expert writer with something to contribute to the organization. Darlene's confidence in herself as a writer became evident to me and to her supervisors during writing review, where she did not hesitate to make comments and suggestions. Furthermore, during writing review Darlene demonstrated her awareness of the purpose of review, that the goal was to produce the best document. Unlike many novices, she did not resent the changes that reviewers made to her text:

> Darlene: Well, I really don't mind changes as long as it makes it better or at the very least does it the same but different.

Her willingness to accept revision was also evident in the discourse-based interviews that I conducted with her. In those interviews, I frequently rewrote sentences or phrases from reports and asked the newcomers how it would affect the report if we substituted my version for the original. Darlene's response after reading one such revised sentence[4] demonstrated her lack of investment in her own prose:

> Darlene: Ooh, I like this one better! (Laughter.)
> Susan: (Laughter.) You like which one better?
> Darlene: This one better.
> Susan: Oh, the one I wrote?
> Darlene: Yeah.
> Susan: Why? What's significantly different about what I wrote that makes it. . . ?
> Darlene: Well, um, it seems to get to the point a little more and you're speaking, saying that [the responses to the concerns itemized in the report] are necessary instead of turning it around and saying without these this will happen, so, that's probably a much better way to word it.

Although this response demonstrated Darlene's openness to revision, she did not hesitate to defend her own text when she felt that suggested changes were inappropriate or unwarranted. In describing a brochure that she had written for the Agency, she told me that she had written a draft, made changes suggested by Harry (the director) and Tom (an assistant director), and then

> About a week later, when she thought the brochure was finished, Harry came and said that Dick [the other assistant director] had made some suggestions for changes. Darlene did not like the suggestions that Dick had made. She prepared written arguments to counter Dick's suggestions. She told me that when she reads comments on her writing, if she agrees, she just says fine. But if she disagrees, she writes out an argument.

In this instance, Harry agreed with Darlene's argument, and she did not have to incorporate Dick's suggestions. Although she did not always get her way, her sense of personal authority allowed her to speak up when she felt strongly about a change. In a review session with Gene discussing the first report that she wrote on her own, Gene suggested that the document would "flow better" if she rewrote a bulleted list

---

[4] The original sentence read: "Without appropriate and detailed responses to the concerns we identified, [provider] risks both the financial stability and programmatic integrity of its [programs]." My revision read: "Appropriate and detailed responses to the concerns we identified are necessary for the financial stability and programmatic integrity of [provider's programs]."

using parallel construction. In fact, what we suggested was that she start each bulleted item with the words "collection rates," instead of

- "Collection rates. . ."
- "Revenue for services. . ."
- "[insurance company name], the largest third party insurer. . ."
- " third party insurers. . ."
- "Collections on uninsured self-pay clients. . ."

Her response to his suggestion was forthright:

Darlene: I think honestly that's what I tried to avoid.
Gene: Yeah?
Darlene: Just because, I mean, it flows better probably, but I just think that it's like, kind of dull if you do that? Like, every bullet starts the same?
Gene: Hmmmm?
Darlene: Not that I would be averse to doing it, but that's
Gene: Okay.
Darlene: just kind of my opinion. An audit report is boring enough as it is without having every sentence starting exactly the same way.

In the final version of the document, where less confident newcomers might have changed the text as Gene suggested, Darlene compromised: She left her text unchanged for the first two items and used "collection rates" for the last three.

Writing review sessions gave Darlene the opportunity to demonstrate her personal authority as a writer because:

- Both Darlene and the supervisors understood the *purpose* of the review as the production of the best possible document.
- The supervisors were willing to *share their authority* over the final product with Darlene in order to meet that purpose.
- The *face-to-face setting* gave Darlene the opportunity to explain and, when appropriate, defend her rhetorical choices. These explanations demonstrated both her writing skill and her interpersonal communication skills.

Darlene and Social Authority

Through sessions such as the ones described here, Darlene's supervisors and colleagues came to recognize her expertise and grant her social authority. Social authority may have been granted fairly quickly because of the nature of the writing review process at the Agency: With multiple reviews of each document at different levels within the organization, Darlene's skills became evident to many different individuals within a very short time. These more experienced individuals described her as someone who could write clear, coherent prose, and (what was more astonishing to this group) they said that she *liked* to write. While she was a competent accountant, the attribution of "expert" that was granted to her was directly tied to her writing ability.

For example, her writing ability was the only specific skill mentioned when she was nominated for the "Employee of the Year Award." The nomination, written by Dick (an associate director) reads, in part:

> For an auditor with only about one year's [sic] of experience, Darlene has progressed faster than any other trainee in the Agency. Darlene's evaluations have all been at the outstanding level.
>
> Darlene is an extremely talented writer who has greatly assisted the Agency in clearing the backlog of reports and issuing reports on a timely basis.

The first paragraph in this nomination describes Darlene's overall capabilities, but it is the mention of her writing ability in the second paragraph that distinguishes her from the other auditors. While the nomination goes on to applaud her efficiency, productivity, enthusiasm, and initiative, no other disciplinary talents are mentioned. The nomination written for another newcomer, Rebecca, contained no mention of writing at all, even though writing is a critical component of the position. Instead, Rebecca was nominated for "her technical ability" (which referred to her auditing ability) and the fact that "she possesses great computer skills."

In addition to Darlene's nomination, evidence for attribution of expertise can be found in conversations, interviews, and meetings with other newcomers as well as many senior members of the Agency.

- My initial interview with Rebecca took place just as she was about to begin an audit where she would be working with Darlene for the first time. She told me that she thought that Darlene was "a really good writer" and that she would "learn a lot about writing from Darlene by working with her on this project."
- In an early interview, Al, the most senior AIC in the Agency, recalled the first time he read something that Darlene had written. He said, "I thought I had died and gone to heaven. She puts everyone else in the Agency to shame."
- In a meeting, Gene referred to Darlene as "the report *maven*" ("*maven*" is Hebrew for "expert").
- In a casual conversation about the importance of writing for auditors, Dick talked for several minutes about Darlene as someone who was "unbelievable." He told me that "it is so rare to find an auditor who can write."

Darlene was granted social authority by this organization because her writing was exceptional; individuals became aware that her writing was exceptional because she participated in writing review sessions. Specifically, because the *process* of writing review at the Agency involved multiple reviews at several levels, Darlene had exposure to most of the senior staff within a few months of her employment. The attribution of expertise was virtually universal and opened the door for her to gain situational authority.

Darlene and Situational Authority

Darlene gained situational authority when Harry, the Director of the Agency, decided that her expertise could help solve a significant new problem. The problem was created when a new governor was elected, defeating the three-term incumbent.

Although most of the employees in the Agency were somewhat protected through their union membership, Harry was a political appointee. The election of a new governor put his position in jeopardy. In addition, the new governor had run on an austerity platform, and ever since his inauguration he had been promising massive cuts in the state workforce. More specifically, there had been discussions of completely eliminating the Department of which the Agency is a small part. Gene explained the situations to Darlene this way:

> I think [Harry's] saying, "If all the reports come out really quick, I [Harry] get to keep my job." . . . Politically, he's in an unprotected job and anybody who's owed anything could be put into it instead . . . I think he has some things to worry about that are very real, and I think that one of the things he's done is to say to himself, one some level, if we get reports out faster then everything will be okay.

Darlene's writing expertise is crucial to the process of getting "reports out faster." Previously, AICs and managers took months, sometimes even years, to draft and polish a report. Harry had hoped that Darlene would be able to intervene and speed up the process. The extent to which she was able to do this came from personal, social, and situational authority based on her expertise as a writer.

Darlene was not hired for her rhetorical expertise. She was hired because she had the right credentials (a B.S. in Accounting) and the right kind of experience (several years as a bookkeeper) to work as an auditor. Unlike Darlene, however, Luke (the marketing representative at the engineering firm) was explicitly hired because of his expertise as a marketing representative.

Luke and Personal Authority

At ECI, marketing representatives were not expected to be technically knowledgeable, but they were expected to have well-developed interpersonal communication skills that would allow them to solicit business opportunities for the firm. With almost six years of experience as a marketing rep for an adhesives manufacturer, Luke had a lot of useful contacts and the ability to understand technical subjects without actually having a lot of technical knowledge. Both Walt, Luke's supervisor, and Luke himself told me that Luke was hired to fill a gap in ECI's marketing department, that ECI needed Luke to help take the firm in a new direction.

Luke's personal authority developed through his experience: He had been very successful working for his prior employer, had prevailed over many other applicants for the job at ECI, and was confident of his ability to succeed in his new position. This confidence was evident in his behavior at staff meetings.

Although the marketing department consisted of eight people, the weekly staff meetings were not attended by everyone. Typically, Walt, (the director of marketing), would meet with the four external reps to discuss their recent activities and future plans. The three internal reps did not attend most of these meetings. Although Luke had only been with ECI for two weeks when he attended his first staff meeting,[5] he did not hesitate to contribute to the meeting: He asked questions, volunteered infor-

[5] There had been no meeting during the first two weeks because one of the reps was on vacation and one was on business in another part of the state.

mation about his background, reported on upcoming events of interest to the group, and instigated discussions about the organization's past experience in specific areas. In fact, although socialization theory suggests that newcomers often spend their first few weeks quietly observing the behavior of their peers, particularly in new situations, Luke took charge of the opening part of this meeting by asking Walt a question that led to a discussion of Luke's plans for the upcoming weeks. In essence, Luke had diverted the meeting away from Walt's planned agenda.

Luke's ability to take control of the meeting demonstrated his self-assurance. Furthermore, the level of detail he provided as he described his plans demonstrated his confidence that he was doing what was expected of him. He told where he would be each day for the following two weeks, including information about a seminar and a two-day conference he planned to attend and the prospective clients he planned to visit during a three-day trip through another part of the state. When questioned by Walt or others in the group about the nature of the business of each of these prospects and his reasons for calling on them, Luke was able to supply answers and information to support his plans. A few days later, I observed him make 17 phone calls in the space of two hours, setting up appointments with prospective clients or gathering information from other ECI employees. He worked diligently, took copious notes, made several appointments, and generally seemed confident in working with little obvious supervision. Luke's ability to work without supervision so soon after joining ECI reflected his sense of personal authority, his sense that he knew what he was supposed to do and how to do it.

Luke and Social Authority

Perhaps because he had such a strong sense of personal authority, his social authority developed rapidly. After Luke's second staff meeting, I spoke with Walt about Luke's progress:

> Walt and I talk about how easily Luke seems to have fit into the group. Luke seems so comfortable, confident. Walt says that is why they hired him—he's genuine, sincere, real.

Six weeks later, Walt was more specific about Luke's performance:

> Walt: I like him, and he's very sensible about how he's going about [representing the firm]. And he's also conscientious about what he's doing. He's concerned about what he's putting down in his call reports, and if it's thorough enough, and if it's politically sensitized.

Walt recognized Luke's expertise as a marketing rep, and he learned about Luke's expertise not only by observing him and listening to him in staff meetings, but also by reviewing Luke's call reports.

Every time Luke met with a client or a prospective client, he wrote a one-page summary of that meeting. The primary audience for the call report was Walt, but the report also went to other relevant personnel in the organization if the situation warranted greater distribution.[6] Through these call reports, and the discussions that Walt

---

[6] For example, if Luke had a discussion with a prospect about the need for a particular service, he would forward a copy of the report to the manager of the group that could perform the service.

and Luke had about the call reports, Walt was able to recognize Luke's rhetorical expertise. That is, Walt could see Luke's concern with audience (from the way he wrote the reports for diverse internal audiences) and purpose (from the information he presented about his meetings with clients). Later in the same interview mentioned above, Walt gave even more details about what Luke wrote in his call reports and how those reports reflected an ideal marketing rep:

> Walt: He's being very objective as he's sitting there [with the client] . . . he's not really forming an opinion one way or another on his own. . . . He's not editorializing it, either. . . . And he wasn't defensive about it.

Walt knew that a good rep is a go-between, serving both client and employer, walking a fine line between the needs of both. Luke's reports reflected his ability to walk that line, to be "objective," to avoid "forming an opinion," to allow the client to speak through his report without "editorializing." In this manner, his reports provided the kind of information that was most useful to Walt and others in the organization who used Luke's reports to make decisions about the future of ECI.

Once Luke had gained social authority, once he had been individuated as an expert, his supervisor granted him some license in the content of the reports. For example, sometimes Luke's reports were vague, but Walt did not ask for additional information. In one such report, Luke explained that a prospect was annoyed with ECI because of some misinformation. He went on to say that he would have to find out the source of the misinformation and report back to the prospect. He ended with the sentence, "I will follow up with an explanation and we will take it from there." Walt's response to this sentence was,

> Well, that's a little unclear, what he's going to take, but at this point, I've developed such a trusting relationship with Luke that I know he'll do the right thing. At first, I would have sat down with him and said, "What do you mean? How are you going to take this and explain it more?" But now, just based on his actions, what he has done, it's fine. So that I feel, I feel that [the report] is perfect.

Others in the organization also gave Luke social authority: His colleagues in the marketing department listened to him with respect in meetings, managers and supervisors in other departments shared information with him and traveled with him to meet prospective clients, and other personnel contacted him for advice when they were asked to take on tasks for their own departments related to marketing.

Perhaps the greatest evidence for Luke's social authority is the very fact that he was still employed at ECI: Only three months after Luke joined ECI, the president began restructuring the organization and eliminating positions. Although marketing did not suffer in the first round of eliminations, it went from eight positions to five only a month later. A year later, two of those positions were eliminated and two were changed. Luke, the newest member of the marketing group, was the only person whose position had remained stable.[7]

---

[7] In the first round of cutbacks affecting the Marketing Group, one internal rep and one external rep were fired, and one external rep was transferred to a different department. In the second round of cutbacks, Walt (the director of marketing) and a second internal rep were fired; the regional rep was brought into the home office and the last internal rep moved into a supervisory position over two new clerical positions.

To a great extent, Luke developed social authority through the review of his call reports. Although the ECI review process is not as multi-faceted as review at the Agency, Luke's reports were reviewed by his supervisor and other relevant personnel. These reviews facilitated the development of social authority because:

- Each text existed within a very specific *context* defined by individuals within the organization and by the particular client discussed in the report. Luke's reports demonstrated his understanding of context.
- Call reports have a very specific *purpose*, and the review of Luke's reports was, in part, an evaluation of Luke's awareness of that purpose.
- The reviewers recognized that often Luke had a greater knowledge of the prospective client than they did, so they were willing to grant him a great deal of *authority* over specific texts.

Over time, as Luke continued to demonstrate his understanding of context and awareness of the purpose of his reports, the authority granted to him over specific texts contributed to a significant level of social authority within the organization.

Luke and Situational Authority

Unlike Darlene, who gained situational authority based on her unique expertise, Luke gained situational authority by default. The president of ECI created a situation where Luke was one of only two people who had the expertise to call on clients and generate business for the firm. While the problem that Luke's expertise can solve is not new or temporary, he gained authority within the organization because of the particular situation that existed at that time. Luke's situational authority came from a situation that was created internally, which was quite different from the Agency, where the organization had the situation thrust upon them from outside. However, the resulting opportunities for Darlene and Luke were very similar: They both could use their expertise to help solve a problem caused by a particular situation. The origin of the situation was not significant to that opportunity.

When individuals gain authority, they can use that authority in many different ways. The newcomers in this study often used their authority to challenge the status quo and to attempt to effect change within the organization.

CHALLENGING THE STATUS QUO

As described earlier, nonacademic writing review takes place in a face-to-face setting, focuses on a particular text, and exists to produce the best possible document for a particular audience and need. These aspects of writing review encourage newcomers to ask questions and test the limits of accepted conventions.

Questioning and testing are recognized strategies for socialization to the organization's conventions, but they can also be seen as attempts to individualize or challenge the status quo. That is, if the questioning and testing serve to enhance the newcomer's understanding of conventions and lead to conformity, then they have served a socializing purpose. If, on the other hand, the questioning and testing result in "creative individualism"—the rejection of some, but not all, conventions or behaviors

(Van Maanen, 1976, p. 109)—then they have served an individualizing purpose. The extent to which this creative individualism is accepted by the organization depends on how the newcomer has been individuated: If the newcomer has gained some level of social or situational authority in the area of the challenge, the organization is more likely to pay attention to and accept the challenge to conventions or behaviors.

Darlene's Challenges

Even before Darlene assumed her role as editor, she felt comfortable challenging a report that had been written by Rebecca and Will (an AIC who had recently left the Agency) and revised by Gene. Since Darlene had no official involvement with this report, I asked Gene why he had given it to her to read. He told me that he was trying to help prepare her for the role of editor that she would be likely to assume at some point in the future. He further told me that he wanted Darlene to read the report so she would see the kinds of decisions he made and to give her an opportunity to tell him how she would have decided what to include.

This review session was atypical for two reasons: Darlene had not written any portion of the document in question and her comments were unlikely to have any effect on the final version. However, she took advantage of the opportunity provided by Gene to read the report carefully. After she had read the report she:

- asked about certain wording that seemed "redundant" and "doesn't seem like the normal way to do an audit,"
- suggested that two paragraphs could have been combined,
- pointed out the need for some specific information,
- said that there was insufficient evidence for the findings, and
- commented on the lack of attention to audience needs.

In each instance, Gene defended his decision and explained why her choice would be less appropriate. Darlene did not sit quietly and accept his explanations in all cases. For example, she felt strongly about the lack of evidence provided in the report. They discussed the reasons behind the paucity of evidence:

> Gene says he would prefer to disclose everything, but there is a benefit to this system. By not disclosing, it gives you ammunition to go back with if [the providers] disagree. Darlene suggests that it's not fair, not right, to hold anything back. Gene says the auditee has several opportunities to ask questions or raise issues, so it's not unfair. Darlene says it's sneaky.

They continued this discussion for several minutes, with Darlene refusing to acknowledge the validity of Gene's explanation. In the end, "Gene tells her she is right," but then he "explains that this provider is a good one, if it were a bad one they might handle things differently." In other words, Gene was explaining how context (the difference between a "good provider" and a "bad one" as the subject of the audit) influences text.

Darlene's strong sense of personal authority and Gene's willingness to share authority over the text by allowing her to question and challenge him created a situation where she could learn about the conventions of the organization, express her opinions about those conventions, and display her expertise as a thoughtful, careful, rhetorically skilled writer. Although in this instance Gene did not change the report based on Darlene's comments, in many cases, particularly after Darlene assumed her role as editor, he and others did accept and act on her suggestions. Her influence on the reports was most obvious in the area of visual presentation, but she also challenged the word choices and phrasings that had become traditional in the organization.

Darlene's use of visual aids to present information may have been connected to her experience using computers. She was more comfortable with computers and more knowledgeable about the variety of features offered by computer software than many of her more senior colleagues. She enjoyed experimenting with the various tools that the computer offers and liked to incorporate visually appealing and effective computer-generated tables, charts, graphs, and bulleted lists as well as highlighted headings and dramatic borders. When she began writing and editing reports, she reduced the amount of narrative by incorporating graphic elements into the text. These challenges to the conventional presentation were supported by her supervisors. In his review of the first report that Darlene wrote, Gene commented on her use of charts:

Gene: Now we have our spiffy productivity chart.
Darlene: Is that all right? You didn't have any questions?
Gene: Peachy!
Darlene: I though maybe we would make it smaller, for the actual report. I mean, we thought it was kind of big. Didn't you think. . . ?
Gene: Yeeaahh. I mean, if you can cut it back to a half page, without losing . . . I just don't know, if it would translate down well, but if you can cut that down to half page, great. And if you can't, it's no big deal.

Although the managers accepted her challenge to the traditional narrative presentation of material, they were not as willing to accept some of the atypical word choices she tried to inject into the reports. The following examples describe some of the changes that supervisors made to Darlene's texts. While the changes may not seem significant when viewed out of context, within the context of the Agency's audit reports they demonstrate Darlene's intentional and explicit challenge to what she saw as boring reports:

- "in years to come," was replaced with "in the future"
- "services were deemed uncollectible" was changed to "services were uncollectible"
- "The effect of . . . cannot be understated" was changed to "The effect of . . . is critical"
- "still more dismal" was changed to "not good"

In review sessions, Darlene frequently argued for her word choices, but she often lost those arguments. As Gene explained, his decisions to change some of her words were based on whether or not she was using "audit words or nonaudit words. 'Privy to,' that's not the kind of thing we say. So I blipped it out of there."

Gene was even more explicit about Darlene's use of language on another occasion, prior to Darlene moving into the editor role. As previously described, Gene had offered her the opportunity to review a report (written by Rebecca and Will) that he had just finished revising. She commented on an unusual word—"disaffiliation"—that he had used, and Gene responded by laughing and saying he was surprised that she would comment on an unusual word:

> They talk about Darlene's word choices: Gene says that she uses interesting words, too. She says there are times when they won't let her use the words she wants. Like, she said the records had "vanished" in one report, and they made her change that because it wasn't an "accounting word." She thinks it makes reports more interesting to read if you use different words than the ones that are expected. Gene agrees, but says the interesting words don't usually make it through the system.

While the supervisors seemed willing, even eager to accept her challenge to the visual presentation, they seemed more conservative when it came to their language conventions: Their reasons for not letting her use "nonaudit" words were usually some variation of Gene's "that's not the kind of thing we say." In other words, the supervisors seemed to think that they had a better understanding than Darlene of what constituted the best possible document with regard to word choice. Their reliance on traditional or standard words and phrases reflected a sense of what was appropriate and acceptable within a particular discipline, a sense that they had a better understanding of context than Darlene.

Luke's Challenges

The personal authority that Luke brought with him to ECI and the social authority that he gained, in part, through writing review, gave him the freedom to challenge some aspects of his job. Luke's challenges had little direct impact on others in the organization; they pertained to the way he performed his tasks but did not significantly alter the way he interacted with others in the organization or the way that others performed their own tasks. For example, Luke rejected the form he thought he was supposed to use for his call reports. In my first meeting with Luke, he told me about the kind of writing he would be doing.

> He showed me a copy of how they do call reports here—in columns, like conversation analysis. He doesn't like [the form], and showed me examples of his call reports (he has only done three so far). He writes across the page, ignoring the column lines. No one has complained about the way he writes up call reports. Everyone else does the script style report. He would prefer to summarize the total conversation than "quote" each party to the call.

Although Luke felt comfortable challenging the use of this form, in fact, he had misunderstood. The form that he had showed me, the "script" form, was used to record telephone conversations as they occurred; "call reports," records of meetings with clients, were filed on another, more complex form, with spaces for different types of narrative summaries to be completed after the visit. Although Luke quickly learned about his mistake, he never accepted the use of the script style form and continued to record his telephone conversations in summary style.

Another of Luke's challenges focused on the way he gathered information about prospects and clients. Luke had complained to me on several occasions about the difficulty he had finding out about ECI's history with clients. In one of our early sessions, Luke:

> closes the door and tells me that the company does a lousy job of documentation. That is, there is no central database with information about the history of the company's relationships with any other company. When he calls on someone, he would like to know whether or not they have ever done previous work, when someone last called on them, what the experience was like, etc. There is no way to get all that information.

Seven weeks later, Luke had found a way to get the information. During one of our observation sessions, I noticed Luke referring to a large orange binder before he made each phone call. I asked him to tell me about the binder:

> Luke explains that this is a list of all clients that the organization has ever dealt with. It includes information on work done, proposals (whether accepted or rejected), contact, address, phone, date of last official contact, etc. . . . Luke had not known of the binder's existence—the binder is an accounting tool. Luke asked for and was given his own set—two volumes, each about 4" thick. . . . Now he always looks up prospects in the binder before calling.

Luke's challenge here was to the process by which the marketing representatives created and maintained relationships with prospects and clients. When the tools at his disposal in marketing proved insufficient for his needs, he had the authority to go outside the department to find other materials that would more adequately meet those needs. Although Luke found this database extremely useful, he had not considered suggesting that others use it as well:

> I asked if he has shared this information with the marketing group. He says he has not talked about it in a meeting, but thinks maybe he should. . . . Luke says he will talk about use of these tools in a meeting some time, just so everyone is aware of what he is doing, how he is using them.

Until I asked Luke if he had shared the database with others, it had not occurred to him to do so. While he had challenged the process and individualized it for himself, he had not tried to extend his influence to the other marketing representatives.

Darlene and Luke demonstrated many instances of challenging their organizations, and both of them seemed concerned with finding "better" ways of performing their tasks. Some newcomers, however, even newcomers with authority, seem to avoid any explicit challenge to their organizations.

Rebecca's Lack of Challenge

Rebecca had considerable personal and social authority and was recognized by the Agency as a remarkably adept auditor. As noted previously, she was nominated for Employee of the Year based primarily on her technical (auditing and computing) skills.[8] Rebecca demonstrated a high level of self confidence at all times. In our first interview:

- She commented on the ineffectiveness of the Agency's orientation program: "They gave me a bunch of stuff to read—useless stuff. [In the future] they should just send [new people] out into the field."
- She told me about her first assignment: "It was very simple, very straightforward. I did a lot of the writing because it was so simple."
- When I asked her how she knew what to write in her first assignment, she said: "I don't pay attention to how the others write, I just write my own way and so far they've told me it was good."

Throughout that interview and all subsequent observations, Rebecca displayed a confidence in her own ability and a lack of respect for the abilities of many of her supervisors that bordered on arrogance. Toward the end of my observations, I conducted a discourse-based interview with Rebecca in which I asked her to choose between two versions of a number of "findings."[9] One version of each finding was taken from an actual report and the second was a version that I had rewritten, but I did not indicate which version was the authentic one. Her responses reflected her attitude toward the word done by others:

- In response to a finding about the lack of signatures on Board minutes: "I think those are silly findings. (Laughter.) I mean, who cares? They didn't sign it, they should have, that's it."
- In response to a finding about claims for interest expenses: "I don't like either of these, because the interest expenses are not allowable, not because of [the reason stated in the findings], but because you can't claim interest. It's against the law. An interest expense would never be eligible for reimbursement."

Rebecca was so adamant about her answer that claiming interest was against the law that I later rechecked the finding with one of the managers.[10] He said that there are certain situations where interest can be claimed, so Rebecca's confidence in her knowledge was somewhat misplaced in this instance.

Despite, or perhaps because of, her independent attitude, the organization perceives her to be extremely capable. Although she was never singled out as an exceptional writer, her writing skills were regarded as more than adequate, and she was asked to take on AIC work long before she was eligible for that role according to civil service classifications. In fact, she was so highly thought of that Gene once told me that "if we had a few teams like Rebecca and Darlene [who had worked together on a

[8] Although Rebecca lost the Employee of the Year award to Darlene, she did earn this recognition the following year.

[9] A "finding" is a statement of a problem or issue that was uncovered during an audit. An audit report is actually a report that lists all the findings and provides the background (evidence) and recommendations for the resolution of each problem.

[10] I had done a pretest of the discourse-based interview with a manager to gain the perspective of an experienced member of the organization as a check against the responses of the newcomers. In that pretest, the manager had not indicated that the interest finding was faulty in any way.

recent audit], we could do dozens of meaningful audits every year and save the state millions of dollars."

Although Rebecca had both personal and social authority, she did not use that authority to challenge the organization and she did not gain any situational authority: Her expertise was not seen as relevant to solving the problem of the protracted report production process. Further, I did not observe her asking questions, making suggestions, or in any way attempting to influence the work processes of those around her. The only obvious outcome of her authority was increased responsibility in the workplace and the possibility of accelerated advancement.

It seemed that Rebecca was unconcerned with attempting to challenge the organization. She did not significantly influence her organization because she was not interested in acting as an agent for change, and we cannot predict how successful she would have been if she had chosen otherwise. While Luke challenged those aspects of the organization that affected his ability to perform his tasks, he seemed content to let others in the organization continue working as they had in the past. Darlene, however, was concerned with changing processes enacted by everyone in the Agency.

BECOMING AGENTS FOR ORGANIZATIONAL CHANGE

Although it is difficult to establish causality, there are situations that seem to indicate that newcomers do, on occasion, push their challenges beyond their own sphere. That is, some newcomers challenge aspects of the organization beyond their own role and attempt to change the organization.

Darlene's Attempts at Change

When Darlene was made editor for the Agency, she moved into a position where she could use her authority to make meaningful changes in the organization. Significantly, she also moved into a situation where she was involved in writing review sessions on an almost daily basis. The increased participation in writing review processes—both as a writer and as a reviewer—may have increased her authority and provided the catalyst for her involvement in organizational change. At the very least, her increased participation provided additional opportunities to display her skills, question conventions, and attempt to change the organization.

The first, and perhaps most predictable, change was in the format and language of the audit reports. As would be expected when one individual is responsible for editing all documents, the reports became more consistent and language began to be standardized.

To facilitate her revision process, Darlene created a list of tips for "Improving Report Writing Efforts." She began this list when she assumed the role as editor in an attempt to resolve what she and some of her colleagues saw as inconsistencies in the director's attention, and "he said I should write things down that he prefers, and then hold him to it." Over the first few weeks in her new role, she added items to the list following review sessions with various managers. The list consisted of information in five categories: General Format; Findings, Background, and Recommendations (auditing terms referring to specific sections of audit reports); and Wording Choices.[11] Eventually, Darlene would have several pages of "tips"—now formalized as "Guidelines for Report Writing"—which she would distribute to everyone involved in writing reports with a cover memo that began, "Attached are guidelines to be followed when preparing our audit reports" (emphasis added).

In addition to attempting to standardize language and format, Darlene suggested a change in the very process by which the reports were drafted, a change that dramatically reduced the amount of time involved in producing a report. Figure 1 describes the report writing process prior to the creation of the editor position. This process shows the report being drafted by the AIC (usually within one week of the completion of an audit), and then moving on to the manager for review. The report would cycle between the AIC and the manager until the manager was satisfied with the document, and then it would be sent up one level in the hierarchy to an associate director. The report would then cycle between the manager and the associate director, be sent on to the director, cycle again, and finally be distributed. While the chart describes the movement of the document, it does not give an adequate indication of the amount of time that this complex process could take. A typical audit might take three to four months; the production of the report for that audit could take anywhere from six months to two years.

When Darlene became editor, it was necessary to change this process to allow for her role. Figure 2 shows the process that Darlene proposed and which, in fact, was accepted by the organization. In this streamlined process, the number of cycles was dramatically reduced. Although all of the same people participated in the process, their review of the document was much more efficient.

One of the major reasons for this efficiency was that Darlene's revised process called for a meeting between the manager, AIC, and editor before writing began. In this meeting, they discussed the content, order of findings, and general tone of the report. Also in this meeting, Darlene reminded the AIC of the "Guidelines for Report Writing" that she had produced and distributed. This planning meeting dramatically improved the quality of the initial draft produced by the AIC. Furthermore, by having the manager and editor review the document simultaneously and meet to discuss their independent reviews, and by having the editor make the actual changes, the number of cycles and the time spent on the revision process was reduced significantly.

Ideally, this new process produced an audit report in about a month. Initially, however, the reports were still taking several months to complete. Darlene suggested an additional change which resulted in a considerable revision of the role of the managers in the audit process. The suggestion was instigated by Darlene's observations at the prewriting meetings.

The prewriting meeting was one of the most important innovations in Darlene's report writing process. While sitting in on these sessions, Darlene realized that quite

[11] Items in the list varied from the simple "Define all acronyms and technical terms used," to the more complex "Make sure the recommendation addresses all points raised in the finding and background sections, and that no new issues are introduced." The first version consisted of 26 items, the second version had 46, and the final version had 64.

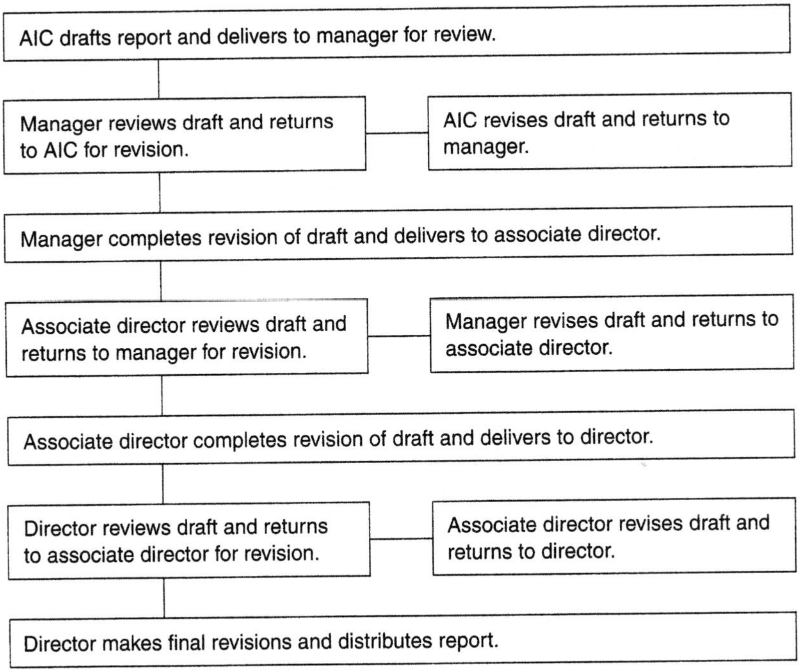

Figure 1. Report writing process at the agency.

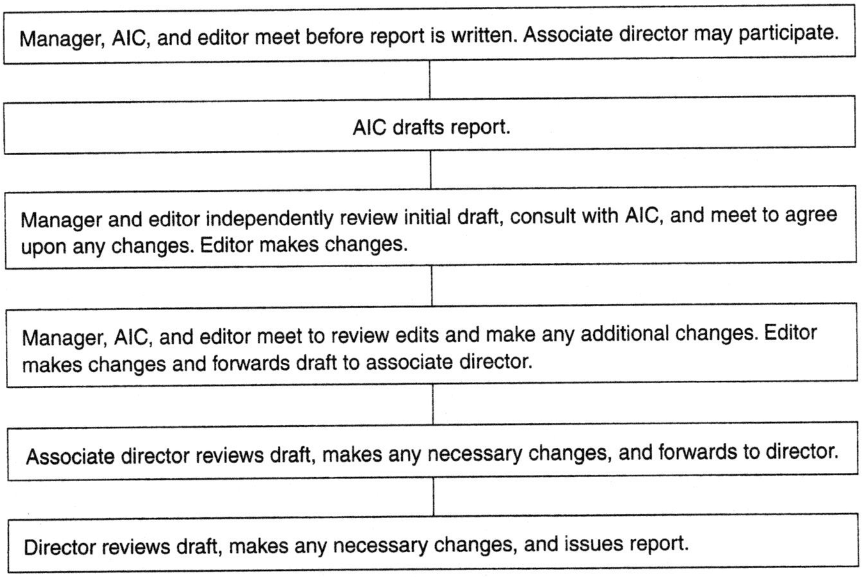

Figure 2. Report writing process instituted by Darlene.

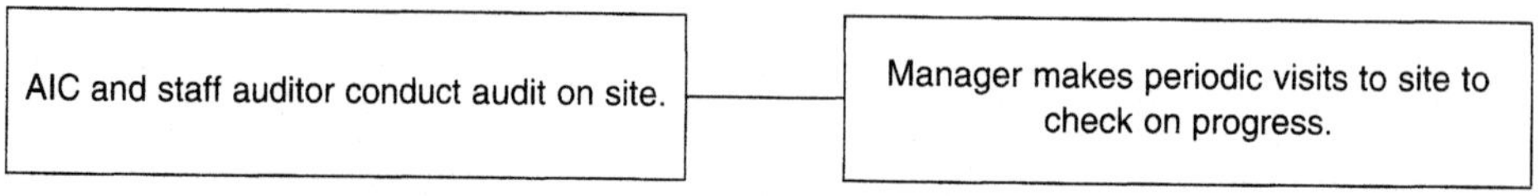

Figure 3. Original audit process.

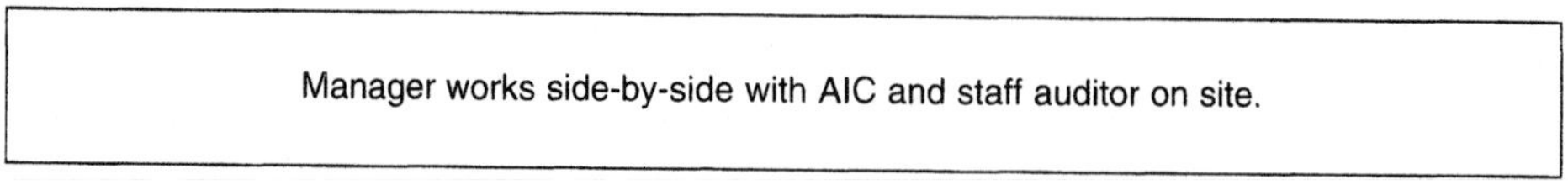

Figure 4. Revised audit process.

frequently the manager would ask for information or evidence that the AIC would not be able to provide. In these cases, the manager would send the AIC back to the audit site to gather more material. When this occurred, it would halt the progress of the report, often for several weeks. Darlene suggested that if the manager were more involved in the audit process itself, these questions would be answered while the AIC was still on site, eliminating the need for return visits. This suggestion was put in practice, and thus Darlene changed the audit process that had been in use for decades. Figures 3 and 4 describe a simplified version of the original and revised processes.

In effect, this revision to the audit process changed the job description of the managers: They no longer supervised from a distance, but instead supervised as part of the audit team on site. The significance of this change is greater than it may seem: The managers, all of whom had grown accustomed to working a typical eight-hour day in a typical office setting, were now required to travel throughout the state on a regular basis. They frequently had to spend several days each week working in borrowed offices dealing with frightened or hostile providers. They frequently had to stay in hotels and eat in restaurants for extended periods of time, returning to their homes and families only on the weekends. While traveling was an accepted aspect of the job for the AICs and staff auditors, one of the benefits of becoming a manager had always been that managers didn't have to travel. By rewriting the managers' job description, Darlene took away one of the most envied and enjoyed benefits of the job.

Darlene's level of authority and impact on the agency are dramatic. However, this does not mean than all newcomers with authority—even those who are interested in challenging and changing their organizations—will be successful in their attempts to foster change.

Peter's Unsuccessful Attempts at Change

Peter had a sense of personal authority based on his previous experience as an architect and as an architectural software specialist. He also gained confidence from his perception of the organization's need for his expertise. Peter was hired because of his unique experience (not many architects were as knowledgeable as Peter about software systems), and he was initially granted social authority based on his employer's recognition of the expertise gained through that experience. In fact, Peter was explicitly hired to be an agent for organizational change: ECI planned to upgrade their internal computer systems, and they were counting on Peter to assist in this major project.

In one of his initial assignments, Peter was asked to help draft a "strategic plan" that would allow ECI "to become the foremost EA [Engineering and Architecture] firm in the area with regard to state-of-the-art technical expertise." Peter spent a great deal of time on this document, which was reviewed by individuals at many levels of the firm. Eight months after the drafting of the plan my notes show that

> I asked what happened to the strategic plans that each group had to write. Peter says he doesn't think anything was done with them. He thinks the problem came because each department did its own plans, but no one had the ability to produce a unified plan from all the different pieces.

Peter's frustration at his inability to effect change was, by this time, causing him to think about changing jobs or possibly going back to school for an MBA. He felt that the firm's inability to make good decisions stemmed from a president who was "an engineer first and a businessman second—and not a very good businessman at that."

Peter's cynicism about the organization developed gradually, and can be seen as stemming from an organizational conflict over espoused goals and values and the actual goals and values adhered to by most of his colleagues. The new president (he became president just a few months before Peter was hired) had a vision for the organization, a vision that he hoped would become reality through the creation and enactment of the strategic plans. To carry out this goal, the president had frequent meetings where he espoused a "new direction" and several times announced a restructuring of the organization.

The restructuring was not well received by many of the employees. As Kurt, the marketing representative for the architecture group, told me, "The groups have, for the most part, not been happy with the new restructuring—there is a lot of resistance. The primary cause of resistance is the refusal to admit that they [the units merged with other work groups] were weak units."

Peter, however, embraced the restructuring. He told Kurt that he interpreted the president's actions as an opportunity to work together and make [the new group formed by the merger of four units: architecture and structural, mechanical, and electrical engineering] a strong group. Peter further saw himself as central to the new direction of the group: "Peter told Kurt that the president referred several times to the approach that he, Peter, has been taking—joint venture projects to help build the expertise and portfolio of the group."

Although Peter had been hired for his computer skills, the president had recognized his experience in "joint venture projects," particularly within the health care field. The president saw Peter's expertise as useful in the fulfillment of his plans for the organization.

Peter's sense of personal authority and his awareness of the social authority being granted to him for his expertise in these areas was frequently reinforced. For example, Marge told Peter that she wanted him to attend a conference on architecture for the health care industry that would be held in Florida in February. "She says that the president's approval [for Peter to attend the conference] is one more sign that he is behind their decision to specialize in this area [health care] and that he really is supportive of the architecture unit." Peter did attend the conference, and he came back optimistically full of ideas to help the firm move in the new direction.

Although Marge seemed to be supporting the president's plan, Kurt told Peter that Marge was "the biggest problem," the person in the unit who would keep the plan from working. The conflict between Marge's view of the organization and Peter's view provide an example of what Van Maanen (1976) calls "inefficacious socialization" (p. 107): Peter was not being effectively socialized to the values and goals of the organization because there were two sets, one espoused by the president and one held by the oldtimers. A theoretical note from the third month of my observation of Peter says:

> One of the few things that still marks Peter as a newcomer is his acceptance of the restructuring. He sees it as a positive step and is embracing it wholeheartedly. People who have been with the firm longer are skeptical at best, fearful at worst. . . . It will be interesting to see if Peter is able to share his enthusiasm for the upcoming challenge with others in the group.

In fact, what happened was that the skeptics refused to work toward the new direction imposed by the president, making Peter the outsider within his workgroup.

Peter behaved according to what he perceived to be the goals of the organization; his manager and those around him continued to behave in accordance with the "old" goals. This became strikingly clear in the discourse-based interviews I conducted with Peter and others in the organization. These interviews were based on four different types of documents drafted by Peter: call reports, job reports, client letters, and a client proposal. I conducted identical interviews with Peter and Marge and, since proposals are typically written by the marketing group, I also conducted the proposal portion of the interview with Walt, the director of marketing. The results of these interviews paint a disturbing, but enlightening, picture. I asked each participant to explain the significance of the differences between sentences written by Peter and sentences taken from the boilerplate proposal Peter had used as a model for his document. While I did not identify the source of any of the sentences during the interviews, Peter immediately recognized his own text. Marge and Walt did not know which sentences had been written by Peter.

- Both Marge and Walt actually preferred the boilerplate. And since they saw it as "better," they both assumed it had been written by Peter.
- Marge and Walt said the preferred version was "more professional"; Peter said, "The significance here is that the company had gone through a reorganization since the boilerplate was written, and it was important to reflect the new organization of the company."

Marge and Walt were continuing to do business as usual; Peter was trying to write to match the new direction of the organization.

Similarly, in the other sections of this interview, Marge and Peter disagreed about the necessity of including specific words or phrases in nine cases out of 14 that I had highlighted.[12] With this level of disagreement, it is not surprising that Peter grew frus-

---

[12] Specifically, of the 14 highlighted passages, Marge said seven were unnecessary; Peter said only three were unnecessary. Since the documents were written by Peter, it is probably understandable that he would say that most of the items were necessary. However, of the three passages that Peter said were unnecessary, Marge said two were necessary.

trated. As noted above, he became disillusioned with the president's ability to make the plan work. The organization never took advantage of the computer expertise they hired him for, his manager never fully supported the joint-venture projects he wanted to pursue, and, within a year of joining ECI, he had left the firm for another job.

THE ROLE OF INDIVIDUATION

Individuation emphasizes the newcomers' characteristics that dominate their role within the organization, and sometimes the results are exactly what the organization would anticipate: As expected, Luke became recognized as an expert marketing rep and Rebecca became recognized as an expert auditor. However, the way a person will be individuated cannot always be predicted: Darlene was individuated as an expert writer, even though she was hired as an auditor. Peter was individuated as someone swimming against the stream, even though he had been hired as an architect and a computer specialist who could lead them in a new direction. The other two newcomers who participated in this study, Katy and John, did not individuate as quickly as the others.

Although I cannot say with certainty why this is so, I would suggest that both Katy and John were less secure—had less personal authority—and thus were less likely to take advantage of opportunities to demonstrate their abilities or challenge their organization than any of the other newcomers in this study. Their lack of personal authority may be attributed to their lack of experience and skill in both the "domain content space" (Katy had only four months previous experience working in the field of auditing, John had none) and the "rhetorical process space" (Geisler, 1994). Katy and John were not perceived by the supervisors at the Agency as being particularly skilled writers, a perception that was fostered in the writing review sessions of their work. In those sessions, not only did their writing require significant revision, but the supervisors also noted that neither of these newcomers questioned the changes to their writing, defended their text, expressed opinions about what they had written or about their supervisor's suggestions, or offered their own suggestions for revision. Neither Katy nor John was individuated as rhetorically skilled; in fact, neither of them demonstrated any unique characteristics that were valued by the supervisors at the Agency.

The way in which people are individuated has significant consequences for their careers as well as for the organizations where they work. However, individuation has not been examined previously from either a personal or organizational point of view. By examining the opportunities for individuation that exist within the writing review process, I hope that I have begun a conversation that will help prepare people to recognize opportunities to question, challenge, and change their organizations.

DEVELOPING YOUR UNDERSTANDING

1. **Identify two of the key features from Darlene's and Luke's organizational contexts that helped them gain organizational authority and explain how these features assisted them in doing so.**
2. **Katz refers several times to "review," the "review process," or "nonacademic writing review." Describe the review process and its functions, referring to the Darlene and Luke examples to illustrate your description.**

3. Of the three kinds of authority outlined by Katz, identify which you believe students are most capable of developing while in school, and which are more dependent on specific workplace contexts and/or experiences. Explain why you believe so. Finally, propose how students can build even the more workplace-dependent authorities while still in school, and develop a personal plan for doing so, according to your particular writing career interests.
4. Identify and illustrate the communication strategies, interpersonal and otherwise, that most helped the subjects in Katz's study become individuated. Refer to several specific examples to support your argument.
5. Evaluate who—Darlene or Luke—you believe gains the most authority within her or his organization. Be sure to clearly define authority and to support your assessment with specific examples.

FOCUSING ON KEY TERMS AND CONCEPTS

Focus on the following terms and concepts while you read through this selection. Understanding these will not only increase your understanding of the selection that follows, but you will find that, because most of these terms or concepts are commonly used in professional writing and rhetoric, understanding them helps you get a better sense of the field itself.

1. orality
2. literacy
3. target readers
4. intermediary readers
5. multiple audiences
6. task representation
7. strategy of isolation

Orality and Literacy in the Workplace: Process- and Text-Based Strategies for Multiple-Audience Adaptation

RACHEL SPILKA
University of Maine

What is the role of interaction, or, more generally, orality, in multiple-audience analysis and adaptation? How does orality relate to literacy in the evolution of corporate documents? A qualitative study of how seven engineers in two divisions of a large corporation wrote for multiple audiences revealed that, in the more rhetorically successful cases observed, interaction was the central means of analyzing and adapting discourse to multiple audiences, fulfilling rhetorical and social goals, and building and sustaining a corporate culture; and orality was more potent than literacy

Source: Spilka, Rachel. "Orality and Literacy in the Workplace: Process- and Text-Based Strategies for Multiple-Audience Adaptation. *Journal of Business and Technical Communication* 4 (1990), pp. 44–67.

in the engineers' composing behavior and the audiences' acceptance of the engineers' ideas and documents.

In recent years, a growing number of theorists have speculated that, in order to communicate effectively in a nonacademic setting, writers need to learn about the social contexts of rhetorical situations they encounter (Faigley; Odell). Correspondingly, more technical-communication researchers have been focusing their attention on how professionals in the workplace apply this social knowledge both to private composing decisions and to public interaction with such influences as collaborators, reviewers, clients, and users (Doheny-Farina, "Writing in an Emerging Organization"; Odell and Goswami; Paradis, *et al.*; Selzer; Spilka).

The advent of these theoretical and empirical probes signals a critical phase in the evolution of technical communication: At this time, more technical-communication specialists than ever before are joining rhetoricians in focusing their attention, not only on the composing processes of writers, but also on the special influence of social contexts and interaction in the invention of documents in naturalistic settings. In particular, they are demonstrating that, as nonacademic writers consider what to say in their documents and how to say it, they benefit from interacting with others (Dohney-Farina, "Writing in an Emerging Organization"; Odell and Goswami; Paradis, *et al.*; Selzer; Spilka).

Yet, how much is really known about the role of interaction, or more generally orality, in the production of documents in the workplace? For example, exactly how do nonacademic writers benefit from interacting with others during invention? To what extent and in what ways does interaction influence writers, readers, and other project participants as a document evolves? How does the role of orality compare to that of literacy in the fulfillment of rhetorical goals in the workplace? I use orality to refer to the process of transmitting ideas via any conversation or message between project participants that involves speech (e.g., conversation in person or on the phone) or via written forms resembling speech (e.g., electronic mail, written notes sent between writers and readers, or comments written in margins). Literacy refers to the process of transmitting ideas via any written materials used primarily in isolation rather than for conversational purposes—materials such as preliminary drafts, final products, source documents, outlines, and planning notes used by writers while working on subsequent drafts.

In a naturalistic study completed in spring 1988, I used converging, qualitative measures to examine how seven engineers in two diversions of a large corporation adapt discourse to multiple audiences. In this study, I explored two questions in some depth:

- What is the role of orality in the composing processes of corporate writers?
- How does orality relate to literacy in the evolution of corporate documents?

While analyzing patterns emerging from this study, I discovered that, in the five more successful rhetorical situations observed, orality played at least four significant roles in the writers' composing processes:

1. *Orality was the central means of analyzing multiple audiences.* The more rhetorically successful writers were those who interacted with readers and

other project participants throughout the composing process to update, revise, and make more accurate their initial impressions about their rhetorical situations (including audience needs, orientations, and preferences) and to expand their knowledge of social contexts characterizing these situations.

2. *Orality was the central means of adapting discourse to multiple audiences*. In the more rhetorically successful situations, the writers interacted with readers throughout the composing process to negotiate toward a consensus about rhetorical decisions (e.g., purposes and content) affecting the document in progress. Writer/reader interaction became the writers' central means of appealing to most audience members.
3. *Orality was the central means of fulfilling rhetorical goals*. In the more successful cases, interacting with readers throughout the composing process was the primary way writers detected incompatible or conflicting perceptions or goals between themselves and readers and among audience segments. With this social knowledge, the writers then interacted further with readers to resolve incompatibilities and conflicts and to respond to those questions and concerns that mattered most to those readers. Interaction was more powerful than drafts in allowing writers to achieve compromises and fulfill rhetorical goals.
4. *Orality was the central means of fulfilling social goals and building and sustaining the corporate culture*. By using interaction to achieve a corporate consensus concerning documents' purposes, contents, and other rhetorical features, the writers helped their corporation fulfill social goals, such as encouraging collaboration and working toward compromise. Simultaneously, these writers also assisted the corporation in developing and sustaining its unique cultural values, conventions, and taboos.

This article will discuss these roles of orality in detail. In doing so, some salient process-based and text-based strategies effective for multiple-audience analysis and adaptation will be introduced, along with a model for audience adaptation that technical-communication specialists might find useful in their own explorations of the roles of orality in nonacademic writing or in classroom instruction. In addition, the article will analyze how the oral culture observed at this corporation reinforces Bitzer's theory of public knowledge, in which he speculates that orality, in the form of continuous interaction, is essential in initiating newcomers into an established community and in aiding a community in the growth and continuation of its culture.

THE STUDY'S SETTINGS AND METHODOLOGY

This study was conducted in two divisions of Forbes Electric (all names used in this article are fictitious), a large northeastern corporation. So that conditions could be optimized for detecting significant cultural and social differences that might affect composing behavior, the study was conducted in divisions with radically different functions. At one division, most engineers conceptualize and communicate ideas. Here, the engineers invent more efficient processes and then train other employees throughout the corporation in the use of these processes. In contrast, the second division has a "line" or production function. Here, most engineers are busy producing nonnuclear energy products.

Also in order that detection of interesting cultural and social differences could be optimized, case studies were conducted of six writers, three in each of the two divisions, as they worked on single documents aimed at multiple audiences. (Later on, one of the six writers handed over his project to another engineer, who became the seventh subject of the study.) The subjects selected were all engineers who had the time to meet with the researcher, were willing to participate in the study and help in data collection, and were scheduled during the next half-year to be the primary writers of documents aimed at multiple audiences. The six documents focused upon were a tutorial, monthly report, operations manual, business plan, recommendation report, and business-unit newsletter.

For each case study, I used methodological triangulation for data collection. Using a variety of research measures was a way, first, of approaching each rhetorical situation from multiple perspectives, which was necessary to compensate for the weaknesses of each research measure used in the study; and, second, of detecting significant patterns occurring within each case study, as well as across several or all of the case studies conducted.

So that the influence of orality on the subjects' composing behavior could be more fully appreciated, the engineers were observed throughout the composing process. This study, therefore, supplemented the traditional pre- and postcomposing process measures (i.e., surveys and interviews) with a group of process measures that had potential to allow for close observations of the engineers during invention. In each of the case studies, I used as many research measures in the following phases as the engineer and division sponsor would allow and as many as would help me detect patterns that would lead to answers for my research questions.

Phase I: Preliminary Observations and Data Collection

In this phase, traditional precomposing process measures—surveys and interviews—were used to gather initial data about the engineers and their environments. First, surveys were conducted of the professionals and managers at each of the two divisions, to gather information about their social and rhetorical environments. Open-ended interviews were then conducted with the subjects to learn about their communication histories, composing behaviors, perceptions of and attitudes toward readers, and predictions about their current rhetorical situations. In addition, as many documents as possible were collected. These documents were ones that the subjects had written in their current or previous jobs at the corporation.

Discourse-based interviews concerning these documents were also conducted. In these interviews, writers were asked about specific composing decisions they had made in texts, such as why they had phrased sentences in certain ways, selected certain words, or chosen particular details. Typically, writers were asked about different choices they could have made at particular junctures of the texts. In the discourse-based interviews conducted in this study, special attention was paid to how composing decisions might have been influenced by social constraints.

Phase II: Observations of the Writers at Work

In this phase, the writers were observed performing normal job responsibilities at work. At the first division, two mornings and two afternoons were spent observing any interaction the writers had with other people. At the second division, I was not allowed to spend a complete morning or afternoon at the site and compensated for this difficulty by training each writer's secretary to be a research assistant who would record the information needed.

Phase III: Observations of the Writers' Composing Processes

Phase III involved using the following process measures to observe the writers' composing processes as they worked on their documents:

- attending as many informal and formal meetings as possible between each writer and project participants, and, soon afterwards, conducting open-ended interviews of all participants about what had happened at those meetings
- asking each writer to record in a small index-card book (kept next to a desk phone) information about each telephone and in-person interaction he had with a potential reader or project participant
- asking each writer to collect printouts of all e-mail concerned with the document, copies of all drafts and notes, and written comments returned to him by reviewers
- conducting discourse-based interviews with each writer of how decisions made in drafts might have been influenced by social constraints characterizing their environments or rhetorical situations
- whenever possible between each major draft, conducting open-ended interviews with each writer and his readers about possible social constraints on composing decisions or readers' reactions

Phase IV: Postcomposing Process Measures

After each writer completed the document (or, in some cases, when documents were aborted for various reasons), open-ended interviews with the writer and intended readers were conducted concerning social constraints that may have affected their perceptions, decisions, reactions, and interactions during the document's evolution. In addition, discourse-based interviews with each writer were conducted concerning perceptions about social constraints that may have influenced production of the latest draft of the document.

Methodological triangulation was also used to analyze the data collected in each rhetorical situation observed. First, study participants were asked, as often as possible, to listen at the end of interviews to my interpretation of what had been said, and, at those junctures, adjustments in the data were made, as needed, to ensure accuracy. Once all the data had been collected and adjusted for accuracy, categories were developed from the data, relationships between categories (themes) were detected, and then relationships between themes (models) were established, in a standard procedure described more fully in Doheny-Farina (*Writing in an Emergent Business Organization*). For example, in my study, one category that emerged from the data was called "writers' perceptions of readers." One relationship, or theme, that was detected between this and other categories was called "writers' perceptions versus reality." Analysis of this theme involved examining instances to see whether the writers' perceptions about readers (e.g., reading behavior, purposes for reading, attitudes toward the text) matched what the readers reported (e.g., methods of reading the text, purposes for reading, attitudes toward the text).

RESEARCH FINDINGS

The model that emerged between the theme "writers' perception versus reality" and another theme called "orality versus literacy" suggested that writers who planned mostly in isolation tended to have inaccurate perceptions about their rhetorical situations. In contrast, writers who interacted continuously during planning tended to emerge with accurate perceptions about their rhetorical situations. In order to relate in detail this important finding, I will first discuss typical rhetorical strategies employed by engineers who produced documents mostly in isolation and then the strategies employed by engineers who interacted continuously throughout the composing process.

The Rhetorical Strategies of Engineers Who Produced Documents Mostly in Isolation

The two engineers in the study who did not communicate effectively with the majority of their multiple audiences were the only ones who made conscious decisions to work on their documents mostly in isolation. These engineers planned their documents and wrote first drafts on their own, and then showed these drafts to just one or two reviewers. (This kind of review was a social expectation at both divisions of the corporation.) The engineers then ignored most of the feedback received from reviewers as they revised their documents—again in isolation—before distributing second drafts to target readers.

Because these engineers worked on their documents mostly in isolation and paid little attention to any feedback that came their way, they virtually ignored what was going on in their current rhetorical situations. They also ignored any social or rhetorical changes that might have taken place since their previous rhetorical situations or since the start of their current projects. As a result, they based their composing decisions primarily on what they could recall about the past. As these writers planned, wrote, and revised their documents, they focused on their recollections of past purposes, past audiences, and past constraints. Therefore, instead of thinking in terms of the particulars of their current situations, these engineers were thinking of the particulars of past situations; instead of considering how to solve current problems, they were thinking of how they had once solved other, quite different problems in the past.

Because these engineers were thinking in terms of past rhetorical situations, they brought to their current situations content and social knowledge that had long since become outmoded and inaccurate. By relying on knowledge rooted in the past, the engineers made composing decisions that their readers considered socially insensitive and rhetorically inappropriate.

This study suggests that if corporate writers apply outmoded content or social knowledge when making composing decisions, they may have

- inaccurate perceptions of their readers and rhetorical situations
- different task representations from those of their readers
- a lack of awareness of changes in social and political constraints

All three of these problems can alienate readers.

Inaccurate Perceptions of Readers and Rhetorical Situations

First, writers working mostly in isolation can bring to their projects inaccurate perceptions about their current readers and rhetorical situations—perceptions that can lead to composing decisions that are insensitive to readers' needs, orientations, or expectations. In one case in the study, Don, working on the second draft of a written tutorial, assumed his intermediary readers (members of a project team) would not care at all about the tutorial and would not read it: "They're equally disinterested," Don said. As he revised the tutorial, Don decided he would encounter no resistance if he ignored most of the ideas presented by the project team at a recent meeting. As it turned out, however, the project-team members did care about the tutorial; most read the second draft quite carefully, and, in interviews, most expressed dismay that Don had decided to incorporate just his own ideas into the draft. These readers felt that, despite Don's revisions, the document remained insensitive to their viewpoints. Some representative comments were as follows:

> It's not a shared view. The document still needs to show more consideration for the ideas of others.
>
> Meetings haven't been an influence on Don. Don has more influence on himself than anyone else can have. This is Don's document.
>
> Don's enthusiastic, but he doesn't perceive others' viewpoints. Too soon old, too late smart. Don has a limited perception of what the real world is. Don doesn't always know what's right, although he thinks he does.
>
> When Don writes alone, his writing is clean, crisp and precise, but not complete. It's a bit parochial in view, and lacks depth of content. Writing successfully here is not intuitive—it comes from training and exposure.

Similarly, in another case, Bill had been writing monthly reports for over two decades on a project he managed. While working on one particular monthly report at the end of 1987, Bill indicated that he was relying on the same content and social knowledge he had used to produce these reports two decades ago, and he also assumed the clients receiving the report were as interested in the research details as his previous readers had been. However, the clients, when interviewed later, indicated they had little interest in research details and felt the report did not address their major concerns regarding the project:

> So, what's the conclusion? If you can't tell me what's happening, why am I reading this?
>
> It doesn't tell me what I need to know.
>
> Bill's reports aren't as well tuned to my needs as the reports I'm getting from other contractors.

According to one of Bill's readers (a manager at another division), Bill tends to alienate current readers by continuing to think in terms of past audiences he used to address several decades ago when working on journal articles:

> Bill's early stuff is famous and quite good. Here, he's still writing for physicists, but he has a different audience, so this doesn't work as well as his journal style. Here, he's writing for program managers and engineers. They do not want to see the scientific method demonstrated. They want to see progress toward commercializing a project.

Different Task Representations

In addition to bringing to writing projects inaccurate perceptions about readers and rhetorical situations, a second problem with a strategy of isolation is that the writer can emerge with task representations that differ substantially from those of readers. A task representations can be defined as the perception someone has about the fundamental features of a rhetorical situation, such as the expected purposes, audiences, and form of a document. If a writer's task representation does not match those of intermediary readers, the writer is likely to produce a document those readers will consider inappropriate and unacceptable.

Again, probably because Don worked mostly in isolation and ignored feedback and ideas from the project team, his task representation matched that of his former boss (who had just transferred to another corporate division) but not those of his current readers. According to Don, the tutorial's major goal was to assemble in one document the project teams' consensus about conceptual models. The audience would consist of just the project team, and the appropriate form would be an outline of major points covered at the meetings. However, unknown to Don, most members of the project team wanted the tutorial to persuade upper management to approve the project team's consensus of the document's message and instruct engineers about the message. Most team members, after reviewing a draft of the tutorial, did not feel it had achieved these purposes or was destined to meet the needs of its future audience. They also considered the document's form inappropriate: They felt a formal report, not an outline, would be the best way to demonstrate to upper management that the team members had taken the corporation's mission seriously and that they had completed a thorough process of coming to a consensus about the corporate message. Because they viewed Don's draft as a major departure from what they had hoped to achieve with the document, they reacted quite negatively to the draft, considering it unacceptable for the purposes and target audience they had in mind for the document:

> We wouldn't send this to the outside. We need to think in terms of the customer [upper-level managers].

> There are no direct links to how the internal customers would use this, or to their concerns.
>
> It won't be clear to future users.
>
> It's not ready for VPs yet.

Lack of Awareness of Changes in Constraints

A third problem with a strategy of isolation is that writers may remain unaware of changes in social and political constraints—conventions and taboos—caused by rhetorical, cultural, and environmental shifts that have evolved over time (since a previous project or during the current project). In the case of Bill's monthly reports, several clients mentioned that Bill seemed out of touch with the economic problems that had emerged during the past decade—problems causing the clients to question whether to continue financing the project. Several clients said Bill's reports typically omitted what concerned them most about the project: evidence of any progress from one month to the next that might convince them to reinvest in the project at the end of the fiscal year. One client said he might not reinvest, but if he did so, he would stipulate changes in the contract that would compel the monthly reports to focus on his major concerns:

> I need a task by task breakdown. When I renegotiate with Forbes, in that contract, I'll specify that I want the report written task by task. I look for difficulties on problem areas that could have an impact on future progress, a forecast of next month's activities, and a monthly incurred cost form, to compare that to what's been done.

Unfortunately, not just Bill's clients, but also the majority of his internal readers—many of them colleagues and managers—felt his reports were focused on the past, not on current problems and concerns. As a result, although most of Bill's readers respected his intelligence and scientific expertise, they usually turned to others at the corporation, not to Bill's monthly reports, to become informed about how the project was going and whether it seemed to be fulfilling corporate or division goals.

Along with these three problems, which led to the failure to satisfy the rhetorical goals of their projects, Don and Bill's strategy of composing in isolation resulted in two social difficulties. First, their strategy of isolation did not meet the social requirements of their corporation. A fundamental social belief at Forbes Electric is that writers need to interact with others during the production of a document for purposes of acculturation and social acceptance. At the divisions observed in this study, continual interaction during a project was considered instrumental to a writer's process of learning about the division's and corporation's social contexts. In several cases in the study, intermediary readers indicated they had been asked by upper management to observe the ability of a primary writer to interact successfully with others during the project. Writer/reader interaction, therefore, can become a sort of initiation ritual, in which writers who interact frequently and effectively are more likely to be socially accepted and promoted in the firm than writers who isolate themselves and ignore what others think, say, and do.

Second, because Don and Bill chose a strategy of isolation, many of their readers expressed personal resentment at having been excluded from the rhetorical process. Many felt socially shunned and were concerned that the writers had not recognized what these readers might have contributed to the document—in particular, to rhetorical decision-making influencing what was said in the document, as well as how it was said. According to several readers interviewed, regular consultation during the composing process signals that the writer cares about what they think, and no consultation during the process probably means the writer doesn't care about their ideas—a conclusion, they said, that contributes to their negative assessment of the writer and document.

The Rhetorical Strategies of Engineers Who Interacted Continuously Throughout the Composing Process

Figure 1, an audience-adaptation model, illustrates the most effective strategies used by the five engineers in the study who met with more rhetorical success than Don and Bill. Although no single engineer used all of these strategies, each engineer in the study except Don and Bill used different combinations during invention. (Note that the steps in this model can be recursive, overlapping, and repeated indefinitely according to what writers deem necessary or most effective.)

This study suggests that interaction can help writers produce effective documents by enabling them

- to revise the initial multiple-audience analysis.
- to adapt discourse to multiple audiences and fulfill rhetorical and social goals.
- to resolve competing or incompatible rhetorical goals, using both process-based and text-based strategies.

Revising the Initial Multiple-Audience Analysis

Steps (A) and (B) in Figure 1 illustrate how corporate writers, after completing the initial rhetorical analysis, can choose to interact with others to revise that analysis. In this way, they make the analysis more accurate and therefore more sensitive to current audience needs, orientations, and expectations, and to the social features characterizing the current rhetorical situation.

Interacting with Readers to Learn about the Rhetorical Situation and to Revise the Initial Rhetorical Analysis

As step (A) in Figure 1 shows, a useful strategy in the first phase of audience adaptation is to delay planning documents until after consulting with readers and other possible project participants to learn about the particular features of new rhetorical situations. This new knowledge helps writers revise, refine, or extend their initial rhetorical analysis, so they can plan intelligent, accurate, and socially sensitive strategies for appealing to readers in particular rhetorical situations.

For example, writers can interact with readers to learn about the people in various locations who are likely to look at the document or be affected by the document's

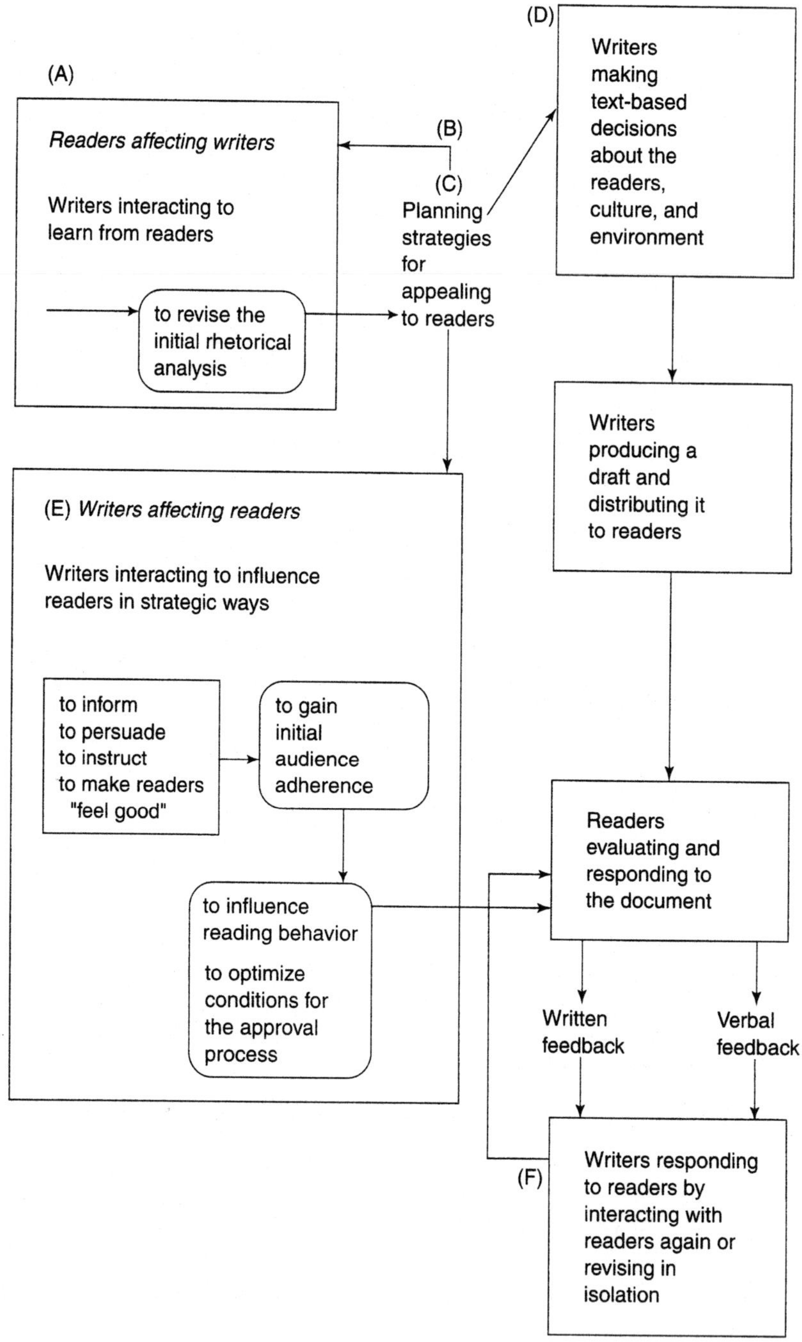

Figure 1. An Audience-Adaptation Model.

message. In consultation with their readers, writers might seek answers to questions like the following:

- What are the unique concerns of these future readers?
- What social, political, economic, and environmental constraints will affect the way they respond to the project or document?
- What are their task representations: What expectations do they have about the documents' purposes, audiences, and forms?

Writers can also inquire about the special constraints, conventions, taboos, and other features of their rhetorical situations:

- How does this situation differ from past rhetorical situations?
- How might it relate to future rhetorical situations?
- How have recent economic, social, or political shifts affected current rhetorical situations, and how might these shifts affect future rhetorical situations?

As shown in the model, readers primarily affect writers during this juncture of the composing process.

Delaying Planning to Interact Further with Readers and Expand Knowledge of Rhetorical Situations

On the basis of what the writers learn from readers (and possibly from others), the writers typically begin building fuller, more accurate constructs of their audiences and rhetorical situations. [See step (B) in Figure 1.] However, in this process, writers may discover gaps in knowledge—information salient to their documents and rhetorical situations—that they would like to understand better before working on their documents. For example, writers might have learned from earlier interactions that more readers will be involved in the communication process than originally expected, that certain constraints (e.g., time, money) are going to be influential, or that certain changes in relationships (e.g., a switch in social roles during the current project) might alter the document's outcome significantly. If writers decide they need to learn more about their rhetorical situations, they might cycle back and conduct more interactions with those people who can teach them what they need to know to make rhetorically and socially sensitive composing decisions.

Adapting Discourse to Multiple Audiences and Fulfilling Rhetorical and Social Goals

As step (C) in Figure 1 illustrates, perhaps while—but most certainly after—the writers conduct their initial rounds of interaction and revise their initial rhetorical analysis, they begin to plan composing strategies. Typically, they decide to take one of two routes: a quasi-independent route or a quasi-dependent (and socially interactive) route.

The Quasi-Independent Route: Text-Based Strategies for Multiple-Audience Adaptation

As step (D) in Figure 1 shows, the quasi-independent route involves working once again in isolation to make composing decisions. However, unlike those writers who have chosen up to this point to work in almost complete isolation, these writers have just completed a series of writer/reader interactions [see step (A) in Figure 1] to learn about the particular features of their current rhetorical situations. Therefore, although the writers are now working mostly on their own, they can use the knowledge gained from the social interactions they have just completed to establish reader-based goals for their documents. With this approach, these writers are more likely than those who have worked in isolation from the very start to produce drafts perceived by readers as appropriate to their rhetorical situations.

During this quasi-independent phase, the more effective writers in this study began by categorizing the expected readers of their documents, based on their new social knowledge of these readers. Then they applied a number of text-based strategies aimed at appealing to those various reader categories. (See Spilka for a fuller account of these strategies.)

1. Categorizing Readers Before Applying Text-Based Strategies

Categorizing readers in a way sensitive to the social or political features of the current rhetorical situation seems an important part of adapting texts effectively to multiple audiences. This study suggests, however, that categorizing readers according to organizational roles (e.g., managers, technicians, engineers) can be a mistake, because doing so can lead to stereotyped impressions that, in turn, can lead to false or misleading generalizations about readers and socially insensitive composing decisions. For example, Don assumed his intermediary readers, categorized by him as "all managers and supervisors," would think and act just like the managers and superiors in his previous Forbes job (at another division), and would be as "disinterested" in his drafts as his former bosses used to be. Similarly, Bill, working on his monthly report, assumed that his managerial readers would be too busy to look at all sections of the report but that the technicians, engineers, and clients would have sufficient time to read all sections slowly and thoroughly.

However, those readers interviewed in the cases of Don's tutorial and Bill's monthly report didn't think and behave the way the writers had assumed they would. Most of Don's managerial readers were extremely interested in his tutorial, because of its importance to the corporate mission. Also, in all seven cases in the study, the stereotype that managers read documents in a certain way, while technicians, engineers, and other types of employees read in other ways, didn't hold true; instead, managers were similar to other types of readers in choosing to skim or read documents thoroughly depending upon their interest in the subjects or individual purposes for reading the documents. For example, if a manager were responsible for reviewing a draft, he or she might read that draft more thoroughly than a client just interested in knowing how project money had been spent in the past month.

Thus, this study suggests that categorizing readers according to their organizational roles can lead to a focus on just one audience segment or feature; to mistaken impressions about readers' needs, expectations, attitudes, and behavior patterns; and to a limited audience analysis that certainly can result in composing decisions destined to alienate readers.

The more successful writers in this study typically categorized their readers in a way that was socially sensitive in their particular rhetorical situations. For example, one engineer found that, in his situation, distinguishing between readers with and without technical knowledge made the most sense, whereas another engineer decided to distinguish between intermediary and final readers while working on a document. Often, these writers would think in terms of the readers' social and political agendas and purposes for looking at the documents; typically, these writers would categorize readers according to the following features:

- the readers' attitudes or levels of caring regarding the documents' contents. (For example, in what way and to what extent are these readers involved in these projects? Which readers seem not to care at all, which ones seem to care just a bit, and which ones seem to care a great deal?)
- the level of feedback on social and political constraints that readers would provide.
- the amount of power readers might hold and the ways readers might think or behave during or after the composing process.
- the readers or types of readers who, for various social reasons, should be avoided or contacted indirectly during the composing process.
- the juncture of the composing process at which various types of readers would be most likely to look at the document. (That is, some readers were likely to look at first drafts, some at second drafts, others at finished versions.)
- the degree to which readers were likely to be thinking in terms of how other, future readers would respond to the documents, and why.

2. Determining Objectives for Each Audience Segment

After establishing reader categories, the more successful writers decided on central objectives or goals they had in appealing to readers in the various categories. For example, one engineer decided he would inform one audience segment, but both inform and persuade another segment.

3. Using the Means/Ends Approach for Choosing Text-Based Strategies

Some of the more successful writers used a means/ends approach to select the strategies that would appeal most effectively to various segments of their multiple audiences. The first step in this approach is to determine a major goal for each important audience segment; for example, a writer might decide to inform an external audience and to persuade an internal audience. The writer then chooses one array of strategies (such as use of metaphor, detail selection, word choice) to meet the first goal of informing the external audience and a second array of strategies (such as detail selection, choosing a certain format, arranging material in a certain way) to fulfill the second goal of persuading the internal audience. Some strategies can help a writer satisfy more than one goal for more than one audience segment.

4. Finding Bridges

Writers can try to include in their drafts features that they might have in common with readers and that their readers might have in common with other readers. For example, one engineer in the study thought all his readers would be interested in a sports-team metaphor and included that metaphor in his draft, while another engineer suspected most of his readers would enjoy physical descriptions of a location they had never before visited and wrote about that location in his document.

5. Designing the Document to Control Reading Behavior and to Focus on Salient Issues

This study suggests writers can manipulate the design of their documents to promote any type of reading behavior (e.g., careful reading, scanning, skimming, skipping over text), according to the goals writers have for readers in their particular rhetorical situations. For example, one writer decided to align headings along the top of a newsletter page to influence nontechnical readers to skim that page, and predicted technical readers would still read all of it carefully. Similarly, another writer chose small type for the first draft of a document to discourage intermediary readers from paying too close attention to the text and to encourage them to approve it with few and minor suggestions for revision. Later, the writer enlarged the type so final readers would read the completed document more closely.

The Quasi-Dependent and Socially Interactive Route: Process-Based Strategies for Multiple-Audience Adaptation

Before writers begin to write or as they write, they might decide to interact further with readers and other project participants, but this time for a new reason. [See step (E) in Figure 1.] Instead of interacting to learn about the audience and social contexts, the writers now interact with others to gain initial audience adherence, not only to their own approach to the project, but also to the documents' intended purposes and contents. For example, before writers show readers drafts (or even start working on these drafts), they can use interaction as a tool for informing, persuading (negotiating with readers toward a consensus on an idea), or instructing readers about material they would like to include in the upcoming document.

Writers can also use interaction during this phase of the composing process as a way to make their readers "feel good" about the project, the writers' ideas, or the documents' contents. One effective method for helping intermediary readers feel good about a document is to ask them to participate in the composing process. In this study, most readers seemed quite pleased when writers asked them to help gather information, write or revise sections of drafts, assist with illustrations, or attend meetings or oral presentations related to documents. For example, Charlie, working on a newsletter, noticed that a particular manager from another division tended to disagree with Charlie's ideas; once Charlie asked the manager to help collect information for the newsletter issues, the manager voiced no further complaints. Similarly, Philip, working on a recommendation report, knew he needed the approval of a newly appointed manager greatly admired by the division asking for the report; Philip thought that if he could please the new manager, he could please several others in that division. By asking the new manager to become a coauthor of the report, Philip succeeded in obtaining her approval of early drafts, which, in turn, helped him secure the approval of others in that division.

Another method that helps readers feel good about a document is to take the initiative to talk informally or formally to them during the composing process about matters that seem to especially concern them (e.g., the project, strategies used during the composing process, drafts, the final product). For example, as Charlie planned what to include in a newsletter issue, he consulted frequently with members of the council sponsoring the newsletter to make sure they agreed with his plans.

Using social interaction as a strategy to gain initial audience adherence seems to have beneficial effects later on when readers receive and evaluate documents. When readers notice from a draft that the writer has listened to what they have said during interaction and has incorporated their ideas into the document, they seem more likely to trust the writer's judgment. This study suggests that readers who trust a writer and who are familiar already with much of what they see in the document (probably because of previous interaction with the writer) tend to read and approve drafts quickly; examine closely just certain document features or sections the writer had mentioned to them earlier; and make just a few suggestions for revision, usually on relatively low-level concerns requiring minor alterations at the word or sentence level. As one engineer in the study pointed out, gaining the trust of readers—both during the project and gradually, over time—is an important strategy for making the entire document-cycling process more efficient.

Besides optimizing conditions for the upcoming review of the document by gaining audience adherence, interacting with readers during invention can fulfill important personal and social goals for writers. For example, when readers receive documents, they might reflect on how writers have handled the projects so far. If social interaction has been successful, the readers might recall how sensitive writers had been in the way they had consulted others at the corporation and how flexible they had been in adapting to audience needs. If people in power at the corporation are pleased by the writers' interaction with others during the project, they might reward the writers by putting them on challenging or prestigious projects or by recommending their work to others looking for talented employees for special projects.

As step (F) in Figure 1 illustrates, interaction can also aid writers after drafts have been produced and distributed to readers. After evaluating a draft, readers typically respond in one of three ways: They remain silent, provide verbal feedback, or provide written comments. Sometimes, readers provide both verbal and written comments or follow a long period of silence with verbal or written responses.

This study suggests that, regardless of the readers' responses or lack thereof, writers at Forbes Electric can benefit from taking the initiative to interact in person once more with readers after giving them drafts. For example, several writers in the study interpreted readers' silence as indicating approval, when, in fact, the readers had responded negatively to the drafts and had decided (for various personal, social, or political reasons) to remain silent about their opinions. In several other cases, writers received verbal or written comments unaccompanied by explanations of the comments or suggestions for strategies the writers might use to resolve the problems commented upon. In one case, the engineer's program manager wrote in the report's margins that the background section needed to be more "reader-oriented." The engineer, who had assumed that section was already reader-oriented, didn't understand why the manager had made that comment, which readers the manager might be thinking about, and how the engineer could revise the section to the manager's satisfaction. In an interview, this manager said he lacked the time needed to write down explanations of his comment but would be willing to talk to the engineer in person about any comments written in the margins. Therefore, corporate writers might benefit from interacting with readers after producing early drafts for clarification of why readers have suggested certain changes, for assistance with strategies for improving documents, and for amplification about the needs of future, intended readers.

Resolving Competing or Incompatible Rhetorical Goals, Using Both Process-Based and Text-Based Strategies

At Forbes Electric, writers often encounter competing or incompatible goals in their rhetorical situations and need to make choices in order to resolve the conflicts or overcome the incompatibilities. For example, a writer might feel pressure to fulfill the corporate goal of being efficient but might also want to delay a document's completion so the audience can have a chance to review and approve each draft produced. Or, perhaps the writer's time frame is incompatible with that of intermediary readers; for example, the writer might want to distribute a draft right away in order to progress to the next phase as soon as possible, but readers might be too busy to look at the draft until much later.

When a writer addresses multiple audiences, the number of competing and incompatible goals is likely to increase. The writer is not negotiating with a single reader, environment, and culture, but rather with multiple readers, environments, and cultures. Certainly, writers need to consider whether their personal goals are compatible with the goals of different readers or types of readers, but writers also need to consider whether the goals of readers will conflict with each other. For example, a writer might receive incompatible advice from readers who perceive the rhetorical situation in completely different ways or who make opposing suggestions for planning, writing, or revising. In addition, writers need to consider whether the goals of their culture conflict with those of the readers' various cultures and whether the goals of the readers' cultures might conflict with each other. For instance, the writer's culture may possess a primary goal of being efficient, while the readers' cultures may have other primary goals, such as ensuring quality in the products, improving customer relations, or producing greater quantities of goods.

How do corporate writers deal with conflicting or incompatible goals when addressing multiple audiences? Markel believes writers might find it impossible to please everyone in multiple audiences by manipulating document features. However, this study suggests writers can resolve many conflicting or incompatible goals by selecting from a variety of text-based strategies enabling them to please different types of readers with different types of goals. For example, writers might aim global strategies at one audience segment and local strategies at another audience segment or aim strategies for informing at some readers and strategies for persuading at other readers. The possible combinations seem endless.

Unfortunately, Markel's belief reflects his focus on text-based strategies of audience adaptation—a focus characterizing most technical-communication textbooks. This study, however, suggests writers can use both text-based and process-based strategies for resolving conflicts or overcoming incompatibilities. In one case, an engineer and an intermediary reader had conflicting ideas about how to arrange a newsletter format so readers would be encouraged to scan certain sections. The engineer used his own method of arrangement (a text-based strategy) and then interacted with the intermediary reader (a process-based strategy) to assure her that even though he had chosen his own text-based strategy over hers, the strategy chosen would succeed in fulfilling the same rhetorical goal she had in mind (i.e., promoting scanning). Both orality and literacy, therefore, can be used to resolve conflicts or overcome incompatibilities; to ensure most audience members will be pleased, writers can draw upon various combinations of process- and text-based strategies.

An Overview: Understanding the Roles of Orality in Multiple-Audience Analysis and Adaptation and in Fulfilling Rhetorical and Social Goals at Forbes Electric

The Primacy of Orality in Multiple-Audience Analysis

In this study, the more rhetorically successful writers were those who talked to readers and other project participants to update, revise, and make more accurate their initial impressions about their rhetorical situations. Interacting with readers early in the composing process also enabled these writers to expand their knowledge about the details of their own social contexts and those of various audience segments. Without interaction, writers at Forbes Electric probably would depend exclusively on their own personal observations and experiences—which typically are rooted in the past—for knowledge about current social constraints and cultural conventions and taboos. Also, these writers probably would depend primarily on guesses, assumptions, and personal observations in evaluating how the various environments and cultures of their readers might affect audience perceptions, attitudes, needs, and expectations. In effect, these writers might try to create knowledge without checking first with those who make and judge knowledge. At Forbes Electric, orality doesn't just help writers identify and learn about their readers; it is the principal medium by which they determine how readers' orientations and impressions differ from their own and then use this knowledge to plan intelligent, socially sensitive audience-adaptation strategies.

The Primacy of Orality in Multiple-Audience Adaptation

At Forbes Electric, orality, not literacy, seems the primary route by which the more successful writers build constructs of readers and rhetorical situations, plan reader-based strategies, write reader-based drafts, and make reader-based decisions during revision. Orality, therefore, contributes substantially to the way documents are produced at the corporation.

More specifically, at Forbes Electric, multiple-audience adaptation can be viewed as an exercise in revising, refining, and extending the initial rhetorical analysis. By interacting with others, writers learn the special features of particular rhetorical situations and then apply that knowledge in the form of process- and text-based decisions enabling them to meet specific rhetorical and social goals. This study suggests, too, that the social contexts of the corporation should have the greatest influence on composing decisions and behavior and that orality is the principal means by which the writers learn about and understand their social contexts and those of their readers. Writer/reader interaction, in particular, enables writers to learn about the special features of social contexts and use this knowledge to gain initial audience adherence to their ideas and documents, thereby optimizing conditions for audience approval of the writers' rhetorical decisions and drafts.

The Primacy of Orality in Fulfilling Rhetorical and Social Goals

This study suggests that, at Forbes Electric, orality is more than just a key part of the document-cycling process; in successful rhetorical situations at this corporation, orality is the primary way people communicate and reach consensus on the ideas and values of the corporate culture. For example, in most cases observed in this study, almost all learning and information exchange took place in writer/reader interaction preceding the production of the first draft, and most of the rest of the information was exchanged at informal or formal meetings or on the telephone after the first draft was produced and distributed. For example, one engineer's intermediary readers found "few surprises" in his recommendation report, because this engineer had made a point of interacting with these readers throughout invention to secure or ensure their initial agreement to the report's contents. In another case, an engineer's business plan was the result of a series of brainstorming sessions about the best marketing direction for the division; all ideas contained in early drafts of the plan had been formulated and developed previously in group discussions, and, if the document were approved by intermediary and intended readers, this approval would mean that the ideas contained in that document would represent the consensus opinion of the division.

These cases illustrate how, even though primary writers at this corporation ostensibly work on their own to produce documents, most writers collaborate with others throughout invention—primarily through social interaction—to formulate ideas; translate them into plans; receive from intermediary readers validation of the ideas and plans; and then produce a final draft which, if approved by readers, reflects the consensus of the corporate culture. At this corporation, the process of interacting is more influential than the product in fulfilling such corporate goals as developing ideas, helping professionals and managers reach a consensus on controversial issues, helping managers decide on favorable products or directions for division operations, helping users learn how to operate machines, inspiring employees to learn useful skills that will lead to improved processes, and instilling certain values in the thoughts and work decisions of employees throughout the corporation. Orality, in the form of writers negotiating with readers and of readers negotiating with each other toward new constructs of social reality, is the primary way that corporate writers, readers, and the composing process influence the culture and environment of this corporation.

Thus, at Forbes Electric, orality is more significant than literacy in fulfilling rhetorical and social goals. To one engineer, documents approved by audiences at Forbes serve mainly as evidence that previous social interaction has been successful in resolving differences between readers on controversial issues and helping them reach a consensus. This perspective seems accurate at this corporation. If a document at Forbes Electric is never read, this might signify that the document's rhetorical and social purposes already have been fulfilled via orality and that readers see no need to look at the document. Similarly, if a document is only partly read, this might signal previous success in persuading a reader, via orality, to take special note of the particular subject covered by the document. If a document is only scanned, perhaps this indicates that readers already know what the document contains because of previous discussions and that they only want to see if any particular subtopic or item might trigger an interest in learning more. At Forbes Electric, therefore, documents typically serve secondary functions. In this study, drafts served most often as backdrops to orality (e.g., previous meetings, formal oral presentations, or informal discussions), as a means of formalizing and confirming the consensus achieved earlier among readers and between writers and readers, and as a means of building corporate knowledge among those who hadn't yet participated in the rhetorical situation.

THEORETICAL IMPLICATIONS OF THIS STUDY: ORALITY AS A PRIMARY MEANS OF BUILDING AND SUSTAINING THE CORPORATE CULTURE

So that we might better understand how, at Forbes Electric, orality can be the primary means of creating corporate knowledge and of building and sustaining the corporate culture, this section will discuss the study's findings in terms of Bitzer's theory of public knowledge.

Bitzer believes public knowledge includes

> principals of public life to which we submit as conditions of living together; shared interest and aspirations; values which embody our common goals and virtues; our constitutions, laws, and rules; definitions and conceptual systems; criticism, philosophy, aesthetics, politics, and science; the accumulated wisdom proffered by our cultural pasts; and, to these, we add the personal facts of our public life.(87)

In terms of Forbes Electric, Bitzer's notion of public knowledge is similar to the notion of corporate knowledge. Bitzer's theory implies that, to become part of (or accepted by) the corporate community, a new employee needs to develop corporate knowledge.

To develop corporate knowledge at Forbes Electric, an employee needs to rely on continuous interaction with the corporate community. This finding is consistent with Bitzer's theory: Public knowledge, he writes, "can only be located in the public" (83).

According to Bitzer's theory, the employee also needs to interact with the public continuously, because the public (and public knowledge) is always changing. As Bitzer puts it:

> Public knowledge will change as new conceptions, values, and principles are added and old ones discarded, and as some of these recede to the background while others become dominant. (89)

As demonstrated in the study at Forbes Electric, with each new rhetorical situation comes a new set of salient features characterizing readers. For example, the issues that concern readers in one rhetorical situation probably won't concern them to the same degree in another rhetorical situation and might disappear completely and be replaced by new concerns. Writers at Forbes Electric, therefore, shouldn't consider initial corporate knowledge to be accurate or even relevant to their current rhetorical situations; instead, to keep up with what concerns readers in these situations, Forbes' writers need to keep interacting with these readers.

Also, as this study illustrates, interaction seems important, not only for gaining knowledge, but also for securing audience adherence to ideas and rhetorical decisions—for gaining a corporate consensus that will have the effect of building or sustaining the corporate culture. This finding is also consistent with Bitzer's theory. "Public transactions" (80), he suggests, are needed because the public will test and confirm public knowledge throughout the rhetorical situation:

> The public is called into existence in response to exigencies produced by public transactions; a growing and rich fund of knowledge, based partly in share means and interests and partly in scientific inquiry, will serve the community as it moves forward in time and the circumstances of its environment; and this fund of knowledge will be subject to testing and confirmation as the public struggles with exigencies throughout its life. Rhetorical communication will enrich the public's information, sustain its experiential knowledge, and provide modes of debate and discussion needed for intelligent decision and action. (80)

As Bitzer suggests, corporate knowledge "is authorized and can only be authorized by the public" (83).

This theory helps explain why writers at Forbes Electric need to discuss their ideas with members of their corporate culture during invention: Doing so helps these writers test and then confirm the ideas with the corporate community. The consensus of the corporate community about certain ideas will determine whether these ideas will be incorporated into the corporate culture or be rejected. Therefore, writer/reader interaction can be viewed as the primary means of building and sustaining the corporate culture.

Corporate writers, this study suggests, can play an enormous role in creating and transmitting corporate knowledge. For them, the process of writing has the potential to shape much of the meaning and direction of the corporation. In effect, the corporation runs on the basis of corporate knowledge. The process of writing, therefore, is more than just another routine task in a corporate environment; rather, it is part of what sustains the community.

At Forbes Electric, the process of writing is also a principal means of learning cultural knowledge. New employees learn about their culture mostly from interacting with others at the corporation, observing what people say and how they say it, and observing other social interaction. Most younger writers have not yet acquired sufficient knowledge about social and political constraints to enable them to make sensitive composing decisions based on this type of knowledge. Typically, they consult with senior mentors: employees who have interacted with others long enough to have gained a fuller appreciation of social and political considerations and to have developed effective process- and text-based strategies. This behavior pattern of consulting senior mentors supports Selzer's hypothesis that the longer an employee works at a company, the more knowledgeable that employee becomes about the company's culture. This pattern also suggests strongly that employees don't learn about their culture or about writing primarily from reading documents, although doing so can help reveal some text-based strategies. Instead, employees learn most about their culture and about writing in the workplace from other employees, via orality.

CONCLUSION

The findings in this study have several implications for the profession. First, we need to broaden our approach to multiple-audience analysis in the workplace: This analysis needs to be considered in terms of the social features of the current rhetorical situation, which can greatly influence readers' responses to the document in progress. At a corporation, conducting an accurate, socially sensitive analysis of multiple audiences probably requires interacting with project participants throughout the composing process to learn about these social features and then using this knowledge to revise, refine, and expand the initial rhetorical analysis.

Second, the profession needs to revise and expand its notion of multiple-audience adaptation. To make socially appropriate composing decisions that please the majority of readers in multiple audiences and that resolve conflicts and incompatibilities between multiple rhetorical participants, corporate writers can do more than implement a wide variety of text-based strategies; they can also influence readers and other project participants in positive ways by interacting with them at strategic junctures throughout the composing process.

Finally, the profession needs to develop a fuller appreciation of the multiple roles of orality in fulfilling rhetorical and social goals. Writing in the workplace is an interactive, ever-changing rhetorical and social process—one that consists of an ongoing chain of verbal and written interactions that, when combined, enable rhetorical participants to build, shape, and reshape their cultures. Writing in the workplace needs to be considered in terms of what takes place over time, across multiple rhetorical situations, as rhetorical participants interact with each other to build and sustain the corporate culture.

This study, therefore, has skimmed the surface of a complex rhetorical and social process. Certainly, the profession needs to explore in greater depth what can happen as rhetorical participants interact with each other, both in writing and verbally, in their attempts to fulfill important rhetorical, social, and cultural goals.

References

Bitzer, Lloyd. "Rhetoric and Public Knowledge." *Rhetoric, Philosophy, and Literature: An Exploration*. Ed. Don M. Burks. West Lafayette: Purdue UP, 1978. 67–93.

Doheny-Farina, Stephen. *Writing in an Emergent Business Organization: An Ethnographic Study*. Diss. Rensselaer Polytechnic Institute, 1984. Ann Arbor, UMI, 1985. DEU 85-00944.

—. "Writing in an Emerging Organization: An Ethnographic Study." *Written Communication* 3.2 (1986): 158-85.

Faigley, Lester. "Nonacademic Writing: The Social Perspective." *Writing in Nonacademic Settings*. Ed. Lee Odell and Dixie Goswami. New York; Guilford Press, 1985. 231–48.

Markel, Michael H. *Technical Writing*. New York: St. Martin's Press, 1984.

Odell, Lee. "Beyond the Text: Relations between Writing and Social Context." *Writing in Nonacademic Settings*. Ed. Lee Odell and Dixie Goswami. New York: Guildford Press, 1985. 249–80.

Odell, Lee, and Dixie Goswami. "Writing in a Nonacademic Setting." *Research in the Teaching of English* 16.3 (1982): 201–23.

Paradis, James, David Dobrin, and Richard Miller. "Writing at Exxon ITD: Notes on the Writing Environment of an R&D Organization." *Writing in Nonacademic Settings*. Ed. Lee Odell and Dixie Goswami. New York: Guilford Press, 1985. 281–307.

Selzer, Jack. "The Composing Process of an Engineer." *College Composition and Communication* 34.2 (1983): 178–87.

Spilka, Rachel. "Studying Writer-Reader Interactions in the Workplace." *The Technical Writing Teacher* 15.3 (1988): 208–21.

DEVELOPING YOUR UNDERSTANDING

1. Use the four roles of orality in the composing process that Spilka outlined to analyze one of the examples in either the Driskill or Katz article. Evaluate how well orality in your chosen example was employed by the writer to achieve effective organizationally contextualized action.
2. Compare and contrast the effects of a strategy of isolation when employed in workplace vs. academic vs. creative fiction contexts. Include a discussion of organizational context in your analysis.
3. Though Spilka's article is not focused on developing a detailed definition of "rhetorical situation," her analysis identifies many features, many writerly-questions, and many issues related to defining what rhetorical situations are. Locate these features, writerly-questions, and issues, and use them to develop your own definition of rhetorical situation.
4. Define organizational context and rhetorical situation. Explain how one relates to the other, and argue why you think it is, or is not, valuable for professional writers to make a distinction between the two terms. Use specific examples to support your argument.
5. Assess the strengths and weaknesses of Figure 1 as a model for guiding professional writers. Based on your assessment, write a one-page plan of revision for the figure. Implement your revision plan, developing a revised figure that you, as well as your classmates, would actually use to guide your professional work. (Your revision must include a visual or visuals, but it can also include accompanying text.)
6. Those in the field of writing and rhetoric have long debated the effects of defining writing as a socially interactive and contextualized practice vs. defining writing as the creation of a product. Though Spilka's study focuses on advocating a perspective and approach to writing that is process-oriented, it acknowledges the functions and values of both writing processes (orality) and writing products (literacy) within organizational contexts. Summarize the functions and values of both, as argued by Spilka, and explain why you agree and/or disagree with her position.

CHAPTER 3
Projects

1. Assume the role of a freelance professional writer specializing in, among other areas, improving internal communications. Write a white paper analyzing the relationships between organizational contexts and internal communications. Use examples from Driskill, Katz, and/or Spilka to support your analysis.

2. *Option 1* Assuming the role of a corporate trainer, develop a half-day workshop for department or division managers aimed at applying what you have learned about organizational and rhetorical contexts to help them improve writing in their departments or divisions. Your workshop should be appropriate for managers in both large and small, as well as profit and not-for-profit organizations.

   Option 2 Assume you are a member of a metropolitan area professional writing group made up of members from a variety of professional writing organizations, like the Society for Technical Communication and the American Medical Writers Association. Your group meets once a month for a "brown bag workshop," of sorts, with the goal of supporting continued education. For these meetings, one writer develops a short presentation, a brief workshop activity, and support materials on a particular issue/topic that will be of general interest to the group. Develop a presentation, workshop activity, and supporting materials aimed at applying what you have learned about organizational and rhetorical contexts to help the other writers become more responsive and strategic organizational authors.

3. As the manager of the technical publications department with 15 staff writers in a growing, mid-sized computer software development firm, you have noticed that only a few of your most senior writers have a clear sense of how their work as writers can change the work of others within the company, as well as that of outside clients. At the same time, you have noticed that a customer-centered approach to product development has increased opportunities for your technical writing staff to participate much earlier in product development processes. With this change in the role of technical writers in product development, along with a company initiative to hire five new staff writers in the next two years, you decide that you need to do something to help your current and newly hired staff understand how their work as writers can impact the whole organization, including clients.

   You decide to write a sort of manifesto, or sub-mission statement for your department encouraging all writing staff members to see themselves as organizational change agents. In your first sketchy notes, you identify two primary purposes for the document. First, you want it to rally your department around a revised sense of identity. Second, you want the document to serve as a constant reminder of and "recipe book" for organizationally situated action. In this second sense, the document should help department members understand the characteristics of organizationally situated action and help them strategize organizational action. Furthermore, you do not want this to be something that gets lost in a drawer; you want it to become a document each writer notices every day, one that your department refers to in staff meetings and while working in project teams. Your job is to write the document and to prepare a presentation you will use to introduce the document into your department's organizational culture.

4. The readings in this chapter have been chosen to open up some of the significant issues related to professional writing as organizationally situated action; by no means do these readings establish a complete and final view of the issues surrounding organizational situation. In fact, there are quite complex arguments about such terms as "organizational context" and "rhetorical situation" across a variety of fields (e.g., philosophy, rhetoric, anthropology, political science, and organizational studies) that, if studied, could yield more valuable insight into the issues surrounding the professional writer's work as organizationally situated.

   Individually, generate a list of possible research questions related to organizationally situated action. As a class, share and expand your list. Individually, choose a question you would like to research and report on back to the class, with the aim of developing a more robust understanding of organizational situation, as well as possible implications for the professional writer. Write a report, publish the reports, and regroup as a class to discuss what you have learned from your reports.

CHAPTER 4

Professional Writing as Ethical Action

Whether I do something or do not do something, I have engaged in action. Even the act of not acting is itself an act. And once I have acted, my actions can be determined to be ethical or unethical.

INTRODUCTION

This may seem like a very abstract way to open a discussion about the ethics of professional writing (and yes, it is), but there is a purpose behind it. The first purpose of this opening is to emphasize that whenever we write, we are in the act of *doing* something, and as a result, it has ethical implications. That probably doesn't surprise you. Most writers can quickly understand how their acts of composing are either ethical or unethical. For instance, if I am a journalist and leave out pertinent parts of an interview that change the spirit of what the interviewee was saying, my writing act is certainly unethical.

What is not necessarily so obvious, though, is that what writers do **not** do is also an action. The second purpose of starting off so abstractly is to highlight this negative aspect of action. As an action, what we do **not** do has ethical implications also. For instance, if I am a journalist and after getting some information from a second party, I do not take time to check my source's information, in my inaction, I may be acting in an unethical manner.

The readings selected for Chapter 4 serve to open up ethical issues related to the practice of professional writing. The following questions can help guide you as you read:

- What is the relationship between rhetoric/writing and ethics?
- What sorts of ethical issues might writers face in professional contexts?
- How can writers assume more active ethical stances?
- What are writers' ethical responsibilities?
- Who do writers serve?
- What defines writers' practices as ethical or unethical?

When writing is understood narrowly as linking words and visuals to produce texts, the ethical dimensions of professional writing are, consequently, quite narrow. But when writing is understood broadly as a social practice, including all the sorts of actions and interactions that are embedded in the writing process, like meetings, brainstorming sessions, researching, and review processes, then the ethical dimensions of professional writing grow quite broad and complex. It is these broader, more complex issues that the readings in Chapter 4 bring to light and address.

The first reading by Ornatowski examines, particularly through one case study, the complexities of balancing being bluntly honest and being politically effective when writing. Ornatowski's selection deals with the kinds of word choice issues that most people associate with ethical issues in professional writing. The final two selections, by Katz and Porter respectively, address less familiar ethical terrain. Katz examines the complex inter-relationships between ethics, rhetoric, politics, and ideology. Porter's selection directs our attention to such issues as the construction of one's self as a writer vis-à-vis the reader and the writer's processes as inextricably connected to ethics.

FOCUSING ON KEY TERMS AND CONCEPTS

Focus on the following terms and concepts while you read through this selection. Understanding these will not only increase your understanding of the selection that follows, but you will find that, because most of these terms or concepts are commonly used in professional writing and rhetoric, understanding them helps you get a better sense of the field itself.

1. selective emphasis
2. bureaucratic rationality

Between Efficiency and Politics: Rhetoric and Ethics in Technical Writing

CEZAR M. ORNATOWSKI
San Diego State University

Traditional textbook rationales for the technical writing course locate the essence of technical writing in objectivity, clarity, and neutrality, and the need for teaching it in its usefulness to employers. Such rationales, however, are unable to accommodate a notion of ethics and responsibility: if the writer merely serves the interests that employ her by reporting facts in an objective way, how can she exercise choice when ethical problems arise? An alternative view is to see technical writing as always rhetorical and involved with potentially conflicting agendas and interests, with objectivity, clarity, and neutrality serving merely as stylistic devices in the writer's rhetorical toolbox. Technical writers are rhetoricians who continually make ethical choices in serv-

Source: *Technical Communcation Quarterly,* Volume 1, Number 1, Winter 1992, "Between Efficiency and Politics: Rhetoric and Ethics in Technical Writing," by Cezar M. Ornatowski, pp. 91–103. Reprinted by permission of The Association of Teachers of Technical Writing.

ing diverse interests and negotiating between conflicting demands. The recognition of the fundamental rhetoricity of technical writing is the first step towards accommodating a meaningful notion of ethics into the technical writing curriculum.

Every semester, I begin the first hour of my basic technical writing class with a brief introduction to "What Is Technical Writing and Why Do We Need It?" I say a few things about what I think technical writing is all about and tell anecdotes from my consulting practice to illustrate some major principles that we are going to cover during the semester. Then I ask students to read Chapter 1 from their textbook, a chapter that is usually called something like "Writing in the Workplace" or "Why You Need This Course."

Here is what my students have found out by reading Chapter 1 in the textbooks I have used in recent years. In the second edition of his *Technical Writing*, Paul Anderson tells the student, "From the perspective of your professional career, one of the most valuable subjects you will study in college is writing" (4). The major argument for the value of writing is implicit in the subheading that follows, "Writing Will Be Critical to Your Success," and the chief premise is usefulness of good communicators to employers. "Besides enabling you to perform your job," Anderson tells the student, "writing well can bring you many personal benefits, including recognition in the form of praise, raises, and promotions" (5). In a similar vein, Steven Pauley and Daniel Riordan, in their *Technical Report Writing Today*, tell the student, "In industry and business today, technical writing is an important part of everyone's career" (4). After citing the usual statistics to show that people at work write a great deal, Pauley and Riordan conclude that "writing is extremely important for moving ahead in any profession" (4). And, they add, when "done well, technical writing is an exciting, fulfilling experience," but when it is "done poorly, it is frustrating, even harmful to career advancement" (3). Finally, in his *Technical Writing*, John Lannon devotes one page to a section entitled "Writing Skills in Your Career." After citing two corporate executives on the necessity of effective communication on the job, Lannon concludes that "good writing gives you and your ideas *visibility* and *authority* within your organization," while "bad writing . . . is not only useless to readers and politically damaging to the writer; it is also expensive" (10–11). Lannon cites the figure of $75 billion spent annually in American business and industry on communication, with roughly 60% of the writing produced being "inefficient: unclear, misleading, irrelevant, or otherwise wasteful of time and money" (11).

Behind the brevity of such rationales seems to lie the assumption that no serious effort—on either the author's or the reader's part—is needed to justify the enterprise, perhaps because the need for "good writing skills" appears self-evident and commonsensical in the terms in which it is presented. These terms are usually, as we saw, those of efficiency, effectiveness, and usefulness. The underlying catechism runs something like this: to work in a social collective (a business organization, a government institution, a cooperative, etc.) you need to communicate. To work effectively, you need to communicate effectively. The principles of effective communication can be derived from the general characteristics (structural, professional, cultural, personal) and needs of the appropriate social collectives and of the people who work in them. More effective communication means a more effective collective. More effective collectives make for a better society. More effective collectives also make more money. An incidental, but not unimportant, benefit is that if you communicate well and make the collective more effective, you advance (and make more money). Ergo: effective communication is useful all around.

Substantively, the major contentions that underlie this catechism are of course true, although the two middle terms, the contentions that the principles of effective communication may be derived from organizations and that more effective organizations make for a better society, have been convincingly criticized by, respectively, David Dobrin and Carolyn Miller. For the rest, however, one must admit that people at work do indeed write a lot, a lot more than most students realize. Communication, and especially written communication, is indeed one of the major modes of operating in any organization or institution, just as the production, circulation, and consumption of information is one of the major activities of modern organizations. Finally, communication does cost money (I was recently told that the average "cost" of a technical report in an aerospace firm to which I was a consultant was around $5000), and poor communication may seriously impede the functioning and effectiveness of an organization.

Recently, however, I have been increasingly uncomfortable with this whole rationale as a way of introducing a course in technical writing. It does not seem to account for what we know about the complex nature of the relationship between language and its social contexts or to provide any meaningful and practical way to talk about the ethical dilemmas faced by writers in organizational contexts. Effectiveness and efficiency, understood in terms of usefulness to employers, as the basic premises for communicative action appear to leave the communicator no provision, at least in theory, for action that does not "efficiently" further the goals of the institution or interests she serves. The clearest illustration of the problem comes from Anderson. In the introduction to his text, Anderson tells the student, first, that "writing well enables you to make a personal impact" (6), while a little later he tells the student that "as an employee, you will communicate for instrumental purposes. That is, most of your communications will be designed to help your employer achieve practical, business objectives, such as improving a product, increasing efficiency, and the like" (7). Not much scope is left in the end of the "personal impact" Anderson promises students.

Several well-publicized cases of communicative "irresponsibility," notably the Challenger disaster and the Three Mile Island nuclear accident, have been widely discussed in the literature (see, for example, Farrell and Goodnight; Winsor). Yet little seems to have changed in the wake of these discussions in terms of how we conceive of the nature of technical writing or how we see its teaching. Could it be that "irresponsibility" is built into our present thinking about the nature of technical writing and into the very foundation of how we go about teaching it?

I shall argue that it may be so and try to articulate the problem by redefining technical writing in rhetorical terms.

THE PROBLEM

In their study of the public rhetoric surrounding the nuclear accident at Three Mile Island, Farrell and Goodnight conclude that "the inadequacies of accidental rhetoric at Three Mile Island point to a failure larger than the technical breakdown of 1979: the failure of technical reason itself to offer communication practices capable of mastering the problems of our age" (271). The "rhetorical crisis" that followed the technical crisis, and that contributed to a sense of helplessness, paralysis, and loss of public confidence in the ability of appropriate agencies to act in the public interest, was caused, according to Farrell and Goodnight, by "systematic failures, emanating from deep-seated contradictions in contemporary theories and practices of communication" (273).

The ethical dilemma that I have articulated at the beginning of this discussion is, I think, but a surface manifestation of one such "deep seated contradiction." The contradiction goes beyond the difficulty of reconciling usefulness to employers with a sense of personal and social responsibility. It is, I think, ultimately the contradiction between two incompatible claims we make at once about the nature of technical writing and two incompatible conceptions of language that these claims imply.

One claim has to do with what technical writing does: it is effective and useful to employers because it accomplishes practical goals. I will use Pauley and Riordan's definition of technical writing as an example. (I realize, of course, that textbooks do not represent the profession and that to cite them as representative of what "we" hold is presumptuous, if not wrong. However, textbooks may be taken, I think, to represent the standard "paradigm" of current classroom practice.) Technical writing, Pauley and Riordan say,

> is the practical writing people do as part of their jobs. . . . [P]eople generate documents as an expected part of their responsibilities. These documents enable businesses; corporations; and public agencies, including governmental units, schools, and hospitals, to achieve their goals and maintain their operations. (7)

The second claim has to do with what technical writing is: it is objective, plain, neutral, clear, and so on. Pauley and Riordan, for example, characterize technical writing as

> written in plain, objective language. Since its purpose is to inform or persuade a reader about a specific matter, technical writing focuses the reader's attention on the relevant facts. . . . As much as possible, the words should not cause the readers to add their own personal interpretations to the subject. (8)

The essence of the contradiction is that two incompatible goals are held out for the technical writer: to serve the interests that employ her effectively and efficiently while being objective, plain, factual, and so on. What she finds in practice is that serving specific interests (any interests, even the most public-spirited) requires at least a degree of rhetorical savvy and that doing so is incompatible with "objectivity," "plainness," or "clarity." The latter, in fact, are not a writer's goals or purposes but stylistic devices that the technical writer employs in serving whatever interests it is she serves. Objectivity may also, of course, describe the writer's attitude (however difficult such an attitude may be to maintain in practice), but the problem is precisely the confusion between attitudes and rhetorical devices. As long as we do not distinguish, and train writers to distinguish, between rhetorical effects and human attitudes, as long as we promulgate the view that there is a language that does not involve "personal interpretation," we have no way of talking about responsibility or ethics that does not appear self-contradictory.

I confront this contradiction every day in my technical communication class. Like probably most of us, I have students with different professional backgrounds. Some are humanities majors, some are engineering majors, and a few are working engineers or other professionals. I cannot help noticing that the working students make adjustments to the stylistic principles we discuss, adjustments that are often as enlightening as they are exasperating: "I can't tell my boss about problems so directly"; "We can't just tell the customer clearly the product failed the test"; "How do I use active voice without blaming my boss?" The honesty and, let's face it, soundness of such comments contrasts with what I can't help feeling is the naivete of the nonworking students, who take at face value—often with a vengeance—the stylistic injunctions to be clear and direct, to state the problem and to make messages explicit. The result may be well-formed reports, but I am forced to admit that they would never "fly" in the complex game of trade-offs and judgment calls that defines the communicative dynamics of real organizations. The student who writes "The engine failed the test and shipping it may endanger the lives of innocent passengers" in a technical report may be ethically well-meaning, but the report would never pass the review cycle. The problem, of course, is how do I explain this to the student without making her feel that this ethics stuff is just school baloney?

RHETORIC AT WORK: AN EXAMPLE

Let me offer an example of the kind of daily ethical and rhetorical decisions writers in organizational contexts typically (in my experience) face. Consider one engineer's reflections on the ethical dilemmas involved in writing a single sentence of an aircraft engine test report. The engineer (whom we shall call Stephen) works as a test engineer in a major U.S. aerospace firm. By education and training, Stephen is a mechanical engineer. He has worked in the firm for over eight years. He is in his mid-30s, intelligent, quick, articulate. He has no formal training in technical writing. What he knows about writing technical reports he learned on the job, mostly from reading other reports, from company manuals, as well as from having his reports "redlined" and discussed by his supervisors. This practical training has made him well acquainted with the rhetorical and political subtleties of functioning effectively in the organization. I observed Stephen (and many other engineers in the firm) for nine months as part of a study of the rhetoric of technical information in the firm. I also conducted numerous interviews with Stephen and with other engineers.

The excerpts below are taken from one of the taped interviews with Stephen, conducted while he was engaged, over a period of several weeks, in testing an electronic controller (which I will call ELCON) for an auxiliary aircraft engine. The tests in question were environmental tests, in which ELCON was subjected to a variety of extreme environmental conditions: cold, heat, water, humidity, sand, and others. At the conclusion of the tests, Stephen began working on a formal engineering test report that would go through an appropriate approval cycle within the company and then be released to the customer (a major airplane manufacturer) and to the Federal Aviation Administration. It is this report, and specifically one sentence in it, that is the subject of the remarks quoted below.

Stephen is trying to express the fruitlessness of repeated attempts to start the engine through the ELCON during the low temperature test. He is trying to express the "hopelessness" of trying to start the engine. He does not want to "lie" (his word) and make it seem that ELCON worked much better than it did, yet he is aware that he cannot simply say "hopeless," because that puts a very important and expensive product in an unfavorable light, and he is sure his supervisors, and especially the program office, would object to his choice of words. The sentence he ended up writing in the official test report reads: "There was a problem encountered during the low temperature test, in which the power-up of the ELCON after cold soak was *fruitless initially*" (my emphasis). Here is how Stephen explained his choice of the word "fruitless." (The quotation is long, but I think worth citing in its entirety so as not to lose the sense of the full dimensions of the rhetorical choices Stephen is making here.)

> Here is a political sentence. "There was a problem encountered during the low temperature test, in which the power-up of the ELCON. . ."After it sits in a box at minus 67 degrees for six hours, then you have to turn it on. Well, when it came time to turn it on, the ELCON, the power supply inside, didn't respond. Kind of like when you turn your car on and the battery's dead and you just hear the click. So, how do you write that up? So, . . . I don't think I actually put the agonizing I did over this sentence in here. But I agonized over this sentence because I didn't want to say, let's see, it still says "fruitless initially," and that's essentially what . . . that's what I ended up saying. But you don't want to say "hopeless," whereas there was no way this thing was going to start, and that's really what it was. We tried for about a half hour and that thing wasn't going to start, the power wouldn't get to it. It had to do with all the circuitry being soaked to that temperature. So we tried for a half hour, which is eons in terms of electricity. You know, you flip a switch the light goes on, that's how electricity is, so half an hour is. . . So how do I write this up? The ELCON failed in power-up. That does not describe it, because we tried again and again and again. Do I really want to say that we tried, you know, forty five times in thirty minutes to power this thing up, you know? So I came up with "fruitless" only because it tells the story without saying that I did it again and again and again. To me it was hopeless, but "fruitless" says it like, "Yeah, we didn't get the power up," and it doesn't say "We tried ninety billion times," and it doesn't say "We tried once," it just says "We didn't get the power up." And that is as vanilla as I can make that, that's as benign a statement as I can make without tipping my hand, so to speak. . . . It's not so difficult to try and document what happened but there is . . . there are points at which you don't want to hurt yourself, you don't want to . . . you don't want to harp on anything in particular. This, the only way I could document what happened without making it sound like the horror show it was, was to just say "Initially we had trouble," which is true. It's a bit of an understatement, but it is true.

Stephen's dilemma is how to tell the technical "truth" while at the same time telling it within the boundaries of outcomes and implications of events acceptable to his supervisors and to the report's customers. Note, however, that even the "technical fact" is largely a matter of personal perception and interpretation: is the attempt hopeless or just fruitless? It may depend on how impatient one feels that day or how long it is till lunch time. Elsewhere, Stephen described the rhetorical strategy he uses in writing his reports as "selective emphasis." In the statements below (taken from the same interview), he explains what he means by "selective emphasis."

> You can change the message that you give with a report by not highlighting certain data. It's probably most apparent in something like a performance report. If I'm documenting engine performance there is [sic] a lot of things that I could look at, to say that it's a good performer or a bad performer. You ask can I change the slant of the report by changing the data. I would say no, the data is what the data is: sacrosanct. What I can do is highlight

> data, withhold data, present it in a different fashion, like, I could tell you this engine was a great performer. It's delivered its shaft power; it's delivered bleed. And you might not know that this engine is high on specific fuel consumption. . . . So, yes, there are ways you can change how a report is perceived, or how a test event is perceived. Actually, if you present everything, it becomes a matter how you highlight it with the prose. And that's the final thing I would say to you: *You can't change the data, but you can certainly change the positive/negative of a given issue*. I even think *it is incumbent on everybody to do just that*, particularly when you are writing for the customer. If you ask anybody in this [test] department, I'm sure you'll get some hard-ass answers about, you know, the data is what the data is. I don't think anybody here [in the test department] has ever cut a corner in that sense. On the other hand, I think between us test [department] and the project [project management, responsible for liaison with customers], the stuff that gets out of here is reasonably, I wouldn't say "censored" but its . . . it's, maybe I will say "censored," yeah, it's been reviewed to the point that as little damaging statement as possible is included.

Stephen is considered a "good writer" among the people for whom he writes: the program managers and the customers. His success appears to be due, in large part, to his rhetorical acumen, to his ability to make the delicate trade-offs and decisions alluded to in the passages quoted above. By contrast, I have talked to, and seen reports written by, engineers who do not enjoy the reputation of being "good" (read "effective") writers and whose reports are always heavily reviewed and often need to be revised by the supervisors themselves before being officially released. The major problem that these writers have (as I gathered from studying their reports and from discussing their reports with their supervisors and co-workers) is not that they do not know what an engineering report should look like (they have learned that on the job and the format is prescribed in departmental Policies and Procedures) or that they cannot write "good" sentences (their sentences are no better or worse than Stephen's). Rather, they are heavy-handed and naive in reporting facts, not politically astute or willing to be astute, or not linguistically accomplished enough to be able to be astute, to negotiate the subtle lines between political prudence and necessary disclosure.

From my research in that firm I concluded that a good organizational writer is one who can negotiate successfully the subtle boundary between, on the one hand, the stylistic and formal demands of clarity, objectivity, neutrality, format, and effective use of visual devices, and, on the other hand, the institutional, social, and situational (read: political) demands placed on the text. In my experience, all organizational communication, including technical documentation, bridges the gap between the physicality of technical phenomena on the one hand and their organizational and social meanings on the other hand, between data as technical and as social facts. It is this inevitable gap—the gap that arises, for instance, when unruly events and the idiosyncrasies of gas turbine thermodynamics are at odds with the political and economic exigencies and stakes—that defines the rhetorical "space" which most technical documents written in real environments must bridge. As Edwin Layton has observed, "organizations like the federal government or a modern corporation have other ends in view than the best and most efficient engineering" (3). It is the difference between these "other ends" and the theoretical ideal of the "best and most efficient engineering" that define the major arena of rhetorical choice and ethical conflict for writers of technical documents in organizations. The ethos of engineering reports, I noted in my research, is always constructed out of uneasy compromises and accommodations, even though many engineers are troubled by the fact that their reports are "censored"—as Stephen hesitantly put it.

THE DILEMMA REVISITED

My problem as a teacher (and theorist) comes down to this: I know that technical writing in real-world contexts is much more political and rhetorical than the textbooks make it out to be and that it is in this political/rhetorical sphere that ethical problems arise. The problem is that the standard textbooks and the standard curriculum offer no way to deal with this knowledge without seeming to throw away the whole shop. If writing documents is only a matter of clearly marshalling objective facts and designing readable texts, "ethical" problems should not arise. Something would either be a fact or not be a fact, be clearly relevant or be clearly irrelevant. My students, the ones who do not work, feel very smug about being honest and get impatient when I talk about ethics. How could one be so stupid or such a crook, they ask when we discuss the Challenger tragedy. They are right. If technical writing is just a matter of designing clear documents, the only way ethical problems could arise is through ignorance of facts or sheer stupidity (the former being, interestingly, the favorite defense of those who are accused of "irresponsible" practices). In fact, the way ethics are alluded to in most textbooks, almost incidentally and parenthetically and usually in the form of a simple injunction to be "honest" or "ethical," shows the inability of the standard rationale to develop a more sophisticated notion of ethics than simple avoidance of outright dishonesty. The mechanisms for "audience analysis," for determining the reader's needs, for analyzing the situation, offer no tools for discerning or discussing the more subtle, and essentially political, choices, pressures, and agendas that one encounters in real environments.

As a teacher, one could agree with Sherry Southard that "being professional means knowing the proper protocol for the corporate world" and that it is such protocol we should be teaching (90). While such a position is no doubt realistic, it amounts to throwing in the towel on ethics, because it contains few provisions for action outside bureaucratic control and outside simply serving corporate interests that pay the bill. One could go to the other extreme and advocate rebellion, a rhetoric of "dissention," as proposed by John Trimbur, who suggested that we should replace "the 'real world' authority of consensus with a rhetoric of dissensus" (615). The problem with this attitude as a premise for teaching and practicing technical writing is that assuming a perennially contestatory stance puts the communicator in danger of becoming, at best, substantially irrelevant (like those proverbial "nyet" men in the Soviet Politburo) or, at worst, unemployed. It is also questionable whether programmatic contestation really leads to a better understanding of the nature of language, of the relationship between language and its social contexts, and of the nature and role of technical communication. I have problem seeing it, for instance, as a coherent rationale for a technical writing curriculum.

Carolyn Miller has suggested that technical writing ought to be reconceived as "a matter of conduct rather than of production, as a matter of arguing in a prudent way toward the good of the community rather than of constructing texts" (23). Although emotionally I am in agreement with this sentiment, I do not see how it solves the practical and pedagogical problem. Employees, after all, are paid to render services to employers and to further their goals "effectively" and "efficiently." Institutional readers "need" a document that does a certain job, and they won't accept one that does not do it to their satisfaction, no matter how lofty the social sentiments of the writer.

EFFICIENCY IS POLITICS
TECHNICAL WRITING IS RHETORIC

To begin to find a way towards dealing with the problem of ethics realistically and effectively, we must, I think, accept two premises. The first premise is that *organizational efficiency is always involved with politics.* Aaron Wildavsky has offered a definition of politics that, I think, is very appropriate in this context. Politics, Wildavsky suggests, is an essentially social process by which an institution "mobilizes resources to meet pressing needs" (94). The second premise is that *technical writers are always inevitably rhetoricians* precisely because they are "useful" to employers and their writing is "effective." In fact, technical writing is a form of rhetoric developed in, and uniquely suited to, the social contexts in which it typically takes place. Let me explain.

The most common contexts for technical writing are business/industrial organizations and government institutions. The dominant purpose of such organizations (as well as the major criterion by which the people who work in them are judged as effective) is the maximization of advantage and increased control of resources in the major operative spheres: economic, political, intellectual. It is the maximization of the symbolic or real "capital."

Communication serves this goal as much as do other institutional activities. Technical documents (engineering reports, model specifications, proposals, annual reports, financial reports, business plans, product brochures, manuals, and other documents) play an important role in the dramaturgy of organizational life through which the organization and, within the organization, the various departments, interests, or individuals, garner resources and advance their causes (for an example, see Barbara Mirel's discussion of the political functions of a user manual). In this dramaturgy, reason (or the appearance of reason) is the fundamental condition of effectiveness, and it is reason, or its appearance, that well-designed technical documentation provides. Richard Ritti has argued, following Max Weber, that the dominant "world-view" (by which Ritti understands "the framework of beliefs, attitudes, and values that provides the basis for classifying, interpreting, and understanding the events of everyday life") of most contemporary organizations, whether private, public, or governmental, is bureaucracy (104). Bureaucracy is characterized by legality, impersonality, and rationality, subsumable under the major functions of planning, organizing, leading, and controlling. Within the compass of bureaucratic rationality, all problems must be (or must appear to be) amenable to technical, rational (note the common collusion of the two terms) solutions comprised of a step-by-step cycle of setting goals, developing a plan, accumulating relevant facts, deciding on the best course of action, and so on.

Because of its implicit and traditional association with science and technology and its "rational" pretensions, embodied in the stylistic attributes of clarity, directness, factuality, objectivity, and neutrality, technical communication is well suited to serve as the rhetorical instrument of organizational-bureaucratic rationality. It is also uniquely, and conveniently, congruent with the cultural self-perceptions of the people who dominate business and industrial organizations. Ritti describes American corporate culture as a culture of "'decision-makers'; of crisp, hard-hitting, two-fisted decision-makers; of bottom-liners; of players of hardball. The metaphors are redolent of *efficiency and effectiveness*. Facts exist, can be weighed, and objective outcomes maximized" (104, emphasis added). Such people want their reports to be factual, objective, and effective—as long as the facts "objectively" and efficiently support their agendas and objectives.

It is largely in this sense that technical writing is "useful" to employers and that is implicitly and ultimately why we teach it.

The traditional catechism has got it wrong—it is not that technical writing is important because communication costs money and may have an impact on one's career, but the other way round—organizations and institutions are willing to spend big money on communication, and communication may have an impact on one's career because it is useful—and that's why it is in demand and that is why we teach it. Here comes the rest of my catechism. It is useful because it is a vital element in furthering the goals (business, political, economic, and other goals) of organizations and institutions. It furthers those goals because it is fundamentally rhetorical, just like any other use of language, and its rhetoric is uniquely suited to organizational and institutional rationality. The essence of its usefulness lies in the fact that it is capable of giving a rational, technical, "scientific" mantle to agendas that are always in some measure political and that it gives the illusion of objectivity to what are always and inevitably interpretations. The fact that we do not commonly think that way about it is its greatest rhetorical asset. Realizing this is the first step towards having the capacity to analyze the trade-offs and bargains that one makes. And that is the first step to talking sensibly about ethics.

Of course, in saying this, I have not solved my initial problem. But I have articulated the problem in a way that lets me talk abut it as an integral part of what I see technical writing to be, and that articulation is in keeping with some other things I know about language, writing, and social contexts. That way of articulating it also engages technical writing with theories of human action and its institutionalizations. Paul Anderson is right in telling students that writing is one of the most valuable courses they will take in college. But I think the reasons we usually give for this value miss the real point.

Works Cited

Anderson, Paul V. *Technical Writing: A Reader-Centered Approach*. 2nd ed. New York: Harcourt, 1989.

Dobrin, David N. "What's the Purpose of Teaching Technical Communication." *The Technical Writing Teacher* 7 (1985): 146–76.

Farrell, Thomas B., and G. Thomas Goodnight. "Accidental Rhetoric: The Root Metaphors of Three Mile Island." *Communication Monographs* 48 (1981): 271–300.

Lannon, John M. *Technical Writing*. 4th ed. Glenview: Scott, 1988.

Layton, Edwin, Jr. *The Revolt of the Engineers: Social Responsibility and the American Engineering Profession*. Baltimore: Johns Hopkins UP, 1971.

Miller, Carolyn R. "What's Practical About Technical Writing?" *Technical Writing: Theory and Practice*. Ed. Bertie E. Fearing and W. Keats Sparrow. New York: MLA, 1989. 14–26.

Mirel, Barbara. "The Politics of Usability: The Organizational Functions of an In-House Manual." *Effective Documentation: What We Have Learned from Research*. Ed. Stephen Doheny-Farina. Cambridge: MIT P, 1988. 277–298.

Pauley, Steven E., and Daniel G. Riordan. *Technical Report Writing Today*. 4th ed. Boston: Houghton, 1990.

Ritti, R. Richard. "The Social Basis of Organizational Knowledge." *Organization—Communication: Emerging Perspectives*. Vol. 1. Ed. Lee Thayer. Norwood: Ablex, 1986. 102–132.

Southard, Sherry G. "Interacting Successfully in Corporate Culture." *Journal of Business and Technical Communication* 4 (1990): 79–90.

Trimbur, John. "Consensus and Difference in Collaborative Writing." *College English* 51 (1989): 602–16.

Wildavsky, Aaron. *The Politics of the Budgetary Process*. Berkeley and Los Angeles: U of California P, 1964.

Winsor, Dorothy. "The Construction of Knowledge in Organizations: Asking the Right Questions About the Challenger." *Journal of Business and Technical Communication* 4 (1990): 7–20.

DEVELOPING YOUR UNDERSTANDING

1. If the value of professional writing is not efficiency, explain what its value is.
2. Analyze at what point(s) in the writing/work process that ethics becomes a matter of practice for professional writers. (Your response will depend on the definition and scope of rhetoric that you assume; therefore, you need to clearly define rhetoric and its scope.)
3. Ornatowski suggests that effective writers "negotiate the subtle lines between political prudence and necessary disclosure." Explain what he means by this, referring to the example of Stephen. Then, assess in what ways you think Stephen effectively and ineffectively acted prudently (you may wish to refer to Miller, Chapter 2, for a definition of prudence). In your assessment, identify some of the consequences that might result from Stephen's rhetorical practice.
4. Ornatowski argues that technical/professional writing is "useful because it is a vital element in furthering the goals (business, political, economic, and other goals) of organizations and institutions.... and its [technical/professional writing's] rhetoric is uniquely suited to organizational and institutional rationality." Explain what Ornatowski assumes is the rhetoric of technical/professional writing, referring to its scope, function, and value. Or, to take a slightly different approach, explain what Ornatowski assumes is the role of the technical/professional writer.
5. Professional writers are regularly engaged in making various levels of ethical decisions; however, these decisions are rarely discussed as part of professional practice. In contrast, hospitals have ethics boards that meet regularly to discuss issues of ethics. Evaluate the idea of instituting an ethics committee for professional writers within an organization (or an online version for independent writers).

FOCUSING ON KEY TERMS AND CONCEPTS

Focus on the following terms and concepts while you read through this selection. Understanding these will not only increase your understanding of the selection that follows, but you will find that, because most of these terms or concepts are commonly used in professional writing and rhetoric, understanding them helps you get a better sense of the field itself.

1. deliberative rhetoric
2. means/ends
3. organizational ethos
4. epistemology of objectivity
5. telos
6. polis

The Ethic of Expediency: Classical Rhetoric, Technology, and the Holocaust

STEVEN B. KATZ

"[T]he stronger this faculty is, the more necessary it is for it to be combined with integrity and supreme wisdom, and if we bestow fluency of speech on persons devoid of those virtues, we shall not have made orators of them, but shall have put weapons into the hands of madmen"—Cicero, *De Oratore* III: xiv. 55.

* * *

Geheime Reichssache (Secret Reich Business)
Berlin, June 5, 1942

Changes for special vehicles now in service at Kulmhof (Chelmno) and for those now being built

Since December 1941, ninety-seven thousand have been processed [*verarbeitet* in German] by the three vehicles in service, with no major incidents. In the light of observations made so far, however, the following technical changes are needed:

[1.] The vans' normal load is usually nine per square yard. In Saurer vehicles, which are very spacious, maximum use of space is impossible, not because of any possible overload, but because loading to full capacity would affect the vehicle's stability. So reduction of the load space seems necessary. It must absolutely be reduced by a yard, instead of trying to solve the problem, as hitherto, by reducing the number of pieces loaded. Besides, this extends the operating time, as the empty void must also be filled with carbon monoxide. On the other hand, if the load space is reduced, and the vehicle is packed solid, the operating time can be considerably shortened. The manufacturers told us during a discussion that

Source: Katz, Steven B. "The Ethic of Expediency: Classical Rhetoric, Technology, and the Holocaust." *College English 54.3* (March 1992), pp. 255–275.

reducing the size of the van's rear would throw it badly off balance. The front axle, they claim, would be overloaded. In fact, the balance is automatically restored, because the merchandise aboard displays during the operation a natural tendency to rush to the rear doors, and is mainly found lying there at the end of the operation. So the front axle is not overloaded.

2. The lighting must be better protected than now. The lamps must be enclosed in a steel grid to prevent their being damaged. Lights could be eliminated, since they apparently are never used. However, it has been observed that when the doors are shut, the load always presses hard against them as soon as darkness sets in. This is because the load naturally rushes toward the light when darkness sets in, which makes closing the doors difficult. Also, because of the alarming nature of darkness, screaming always occurs when the doors are closed. It would therefore be useful to light the lamp before and during the first moments of the operation.

3. For easy cleaning of the vehicle, there must be a sealed drain in the middle of the floor. The drainage hole's cover, eight to twelve inches in diameter, would be equipped with a slanting trap, so that fluid liquids can drain off during the operation. During cleaning, the drain can be used to evacuate large pieces of dirt.

The aforementioned technical changes are to be made to vehicles in service only when they come in for repairs. As for the ten vehicles ordered from Saurer, they must be equipped with all innovations and changes shown by use and experience to be necessary.

Submitted for decision to *Gruppenleiter* II D,
SS-Obersturmbannführer Walter Rauff.

Signed: Just

THE FINAL SOLUTION: AN ETHICAL PROBLEM IN RHETORIC

This is a real memo, taken verbatim from the published transcript of *Shoah*, a 9-hour documentary film on the holocaust directed by Claude Lanzmann (103-05). In this memo, the writer, Just, attempts to persuade his superior, Walter Rauff, of the necessity for technical improvements to the vans being used in the early Nazi program of exterminating the Jews and other "undesirables," just months before the Final Solution of gas chambers and death camps was fully operationalized. In this earlier stage of the Final Solution, four Einsatzgruppen, or "Special Action Groups," A, B, C, and D, had been organized by Himmler to carry out executions by firing squads (Shirer 1248–49). Group D, whose field of operations included the southern Ukraine, was from June 1941–June 1942 headed by Otto Ohlendorf, in the R.S.H.A., Himmler's Central Security Office (Shirer 1249). In 1942, Himmler ordered gassing vans to be used for executing women and children, because it was more efficient, "humane" (see Shirer 1250–51, 1254n.). The Wannsee Conference, in which the details of the Final Solution were worked out, had been held on January 20, 1942.

To begin to get at the ethical problem in rhetoric here, let's do a brief rhetorical analysis of this memo from the standpoint of technical communication, argumentation, and style. By any formal criteria in technical communication, it is an almost perfect document. It begins with what, in recent composition theories and technical writing practices, is known as the problem or "purpose statement." According to J. C. Mathes and D. W. Stevenson, this statement should invoke an assumption or goal shared by the audience—here the statistic that 97,000 have been processed without incident—and then introduce a fact that conflicts with that assumption or goal—technical changes are needed—thereby effectively setting up the problem to be solved (29–38; see also Olsen and Huckin, *Principles* 94–104). In keeping with some of what today are recognized as the rules of good document design, the memo is also divided into three numbered sections that are clearly demarcated by white space for easy reading. And most importantly from the standpoint of technical writing, this recommendation for modifying the vehicles is technically accurate and logically argued.

Indeed, in this memo one can find many of the *topoi* first defined by Aristotle in the *Rhetoric* (II, xxiii. 1397a6–xxiv. 1402a29) that are used to investigate any situation or problem and provide the material for enthymemic arguments. For example, in the first section the writer uses the common topic of relationship: cause/effect arguments, in conjunction with the topic of comparison (difference) and the topic of circumstance (the impossible), are used to investigate the problem of maximizing the use of space, to refute the manufacturer's claims that the problem is one of overloading, and to conclude in an enthymeme that a reduction in the load space is necessary. Just further supports his conclusion by cause/effect arguments embedded in the topic of contraries that reducing the number of "pieces" loaded would extend operating time because the empty space would have to be filled with carbon monoxide, while reducing the load space would actually shorten the operating time. Finally, Just argues by cause/effect and contraries to refute the manufacturer's claim that reducing the load space would overload the front axle by arguing from precedent (example) that "the merchandise . . . displays during the operation a natural tendency to rush to the rear doors, and is mainly found lying there at the end of the operation. So the front axle is not overloaded." Thus, in a series of enthymemes that make use of the *topoi*, Just investigates and proves his case for a reduction in load space.

But of course, this is not the problem with this memo. In fact, given the subject matter, we might wish to claim that this memo is *too* technical, *too* logical. The writer shows no concern that the purpose of his memo is the modification of vehicles not only to improve efficiency, but also to exterminate people. This is the ethical problem in rhetoric I wish to discuss. Here, as in most technical writing and, I will argue, in most deliberative rhetoric, the focus is on expediency, on technical criteria as a means to an end. But here expediency and the resulting *ethos* of objectivity, logic, and narrow focus that characterize most technical writing, are taken to extremes and applied to the mass destruction of human beings. Here, expediency is an ethical end as well.

This "*ethos* of expediency" can be seen in the style of Just's memo, particularly the euphemisms and metaphors used to denote, objectify, and conceal process and people—"observations," "load," "pieces," "operating time," "merchandise," "packed solid," "fluid liquid," "large pieces of dirt"—as well as use of figures of speech such as ellipsis ("97,000 have been processed") and litotes ("alarming nature of the darkness," "displays a natural tendency to rush to the rear doors"). What concerns me most here is how, based on an ethic of expediency, rhetoric was made to serve the holocaust.

It is well known that to perform well in a professional organization, writers must adopt the *ethos* of that organization. Barring errors in translation or differences in language structure between German and English, the *ethos* of Just's memo is created and supported by a grammatical style that Walker Gibson has labelled "stuffy" (90–101): the heavy use of polysyllabic words, modified nouns ("natural tendency," "full capacity," "sealed drain," "fluid liquid," "technical changes"), of a passive voice that obscures the role of the agent, and of subordinate clauses that separate subject from verb. As Gibson points out, in this style responsibility is shifted from the writer (and reader) to the organization they represent, the organization whose voice they now speak with, in whose interest they act, whose *ethos* they have totally adopted as their own. All the stylistic features I have pointed out communicate and reveal a "group think," and officially sanctioned *ethos* grounded in expediency.

Indeed, this brief analysis reflects the rhetorical problem with Just's memo: it is based *purely* on an ethic of expediency. This claim at once corroborates and goes beyond Hannah Arendt's controversial conclusion that Eichmann, the inventor of "the final solution," was not a psychopath but a bureaucrat simply doing his duty. For Just is not merely performing his function; in order to perform it effectively, he has adopted the *ethos* of the Nazi bureaucracy he works for as well. But in Nazi Germany, that *ethos* also involved an entire nation of people, a whole culture. Thus, I believe the ethical problem is even deeper and more widespread than the *ethos* of a single bureaucracy. In this paper I will attempt to show that what I have called an ethic of expediency underlies technical writing and deliberative rhetoric (see Olsen and Huckin, *Principles* 70), and that this ethic, which is so predominant in Western culture, was at least partially responsible for the holocaust.

Thus it will be my contention that the ethical problem represented in Just's memo to his superior, while an extreme case, is not an anomaly nor a problem in technical writing only, but a problem of deliberative rhetoric—defined by Aristotle as that genre of rhetoric concerned with deliberating future courses of action. I will argue that the ethic of expediency in Western culture which Aristotle first treated systematically in the *Rhetoric*, the *Nicomachean Ethics*, and especially the *Politics*, was rhetorically embraced by the Nazi regime and combined with science and technology to form the "moral basis" of the holocaust. While there is a concern for ethics in the field of technical communication, and while few in our society believe expediency is an adequate moral basis for making decisions, I will suggest that it is the ethic of expediency that enables deliberative rhetoric and gives impulse to most of our actions in technological capitalism as well, and I will explore some of the implications and dangers of a rhetoric grounded exclusively in an ethic of expediency. In doing so, I hope to mount a critique of the ethic of expediency that underlies technical communication and deliberative rhetoric, and by extension writing pedagogy and practice based on it.

In "The Rhetoric of Hitler's 'Battle'" Kenneth Burke has already demonstrated the importance of rhetorical analysis for understanding the source of Hitler's power, and the significance of his misuse of the rhetoric of religion. However, despite Burke's warning, we have tended to understand the holocaust from a nonrhetorical, Platonic standpoint, which amounts to a refusal to understand it at all. Sometimes this standpoint is justified. Elie Wiesel, for example, eloquently argues for the sacredness of the memory of the holocaust against the attempts to absorb it into popular culture and so trivialize it. But for Wiesel, and I would suggest, most people, the holocaust appears as a breach in the Platonic wall of Virtue, an aberration in Western civilization, and so lies outside human culture: "Auschwitz is something else, always something else. It is a universe outside the universe, a creation that exists parallel to creation" (Wiesel 1). In this Platonic realm of anti-Forms, the holocaust lies beyond rhetorical analysis. For Wiesel and many other survivors and scholars, the holocaust can best be comprehended by the reverence of silence that surrounds a mystery.

However, as George Steiner intimates throughout *In Bluebeard's Castle*, the holocaust may not be so much a breach of the Platonic wall of Virtue, an aberration of Western culture, as an outgrowth of it, the final development and manifestation of something deeper and more problematic in Western civilization itself. In this view, the holocaust falls under the purview of rhetoric. Although Steiner points to the Platonic utopianism inherent in Western culture rather than to expediency as the root of the holocaust, I will show that much of Hitler's ethical and political program is also directly or indirectly based on the ethic of expediency first treated by Aristotle, and is thus amenable to analysis from an Aristotelian point of view. While I agree with Wiesel's argument against the trivialization of the holocaust through popularizations and respect him immensely, an exclusively Platonic stance toward the holocaust prevents us from fully understanding how it happened, and from understanding the relationship it reveals between rhetoric and ethics.

ETHICS IN DELIBERATIVE DISCOURSE: EXPEDIENCY

Let's start with the issue of objectivity in technical writing. While the fallacy of the objective stance in technical writing has been discussed extensively from an epistemological standpoint (see Miller, "Humanistic Rational"; Dobrin), it has not been discussed enough from an ethical one. The concept of *ethos* in rhetoric might help us here. In rhetorical theory, the role of *ethos* ("the moral element in character") in enthymemic arguments has been demonstrated by William Grimaldi, for example, who, interpreting Aristotle, argues that it is an essential link between deliberation and action (144–51). Virtue for Aristotle involves choice informed and led by both intellect *and* natural disposition or appetite (*Nicomachean Ethics* VI. Xii. 1143b16–xiii. 1145a14). Thus Grimaldi argues that while *logos*, or reason "considers the means necessary" to reach some end in the deliberative rhetoric, it is *pathos* and *ethos* that provide the impetus to act.

In this sense, ethics, defined as human character manifested in behavior, is an important consideration in deliberative rhetoric. All deliberative rhetoric is concerned with decision and action. Technical writing, perhaps even more than other kinds of rhetorical discourse, always leads to action, and thus always impacts on human life; in technical writing, epistemology necessarily leads to ethics. The problem in technical communication and deliberative rhetoric generally, then, is not only one of epistemology, the relationship of argument, organization, and style to thought, but also one of ethics, of how that relationship affects and reveals itself in human behavior.

It is easy to see how the epistemology of objectivity would lead to an ethic of expediency (or how the ethic of expediency would lead to an epistemology of objectivity) in so far as the viewing subject and the viewed object are technical means to some "higher" end—that is, "truth." But even discussions based on the principals of problem statements, audience adaptation, and rhetorical argumentation—upon which the more sophisticated teaching (and practice) in technical writing as well as rhetoric are based—only begin to get at the fundamental issue that thrusts itself upon our attention in Just's memo. As we will see, based on the ethic of expediency that underlies not only technical writing and rhetoric but also most behavior in Western civilization (see Olsen and Huckin, *Principles* 70), those same principles were used to form the "moral" basis of Nazi society, to create the *ethos* of that entire culture, and to provide the necessary warrant for the holocaust. As Olsen and Huckin suggest in the second edition of their textbook (*Technical Writing* 40–41; 91–94), we need to consider technical writing based on deliberative rhetoric from the standpoint of both rhetoric *and* ethics.

From the debates between the sophists and Plato to present-day criticism of advertising and political propaganda, there has always been an uneasy relationship between rhetoric and ethics. Perhaps nowhere is that relationship more clearly treated—and the strain more evident—than in Aristotle's *Rhetoric*. In the *Rhetoric*, Aristotle states that "rhetoric is a combination of the science of logic and of the ethical branch of politics" (I. Iv. 135b10)—of logic *and* ethics. According to Aristotle, then, ethics in political discourse is a matter of Goodness as well as Utility. However, in his discussion of deliberative discourse in the *Rhetoric*, Aristotle elides Goodness and Utility: "the political or deliberative orator's aim," he says, "is utility: deliberation seeks to determine not ends, but means to ends, i.e., what it is most useful to do" (I. vi. 1362a17–20).

In the *Rhetoric* Aristotle thus seems to collapse all ethical questions in deliberative discourse into a question of expediency. As he says, "all other points, such as whether the proposal is just or unjust, honourable or dishonourable, he [the political orator] brings in as subsidiary and relative to this main consideration" (I. iii. 1358b23–25). Nan Johnson argues that it did not seem to matter much to Aristotle whether the ends of deliberative rhetoric were ultimately just or unjust, true or false, as long as the means were expedient. However, several scholars have argued that Aristotle's conception of rhetoric as *praxis* (social action) is not amoral, but rather ethical insofar as *praxis* involves *phronesis* (practical wisdom or prudence) as an end in itself (see Sullivan 377–78; Kallendorf and Kallendorf 55–57; Rowland and Womack). But it is precisely because rhetoric is a practical art rather than a theoretical science, one located in *praxis*, in the contingent realm of action, that deliberative rhetoric can be understood to be primarily based on an ethic of expediency. If *praxis* depends on *phronesis*, on the practical wisdom or prudence of the speaker to reason about "the good," that wisdom, that prudence, is itself a means to an end, that end being *praxis*.

Further, as Dale L. Sullivan points out, "the good," and thus what counts as practical wisdom or prudence, is defined by society (378). Thus *phronesis*, like ethical appeal for Aristotle (*Rhetoric* I. Ix. Esp. 1367b10), can also be considered an expedient, a means to an end of rhetoric as *praxis*—determining the "right" course of action in the first case, finding the available means of persuasion in the second. (Eugene Garver, however, argues that this understanding of *phronesis* depends on whether one defines it as "prudence," which is rooted in character as an end in itself, or as "practical reason," which is detached from character in modern political thought and thus more "technical" [xi]. But as I will show, prudence, like virtue itself, can be redefined by society, become a means to another end, as was the case in Nazi Germany.) In Aristotle's treatment of deliberative rhetoric, then, expediency seems to become an ethical end in itself. Expediency is always the good—"utility is a good think" Aristotle says (I. vi. 1362a20), concluding: "any end is a good" (I. vi. 1363a5). This is a conclusion which, in light of the holocaust, we may want to reconsider. For following Aristotle, in deliberative discourse, including technical communication, we are in the habit of giving expediency too much free reign.

In fact, most technical communication is deliberative. (Indeed, in a scientific and technological society, much deliberative discourse is technical.) As Olsen and Huckin teach, technical writing is concerned both with arguments of fact *and arguments of policy*—with what should or should not be done (*Principles* 67). But as they also point out, since most technical communication is deliberative, it is based primarily on arguments of expediency rather than worth or goodness (*Principles* 70). What Aristotle gives us in the *Rhetoric*, then, is a practical ethic for technical writing and deliberative discourse, an ethic based almost exclusively on expediency. Most arguments of worth and goodness, if they are present at all, are subsumed under expediency, becoming another means to a desired end, becoming expedient in themselves (like appeals to give to charity based on the advantage of a tax break).

However, Aristotle's treatment of ethics is not as simple as that. Alasdair MacIntyre argues that in Aristotle's *Nicomachean Ethics* the relationship between means and ends is ambiguous (148). On the one hand, it does seem that for Aristotle virtue is a means to an end, that end being happiness. In the *Nicomachean Ethics* Aristotle says that "Happiness . . . is something final and self-sufficient, and is the end of action" (I. vii. 1097b21; see also X. vi. 1176a30–viii. 1179a34). It is not erroneous, says MacIntyre, to see that in positing "the good" as the *telos* or goal of human life and defining that *telos* as happiness or pleasure, Aristotle renders happiness the ideal object of all virtue (148). In fact, G. E. R. Lloyd suggests that Aristotle waxes positively Platonic in his discussion of happiness (239).

On the other hand, according to MacIntyre, Aristotle does not clearly separate means and ends as we do. MacIntyre argues that in Aristotle's teleological philosophy, happiness as "the good" is not only an end of virtue but a part of virtue, the result of virtue as an activity of the soul: "The enjoyment which Aristotle identifies is that which characteristically accompanies the achievement of excellence in activity" (160). Lloyd too points out that there is no ideal form of the Good as such, but rather individual goods associated with particular activities or subjects (208–13). Thus, says MacIntyre, "the enjoyment of itself provides us with no good reason for embarking upon one type of activity rather than another" (160).

Further, if there is no ideal form of the Good, virtue (like knowledge without the ideal form of Truth) is communal in nature, and is at least partially determined by the society in which one lives. That is, virtue, like knowledge, is socially constructed, culturally relative, an awareness of a condition of our civilization from which, as Steiner laments, there is no turning after the holocaust (59–93). In fact, according to MacIntyre, virtue was not a matter of individual moral authority for Aristotle, as it is for us, but was always directed toward and made possible by the *polis* (148–64). Thus, MacIntyre suggests, it is probably incorrect to consider happiness or pleasure the *telos* of human life for Aristotle; rather, it was the excellence of activity (160).

And of course, the highest activity resulting in supreme happiness was philosophical contemplation. For Aristotle, the reason for the *polis* to exist is to make possible the pursuit of excellence and the happiness that is concomitant with it (*Ethics* I. ii. 1094a20–1094b10; *Politics* VII). Indeed, to reduce MacIntyre's thesis to its simplest terms, the decline of both the philosophy of ethics and of virtue itself is marked by the breakdown in Western culture of a communal teleology and the shift to an individual moral authority and utilitarianism that can be seen, for instance, in the philosophies of Nietzsche and Bentham (MacIntyre 62–78; 256–63). This last point may be important when we consider some of the implications for rhetoric of the ethic of expediency in a capitalistic culture.

Thus, although the roots of totalitarianism have been perceived in Aristotle's conception of the *polis* as well as in Plato's conception of the republic (see Popper 1–26), and the darker side of the Greek *polis* itself has come under some scrutiny from rhetorical quarters (Miller, "Polis"), we may wish to locate the ethic of expediency that culminated in the holocaust not in Aristotle's corpus, but rather in the trace of subsequent history. For if MacIntyre is correct, not only Aristotle's concept of ethics but virtue itself has "deteriorated" under the pressure of individualism and the utilitarianism that individualism gives rise to. In any case, it is important to understand how the ethic of expediency that evolved in Western culture and underlies most deliberative discourse also at least partly formed the moral basis of the holocaust. And Aristotle's treatises can provide a clear point of reference.

It is not my purpose in this article to establish a direct connection between Aristotle and Hitler. There is little evidence in *Mein Kampf* to suggest that Hitler actually read Aristotle either when he "studied" in Vienna or while he was an inmate at Landsberg Prison, where he wrote *Mein Kampf*, although he almost certainly read or had secondhand knowledge of the work of Plato, as well as Fitche, Nietzsche (see *Mein Kampf* 579–81n.), and other German philosophers and historians (see Shirer 142–64). Indeed, in his early days in Vienna, Hitler "was a voracious reader" (Shirer 40), and throughout his life possessed a keen if selective passion for political writing and biographies of powerful leaders (see Shirer 1439). But it is my belief that Hitler, like those around him (see Speer 246), was at least familiar with Aristotle's work, especially the *Politics*. Machiavelli, Renaissance statesman, student of politics, and author of Hitler's "bedtime reading" (Gauss 8), almost certainly was (cf. Garver).

But it is crucial that we examine Hitler in conjunction with Aristotle's *Rhetoric*, *Nichomachean Ethics*, and *Politics* to see how Hitler used the ethic of expediency rhetorically to create a "moral" warrant for Nazi action. To do so, it will be necessary to turn to Hitler's writings, speeches and conversations (as collected, edited, and in some cases translated for the first time in the short but incisive *Hitler* by George H. Stein). For it is in his writings, speeches, and conversations that Hitler lays bare not only his political program, but the ethic of expediency that guided it.

HITLER'S "ETHICAL" PROGRAM?

Although the characterization seems hard to swallow, Hitler's *was* an "ethical" program in the broadest sense of that term. As Stein writes in a prefatory remark, "In *Mein Kampf*, Hitler set down clearly and systematically his principles for political action" (45). Indeed, in *Mein Kampf* Hitler asks: "Can spiritual ideas be exterminated by the sword? Can 'philosophies' be combated by the use of brute force?" (51)[1] If Aristotle maintains in the *Nicomachean Ethics* (VI. xii. 1143b16–xiii. 1145a14) that "practical wisdom" must be accompanied by "moral virtue" to supply the right end, that "it is not possible to be good in the strict sense without practical wisdom, nor practically wise without being good" (VI. xiii. 1144b30), Hitler maintains that the application of technique and power must be based on a "spiritual idea," a philosophy, to be successful. Hitler understood—all too well—that his political program for world war and mass extermination would not be accepted without a moral foundation. While "the continuous and steady application of the methods for repressing a doctrine, etc., makes it possible for a plan to succeed," Hitler proclaims, "this persistence . . . can always and only arise from a definite spiritual conviction. Any violence which does not spring from a firm, spiritual base, will be wavering and uncertain. It lacks the stability which can only rest in a fanatical outlook" (52).

For Hitler, as for Aristotle—at least in his discussion of deliberative rhetoric—there seems to be no distinction between "practical wisdom" and "moral virtue," between expediency and the good, as long as rhetoric serves its end, that is, the State. Thus Hitler asserts: "Conceptions or ideas, as well as movements with a definite spiritual foundation, regardless of whether the latter is false or true, can, after a certain point in their development, only be broken with technical instruments of power if these physical weapons are at the same time the support of a new kindling thought, idea, or philosophy" (51). In Hitler's rhetoric, expediency is the necessary good that subsumes all other goods, and becomes the basis of virtue itself.

And depending on how one interprets the word "support" in the previous quotation, there were two possible ways in which expediency might become the basis of virtue for Hitler: politically and technologically. In the first interpretation, "support" can be read to mean that the technical instruments of power must be used in the service of (must implement and enforce) a new political philosophy. In the second interpretation, "support" can be read to mean that the technical instruments of power must themselves become the basis of (must embody and engender) a new "technological philosophy." In other words, for Hitler there seem to be two kinds of expediency that can be used to supplant an existing morality: political expediency, motivated by a "concern" for the State (at least ostensibly), and technological expediency, motivated by technology itself.

Thus, to see how Hitler "takes" the Aristotelian notion of expediency and combines it with technology to create a new moral order, it is useful to make a distinction here between expediency based on politics and expediency based on technology. I have already mentioned that for Aristotle, if the end or "good" in deliberative discourse is political expediency, the function of the "ideal" state is to supply the material means necessary to secure "happiness" and the "good life" for *its* citizens—their moral and intellectual development. These material means included enough people and land to be self-sufficient (*Politics* VII. iv. 1326a5–v. 1327a10), a defense against enemies, both external and internal, both in the present and in the future (V; VII. vi. 1327 all–1327b15; xi. 1330b35–1331a17), and a large slave class (I. v. 1254a18–vi.

[1] All page numbers following Hitler quotations are from *Hitler* by George H. Stein.

1255b15; VII. ix. 1328b25–x. 1330a34). (Based on the ethic of expediency, it also included killing deformed children or mandatory abortion to control the population of the state! [*Politics* VII. xvi. 1335b20–28].)

Hitler almost seems to put Aristotle's observations into practice. In his political speeches and writing, Hitler continually proclaimed the political (i.e., "ethical") need and practical utility of conquering Europe and enslaving its farmer peasants, turning Russia into "Germany's India" (63), and exterminating the Jews and other "inferior, subhuman species" in order to eradicate "social disease" and facilitate the moral, material, and intellectual development of the German people. In Hitler's oratory and mind run amok, the Final Solution was necessary because neither exile nor quarantine of the Jews could guarantee the purity, safety, and well-being of the Aryan race.

But Hitler unfortunately also understood that the moral grounds for war and mass extermination could be rhetorically founded on science and technology themselves. Science and technology as moral expedients could be used to generate a "new philosophy," a "spiritual foundation," a "fanatical outlook." There was the belief in genetic hygiene and Germanic superiority grounded in racial biology as well as natural selection (see Proctor). But in addition, grounded in the ethic of expediency, "the technical instruments of power" themselves, "the physical weapons" as well as the political program they served, also could be the rhetorical basis of the spiritual element.

In Nazi Germany (and I will suggest, in our own culture) science and technology become the basis of a powerful ethical argument for carrying out any program. Science and technology embody the *ethos* of objective detachment and truth, of power and capability, and thus the logical and ethical necessity (what Winner has called the "technological imperative" [*Autonomous* 100–06]) for their own existence and use. Sullivan arrives at a similar conclusion (379). But in Just's argument for technical improvements to the gassing vans, we see the technological imperative at its worst. Technological expediency actually subsumes political expediency and becomes an end in itself. Progress becomes a virtue at any cost.

Thus, the theoretical distinction I just made between technological and political expediency breaks down in practice. Technology is political (see Winner, "Artifacts"; *Autonomous*). Both technology and politics can become the basis of ethics; both lead to power. But technology can become the basis of political as well. Based on what we now know about the holocaust, there can be no doubt that Hitler believed in the efficacy of science and technology, no matter how perverted, as the basis of ethics and politics. "A movement like ours mustn't let itself be drawn into metaphysical digressions." Hitler states; "It must stick to the spirit of exact science" (69).

The result: Just's memo. Mass extermination. Horrible biological and technological experiments on those considered subhuman. A cold-blooded methodology the standard for dealing with the Jews, as well as with the conquered. A cold-blooded method the *ethos* of an entire country. Gas chambers replacing vans, systematically "processing" hundreds of thousands of "pieces" a day. New and improved methods for administering pain and eliminating people. The whole society organized into a death machine for the efficient extirpation of millions, lauded by the Nazis as a hallmark of organization, elegance, efficiency, speed, all of which became ends in themselves for those planning and those executing the procedures.

For Hitler, technological expediency served to make mass extermination seem not only necessary, but just and honorable: "every persecution which occurs without a spiritual base seems morally unjustified," says Hitler (51–52). It is the ethic of technological expediency that we sense in the memo by Just to the SS—if we sense any ethic at all. Underlying the objectivity, detachment, and narrow focus of this memo (and of Nazi rhetoric in general) is an assurance that the writer's "action" is *technically* justified and correct, and thus morally right, an assurance that is grounded not in the arrogance of a personal belief in one's superiority, but rather in a cultural and ethical form of technology as well as Party. The ethic of technological expediency that underlies this memo and constitutes its *ethos* at least in part provided the warrant that propelled Nazi Germany into the forefront of war and of infamy. Perhaps this ethic can explain the cold logic with which Just addresses the gassing of innocent people. Perhaps the ethic—as well as apathy, and fear, and hatred—can explain the complicity of millions.

THE TECHNOLOGICAL *ETHOS* AND NAZI RHETORIC

To further understand how the ethic of expediency based on technology partially formed the moral basis of the holocaust, and to begin to realize the implications of this for rhetoric, it would be useful to understand the *ethos* of technology a little more, how rhetoric was used to create it, and what its effect on rhetoric was. While I don't mean to suggest this is the "final answer" to that question murmured so many times before—how could the holocaust have happened?—the imperatives of science and technology as moral expedients create a powerful *ethos* that may partly explain what occurred. As Jacques Ellul discusses at length in *The Technological Society*, technology, the embodiment in techniques and procedures as well as machines of scientific method, becomes its own *raison d'etre* and driving force in culture. Technology becomes both a means and an end in itself.

In addition, Jurgen Habermas argues that in late industrial capitalism, technological values do indeed subsume political/economic ones, and that this "purposive-rational subsystem" of industrial capitalism quietly usurps the "traditional-institutional framework" of social customs, values, and beliefs (90–107). That is, a "technological rationality" that calculates the value of everything in terms of its own technical criteria and use (and that drives postmodern economics, for example), supplants and replaces the traditional values of the society. In Just's memo, we see that technical improvements to the vans become the only criteria necessary to consider.

Obviously, "technological rationality" is based on expediency. Unlike honor or justice, which are based on higher, more abstract moral principles, expediency is the only "technical" ethic, perhaps the only ethic that "pure rationality" knows. (Stein even calls Hitler a "religious rationalist" [67].) With expediency, the only ethical criterion necessary is the perceptible movement toward the technical goal to be achieved—including expediency itself. Indeed, expediency is the only ethic that can be "measured," whether that measure be a cost-benefit analysis employed by an industrial engineer to argue for the automation of a plant, or the number of people exterminated in one day—"pure" expediency (undiluted and uninhibited by other ethics) recognizes no boundaries, no degrees of morality or other ethical limits. While expediency can be the basis of desire and emotion (like greed or the lust for power), the ethic of expediency is an exclusively logical, systematic, even quantifiable one, can lead to a rationality grounded in no other ethic but its own, and is symptomatic of a highly scientific, technological age.

And of course, technology is the embodiment of pure expediency. Thus, "the spiritual element," the *ethos* of technology, is expediency: rationality, efficiency, speed, productivity, power. It is in this way that technology creates the "ethical appeal" I mentioned earlier. Both science and technology are "a good" not only because they are a rational means for accomplishing a task and/or achieving leisure and thus happiness (the virtues heard most in regard to scientific and technological progress), but because they are ethical ends in themselves as well. As Carolyn Miller points out, the *ethos* of technology can even become a form of consciousness (see "Technology"). And as Heidegger expounds, the essence of science and technology is "enframing," a manifestation and mode of perception and of being that arrests, objectifies, turns everything into a "standing-reserve" for use (14–49).

In Nazi Germany, where gold fillings were extracted from the teeth of the victims of the gas chambers and melted down and the hair of victims was used "in the war effort," we see the ethic of expediency taken to extremes. Germans under Nazi rule were an efficient people of an industrious nation who totally lost themselves in the *ethos* of technology. The holocaust reminds us not only of the potential brutality and inhumanity of the ethic of expediency, but of a rationality taken to such extremes that it becomes madness.

How did this *ethos* come about? If Hitler used the ethic of expediency as first treated in Aristotle's *Politics* as part of the moral basis of his political program (significantly, his fervent appeals to the "Platonic" right of the Third Reich were the other part), he used the ethic of expediency first treated in Aristotle's *Rhetoric* to create the technological *ethos* of Nazi consciousness and culture. Based on that ethic of expediency, Hitler can be understood to have turned Aristotle's concept of deliberative rhetoric inside out, exploiting the ethic of expediency that underlies and enables it and essentially turning deliberative rhetoric against itself. To understand how Hitler perverted Aristotle's concept of deliberative rhetoric to create the *ethos* of Nazi Germany, we must look more closely at Hitler's conception of rhetoric.

We have seen that Just's memo is based purely on expediency; the memo itself is a technical instrument (like the vans themselves) for carrying out the organizational "task." I have also already pointed out how in Aristotle's conception of deliberative rhetoric, expediency seems to be the primary virtue. Deliberative rhetoric is expedient when it serves its end, that is, political persuasion. The test of success in Aristotelian rhetoric is in the persuasion of the audience (the so-called "audience criterion"). As "the art or faculty of observing in any given case the available means of persuasion" (*Rhetoric* I. ii. 1355b26), then, rhetoric could be considered a means to an end, an expedient, a *techne* (although as Grimaldi and others have shown, for Aristotle it was much more than this; for Aristotle rhetoric was also an *episteme* or faculty for discovering social knowledge).

Hitler takes the ethic of expediency underlying deliberative rhetoric to its logical extreme. For Hitler, propaganda, the truest form of "technical rhetoric," replaced deliberative discourse as the preferred mode of communicating with the masses:

> The function of propaganda does not lie in the scientific training of the individual, but in calling the masses' attention to certain facts, processes, necessities, etc., whose significance is thus for the first time placed within their field of vision.
>
> The whole art consists in doing this so skillfully that everyone will be convinced that the fact is real, the process necessary, the necessity correct, etc. (46)

Based on the ethic of expediency, rhetoric for Hitler was pure technique, designed not to encourage debate, but rather to indoctrinate: "all effective propaganda must be limited to a very few points and must harp on these slogans until the last member of the public understands what you want him to understand by your slogan"; the reason, Hitler adds, is that "As soon as you sacrifice this slogan and try to be many-sided, the effect will piddle away, for the crowd can neither digest nor retain the material offered. In this way the result is weakened and the end entirely cancelled out" (47). Even in these abbreviated quotations we see not only a greater (political?) distrust of the masses than we find in Aristotle (*Rhetoric* I. ii. 1357a5), but also a greater "technical" preoccupation with the end to be achieved, both of which tend to work against free discussion, true deliberation.

In fact, founded on the ethic of expediency and taken to extremes, rhetoric itself becomes a kind of technology, an instrument *and an embodiment* of the end that it serves. In *Mein Kampf* Hitler asks, "Is propaganda a means or an end? It is a means, and must therefore be judged with regard to its end. It must consequently take a form calculated to support the aim which it serves" (45). In Nazi Germany, propaganda served the function of creating the technological/political basis for the new order, which, given the ethic of expediency, becomes the moral basis for it as well. As Hitler states, "The first task of propaganda is to win people for subsequent organization; the first task of organization is to win men for the continuation of propaganda. The second task of propaganda is the disruption of the existing state of affairs and the permeation of this state of affairs with the new doctrine" (49).

Propaganda thus served to create the technical *ethos* of Nazi consciousness and culture: rationality, efficiency, speed, productivity, power. In fact, as a technology, propaganda itself embodies this *ethos*, actually becomes personified in Hitler's rhetoric as existing for those ends only. If Aristotle observes that deliberative discourse is based on questions of expediency rather than just or honor, Hitler declares that "The function of propaganda is . . . not to weigh and ponder the rights of different people, but exclusively to emphasize the one right which it has set out to argue for. Its task is not to make an objective study of the truth, in so far as it favors the enemy, and then set it before the masses with academic fairness; its task is to serve our own right, always and unflinchingly" (47).

For Hitler, this technical *ethos* was necessary to create the rhetorical/moral basis for the violence and brutality to which he incited the German masses. If Aristotle observes that for political orators "all other points, such as whether the proposal is just or unjust, honourable or dishonourable, are subsidiary and relative and have little place in deliberative discourse. . . . [W]hether it is not *unjust* for a city to enslave its innocent neighbors often does not trouble them at all" (*Rhetoric* I. iii. 135b25; 1358b35), Hitler insists that in questions of political struggle, "all considerations of humanitarianism or aesthetics crumble to nothingness . . ." (45).

Finally, if the purpose of Hitler's propaganda was to instill in the German people an *ethos* of detachment and power by which the Aryan race would build the Third Reich, as leader of this race Hitler sought to embody this *ethos* himself: "the masses love a commander more than a petitioner and feel inwardly more satisfied by a doctrine, tolerating no other beside itself . . ." (42–44). If ethical appeal, the most important of the three appeals for Aristotle (*Rhetoric* I. ii. 1356a4), is created when the speaker convinces the audience that he or she possesses sound sense, high moral character, and good will (II. i. 1378a9), Hitler redefines these ethical categories based on the ethic of expediency, reducing them to their basest, "technical" level. In the ethical system Hitler rhetorically created for "the master race," sound sense is reduced to expediency, high moral character is reduced to courage to use brutal force, and good will is reduced to "benevolent violence" against those considered inferior: "When I think about it, I realize that I'm extraordinarily humane. . . . I restrict myself to telling them they must go away. If they break their pipes on the journey, I can't do anything about it. But if they refuse to go voluntarily, I see no other solution but extermination" (72).

In word and act, Hitler created an *ethos* of expediency in order to carry out his pogrom for the greater good of Germany: "The people at all times see the proof of their own right in ruthless attack on a foe, and to them renouncing the destruction of the adversary seems like uncertainty with regard to their own right if not a sign of their own unright" (50). It was an *ethos* that Hitler thought necessary for the German people to embrace and adopt as well: "Close your hearts to pity. Act brutally. Eighty million people must obtain what is their right. Their existence must be made secure . . ." (76).

It is clear that Hitler combined the ethic of expediency embedded in rhetoric with technology to create the *ethos* of Nazi Germany. That is, Hitler used technological expediency to create the *polis* necessary to carry out world war and mass extermination. In addition, the ethic of expediency then served as the *telos*—"the will to power"—of that *polis*. It is therefore also clear that the *telos* within a *polis* is not universal but socially constructed and relative, and renders ethics that are based on and serve them relative as well. MacIntyre too recognizes this (159). In fact, if we understand Aristotle's acceptance of slavery as a reflection of "the blindness" of his culture (MacIntyre 159), then perhaps we can also understand the holocaust as a reflection of "the blindness" of Nazi culture as well—a political and technological blindness deliberately created in and through rhetoric.

This is in no way meant to diminish or forgive the profound tragedy of the holocaust. Nor is it meant to devalue rhetoric. Rather, it is to bring home the significance of the holocaust for our understanding of the essential relationship between rhetoric and ethics. In considering that relationship, we must always look at rhetoric in the context of historical, political, social, and economic conditions which govern the nature of and use of rhetoric in culture. In Just's memo to the SS we clearly see the view of human beings that can result when technology becomes an *ethos*, when a *polis* embraces a "pure" ethic of expediency as its *telos*. To understand the holocaust from a rhetorical point of view is to understand the *extreme* limits and inherent dangers of the prevailing ethic of expediency as ideology in a highly scientific and technological society, and how deliberative rhetoric can be subverted and made to serve it.

EXPEDIENCY IN TECHNOLOGICAL CAPITALISM: THE "FINAL PROBLEM" FOR US

Having said this, I think it is important in the conclusion of this paper to briefly explore the implications of the ethic of expediency manifested in Nazi Germany for rhetoric in our capitalistic culture. Certainly, our *polis* is as different from Nazi Germany's as Nazi Germany's was from ancient Greece's. While the *telos* of the ancient Greek *polis* was the intellectual development of the mind (for its few "citizens" anyway), the *telos* of the Nazi *polis* was the development of the power of the State itself, as embodied in technology, Party, and Führer. And while the *polis* in both ancient Greece and Nazi Germany can be understood to have had a communal *telos*—the development of the State (though for different ends)—the *telos* of our *polis* is understood to be the individual. Individualism is the basis of both democracy and capitalism.

I said earlier that MacIntyre believes that Aristotle's concept of ethics and virtue itself have "deteriorated" under the pressure of individualism and the utilitarian ethic that individualism spawned. As MacIntyre suggests, we probably can't understand happiness as Aristotle did. We may not understand Aristotle's concept of expediency either. Whether ethics have actually "deteriorated" or not, with the shift in moral authority from the State to the individual, *personal* happiness has become the goal of life in the United States. And that happiness has come to be defined primarily in economic terms. I think it can be asserted without too much argument that the *telos* of life in the United States is economic progress. In the United States, success and happiness, both personal and communal, are measured in monetary terms. In a capitalistic culture, it is "economic expediency" that drives most behavior.

Further, that expediency is both political *and* technological. I have already mentioned how Habermas believes that in postindustrial societies technological and political values unite and subjugate the traditional values of those societies with a technological rationality that calculates the worth of everything in terms of its own "technical" aims. In our capitalistic society, economic rationality, facilitated by and dedicated to the development of new technologies, is one manifestation of this. The danger, then, is that technological expediency in the guise of free enterprise can become de facto both a means and an end. That is, in our culture, the danger is that technological expediency (unlike happiness for Aristotle, which appears to be only a part and result of virtue) can become the only basis of happiness, can become a virtue itself, and so subsume all ethics under it, making all ethics expedients and thus replacing them. According to Habermas, this has already occurred.

The ethic of expediency *in extremis* and combined with technology underlies the rhetoric of Just's memo to the SS and the holocaust in general. But *to some extent*, technological (i.e., economic) expediency is the "moral" basis of many decisions/actions in our society that sometimes harm human welfare or imperil human life. A recent example would be the decision not to notify the public of the bomb threat to Pan Am Airlines to keep the airlines operating; in December 1988, Pan Am Flight 103 from London to New York exploded over Locherbee, Scotland, killing all two hundred and seventy people on board. Ethically speaking, the difference is only one of degree, not kind. The decision not to notify the public was a "systems decision," concerned more with the "efficient" operation of the transportation system than with the people the system is supposed to serve. In any highly bureaucratic, technological, capitalistic society, it is often the human being who must adapt to the system which has been developed to perform a specific function, and which is thus always necessarily geared toward the continuance of its own efficient operation (see Winner, *Autonomous*, especially 238–48). In a capitalistic society, technological expediency often takes precedence over human convenience, and sometimes even human life.

Now, I am not saying that science and technology are inherently fascist, or that we are becoming like the Nazis. Nor am I saying that expediency is all bad. It can be and is used to argue for increased safety or to otherwise enhance human welfare. What I am saying, however, is that expediency as we understand it in our culture in the twentieth century, as a technological end in itself, is problematic. The ethic of expediency that provides the moral base of deliberative discourse used to make decisions, weigh consequences, and argue results in every department of society, also resulted in the holocaust—a result that raises serious and fundamental questions for rhetoric. (This is especially important when so many of our decisions, so much of our discourse, both public and professional, is technical in nature, and is therefore most likely to be dominated by the ethic of expediency.)

If technology can become a form of consciousness, as Miller suggests, and technological expediency in the guise of economic rationality can become our *telos*, then deliberative rhetoric—devoted to the use of reasoned debate to arrive at informed consensus and decisions in a democracy—could become nearly impossible, at least as far as technological/political issues are concerned. Although in "Rhetoric of Decision Science" Miller holds up deliberative rhetoric as a form of reasoning that is opposed to decision science—a technique based on technological rationality that is used to make managerial decisions by quantifying all the variables—we have seen that based on the ethic of expediency that underlies and enables it, deliberative rhetoric can be made to serve exclusively the technological interests of "the State." Although not a decision "science," deliberative rhetoric could become technological, replacing the democratic decision-making process with *techniques* of persuasion and audience adaptation calculated to serve their own end only. Some would argue it already has.

Although I can't explore it here, there are many parallels between Hitler's propaganda techniques and contemporary political campaigns and commercial advertising in the United States. Rhetoric, especially the "rhetoric of science and technology," is increasingly being called upon and used to make or justify decisions based on technological expediency—to create the necessary technological *ethos* for accepting actions or events, especially in military procurement and operations, and in the management of risky technologies such as hazardous waste disposal facilities or nuclear power plants.

The question for us is: do we, as teachers and writers and scholars, contribute to this *ethos* by our writing theory, pedagogy, and practice when we consider techniques of document design, audience adaptation, argumentation, and style without also considering ethics? Do our methods, for the sake of expediency, themselves embody and impart the ethic of expediency? If *telos* is politically constructed and ethics are culturally relative, we must realize the role our rhetoric plays in continually creating, recreating, and maintaining not only knowledge, but values as well—including the value of technological expediency—through how we teach rhetoric, and how we use it.

And if we do contribute to this *ethos*, what can we do to counter it? We can begin by recognizing the essentially ethical character of all rhetoric, including our writing theory, pedagogy, and practice, and the role that expediency plays in rhetoric. We no longer have the luxury of considering ethics outside the realm of rhetoric, as in the Platonic model of knowledge, for the holocaust casts serious doubt upon this model. And Aristotle's division of ethics in rhetoric according to audience and function (deliberative, forensic, epideictic), is appealingly *useful* but problematic and ultimately limited. For based on that division, and the ethic of expediency in deliberative rhetoric under which we have operated, Aristotle does not seem to consider other ethics, such as honor and justice (or kindness and humility) important in deliberative discourse—at least not for their own sake.

In the gruesome light of the holocaust, then, we should question whether expediency should be the primary ethical standard in deliberative discourse, including scientific and technical communication, and whether, based on Cicero's advocacy of a rhetoric grounded in a knowledge of everything and Quintilian's definition of the orator as "a good 'man' skilled in speaking," we can and should teach the whole panoply of ethics in deliberative discourse in our rhetoric and writing courses. We could start with Just's memo. Perhaps we should even begin to question whether "happiness"—as *we* understand it in our individualistic and utilitarian culture, as personal or corporate gain grounded in economic progress—should be the only basis of virtue and the primary goal of human life. For when expediency becomes an end in itself or is coupled with personal or political or corporate or scientific or technological goals that are not also and ultimately rooted in humanitarian concerns, as is often the case, ethical problems arise. (Of course, this presumes that we can define and agree upon what these "humanitarian concerns" are—a presumption which is not at all certain, given the "true" relativity of values, the multiplicity of needs, and the current climate of personal and corporate greed.)

But I trust we can agree that Hitler's rhetoric, politics, and ethics are not based on "humanitarian concerns." I also hope we can agree that Hitler's rhetoric, politics, and ethics are not only based on insane hatred and racial prejudice, but also on the ethic of expediency carried to extremes and unchecked by any other ethical concerns, on science, technology, and reason gone awry. For in an age when it is sometimes considered "economically rational" to accept high insurance costs on plane crashes rather than improve the safety of planes; when Ford Motor company decided that it would be more cost-effective to incur the law suits (and loss of life) caused by the placement of the gas tank on the Pintos rather than fix the problem, and only changed its mind when an equally expedient solution was found; when *person*nel are now referred to as Human Resources, like shale or oil, with the metaphorical implications that they (we) can be used up and disposed of or replaced when need be; when launch dates are more important than the safety of astronauts and production quotas more important than the safety of workers and residents alike; when expediency outweighs compassion in government and cost/benefit analyses are applied to human welfare and technical considerations outweigh human considerations in almost every field of endeavor, even in the social sciences and humanities—when every field strives to be scientific and technical and decisions are made and consequences weighed and value argued on the ethic of expediency only—the holocaust may have something to teach those of us in technical communication, composition, and rhetoric.

Works Cited

Arendt, Hannah, *Eichmann in Jerusalem: A Report on the Banality of Evil*. New York: Viking, 1963.

Aristotle. *Nicomachean Ethics*. Trans. W. D. Ross. *Introduction to Aristotle*. Ed. Richard McKeon. New York: Modern Library, 1947.

—. *The Politics*. Ed. S. Everson. Cambridge: Cambridge UP, 1988.

—. *The Rhetoric*. Trans. W. Rhys Roberts and I. Bywater. *The Rhetoric and Poetics of Aristotle*. New York: Modern Library, 1954.

Burke, Kenneth. "The Rhetoric of Hitler's 'Battle'." *The Philosophy of Literary Form: Studies in Symbolic Action*. 3rd ed. Berkeley: U of California P, 1973. 191–220.

Cicero. *De Oratore*. Trans E. W. Sutton and H. Rackham. Vol. 2. Cambridge, MA: Harvard UP, 1942. 2 vols.

Dobrin, D. N. "Is Technical Writing Particularly Objective?" *College English* 47 (1985): 237–51.

Ellul, Jacques. *The Technological Society*. Trans. John Wilkinson. New York: Knopf, 1964.

Garver, Eugene. *Machiavelli and the History of Prudence*. Madison: U of Wisconsin P, 1987.

Gauss, Christian. Introduction. *The Prince*. By Niccolo Machiavelli. Trans. Luigi Ricci. New York: New American Library, 1952. 7–30.

Gibson, Walker. *Tough, Sweet and Stuffy: An Essay on Modern American Prose Styles*. Bloomington: Indiana UP, 1966.

Grimaldi, William M. A. *Studies in the Philosophy of Aristotle's Rhetoric*. Weisbaden, Germany: Franz Steiner Verlag GMBH, 1972.

Habermas, Jurgen. "Technology and Science as 'Ideology'." Trans. Jeremy Shapiro. *Toward a Rational Society: Student Protest, Science, and Politics*. Boston: Beacon P, 1970. 81–127.

Heidegger, Martin. "The Question Concerning Technology." *The Question Concerning Technology and Other Essays*. Trans. William Lovitt. New York: Harper, 1977. 3–35.

—. "The Turning." *The Question Concerning Technology and Other Essays*. Trans. William Lovitt. New York: Harper, 1977. 36–49.

Hitler, Adolph. *Mein Kampf*. New York: Reynal, 1941; Houghton, 1939.

Johnson, Nan. "Ethos and the Aims of Rhetoric." *Essays on Classical Rhetoric and Modern Discourse*. Ed. Robert Connors, Lisa S. Ede, and Andrea Lunsford. Carbondale: Southern Illinois UP, 1984. 98–114.

Kallendorf, Craig and Carol Kallendorf. "Aristotle and the Ethics of Business Communication." *Journal of Business and Technical Communication* 3 (1989): 54–69.

Lanzmann, Claude. *Shoah: An Oral History of the Holocaust*. New York: Pantheon, 1985.

Lloyd, G. E. R. *Aristotle: The Growth and Structure of His Thought*. Cambridge: Cambridge UP, 1968.

MacIntyre, Alasdair. *After Virtue: A Study in Moral Theory*. 2nd ed. Notre Dame, IN: U of Notre Dame P, 1984.

Mathes, J. C., and D. W. Stevenson. *Designing Technical Reports: Writing for Audiences in Organizations*. Indianapolis: Bobbs-Merrill, 1976.

Miller, Carolyn R. "A Humanistic Rationale for Technical Writing." *College English* 40 (1979): 610–17.

—. "The Polis as Rhetorical Community." Paper presented at the International Society for the History of Rhetoric. Gottingen, West Germany, 1989.

—. "The Rhetoric of Decision Science, or Herbert A. Simons Says." *The Rhetorical Turn: Invention and Persuasion in the Conduct of Inquiry*. Ed. Herbert W. Simons. U of Chicago P, 1990. 162–84.

—. "Technology as a Form Consciousness: A Study of Contemporary Ethos." *Central States Speech Journal* 29 (1978): 228–36.

Olsen, Leslie A., and Thomas N. Huckin. *Principles of Communication for Science and Technology*. New York: McGraw-Hill, 1983, (Rpt. as *Technical Writing and Professional Communication*. 2nd ed. New York: McGraw-Hill, 1991.)

Popper, Karl R. *The Open Society and Its Enemies*. Vol. 2. New York: Harper 1962. 2 vols.

Proctor, Robert. *Racial Hygiene: Medicine Under the Nazis*. Cambridge, MA: Harvard UP, 1988.

Rowland, Robert C., and Deanna F. Womack. "Aristotle's View of Ethical Rhetoric." *Rhetoric Society Quarterly* 15 (1985): 13–31.

Shirer, William L. *The Rise and Fall of the Third Reich: A History of Nazi Germany*. Greenwich, CT: Fawcett, 1959.

Speer, Albert. *Inside the Third Reich*. Trans. Richard Winston and Clara Winston. New York: Avon, 1971; Macmillan, 1970.

Stein, George H., ed. *Hitler*. Englewood Cliffs, NJ: Prentice Hall, 1968.

Steiner, George. *In Bluebeard's Castle: Some Notes Towards the Redefinition of Culture*. New Haven, CT: Yale UP, 1971.

Sullivan, Dale L. "Political-Ethical Implications of Defining Technical Communication as a Practice." *Journal of Advanced Composition* 10 (1990): 375–86.

Wiesel, Elie. "Art and the Holocaust: Trivializing Memory." *New York Times* 11 June 1989, sec. 2: 1+.

Winner, Langdon. *Autonomous Technology: Technics-out-of-Control as a Theme in Political Thought*. Cambridge, MA: MIT P, 1977.

—. "Do Artifacts Have Politics?" *Daedalus* 109 (1980): 121–36.

DEVELOPING YOUR UNDERSTANDING

1. Define what Katz means by an ethic of expediency.
2. Based on Katz's argument, explain the relationship between rhetoric and ethics, or more specifically the relationship between deliberative rhetoric and an ethic of expediency.
3. Katz argues that "it is the ethic of expediency that enables deliberative rhetoric." Identify what one or two alternatives to an ethic of expediency might be. For each, explain the types of rhetoric they would enable.

4. In several instances, Katz states clearly what he *is* and is *not* trying to argue. Locate and list these instances, and use them to develop an abstracted view of his argument. Then, forecast what value his argument might have for professional writers.
5. Develop two or three alternative scenarios describing what Just, the Nazi memo writer, might have done instead of writing the memo he wrote. Identify the ethical choices he would have had to make, the constraints under which he would have had to make them, and the probable consequences of his alternative actions.

FOCUSING ON KEY TERMS AND CONCEPTS

Focus on the following terms and concepts while you read through this selection. Understanding these will not only increase your understanding of the selection that follows, but you will find that, because most of these terms or concepts are commonly used in professional writing and rhetoric, understanding them helps you get a better sense of the field itself.

1. phronesis
2. deliberation
3. heuristic
4. rhetoric/writing

FRAMING POSTMODERN COMMITMENT AND SOLIDARITY

JAMES E. PORTER

I am arguing for politics and epistemologies of location, positioning, and situating, where partiality and not universality is the condition of being heard to make rational knowledge claims.

—Haraway, 1991, p. 195

The position I am developing is a postmodern pragmatic one that involves a lot of deliberation and balancing of competing perspectives. You might consider it a kind of neo-Aristotelian brand of postmodernism because of its revival of *phronesis* as an important faculty of the art of rhetoric—and yet it is also very much sophistic. I am less interested, however, in naming it than I am in describing its operation.

This rhetorical ethics begins by seeing values and language use as inextricably bound. It sees writing as an act fraught with ethical issues. The stance I am advocating is respectful of law, but does not see law as a determining or ultimate authority. It puts a high value on community customs, conventions, and beliefs (*nomoi*), but recognizes that communities can be wrong, hegemonic, or oppressive. It puts a high value on the right of individual free expression and the importance of protecting that right, but also notes that individuals can wreck havoc on communities, and that,

Source: *Rhetorical Ethics and Internetworked Writing,* James E. Porter, "Framing Postmodern Commitment and Solidarity," 1998, pp. 149–166.

ultimately, the individual is not a very solid or trustworthy ground for moral authority. It places a high value on the spirit of caring, of brotherly and sisterly love, of friendship—of a number of related principles that have status in numerous religious and cultural traditions, which we might sum up with the phrase "love and respect for others" (although that notion admits to a wide range of interpretations). It assigns a preferential option for the poor, oppressed, and marginalized, saying in effect that any ethical system based mainly on the principle of operative equality will fail to account for systemic inequities and, thus, will risk perpetuating those inequities. It places a high value on involvement and collaboration with others (and with The Other) and on developing moral positions within a collaborative and cooperative framework of dialogue. It places a high value on deliberation (a key principle in Aristotle's [1976] *Nicomachean Ethics*), the importance of carefully considering each situation in an antilogical manner before pushing ahead. It places a high value on action: Deliberation is necessary, but to overindulge in it is to defer action and, ultimately, to freeze oneself into inaction. It stresses the operation of *phronesis*, the quality of being able to determine what one should do, how one should act, and what one should be even when the circumstances are murky, the issues complex, and the right action hard to determine. It places a high value on the constraints of the particular case.

A list of counterprinciples is not very satisfying, I realize, because such an approach does not generate clear and concrete answers to pressing problems. What it does is suggest a process by which problems might be addressed. It does suggest some key principles one should bring to bear on rhetorical/ethical problems, as suggested earlier. It does not, however, tell us which principle ought to control a decision in any given case or how to reconcile opposing principles. It does suggest criteria for judging the process—and in that sense is Habermasian.

The whole point of my position is that we should not attempt to develop a moral geometry—a scientific, rigorous, rules-based approach to settling ethical decisions. In this respect, my rhetorical ethics differs sharply from the approach of Kantian-type ethical theories (like Rawls', 1971). My approach is closer to Aristotle's in its insistence on the necessity of developing "the art of moral judgment" (Benhabib, 1992, p. 53). This art of moral judgment requires a strong sense of values and priorities and principles, but does not leverage those values and priorities in top-down, rules-oriented procedure. It places a high value on reflection, deliberation, and "theorizing," but does not place much value in Theory, as the abstract formulation of determining rules and principles. Rather, it places values and priorities into a heuristic tension (a) with competing values and priorities, and (b) with the concrete particulars of ethical and rhetorical situations involving real people and particular technologies.

My approach is also an effort to develop a procedural heuristic that will assist the rhetorical ethical *process*, that is, that will help writers and writing teachers in the act(s) of producing discourse. Too often ethical systems of the philosophical sort fall into a backward-looking hermeneutic rut: they focus too much on post facto moral justification (or critique) of past action and do not provide us with much help figuring out what to do *now* as we write *today*. When one writes, one decides. Writing is an action involving an ethical choice about what one is to be and what one is to do. At the point when you begin to write, you begin to define yourself ethically. You make a choice about what is the right thing to do—even if that choice is a tentative and contingent one.

As I argued earlier (in Chapters 2 and 3), adopting this point of view requires seeing rhetoric *as* both productive and practical art, not merely as one or the other. It requires seeing rhetoric and ethics as distinct yet also overlapping arts. Writing is not only a making (*poesis*), but it is also a doing (*praxis*). Writing is an action with ethical and political consequences. Writing is a praxis that is both an act and a relationship (Dussel, 1988), that is, it does something to someone, one part of which is establishing power relations (e.g., between writers and readers, among various audiences).

Paulo Freire (1970/1993) echoes this important point as well in his description of the two dimensions of the word: reflection and action. He argues that:

> When a word is deprived of its dimension of action, reflection automatically suffers as well; and the word is changed into idle chatter, into *verbalism*, into an alienated and alienating "blah." It becomes an empty word, one which cannot denounce the world, for denunciation is impossible without a commitment to transform, and there is no transformation without action. (p. 68, emphasis in original)

The view of language and writing as *poesis* only ("verbalism") renders language into impotent, "idle chatter." Viewing language and writing as also action allows for commitment and transformation. The praxis that my rhetorical ethics promotes is the coordinated and dialogic view that Freire espouses: viewing the full capacity of language/writing as both reflection and action in dialogic tension, as both a making and a doing.

So what does all this mean on an operational level? I mention now several procedural principles that I see deriving from this approach. Despite the many deep and murky ethical problems we seem to face on electronic networks, we can cull some principles and advance some arguments for a critical rhetorical ethics for internetworked writing, that is, we can make some positive statements about how one ought to act as a writer (and secondarily as a reader). I divide these principles into two categories: writer posture toward audience and procedural strategies.

The question of whether these principles do, or should, have any legal force is a legal question that I do not address here—although, of course, several of the discussions here have become intertwined with legal issues. The ethical cannot ignore the legal, nor the legal the ethical.

THE WRITER'S RELATIONSHIP FOOTING VIS-À-VIS AUDIENCES/READERS

Respect Audience/Respect Differences

If there is a single main principle of postmodern ethics, it is probably this—acknowledge differences. Postmodernism's primary critique of modernism is that in its move to rationality, impartiality, and objectivity, modernism obscures difference, and difference as an operating principle is seen by many postmodernists as fundamental. Show respect for—care for—others. Respect their differences. Heed the values of your audience, both the values of the community/culture at large and the differences among individual audience members.

What do we mean by "audience"? The postmodern injunction to show respect for difference extends beyond audience in the limited sense (that is, those whom one directly addresses in a discourse) to pertain to all those who are represented in or affected by such discourse as well. Audience in this broader sense refers to an entire community, not simply to a physical flesh-and-blood assemblage, not simply to the evident or immediate readers or listeners of a discourse (see Porter, 1992a). When we start thinking about the vastness and diversity of the potential audience(s) involved in the Internet and World Wide Web, then we can see that internetworked writing poses a challenge for conventional (and more constrained) print notions of audience.

A postmodern rhetorical ethics says that differences among audiences must not be obliterated. Distinct identities must be recognized. In a moral discourse readers should not be treated as an homogenous whole or as if they were all members of a universal collective. Difference must be respected, but we should move beyond merely "respect" for difference, which could be seen to promote a kind of distant, begrudging tolerance. Luce Irigaray (1984/1993) suggest that we should move actively to *embrace* difference and to *celebrate* it, and that we should certainly be in awe of the mystery of difference. When I was first joining academic LISTSERV groups in the late 1980s, I remember how such groups had to adjust to the presence of "outsiders," that is, people from other fields and disciplines would frequently participate in such groups and would occasionally invite the ire of the "insiders" by doing so. The insiders in the field would sometimes wave the flag of disciplinary expertise as a way to stifle the contribution of those contributing other viewpoints or working out of alternate disciplinary paradigms. (For example, sometimes lawyers on the CNI-COPYRIGHT group would express impatience with the nonlawyers. One speech communication theorist on the IPCT list, Interpersonal Computing Technology, would trot out his academic credentials whenever the discussion was threatened by a nonacademic viewpoint.) What I have noticed more recently, at least in the academic electronic groups I belong to, is that there is greater tolerance for disciplinary difference and often even an appreciation for multidisciplinary difference. I cannot say, however, that such a respect for difference extends much beyond academic notions of multidisciplinarity.

The importance of differences among audiences is a key, although seldom discussed, principle in Aristotle's *Rhetoric*. Yes, Aristotle's conception of audience is exclusively male, but what Aristotle builds into his art is the essential fact of difference: Considering differences between audience types (e.g., age, gifts of fortune, emotional disposition) is a crucial part of the art (Book 2). Unless the rhetor knows quite specifically the various characters and positions of his audience, he cannot begin to understand how to talk to them. Furthermore, Aristotle posits that rhetorical argument begins with audience belief, the basis for enthymemic reasoning. So in this sense, respect for audience standpoint is integral to Aristotle's conception of the art.

Irigaray (1984/1993) provides a direct critique of Aristotle's sexual monism. In her deconstruction of Aristotle's *Physics* (in *An Ethics of Sexual Difference*), she argues for physical sexual difference as the foundational ground for ethics. Both Irigaray (1984/1993) and Iris Marion Young (1990) point to difference as a foundational principle. For Irigaray, sexual difference is the foundation of a relational ethics. For Young, acceptance of the validity of differences among social groups must be the basis for political justice. Irigaray and Young both point to how traditional philosophy—or what Young calls "normative social philosophy and political theory" (p. 149)—has worked to obscure difference. The result of such an obfuscation is domination by a privileged group that hides its "groupness" (e.g., whiteness, masculinity) under the coordinated claims of objectivity, impartiality, neutrality, and universality. The first step, then, in achieving political or ethical justice is the process of acknowledging differences, including exposing the hidden or obscured differences that often end up dominating. The critical turn that Young makes is to view differences not as undesirable, or alien, or the Other (p. 170), not as the subordinate second term of a deconstructed binary, but as a value in its own right—that is, as "specificity, variation, heterogeneity" (p. 171). The turn Irigaray makes is a theological one: Embracing and celebrating difference brings us to a transcendent spirituality.

Charles Scott (1990) reminds us of the importance of being "alert to exclusions and to forgotten aspects in a people's history" (p. 7). Our uses of discourse must work to insure that the values we invoke do not drown out alternative voices. Thus, a critical rhetorical ethic does more than merely "respect" audience in a passive or detached way. It works to liberate the Other by naming and making space for that which is taken for granted or "routinely excluded and silenced" (p. 8).

Care for Audience/Care for Concrete Other (Versus Generalized Other)

In Book 2 of *Rhetoric*, Aristotle introduces several characteristics of speakers/writers as ethical principles. These are especially powerful principles, as Aristotle articulates them, because they combine *ethos* and *pathos:* They are feelings as well as elements of character. Aristotle emphasizes the importance of expressing kindliness, of having good feeling toward others (*karis*) (see Grimaldi, 1988, pp. 127–128). Also important is entering discursive relations in a spirit of friendship and mutual caring (*philos*) as well as a generosity of spirit (*caritas*). Cicero (1942) pushes "respect" for audience even further, into the realm of emotional and psychological commitment, into what we might consider the realm of *caring* for audience (*De Oratore*, 2.44.189). We see the principle of "commitment to other" as a principle in Greek culture as well as classical rhetoric. The "guest friend" relationship is an important ethic, perhaps the most important, in Greek tragedy: The principle says that as host or hostess you have a sacred duty to protect and care for the Other, the visiting alien. In Greek tragedy, violating this ethic leads to war, death, destruction, and chaos. (Usually only the intervention of the gods, *deus ex machina*, can restore order.) If we think about such a principle as applied to the management of LISTSERV groups, we can see that it would call the listowner to take on some responsibility, as host or hostess, for those "guests" participating in any particular electronic community.

The caring ethic, as we saw in Chapter 4, gets its most thorough treatment in the feminist ethical articulations of Nel Noddings (1984) and Carol Gilligan (1993), who view the ethical stance of the "caring one" as a distinctly feminist position that has been historically subordinated to the masculine deontological and abstract ethic of rights. Rather than viewing abstractness and impartiality as the revered characteristics of the ethical stance, the caring ethic views personal and emotional concern for the distinct person as a superior ethical posture.

In her critique of the abstract impartiality of Habermas's discourse ethic, Benhabib (1992) stresses this point in particular. One must avoid "generalizing the other," but must engage the concrete and particular features of one's audience, who is actually "there," as far as that can be determined (although that becomes particularly difficult in electronic discourse): "The standpoint of the concrete other . . . requires us to view each and every rational being as an individual with a concrete history, identity and affective-emotional constitution" (p. 159). Stereotyping of audience differences—for example, referring to "women" as a political or social collective—is what Benhabib means by treating the other as a generalized type (see hooks, 1990; Rorty, 1991). Benhabib critiques Habermas's discourse ethic for not stressing the importance of "the concrete moral self" (p. 146). Similarly, Feenberg (1991) sees the kind of formal abstraction represented by Habermas's ethic as too remote, both from specific material technological settings as well as from the experiences and identities of real people.

We see a theological take on this principle expressed in liberation theology. The ultimate sin, according to liberation theologian Enrique Dussel (1988), is instrumentalizing people as things: "Sin is domination over the other" (p. 61); it is failing to take into account distinct personness (needs, identity, character, gender, background, experience). From this perspective, formal abstraction does not simply result in bad theory or bad rhetoric, it is unjust and sinful.

Do Not Oppress/Do No Harm

The negative corollary of the principles "respect audience difference" and "care for audience" is do not oppress them or dominate them. Do them no harm. But what exactly does this common ethical principle mean in terms of discourse?

Invoking Habermas's discourse ethic, Benhabib (1992) identifies two key principles of communicative ethics that provide some criteria for determining whether a discursive setting is indeed "doing harm"—the principles of universal moral respect, and the principle of egalitarian reciprocity:

> The "universal and necessary communicative presuppositions of argumentative speech" entail strong ethical assumptions. They require of us: (1) that we recognize the right of all beings capable of speech and action to be participants in the moral conversation—I will call this *the principle of universal moral respect*; (2) these conditions further stipulate that within such conversations each has the same symmetrical rights to various speech acts, to initiate new topics, to ask for reflection about the presuppositions of the conversation, etc. Let me call this *the principle of egalitarian reciprocity*. The very presuppositions of the argumentation situation then have a normative content that precedes the moral argument itself. (p. 29, ital in original)

These principles extend from Habermas's (1990) distinction between strategic action and communicative action. Strategic action is manipulative persuasion or public speech acts that attempt to coerce or manipulate audiences. Communicative action is Habermas' ideal ethical discourse action, the criteria for which are expressed in the two principles stated earlier. In ethical communicative action, the audience is treated not as a passive decoder or receiver, but as an equal interlocutor with reciprocal rights. In this sense, Habermas sets up a discourse ethic that reaffirms the old rationalistic privileging of dialectic (and argument) over rhetoric (and persuasion). The limitations of this ethic are: (a) it does not provide us with pragmatic criteria, as almost no real public discourse are capable of meeting these criteria; and (b) it is not sensitive enough to the inequities of discursive arrangements or to the presence of power and oppression in discursive settings. (In her revision of the Habermasian discourse ethic, Benhabib recognizes these limitations and works to address them.)

For alternate criteria, we can look to postmodern political commentary that begins with the consideration of power and oppression. Foucault (1987) makes a crucial distinction between "relationships of power" and "states of domination." Although all acts of writing or rhetoric might be said to be acts of power, they are not all acts of domination. Power is more general. Foucault says that relationships of power "have an extremely wide extension of human relations. There is a whole network of relationships of power, which can operate between individuals, in the bosom of the family, in an educational relationship, in the political body, etc." (p. 3). Domination, on the contrary, refers to invariable relations of power, that is, to "firmly set and congealed" settings that "block a field of relations of power" and "render them impassive and invariable" by preventing "all reversibility of movement" (p. 3). Both individuals and social groups can dominate.

This distinction is important for an ethics of rhetoric as it provides a criterion for critiquing rhetorical acts. When we write electronically (or otherwise), we are engaging in an act of power, according to Foucault's notion of power: Our effort to inform, persuade, or entertain is an effort to establish an authority vis-à-vis some reader(s). Rather than wring our hands in dismay over the arrogance of wielding this power (what I think of as a spasm of postmodern liberal guilt), we should see the exercise of discursive power as common, frequent, unavoidable, and not necessarily unethical. The distinction that is important pertains to acts of domination and oppression. For Foucault, domination refers to those acts of visible or invisible power that block "reversibility of movement," that is, they prevent a group or individual from expressing alternatives or exercising alternate choices. Although we cannot avoid the obligation of power when we write, we can try to avoid writing that dominates or oppresses.

Young (1990) further distinguishes between domination and oppression, viewing both as institutional forms. Domination refers to "institutional conditions which inhibit or prevent people from participating in determining their actions or the conditions of their actions" (p. 38). All of us are subject to various forms of domination, but that does not mean that domination affects us all in the same way: "[N]ot everyone subject to domination is also oppressed" (p. 38). Domination, in fact, may work to the advantage of some social groups. A social group is oppressed, however, when the domination works to systematically inhibit people's abilities, freedoms, choices (p. 38). Thus, domination means that people are excluded from participating in the systems and institutions that guide or determine their actions. Oppression means that, because of this exclusion, a people's freedoms are inhibited. This distinction is important for an ethics of rhetoric as it provides a criterion for distinguishing competing social groups' claims to being "oppressed." (See Sullivan & Porter, 1997, Chap. 5, for further discussion of the implications of oppression.)

Does your act of rhetoric allow for what Foucault calls "reversibility of movement" and Benhabib (1992) refers to as "reversibility of perspectives" (p. 146)? If you are a LISTSERV manager, for instance, are you autocratic in terms of how you exercise your authority on the list? (The liberal individual theory of property supports such autocracy, preferring to focus on the question of who "owns" the list rather than assigning at least some communal ethical authority to participants.) Or do the members of the list have some say in determining the status of their condition? Does the list operate by a "love it or leave it" philosophy, which says in effect that if you do not want to abide by the owner's rules, then your only choice is to leave the list? This is an act of domination and exclusion, and the presumed "choice" to leave a list is a forced choice and maybe a professionally harmful one for an academic whose area of specialization is represented on the list.

Again, the issue should not be reduced to this kind of either–or false dilemma: If the listowner does not establish authority (i.e., create and enforce strict rules), then the list participants will endlessly debate what the rules should be, and chaos will result. Certainly, with the way LISTSERV operations work, the listowner has every right to set up a list as he or she sees fit. The listowner has power, but with that power comes the obligation and responsibility to list participants. The problem, of course, for listowners or for writing teachers who sponsor electronic conferences is addressing the dilemmas that can arise (as we saw in Chapter 5) between competing interest groups or between individuals and the majority.

THE WRITER'S PROCEDURAL STRATEGIES

Consult Dialogically with a Diversity of Sources

The law should always be considered very seriously in decisions about the status of intellectual property, but the law cannot be relied on, in every instance, to provide clearcut advice or protection. The law is a complex and diverse set of decisions, statutes, opinions, and judgments that intersect with a wide range of human practices. The ethical writer has to consider a plurality of choices, including diverse and often conflicting laws, policies, and theories. The principle of ethical pluralism does not say that "everybody is right," but it does say that there are competing ethical standards that may have application and validity in a given situation. How do you choose among them? First, you do not do it alone. In rhetorical ethics, decision making is social, collaborative, and dialogic. Numerous theorists have recognized openness to dialogue (Haynes-Burton, 1990) and "tolerance for differences" (Lang, 1991, p. 5) as important qualities of ethical writing. Rorty says that responding to ethical dilemmas often involves "interweaving" opposing principles.

This process resembles what Jonsen and Toulmin (1988) describe as the key operating procedures of *phronesis* in Aristotle's *Nicomachean Ethics* (1976). The ethical person/writer must apply *circumspectio* ("looking around") in order to determine options and choices. He or she must be careful to judge *circumstantiae* ("what is standing around"), that is, the current conditions. He or she must apply *cautio* (carefulness) in making any decision, because human action is *multiforma* (takes many forms) (pp. 130–131). In short, ethical writing requires some careful deliberation, some marshalling of options and alternatives, some care in determining the conditions of one's rhetorical setting, and some respect for the variances of human action: We might regard such principles as the basics of an ethical approach to rhetorical invention.

You start this process by collecting a range of authoritative principles from a range of sites. Rhetorical ethics does not assign a priori authority to any site or source, but rather places competing sources of authority in dialectical tension. The list of resources in the Appendix includes a number of online discussion group that consider ethical and legal issues in electronic networking. Writers and writing teachers can use this list both to collect sources of authority and as a forum for ethical dialectic. They can look for ethical guidance to acceptable use policies established by government, education, and private network lists. Information about the examples of various acceptable use policies (as well as records of relevant legal decisions) are stored in an electronic database maintained by the Electronic Frontier Foundation. In short, these network sites serve as a concrete basis for dialogic exploration of rhetorical/ethical issues.

If I have a question about the ethics of the use of electronic text, for instance, an electronic copyright question, I know that there are several sites of authority I can consult. For legal authority and for publishers' and librarians' perspectives, I would consult the CNI-COPYRIGHT electronic list. For a discussion of these issues from the point of view of writing teachers, however, I would consult participants on ACW-L, the list for the Alliance of Computers and Writing. Different lists provide different disciplinary viewpoints and different forms of expertise. For truly problematic cases, I might consult several lists by posting my dilemma in the form of a question to each list. Thus, the electronic network provides the technological mechanism for heuristic invention of rhetorical/ethical issues.

Reading is obviously important. Read suspiciously, read critically, but nonetheless read. Consult a variety of sources for guidance when necessary, especially electronic sources. Cyberwriters can learn more about copyright law, for instance, by consulting print sources on intellectual property law in general and on how it applies to cyberspace in particular (e.g., Branscomb, 1994; Ermann, Williams, & Gutierrez, 1990; Forester & Morrison, 1994; Miller & Blumenthal, 1986; Oz, 1994; Woodmansee & Jaszi, 1995). Cyberwriters should consult electronic publishing guides that develop policy for use of electronic text (e.g., American Library Association, 1971; Association for Computing Machinery, 1995; Bill of rights, 1993; Corporation for Research and Educational Networking, 1993; Duggan, 1991; Kadie, 1991; Kahin, 1992; National Science Foundation, 1992; Oakley, 1991). The work of Paula Samuelson (who writes a regular column for the *Communications of the* ACM) addresses the issues of intellectual property in the digital age—in fact, Samuelson (1996) is one of the leading authorities on electronic copyright. The network itself is a useful source for collecting up-to-date information. The internetworked writer can browse some of the intellectual property Web sites listed in the Appendix and participate in LISTSERV discussion groups that focus on copyright and other legal/ethical issues.

Figure 4.1, the shot chart of ethical positions, can also serve as a rhetorical/ethical heuristic for writers. The diagram problematizes ethical positions. If anyone is inclined to rely on simplistic ethical pronouncements, that diagram serves to identify a range of competing ethical frameworks that can serve to critique and problematize such pronouncements.

Situate and Contextualize at the Local and Particular Level

A postmodern rhetorical ethics emphasizes the authority of contextualized elements and of the situated moment (*kairos*). It says the particular historical and situated moment and the particular details and circumstances of that setting (the "facts," values, audiences, timing, historical circumstances, technologies) are critical. In their neo-casuistic rhetoric ethic, Jonsen and Toulmin (1988) stress the importance of authorizing the details of the case, as well as the particular situated circumstances in which any ethical issue resides.

The situated side of critical rhetorical ethics requires that we make judgments sensitive to the constraints of local policy and practice, particular settings and cases, and, in the case of electronic writing and publishing, to the particular technologies and customs of use for those technologies. Feenberg (1991) in particular says that we have to avoid "the *decontextualizing practice* of formal abstraction [which] transforms its objects into mere means" (p. 170), by which he means, in part, the kind of rationalistic, philosophical approach to ethics and technology advocated by those in the Western deontological tradition of philosophical ethics (including Habermas). Contextualized awareness is sensitive to the constraints and norms of the particular form of technology. In concurrence with Feenberg, Martha Cooper (1991) says that a postmodern ethics must consider (a) "rhetorical situation" more broadly than in the conventional sense to include the medium and the effect of the medium on power relations (questions about access, for instance, and who controls and maintains the medium), and (b) the institutional bureaucracies and professional settings in which political advocacy and decision making occur.

A position sensitive to a technological setting would notice, for instance, that notions of netiquette on LISTSERV groups are different from that on MUDs or newsgroups. LISTSERV groups that have a designated "owner" tend to be more orderly and academic in the way they carry on discussion. Participants know (or should know) that their presence there depends on owner tolerance. In this respect the ethical "law" of the LISTSERV group is fairly absolute and undemocratic: The owner rules. (Even though enlightened owners can choose to share their power if they wish, the choice is that of the owner). Newsgroups and MUDs are much more freewheeling forums that do not have owners in the same sense (although they may have initiators and moderators).

Notions of ownership for print publications are different than for electronic archives. To questions about the ethics of reposting electronic messages, a rhetorical ethic would inquire about the conventions of the specific community, about the nature of the message, about the norms for reposting that may have relevance to the case. In other words, such an ethic would grant authority to the specific situation, to the specific technology, to the exact nature of the message, and to the norms and conventions (*nomoi*) of the electronic communities involved. It is, however, also aware of the danger of letting any single community isolate itself from principles outside that particular community—principles of universal human respect and reciprocity, for instance, from discourse ethics.

Figure 1, which served to identify the various agencies influencing the act of electronic writing in the composition classroom, can also double as a visual heuristic to identify the agencies involved in rhetorical/ethical writing situations. That is, the diagram represents the matrix of forces that one might consider as operative in the act of internetworked writing. The answer to the question of "what policies should guide my writing in this electronic group?" lies not in one place but in several: There are likely to be competing policies and guidelines to be negotiated and reconciled. The visual serves as a prompt to writers to "look in several places" and alerts them to

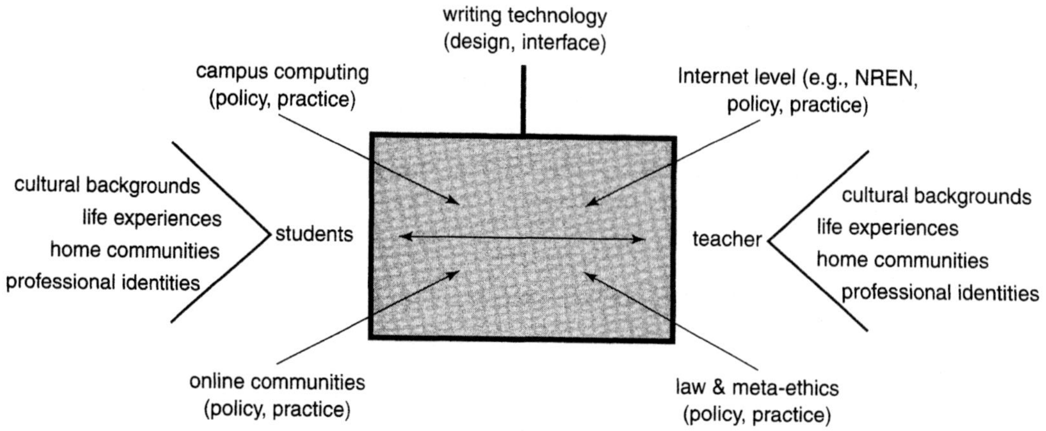

Figure 1. Sites influencing internetworked writing.

the need to reconcile possibly competing sites of ethical authority. The heuristic is general enough to function across a spectrum of electronic writing situations, yet its terms focus the writer on concrete agents and sites. (This is the challenge facing the designer of any heuristic: designing the heuristic for application across a relatively wide range of writing situations, yet not making the terms so general and abstract as to lose pragmatic and situated value in prompting writers.)

Acknowledge Ethical Complexities and Ambiguities

If I offer an "answer" in this book it is that problems (real problems, that is, not just inconveniences, but real problems of incommensurability and ethical dilemmas) are best worked out through the operation of a local and patient rhetorical pragmatism that is deliberative and reflective and that pays some attention to the processes involved, to the particulars of each case and its material conditions; and to the individual persons affected. I admit that at points we may have to resort to abstractions and slogans—yes, by all means, Free Speech! and hurrah for Democracy!—but remember how easy it is to misuse such slogans. Do not use them to shut off or polarize discussion. Be aware of how resistant the really difficult cases are to simple solutions by abstraction.

As I say, this is not the answer that most want to hear. As Stanley Fish (1994) says, "Complexity does not play well in Peoria or anywhere else" (p. 58), including Washington, D.C. But then, that leaves something for educators to do, does it not?: to acknowledge the complexity of public issues, to promote careful thinking and deliberation, to promote fair consideration and representation of all sides, to resist the "single story" approach that the newspapers and television media and, increasingly, the electronic networks promote—and that both right- and left-wing demagogues favor.

As soon as you make claims like the ones in the previous paragraph, you run into trouble, of course, especially with postmodern ironists. The abstractions you casually dismissed, then yourself used, double back to trip you up: Whose idea of "careful thinking"? Which notion of "fair representation"? You can never completely evade the abstractions (a.k.a. theorizing), but you can be wary of them and alert to their operation. You can be an advocate of "free speech," but be wary of those cases that push the free speech principle to its limits and be capable of asking the critical questions ("free" for whom? at whose expense?).

Jonsen and Toulmin (1988) remind us that complexity is the nature of ethics. For some difficult cases and situations there will *never* be clear and easy answers—and to shout at each other as if there were, or could be, is the ultimate act of stupidity. For tough cases there will always be moral ambiguity (this is the de facto nature of tough cases). It is in the nature of human experience to be *multiforma*, complex to the degree that in its distinct situatedness it will always manage to remain one step ahead of our theorizing and our categories for talking about it. Rather than being dismayed by this complexity, we might try to appreciate it.

MOVING TOWARD SOLIDARITY

There is a point at which continual open-ended heuristic thinking is dangerously disabling. Reflection, deliberation, careful interrogation, practiced skepticism—are all valued postures in this critical rhetorical ethic. At some point, however, the questioning process has to stop to make a commitment, even if only a contingent one. (If you deconstruct the idea of "computer classroom" to death, you just are not gonna have a computer classroom to teach in.)

If the position I am advocating tips in a direction, it tips in favor of solidarity—in favor of the position that we have to use rhetoric, ultimately, to enable solidarity: a solidarity that does not build itself on a rockbed of universal strictures, but that is a symbolically constructed solidarity of opinions that we build for our common good because we need a common good. This position tips in the favor of solidarity and the communal, too, in part for an ironic reason: Because in U.S. society the liberal-individualist position has for so long been dominant, I adopt the communal standpoint because it provides the most direct counter to the dominant—and often, at times, oppressive—liberal-individual ethic.

The great fear is that communal identification represents a return to a Western logocentric masculine rhetoric. Critics worry about solidarity and community as operating principles. They say we have been "identified" for too long and that it is time to recognize diversity. Steve Katz (1992, 1993) and others warn us about the problem of the common good becoming a totalitarian evil or a majority oppression. Yet my worry is that we are circling back to a place we have already been. Are we teaching writers to hurl their differences at one another, just like rhetoric used to teach its practitioners how to hurl arguments at one another, without actually *listening* to one another? (This is what appears to happen frequently in newsgroup arguments and on the floor of the U.S. House of Representatives.) Are we returning to a debate model assuming equal and opposite participants in a world where equality is not the common condition? Does the resistance to a community or solidarity model lead us to another version of the old liberal model, perhaps a postmodern version of it? In such a model—what we might call the rock-hurling, debate model of public discourse—the more powerful are bound to win, because they can afford more and bigger rocks. Young (1990) warns against the danger of interest group pluralism that in its stridency can disable dialogue, and thus, block effective change. If interest groups do not work toward solidarity, then they will diffuse political potency and leave untouched and unchanged the systems and institutions that dominate us all.

The community I have in mind is not a melting pot or blender. The solidarity of this community does not require the effacement of difference, but rather builds on difference in the manner suggested by both Irigaray (1984/1993) and Young (1990). This community requires both feminine and masculine modes of thinking. For Irigaray, we achieve transcendence, spiritual fulfillment, in the mutual embrace that is a celebration of our difference. This commonality achieves its strength through and celebrates diversity. For Young, diversity and variation in a society should be viewed as a source of strength (although I do not share her optimism in the city as the ideal model of communal diversity).

There is a tendency in the field of rhetoric/composition, especially in cultural studies discussions, to mislabel community metaphors mainly in a negative way (e.g., as hegemony, exclusivity, the dominant culture) and to set up a straw binary: diversity as hero, hegemony as villain—which is an unassailable position, like saying "good" and "bad," and so is ultimately not very helpful.

There is another set of terms one can invoke to stress the positive notions of community and solidarity: unity, identification, alliance. Rather than throwing our competing sets of god terms at one another, take an ironic viewpoint and, like Kenneth Burke, see these two sets of terms as operating in a kind of binary tension. Burke saw all rhetoric as always involving both Identification and Division: As we strive for likeness and similarity (consubstantiality) we also affirm our differences, and the two notions are in constant oscillating tension. Any act of writing involves an identification—a positioning of a "we" and an establishing of a set of values that are over here, with "us." There is always an embracement: The writer engages an audience, identifies with an audience, and in that process a community is formed or changed. In the best of worlds, we always hope the writer's hug is an embrace, not a strangulation. We hope the embrace is mutual—dialogic if you like. However, you can not have rhetoric/writing without it, or without its counterpart—division.

Thus, the exercise of rhetoric is always at its core ironic, and the trick from the point of view of rhetorical ethics is to not let the irony freeze us into inaction, but to try to negotiate a nonoppressive committed standpoint in the midst of this irony.

At this point, my position may sound like Richard Rorty's (1989) description of "liberal irony" in *Contingency, Irony, and Solidarity*. Rorty assumes an ironic stance through most of the book, yet argues at the same time for liberalism as a utopian ideal. He explicitly situates himself between Foucault ("the ironist unwilling to be a liberal") and Habermas ("the liberal who is unwilling to be an ironist"; p. 61).

The stance Rorty advocates is that the "committed ironist," who looks at all sides of an issues but then takes a stand. The stand may have doubts; it may be contingent, but for the time being it is a strong stand. Like me, Rorty is suspicious of the use of common-sense terms (e.g., *democracy* and *liberalism*) to address problems: Too often they are instruments used to end dialogue by shutting off the Other. Like me, Rorty does not see much promise in extreme postmodern ludicism as a way to address social problems. This will lead to too much self-doubt and lack of action and character on the part of the community.

Unlike me, Rorty views irony as something private, to be carried out among expert philosophers and literary critics of irony, who effect change in small increments within disciplinary groups (e.g., Nabokov). This is where I have trouble with Rorty's conception of the "committed ironist" or "liberal ironist." Such a position ends up reinstatiating the academic male philosophical liberal humanist, the elite intellectual who publicly advocates what he privately and ironically despises. This posture is, ultimately, one of despair that public discourse can ever be anything other than stupid and rhetorically useless. Rorty's position ends up despairing of any productive role for mass education in public life. The best that education can aspire to, in Rorty's model, is to train a few elite and aesthetically sensitive ironists, who will celebrate their critical acumen in isolated academic enclaves—the ethical ivory tower.

My position on irony owes more to Burke and especially to Lyotard and Thébaud's (1985) discussion of rhetorical ethics in *Just Gaming*. It views certain paired opposites as having a rhetorical/ethical value in helping us construct writing stances. (Such pairs include individual/community, cooperation/competition, critique/construct, and static/dynamic.) Rorty falls victim to the philosophical temptations of ironic formal abstraction: He never contextualizes his discussion. However, the Lyotardian position, although ironic and theoretical, is a casuistic one that advocates a pragmatic case-by-case basis. It is ironic because it is constantly questioning the commonsensical, overturning the expected, saying "no" when everybody else says "yes." However, it does manage to locate commitment at a situated case level. Lyotard accepts, finally, the inevitability of *phronesis*, a concession that Rorty is not willing to make.

My cautious endorsement of solidarity also contrasts with Victor Vitanza's (1991a) rejection of Aristotelian pragmatics, which he views as a system advocating a strong view of consensus. Although I agree with Vitanza's critique of strong consensus, I cannot ultimately endorse his construction of ethics as "drifting." To me the ethics that Vitanza ultimately affirms embodies two constructs. One construct is the free-floating individual system resister, which is simply an angrier, more radical version of liberal individualism. The other construct is itself a system: Although Vitanza says he wants to resist all systems, he also clearly wants "to drift toward some notions of a Third Sophistic" (p. 121; see also Vitanza, 1991b, 1994).

The chief components of this situated ironic communalism are pursuit of the common good; respect for individual *and* community, for diversity *and* unity; and constant questioning and revision. I see the community versus individual, consensus versus dissensus, controversy as a false binary. Why not hold both together? In *The Politics* (1962), Aristotle makes an important point about unity versus diversity:

> Certainly there must be some unity in a state, as in a household, but not an absolutely total unity. There comes a point where the state, if it does not cease to be a state altogether, will certainly come close to that and be a worse one; it is as if one were to reduce concord to unison or rhythm to a single beat. As we have said before, a state is a plurality, which must depend on education to bring about its common unity. (2.5 1263b29)

Aristotle says, in effect, that in developing unity we must not obliterate difference, we must not "reduce concord to unison or rhythm to a single beat." Other unity metaphors are more problematic, insofar as they do not so readily grant diversity its due. For instance, in 1 Corinthians 12, Paul talks about the relation of parts to whole and gives us the metaphor of the faithful as one unified body: "There are many parts, yet one body." Postmodern ethics has expressed strong reservations about this kind of single body image (as numerous feminists have said, whenever there is just one body it is bound to have a penis). Irigaray (1984/1993) offers us an alternative scriptural metaphor in her consideration of the two bodies, of the sexual difference that she constructs as the foundation for an ethic or relations. Her account, in fact, shows us how, through embrace of the Other (mutual embrace, not a "disguised version of the master–slave relationship," p. 17), we can achieve a kind of celebratory transcendence.

Is it possible to have a postmodern view of community? Carolyn Miller (1984/1993), Iris Marion young (1990), and Lester Faigley (1992) have all pointed the way to such a vision of community. In a discussion linking communitarian and rhetoric theory, Miller (1993) develops a vocabulary for talking about community in a way that highlights difference: Concepts like "the bazaar, the collage, the carnival, bricolage, pastiche, eccentrism, nomadism, diaspora, migration" (pp. 89–90) all suggest ways of thinking about social/communal events in a post-modern light. The community does not have to be a Governing State, or a Majority consensus, but can be a gathering for the purpose of festival and play, a celebration of difference. The Internet could potentially be the site for such electronic communities, but only if we begin the process of constructing a cyberspace ethic that promotes it as such a site and recognizes that such a festival cannot proceed without some regard for the ethical.

Miller (1993) does worry that a postmodern conception of community might "leave us with a community that is so fragmented, perforated, intermittent, and attenuated that it no longer performs any rhetorical work" (p. 91), but the answer to that is to work to construct, and reconstruct, communities (plural). The effort is ongoing. She thinks that rhetoric must foreground community because it "is necessary both for emotional solidarity and for political action" (p. 91). I very much agree.

Aristotle's musical metaphor is a significant one. The terms *community* and *unity* suggest consensus, but that consensus should not be conceived as homogeneity. Consensus is to be understood here in the sense of harmony: The different sounds aligned and blended in a particular way will produce a pleasant sound, a more aesthetically pleasing state than any of the individual parts. This kind of harmony is not the same as everyone playing the same sound, the same note, in the same key. All that produces is one loud blare, not music. Harmony is not homogeneity.

So, let us work toward this: an electronic network promoting a social harmony that values the diversity of sounds, that celebrates differences for a common good we have to work to construct. If we can use writing to design a network that achieves the spiritual transcendence that Irigaray (1984/1993) hopes for, the third age of the "incarnation of the spirit . . . the era of the wedding and the festival of the world" (pp. 148–149), let us do it.

DEVELOPING YOUR UNDERSTANDING

1. Summarize the distinction between power and domination.
2. According to Porter's argument, explain what the relationship is between rhetoric and ethics.

3. Describe the role of collaboration and interaction in Porter's ethic of postmodern commitment and solidarity.
4. Porter's ethical principles for professional writers are organized by two main categories: (a) the posture, position, or stance a writer takes in relation to her audience/readers/clients and (b) the procedures, processes, or ways the writer goes about her work. Use these two categories to plan how you will go about your work on one of the larger writing projects you have been assigned, or are working on elsewhere (e.g., an internship, another course, or at work).
5. Identify an example of domination in writing. (The example does not have to represent total domination, but can instead reflect aspects of domination.) Explain how it is an example of domination, referring to contextual, textual, and/or procedural details from the example. Then, discuss what a professional writer could do to eliminate, or at least significantly reduce, the domination.

CHAPTER 4

Projects

1. Talking about ethics, even in relation to cases others have faced, is not the same as taking ethical action. But being prepared and able to take effective ethical action requires that we take time to examine it from a distance, so to speak. One way to examine ethical action is to talk about it in abstraction and as it relates to other instances. We can also learn a great deal about ethics and ethical action, without actually being faced with the concrete realities, by assuming the roles of others who are taking such action. This project asks you to take that intermediary step, to put yourself in the place of other writers faced with negotiating appropriate ethical action.

   Assume the role of Just, the memo writer in Katz's opening example, as well as one other writer from any of the cases discussed in the articles from this chapter. Also, assume that you have adopted and are putting into practice Porter's ethical principles for professional writers. Speaking in the first person narrative, describe how you put Porter's ethical practice into action in both cases. In each narrative, examine what you have to do, the constraints and limitations you face, and the emotional responses (e.g., fear, worry, or pride) you have toward the actions you are taking. Thus, you are writing a story of your ethical/rhetorical action. Then, write a separate commentary that analyzes how your revised professional writing practice is similar to and different from the original case.

2. If we take it seriously that professional writing is ethical action, we should pay as much attention to developing phronesis and praxis as we do to developing technē and poiesis. All writing requires both, but too often we direct our attentions toward technē/poiesis and let unreflective habits drive phronesis/praxis. The following client project requires you to shift your attention toward phronesis/praxis.

   In groups of three or four, locate a client on your campus or in your local community who has a writing project you could work on for them. If you can, locate projects that are well within your reach in terms of genre knowledge, rhetorical strategies needed, technical skills required for document production, and overall time needed to complete the project. Do, however, look for projects that require you to interact with several parties. For instance, constructing a whole Web site for a campus organization would probably not be a good choice. Depending on your experience with Web page design and construction, though, creating a splash page only for the same campus organization may be a good choice.

   - At the outset of the project, discuss and develop an ethical action plan delineating the principles you agree will guide your interactions with the client and involved parties, as well as among yourselves as a co-authoring group (these will probably be two related but different plans).
   - As you work through the project, each group member should keep a journal in which you discuss judgments you have made, actions you have taken, and the reasoning behind them. (Do not look for and focus on things that you see as "ethical dilemmas," only; consider actions such as discussing design options with an audience representative as an important action which you have judged to be good, based on a certain line of reasoning.)
   - At the end of the project, write a co-authored report that summarizes and reflects on your group's ethical actions, judgments, and reasoning throughout the project. In addition to turning in your report, the final client project, and your journals to your instructor, be prepared to summarize and discuss your report with the class.

CHAPTER 5

Professional Writing as Technologically Situated Action

INTRODUCTION

The readings in Chapter 5 highlight some of the issues related to the technological contexts in which professional writing practice takes place. When writing is understood as only a way of producing something, technologies are understood as mere "tools" to help us get things done (or tools that get in the way of getting things done). Approaching writing from this direction, technological issues focus on how to use X tool, what Y tool is good for, and when a writer might want to use Z tool.

When writing is understood, also, as a social practice, technologies become more than mere tools; they become part of the context in which writing takes place. Approaching writing from this second direction, technological issues broaden. Tool-focused issues are still important. For instance, one still must address how to use HTML, what HTML is good for, and when a writer might want to use HTML. But other issues also arise, like how does writing in HTML change the writing process, the writer's relationship with a reader, and what we mean by "written texts." From this broader perspective, professional writing does not merely involve using technological tools; professional writing becomes a social practice that is situated within technological contexts. In order to increase your awareness of and sensitivity to the impact of technological contexts, keep the following questions in mind as your read the selections in Chapter 5:

- What is the relationship between technological context and writing processes?
- What is the relationship between technological context and writer-reader interactions?
- In what ways do technological contexts affect what we understand as "written texts"?
- How do technological contexts affect our definitions of what writers do and what writers must know?

One of the best ways to increase awareness of the impact technological context has on professional writing is to defamiliarize the technological contexts that have become "natural" to us. James Kalmbach's "Publishing Before Computers" helps to defamiliarize our current contexts by describing a familiar process—writing and publishing—in unfamiliar contexts. Reading about how writing and publishing were practiced in technological contexts that pre-date desktop computers clarifies how our current technological contexts affect things like writing process and writer-reader relationships.

In addition to looking back at how technological contexts affect the work of professional writers, we can and should look forward to explore how future technological contexts may influence who writers are and what writers do. Instead of putting all of the bulleted questions above in the present tense, consider them in the future tense (e.g., what will be the relationship between tomorrow's possible technological contexts and the writing process?). Professional writers need to look ahead in this way, not only so they are prepared for what is to come, but also because by looking ahead and understanding the impact technological contexts have on writing, professional writers can be active in shaping the technologies that will later shape them and their work. Stephen Bernhardt models this looking-ahead in "The Shape of Texts to Come: The Texture of Print on Screens," even though what he wrote about circa 1992 seems quite natural to us today.

One additional article, by Tharon Howard, examines one way technology complicates authorship. Copyright is often misunderstood as a way to protect authors' individual ideas, and often seen as a principle uncomplicated by context. Howard first corrects our historical understanding of copyright and then examines how electronic media complicate who is the "author" of electronic texts.

FOCUSING ON KEY TERMS AND CONCEPTS

Focus on the following terms and concepts while you read through this selection. Understanding these will not only increase your understanding of the selection that follows, but you will find that, because most of these terms or concepts are commonly used in professional writing and rhetoric, understanding them helps you get a better sense of the field itself.

1. publishing as action
2. letterpress
3. typography
4. lithography
5. xerography
6. desktop publishing
7. economies of scale

Publishing Before Computers

JAMES R. KALMBACH

Writing and writing systems are forms of technology (Ong, 1982). They are technologies in terms of how written symbols represent the sound system of a language, and they are technologies in terms of how those symbols are formed, preserved, and transmitted.[1] Publishing as action is, in its essence, an interplay between social forces and technological resources. The nature of publishing at any point in history is shaped by the actions people wish to take and the technological resources available to express those actions in writing. Over the years, publishing and the technologies for publishing have evolved from a limited, conserving activity to a pervasive form of social transaction. This evolution has always been in the direction of increasing expressiveness of written text, increasing availability of technology, and increasing interactivity.

This chapter traces the evolution of publishing technology from the invention of the alphabet in ancient Greece through to Xerox's introduction of the plain paper copier in 1960. It compresses several thousand years of innovation into a few brief pages. The goal of the argument is not to explicate these technologies in detail, but to show that even though the technologies change the ways in which these technologies are used, the social forces they respond to have remained fairly constant.

THE INVENTION OF THE ALPHABET

The alphabetic writing system, first developed in Greece around the 9th century (Gelb, 1963), is the original technology of publishing. Prior to the alphabet, writing systems were limited in the range of meanings they could communicate. Havelock (1974) suggests that the prealphabetic syllabic writing systems, which attempted to represent in print the actual, physically heard syllables of a language, could only communicate a limited range of texts. The inherent ambiguity of these writing systems made it difficult for readers to decode unique ideas.[2] Instead, topics were limited to those that were familiar and predictable to the reader:

> The record of a culture which is composed under these restrictions is likely to center upon religion and myth, for these tend to codify and standardize the variety of human experience so that the reader of such scripts is more likely to recognize what the writer is talking about (p. 35)

Source: *The Computer and the Page: Publishing, Technology, and the Classroom,* James R. Kalmbach, "Publishing Before Computers," 1997, pp. 55–69.

[1] "The term 'writing' describes a series of technological devices which . . . assist the user in an act of recognition" (Havelock, 1974, p. 16).

[2] "One sign [in syllabic writing systems] has to represent several sounds and the open choices left to the reader . . . become extensive" (Havelock, 1974, p. 31).

In cultures dominated by these prealphabetic scripts, reading was restricted to specialists:

> Expected and recognizable discourse [in syllabic writing systems] becomes highly traditional both in form and content. Such traditionalism is characteristic of a craft, the secrets of which are carefully nurtured by its practitioners. The scribes who used these syllabaries were practitioners of this sort. The so-called literacy that they represent was craft literacy. (p. 37)

Rather than a means of achieving social action, the limited access to literacy in prealphabetic writing systems and the difficulty of interpreting unique texts meant that literacy was primarily a means of acquiring and preserving power.

The Grecian writing system's central technological innovation was to abstract away from the real sounds that syllabic writing systems attempted to represent and instead use written symbols to represent an abstract unit of phonological structure: the phoneme.[3] It was a remarkable conceptual leap. Phonemes such as *b*, *p*, *d*, *t*, *g*, and *k* only exist in the abstract. They cannot be pronounced without an accompanying vowel.[4] Separating vowels from consonants enabled the Greeks to limit the number of unique symbols in their writing system to under 30, while at the same time decrease the ambiguity of the original texts. With a writing system that represented abstract phonological structure rather than real speech, it was possible to write down (and to read) original texts. Writing systems were no longer solely a means of preserving traditional texts. They could be used to create new ideas and to communicate those ideas to others. New technologies enabled the Greeks to use written language in new ways. The result, Havelock (1974) argues, was literacy and the literate basis of modern thought.

As Lanham (1993), Ong (1982), and others have argued, alphabetic writing also introduced a fundamental dichotomy into Western civilization. On the one hand, alphabetic texts were an even better substitute for memory (Ong, 1982) than were earlier forms of writing. As a result, longer more diverse texts could be recorded and preserved. On the other hand, the fact that alphabetic writing and reading were relatively easy to learn and could be used to record and communicate original thoughts made alphabetic writing a means of resisting this conserving tendency. We have struggled ever since with this tension between the conserving and creative functions of writing (Perkinson, 1995).

FROM ORAL TO SILENT READING

The scribes in the Middle Ages who inherited the Greek tradition of alphabetic writing were clearly in the conserving camp. They saw themselves not as creators or publishers of new texts, but as keepers of great texts from past eras; "One of the main functions of the monastic institutions of the Middle Ages was to preserve manu-

---

[3] A *phoneme* is the minimal unit of sound that changes the meaning of a word. Phonemes differ from language to language. *Spit* and *pit* are the same *p* phoneme in English even though one *p* is pronounced with a puff of air and the other is not. In Arabic languages, these two *p* sounds are separate phonemes.

[4] These sounds are called stops because they "stop" the vibrating column of air created by a vowel or semivowel. Stop consonants cannot be pronounced except in combination with other sounds because without the vibrating air there is nothing to stop.

scripts and produce good copies, cleaned of the errors made by weary copyists of the past" (Smith, 1980, p. 5). Their task was to preserve, reproduce, and interpret these texts. Perhaps because of this role as guardians of great texts, scribal publishing also elevated the book-as-a-work-of-art to its highest level.[5]

This conserving role began to break down during the Middle Ages as the act of reading gradually transformed from an oral, communal act into a silent, individual process (Saenger, 1982). Reading for the Greeks and later the Romans meant reading out loud: a slave reading to a master or one person reading to a group. Oral reading was largely mandated by the technology of the writing system. Visible texts were rendered in all-capital letters with no spaces between words and no punctuation. Such texts, Saenger argues, were designed to be a transcription of spoken language that was best read orally:

> The Roman reader, reading aloud to others or softly to himself, approached the text syllable by syllable in order to recover the words and sentences conveying the meaning of the text. . . . A written text was essentially a transcription which, like modern musical notation, became an intelligible message only when it was performed orally to others or to oneself. (p. 371)

Oral reading of such texts must have been physically exhausting and reading for meaning a constant challenge. Even with contemporary typefaces, one can get a sense of the challenge they faced when lower case letters, word spaces, and punctuation are stripped away from a text. The following passage is the opening of Jay Bolter's *Writing Spaces*, set in a Roman style:

> INVICTORHUGOSNOVELNOTREDAMEDEPARIS1482THEPRIESTREMARKEDC-ECITUERACELATHISBOOKWILLDESTROYTHATBUILDINGHEMEANTNOTON-LYTHATPRINTINGANDLITERACYWOULDUNDERMINETHEAUTHROITY-OFTHECHURCHBUTALSOTHATHUMANTHOUGHTWOULDCHANGEITSMO DEOFEXPRESSIONTHATTHEPRINCIPALIDEAOFEACHGENERATIONWOULD-NOLONGERWRITEITSELFWITHTHESAMEMATERIALANDINTHESAME-WAYTHATTHEBOOKOFSTONESOSOLIDANDDURABLEWOULDGIVEPLACE-TOTHEBOOKMADEOFPAPERYETMORESOLIDANDDURABLE

Could you read the passage silently, or did you read it quietly to yourself, pronouncing the text syllable by syllable, sacrificing meaning to make sound–letter correlations? Do you remember any of what you read, or does your head ache and your eyes hurt? Such text clearly privileges oral, linear reading.

The principle innovation that made silent reading possible in the Middle Ages was the insertion of spaces between words (Saenger, 1982). Notice how much easier the Bolter passage is to read with these spaces:

> IN VICTOR HUGOS NOVEL NOTRE DAME DE PARIS 1482 THE PRIEST REMARKED CECI TUERA CELA THIS BOOK WILL DESTROY THAT BUILDING HE MEANT NOT ONLY THAT PRINTING AND LITERACY WOULD UNDERMINE THE AUTHROITY OF THE CHURCH BUT ALSO THAT HUMAN THOUGHT WOULD CHANGE ITS MODE OF EXPRESSION THAT THE PRINCIPAL IDEA OF EACH GENERATION WOULD NO LONGER WRITE ITSELF WITH

[5] *The Book of Kells*, a gorgeously illustrated biblical codex completed around 800, is a national treasure for the Irish (Olmert, 1992, p. 92).

THE SAME MATERIAL AND IN THE SAME WAY THAT THE BOOK OF STONE SO SOLID AND DURABLE WOULD GIVE PLACE TO THE BOOK MADE OF PAPER YET MORE SOLID AND DURABLE

The addition of word spacing enabled readers to group words into meaningful units, to skim texts to locate specific passages of interest, and to predict what would come next. The use of upper and lower case letters and punctuation as clues to grammatical and semantic relationships, which made it easier for readers to infer syntactic and discourse-level structures, further encouraged the move toward silent, nonlinear reading (Levenston, 1992).[6]

LETTERPRESS PRINTING

One consequence of silent reading was a dramatic increase in the amount of text that a reader could process. Saenger (1982) reports that in the 12th-century cloister libraries, books were lent out at Easter for a period of one year; "The lengthy loan period had reflected the slow pace of reading orally either to oneself or to others in small groups" (p. 396). As silent reading became more widespread, libraries had to increase their holdings to meet the demand for more texts: "In 1450, a university population which was approximately forty per cent smaller than it had been in 1300 was reading a much larger corpus of scholastic writing than had existed one hundred fifty years before" (p. 398). The traditional means of producing books by hand could not meet this demand.

These parchment books were large and difficult to use. They were rare and closely guarded, often stored in enormous metal boxes (to keep the parchment pages flat) and chained to library shelves.

Moreover, creating books by hand was slow and labor intensive. Manuscript books were hard to find and harder still to purchase. Commercial trade in books was rare. Instead, the way to acquire books was to hire copyists to create them from scratch.[7]

The invention of letterpress printing by Johann Gutenberg in the mid-15th century came at a time when manuscript books were at their artistic height. When Gutenberg produced his 42-line Bible in 1455, he took fine hand-drawn letterforms as his models because he was competing with copyists, and quality letterforms were

---

[6] The consequences of this shift from oral to silent reading can still be seen today. There are more than 1,800 words in English that are pronounced the same but spelled differently (*there*, *they're*, *their*, etc.), but only about a half dozen words are spelled the same but pronounced differently (*read* and *read*). Homophones complicate sound–letter correspondences, but they make the process of identifying words in silent reading easier by signaling differences in meaning through differences in visual shape. Similarly, spelling patterns often preserve meaning at the expense of phonemic accuracy. The second *o* in photograph or photography represents two different phonemes (/o/ and /ae/), but keeping the spelling the same preserves the common meaning of the root word, which is an advantage in silent reading. Indeed typography and typographic variation also support silent reading. Whether a document is set in Times Roman or Helvetica has little effect on our ability to read that document out loud.

[7] Consider this description of how a 14th-century library was created: "When Cosimo de'Medici was in a hurry to form a library . . . he sent for Vespasiano, and received from him the advice to give up all thoughts of purchasing books, since those that were worth getting could not be had easily, but rather to make use of the copyists; whereupon Cosimo bargained to pay him so much a day and Vespasiano, with fifty-five writers under him, delivered 200 volumes in twenty-two months" (quoted in Chappell, 1970/1980, pp. 36–37).

what would appeal to patrons.[8] However, at the same time, he was developing a technology that could meet the growing demand for printed materials.

Although the scribes no doubt complained bitterly that these newfangled "printed" books were degrading the visual and tactile quality of handwritten books, they could not compete with the speed, precision, and economy of the letterpress. In the first 50 years after the invention of movable type, more books were printed (approximately 10 to 20 million) than had ever been produced before Gutenberg (Craig & Banton, 1987). Because of movable type, books became a commodity, an object of commerce, increasing the availability of printed documents for everyday users.

Letterpress printing may have established bookmaking as a commercial rather than an artistic or conserving activity, but printing was still a slow, labor-intensive process. Letterforms had to be cut in steel, matrices built, and then typecast from molten lead. Signatures had to be hand-composed, the press made ready, and paper hand-fed into and then pulled off the press. As a consequence, only the most significant manuscripts found their way into print—religious or political documents that printers thought would sell or that a patron was willing to commission.

Before the invention of movable type, all documents were created by hand; after the invention of printing, most documents were still produced by hand.[9] Letterpress printing was a technology best used to create many copies of a few documents rather than a few copies of many documents. Letterpress printing helped establish publishing as a commercial activity and made mass literacy possible by producing the books, newspapers, and pamphlets that gave people something to read, but it continued to limit access to the tools of publishing.

Letterpress printing also reinforced the dichotomy between "the book" as something to be bought, sold, and valued and the "everyday document" as something ephemeral and handwritten, to be used and discarded. Between the typeset, printed book and the everyday handwritten document lay an enormous station of opportunity.[10] Not surprisingly, inventors tried to fill that gap with machines that would be fast and simple enough to set type by hand. The first of those machines was the typewriter.

THE TYPEWRITER[11]

From its beginning, the typewriter had been conceived as a publishing device, an alternative to handwriting for everyday publishers. This intent is clear from the first patent application in 1714 by Henry Mill in England: "An artificial machine or

[8] As a result the quality of that first Bible was remarkably high. Chappell (1970/1980) has argued that "it is possible to put the best piece of contemporary printing beside the Gutenberg Bible, and to compare the two without any concession being asked for the latter because it was produced more than five hundred years ago" (p. 19).

[9] Olmert (1992) devotes a whole chapter, "Debits and Credits: The Keeping of Accounts," to the topic of medieval everyday handwritten documents, including ledgers, court records, parish registers, wills, and insurance inventories.

[10] This dichotomy continues in contemporary discussions of electronic publishing that single out the book (rather than the newspaper or the magazine) as the primary artifact of print culture (see Bolter, 1991; Landow, 1992).

[11] The material in the next two sections is adapted from Kalmbach (1988a).

method for the impressing or transcribing of letters . . . whereby all writings whatsoever may be engrossed in paper or parchment so neat and exact as not to be distinguished from print" (quoted in Blanchard, 1981, p. E-2).

No drawing or models exist of Mill's machine. It would be another 150 years before typewriters were commercially feasible. The first American patent for a typewriter was issued in 1829 to William Austin Burt. He called his machine "Burt's Family Letter Press." The first commercial typewriter was marketed in 1852 by John Jones. His advertising copy suggests that he was going squarely after what we today would call the desktop publishing market:

> It may be advantageously used in localities remote from Printing Offices, for printing advertisements, handbills, circulars, cards, etc, . . . It is easily managed, and any child that can read will, with a few hours practice, print accurately. (quoted in Blanchard, 1981, p. E-9)

Although these 19th-century entrepreneurs hoped to bring the power of typography and letterpress printing to everyday users, they could not deliver a machine that matched their vision. Christopher Latham Scholes developed the first commercially successful typewriter in the 1870s by placing individual letters on separate bars. Each bar was activated by its own key—a move that improved speed but severely limited the quality and variety of letterforms.

Scholes's machine could use only one typeface, and that first model offered only all-capital letters. (The shift key, which made possible the inclusion of upper and lower case letterforms on a single typebar, did not appear for another two years.) For the typebars to function without jamming, each character had to be allotted an equal amount of space, and the characters had to be made visually simple so that they would reproduce accurately. Even so, image quality was uneven and spacing erratic. Those early typists could not even see what they were typing. The typebars were arranged in a circle beneath the platen, and text did not emerge until six lines after it was typed. Although the early typewriter entrepreneurs may well have dreamed of a desktop publishing market, it was Scholes's ability to deliver inexpensive, consistent speed rather than visual quality that made the typewriter a success (see Walker, 1984).

Like today's proponents of computers, desktop publishing, and hypertext, proponents of the typewriter were quick to get carried away in promoting this new technology. As early as 1872, Ely Beach announced the following in the pages of *Scientific American:*

> [The Scholes typewriter] requires no especial skill in its manipulation. A child knowing its letters may use it after an hour's instruction, and indeed any one, after short practice, can easily become able to write from 60 to 80 words per minute. (quotes in Blanchard, 1981, p. E-26)

Along the same vein, in 1922, James Collins speculated about the typewriter's likely effect on printers:

> The first inventors thought the typewriter would take the place of a pen—write letters and copy documents faster. But people quickly saw that, by using carbon paper, they could write several copies of a letter or document. That proved to be a fine thing. Thinner paper gave more copies, but not as many as its users wanted. Then Edison invented the mimeo-

> graph, by which the typewriter could write a stencil on waxed paper, and from that, thousands of copies were made. The printers were frightened! If a girl with a typewriter could make thousands of circulars, who would want printed circulars? (p. 493)

Unfortunately, no matter how much inventors and futurists promoted the typewriter, it could still only produce one page at a time. Typewriters were limited publishing devices. They offered an alternative to handwriting only when one copy was needed (or, with carbon paper, a few copies), but typewriters were not an alternative when multiple copies were needed. The gap between the high-end publishing on a letterpress and the low end publishing of a typewriter-driven office continued to be large and largely empty. Not until after World War II did the technologies of offset printing—photo-offset lithography and its electrostatic cousin xerography—fill that gap by moving reproduction from a mechanical to a chemical and eventually an electrostatic basis.

LITHOGRAPHY

Lithography was invented in 1796 by Aloys Senefelder, a struggling Bavarian playwright and musician. No one would publish his stuff so he decided to print it himself. Letterpress printing was beyond his means, so he began to experiment with alternative printing methods. According to the no doubt apocryphal story, Senefelder had just prepared a piece of limestone for etching when his mother asked him to write a laundry list for her. The laundress was waiting and there was no paper or writing ink nearby, so he wrote the list with lampblack on the clean stone. Later, he discovered, largely by chance, that the waxy lampblack could be transferred directly from the stone to paper. Senefelder had discovered an alternative to letterpress printing. Instead of creating images mechanically by pressing paper against type, he could transfer an image chemically (Lawson, 1963).

Senefelder named his new process *lithography* from the Greek for "stone writing" and abandoned the arts to develop and promote his invention. During the 19th century, lithographic stones were used to produce color prints and other illustrations. Mechanized lithographic presses were introduced in the 1840s. Photography was first used to transfer images to stone in 1857, and the offset principle of first transferring an image to a rubber blanket and then from the blanket to paper was introduced in 1875. These different innovations were first combined in 1905 with the rotary offset press (Lawson, 1963).

Pocket Pal (1984), the classic reference guide for the print industry, describes the modern process of offset lithography as follows:

> Lithography uses the *planographic* method. . . . Printing is from a *plane* or flat surface, and there are two basic differences between offset lithography and other processes: (1) it is based on the principle that grease and water do not mix, and (2) ink is *offset* first from the plate to a rubber blanket, and then from the blanket to the paper.
>
> When the printing plate is made, the printing image is rendered grease receptive and water repellant, while the non-printing areas are rendered water receptive and ink repellant. On the press, the plate is mounted on the plat cylinder which, as it rotates, comes into contact successively with rollers wet by water or dampening solution, and rollers wet by ink. The dampening solution west the non-printing areas of the plate and prevents the

> ink from wetting these areas. The ink wets the image areas which are transferred to the intermediate blanket cylinder. The paper picks up the image as it passes between the blanket cylinder and the impression cylinder. (p. 28)

This offset process involves three essential steps:

1. A plate containing the image to be produced is treated in some manner so that ink is attracted by the image area and repelled by the nonimage area.
2. The image is offset from the plate to a secondary medium such as blanket.
3. The image is transferred from the blanket to paper.

Because an image is transferred from plate to blanket to paper, the plates used for offset lithographic printing can be made from materials that could not otherwise withstand the stress of imaging directly onto paper. In particular, paper plates for offset lithography were developed that could be imaged by hand or with typewriters. More popular and more effective, however, were metal or bi-metal aluminum plates that were imaged by various photographic processes.

Creating plates by photographic reproduction (rather than by hand or with a typewriter) had a number of advantages. When a plate wore out, an exact duplicate could be created photographically. Errors could be corrected on the original copy rather than on the plate or on the press. Perhaps, more significant, photographic reproduction freed graphic designers from the need to square lead type and slugs into a letterpress form. Anything that could be put on a piece of paper could be photographed and then printed.

Despite these advantages, lithographic presses were not widely used until after World War II. During the early 20th century, offset printers had to use a linotype machine to set type, pull a proof of that type off a proofing press, and then use that type to create a page (Kleper, 1976).

One alternative to hot lead were typewriter-like systems, often referred to as "cold type." These systems were based on a technology of interchangeable type elements that had originally been introduced in the 1870s to compete with Schole's typebar machine.[12] The cold-type typewriters that inherited this technology offered excellent typographic controls and a reasonable selection of typeface, although with limited type sizes.

Of these cold-type typewriters, one of the best was the IBM Composer. Introduced in 1966, the Composer offered excellent intraletter and interlinear controls with 11 true typefaces ranging from 6 to 12 points and an image quality that was outstanding: dense, black, consistent type that was easily comparable to or better than an all-text page from a laser printer.[13]

Regardless of this quality, cold-type typewriters were limited typesetting devices, and the potential of offset printing remained largely untapped until inventors figured out how to image letterforms directly onto photographic paper, a process called *phototypesetting*. The first phototypesetters were introduced in the late 1940s. They were slow, mechanical devices that operated much like their hot lead parents. The

---

[12] Typewheel typewriters generally offered better image quality than typebar typewriters, but until IBM perfected their Selectric electric typewriter, type wheels could not compete with typebars on speed.

[13] Carte (1974), for example, advocated using an IBM Selectric Composer for "final repro typing" of technical manuals to provide readers with much more professional looking documents.

second generation of machines began appearing in the mid-1950s with the Photon 200. The third generation, using cathode tube technology, began appearing 10 years later.

The marriage of photographically produced type with photographically imaged printing plates has made offset lithography a dominate commercial printing technology. Letterpress, gravure, and other forms of printing continue largely as niche technologies. Photo-offset lithography, in the form of small table-top duplicators, also found its way into commercial and nonprofit organizations for in-house print shops. Today, however, such in-house printing has been all but replaced by a different form of offset: electrostatic printing.

XEROGRAPHY

Electrostatic printing (more commonly known as *xeroxing*) is the process used in most photocopy machines and laser printers. The Xerox machine's story is a classic tale of American business. An inventor working in his kitchen, a small company persevering in the face of skepticism and outright disbelief, and a product—the Xerox 914—combined to change the way we do work.

The process of electrostatic printing was invented by Chester Carlson in 1938.[14] Carlson had graduated from Cal Tech with a degree in Physics in 1930 just as the depression was beginning. He could not find employment as a physicist so he moved to New York, where he worked as a clerk in the patent department of a law firm. His primary tasks involved recopying manuscripts and preparing photostats of drawings for patent applications. Convinced that there had to be a better way, he converted his kitchen to a laboratory and set out to find a way to use photo conductivity to reproduce an image. The process Carlson discovered was complicated and tedious:

> The Carlson process essentially broke down into five steps. First, a special photo conductive surface was given an electrostatic charge [in early demonstrations Carlson rubbed the material with a piece of fur] which it could hold only in the dark. Once exposed to light, the charge would disappear. Next a printed page was placed in close proximity to this surface and light was shone on it so that an image of the printing was projected onto the surface. (Because of the light, the surface kept its charge only in those places occupied by the dark ink.) The third step was to dust the surface with powdered ink, which stuck to the charged portions, creating a mirror image of the printed page. This image was then transferred to a blank sheet of paper. Finally, to make it permanent, heat was applied which melted the ink and fused it to the page. (Kearns & Nadler, 1992, p. 18)

Carlson worked for years refining and promoting his invention. He convinced the Battelle Memorial Institute to invest in the process, and they in turn interested Haloid, a small photo paper company from Rochester, NY to invest.

In 1948, Haloid produced its first product based on Carlson's process: a large flatbed copier. The machine, however, was difficult to operate. A skilled operator had to go through over a dozen separate steps to complete a single copy. Haloid sent evaluation versions of the copier out to large companies. The machines all came back

---

[14] The first words ever photocopied were the date and place of the event, "10–22–38 Astoria." This landmark copy is reproduced by Kearns and Nadler (1992).

with notes of apology to the effect that the process was just too difficult and messy for office use. Things looked bleak until someone from Battelle suggested that the machine might be used to image paper plates for offset lithography as press operators were used to running complicated, messy machinery. Haloid sent more prototypes out, this time to press rooms. Soon a major automobile manufacturer reported that whereas it cost them $3.12 to produce an offset page using a zinc plate, that page could be printed for 37¢ using a paper plate created with a Xerox plate maker, and that plate could image up to 20,000 copies.[15]

Haloid began marketing the Xerox paper plate maker, and throughout the 1950s, profits from the plate maker financed their efforts to bring out a plain paper office copier. That machine, the Xerox 914, was introduced in March 1960, 22 years after Carlson had invented the xeroxgraphic process. Haloid had learned from their difficulties with flatbed copiers. To use the 914, a user had only to put a sheet of paper face down on the glass, twist a dial for the number of copies (up to 15), and then press a button. Fifteen seconds later a copy emerged.[16]

In addition to making their copier easy to use, Haloid's most important innovation was to lease rather than to sell the 914 and to charge a per-page cost for copies. The machines were extremely expensive to make and would have been difficult to sell but for the fact that for only $100 per month anyone could lease one and make 2,000 free copies. After that, copies cost 4 cents. The plan was truly elegant; both large and small businesses could afford the same machine and pay according to their usage. Even better, the need to make copies seemed to feed itself. Unlimited copies of documents was one of those things people did not know they needed until they had a copier in their office. Haloid originally projected that each 914 copier would generate 10,000 copies a month. The actual number proved to be closer to 40,000. A cash cow was born. Largely on the back of a single product, Haloid was transformed from a small Rochester photographic paper maker into a major international technology company. From 1959 to 1968, Xerox (they changed the name in recognition of the success of the product) went from annual sales of $32 million to $1.25 billion.

The critical innovation in xerography was eliminating the plate (and, of course, plate preparation) from the reproduction process. With a Xerox machine, the original paper document also served as the plate. The operator simply puts the original on the glass and presses a button. The image on the paper is then optically translated into a negative electrostatic charge on a drum. This drum rotates through a bath of positively charged toner. The toner is attracted to the negatively charged areas of a drum. The drum then transfers this toner to a piece of paper, and the toner and paper are heat-sealed together.

Plate preparation is a tedious and expensive component of traditional offset printing. The original document has to be photographed to create a line negative (in which any black areas of the original page are clear and any white areas are black). Flaws in the line negative are opaqued. The negative is then "stripped" or taped onto a sheet of golden rod paper (from which rectangular boxes have been cut for each

---

[15] It was perhaps the first report of using desktop publishing technology to save money by sacrificing quality.

[16] To prove the machine was easy to use, Xerox produced a television commercial in which a businessman asks his 6-year-old daughter to make a copy for him. She happily skips off. When she comes back, he asks her which is the original and which the copy, she scratches her head and says, "I forget."

type block). The stripped negative is then attached to a photosensitive plate, and the two are exposed to an intense light. The image that the light has etched onto the plate is developed with chemicals, and the plate is coated with lacquer. All these steps have to be done for each plate before that plate can be mounted on a press and adjusted for printing.

The advantage of a plate, however, is that it can produce thousands to hundred of thousands of copies. Consequently, the cost of preparing that plate can be prorated across the number of copies needed. As the number goes up, the plate's cost per copy goes down. Because a photocopy machine does not use a plate, the cost to create a copy never changes. If the effective cost of creating one copy is 5¢, then the effective cost of creating 100 copies is $5 and the effective cost of creating 1,000 copies is $50. The cost of creating a few documents is quite reasonable, but at some point it becomes cheaper to print documents by offset lithography than it does to use a photocopy machine. Because of these differing economies of scale, chemical-based offset printing and electrostatically based photocopy machines are complimentary rather than competing technologies. Electrostatic printing can produce small numbers of copies inexpensively, but costs mount as copies increase. Offset lithographic printing can produce large numbers of copies inexpensively, but its cost per page to create small numbers is high.

The economic balance point between offset and xerographic printing has, however, been shifting. When I was in college, in the early 1970s, the quality of photocopy machines could not compare with offset printing. We could justify the cost of offset printing if we needed 100 or more high-quality copies. Today, unless special paper, halftones, or color are needed, the decision point is closer to 10,000 copies. Current machines produce not only black, text-only pages of quality comparable to offset printing, but they can print on both sides of the page and collate and bind multipage documents. A set of master pages is put in one end and a bound booklet emerges from the other.

Together the technologies of photo-offset lithography and electrostatic photocopying have helped to shape the information age we now live in. Before, if someone wanted copies of a document, he or she could copy or retype the document by hand, go to a printer and have the work reset and then printed. There was little choice in between. Today anything can be inexpensively printed, and virtually anyone who can put words or images on paper can get those words or images reproduced at a library, grocery store, or laundromat.

A number of information processing occupations (such as technical writing, proposal writing, public relations, etc.) owe their existence in large part to the offset revolution. The major genres of documents that today drive desktop publishing—newsletters, brochures, fliers, booklets, and so on—all became popular because of offset's flexibility. How many brochures would be received in the mail if each was set in hot lead type? How many manuals would be produced, and how useful would they be? Just as letterpress printing democratized literacy by making written texts available to a larger number of people, offset printing—in the form of small duplicators, photocopy machines, and laser printers—has made publishing as social action available to virtually anyone.

New purposes for published have come to supplement older ones as the technologies for creating and distributing copies have changed. Publishing evolved from a preserving activity to a commercial activity as literacy levels increased and access to printed texts became widespread. Offset technologies such as lithography and xerography made the tools for creating copies even more widely available, but it would take an additional technology—the computer—to make publishing part of society's everyday transactions through writing. . . .

DEVELOPING YOUR UNDERSTANDING

1. Analyze how technology affects who reads, what gets written, and why texts are written. Use three technologies, including the Web, as examples.
2. Kalmbach, in this article, does not pay particular attention to the ways technologies affect the writer's work (e.g., where the writer works, the writer's function/role, who the writer writes to/for, who employs the writer, the prestige of the writer, and the writer's routine tasks). Briefly trace, through the technological history represented by Kalmbach, the ways technologies have affected the professional writer's work, and predict some of the ways near-future technologies might affect your work as a professional writer.
3. Describe the ways economic issues like cost, supply, demand, and economies of scale have driven the development and use of writing technologies, as well as the decay and elimination of others. Explain how these economic variables and the availability of such automated features as brochure design templates, grammar check, and fit-to-page commands can both enhance and threaten the professional writer's work.

FOCUSING ON KEY TERMS AND CONCEPTS

Focus on the following terms and concepts while you read through this selection. Understanding these will not only increase your understanding of the selection that follows, but you will find that, because most of these terms or concepts are commonly used in professional writing and rhetoric, understanding them helps you get a better sense of the field itself.

1. task-oriented reading
2. interactivity
3. functional mapping
4. modularity
5. navigational strategies

THE SHAPE OF TEXT TO COME: THE TEXTURE OF PRINT ON SCREENS

STEPHEN A. BERNHARDT

Changes in the technology of text invariably trigger changes in the shape of text. Texts are undergoing monumental transformation as the medium of presentation

Source: Bernhardt, Stephen A. "The Shape of Text to Come: The Texture Print on Screens." *College Composition and Communication 44* (May 1993), pp. 151–175.

shifts from paper to screen. We need to constantly appraise the broad drifts in the shape of text—to anticipate what now constitutes and what will soon constitute a well-formed text. We need to think about how readers interact with text—what they do with it and how. We need to anticipate where text is going: the shape of text to come.

This paper suggests some of the dimensions of change in how text is structured on the page and on the screen. It is necessarily speculative, since the topic is just beginning to receive systematic attention (Bolter; Brockmann; Horton; Kostelnick; Merrill; Rubens, "A Reader's View" and "Online Information"; Rubens and Krull; Special Issue of *Visible Language* 1984).

We have a good theoretical understanding and a highly developed practical art of the rhetoric and text structure of paper documents, and this praxis exerts a strong shaping influence over texts produced via electronic media. We are in a state of rapid evolution, with heavy borrowing on the history of text on paper, applied sometimes appropriately and sometimes inappropriately to the new medium. Because electronic text does not create a totally new rhetoric but depends for its design on the strategies of paper texts, the starting point in this analysis is not "How do screen-based texts differ categorically or essentially from their paper-based counterparts?" but "What is a framework for understanding dimensions of variation in texts across the two media?"

This paper uses a text analytical approach to identify nine dimensions of variation that help map the differences between paper and on-screen text. Screen-based text tends to exploit these dimensions to a greater degree than does paper text.

To a relatively greater extent, then, on-screen text tends to be:

Situationally Embedded: The text doesn't stand alone but is bound up within the context of situation—the ongoing activities and events that make the text part of the action.

Interactive: The text invites readers to actively engage with it—both mentally and physically—rather than passively absorb information.

Functionally Mapped: The text displays itself in ways that cue readers as to what can be done with it.

Modular: The text is composed and presented in self-contained chunks, fragments, blocks.

Navigable: The text supports reader movement across large pools of information in different directions for different readers and purposes.

Hierarchically Embedded: The text has different levels or layers of embedding; text contains other texts.

Spacious: The text is open, unconstrained by physicality.

Graphically Rich: The text exploits and integrates graphic display to present information and facilitate interaction.

Customizable and Publishable: The text is fluid, changing, dynamic; the new tools of texts make every writer a publisher.

As academics with a commitment to certain kinds of discourse, we may not see as desirable all of these developments in the ways text is structured, but they appear to be inevitable. We need first to understand the directions that computers are taking written language, and then to consider these changes as we teach our students strategies for reading and writing text in a new age.

SITUATIONALLY EMBEDDED TEXT

When people voice doubts about whether computers will take the place of books, they are generally expressing doubts about readers' tolerance for extended reading on screen. Reading from screens tends to slow people down and fatigue them, in part because the contrast of print on page is much better than that of text on screen. But when reading is viewed as a sub-task within a larger task environment, the issue of fatigue is not so critical. Extended reading will continue to rely on print, while other functional sorts of reading will rely on screen-based text.

Screen-based text differs from paper text in many ways, and not just because the two media are different. We use text on screens under different conditions and for different purposes than we do paper texts, and it is these differences in use and purpose that will ultimately determine the key points of difference between the two media (Barton and Barton, "Simplicity"; Duchastel).

A real virtue of paper text is its detachment from the physical world. We can read on planes or in the car; we can put books in our backpacks or leave them at home. We can pick up a book or magazine or a newspaper and read in every imaginable situation, no matter what else is going on about us. In fact, reading allows us to escape the immediate situation, to enter other worlds.

In comparison with paper text, screen-based text tends to be more tightly embedded in the context of situation; it is more likely to be bound up as a part of ongoing activities. Reading screen-based text is often integrated with other forms of action—learning to use software, constructing texts from separate files, or searching a database. A reader might search for relevant text, retrieve information in the form of procedures or syntax, and then return to the task environment. In such situations, reading becomes a second-level activity, resorted to when the higher-level task activity hits a snag.

This kind of task-oriented reading stands in contrast to, say, reading imaginative literature or magazines, where readers enter a world of text that impinges little on their real-time situation. Readers of screen-based text are not so much *readers* as *doers* or *seekers:* they read to find out how to do something or to retrieve some bit of information. People tend to read screen-based text to play games or to program; they read-to-write, or read-to-operate, or read-to-look-up. We don't really have language for this kind of reading—it's more like *using* text than *reading* it. Such reading-to-do is more like making raids on print than having extended engagements with a writer's ideas or arguments. It is driven by the pragmatic situation; it is exploitative; it is manipulative (see both Sticht and Redish).

The shape of screen-based texts is influenced heavily by one specific development: *help systems*—those word files that attempt to rescue computer users who encounter difficulties. Nobody reads this kind of text in anything like linear order, but many users make incursions on it as they struggle to work with their machines, reading bits and pieces as needed. The help text is simply part of the machine.

But reading that is situationally integrated with other activities is typical not just of help systems for using computers. The electronic writing classrooms of Project Jefferson at the University of Southern California (Chignell and Lacy; Lynch) or Project Athena's Educational Online System at MIT (Barrett, "Introduction," *Society*) exist to support not so much reading activities as writing activities: researching, keeping track of information, drafting and revising text, sharing text through collaboration, or sharing texts and notes with others in the class. Pieces of text get used, copied, borrowed, annotated, clipped, revised, and passed around in the interest of some governing activity—in this case, the improvement of writing. Text is read throughout the process, but reading is not the primary or ultimate goal. The computer structures an environment, where the writer, a set of texts, and a group of people interact in desirable ways. The screen-based text is interwoven with the larger activity of producing work in a group setting. Text is inseparable from the situation.

Such applications of technology take reading and writing beyond simple interaction with the computer; the computer scaffolds social interaction within an electronic environment. In his Introduction to *The Society of Text*, Barrett describes the use of computers to structure interaction (as opposed to modeling cognition) in MIT's Athena-supported network:

> [T]he internal workings of the mind were not mapped to the machine; instead, we conceived of the classroom as a "mechanism" for interaction and collaboration and mapped those social processes to the computer. In essence, we textualized the computer: we made it enter, and used it to support, the historical, social processes that we felt defined the production of texts in any instructional or conferencing environment. (xv)

In such situations of use, text is embedded within systems—it is not separate like a book or a magazine. Its texture is shaped by both the machine and the instrumental purposes and social interactions to which the text is put. Screen-based text becomes part of a physical system that governs where it can be used, who can access it, what is needed to access it, and so on. Text is inseparable from the machine.

Notice how different this tends to make screen-based text from paper text. While books are self-contained, portable, and usable within almost any situation, screen-based text becomes dependent on a larger technological and social environment, to be used under delimited circumstances, typically as an integral part of other ongoing events. This is an important contrast in the pragmatics of paper vs. screen, underscored by the contrast of *text-intensive* books vs. *situationally embedded* screen-based language.

INTERACTIVE TEXT

It is commonplace to characterize the reader's role in a text as being active or transactive, construction or constitutive. In this view, readers construct or reconstruct a text in their own image, bringing as much to a text as they take from it. When we talk in these ways, we often have in mind private encounters with text in physically inactive settings. We are talking primarily about mental processes, or language processes, or sometimes social processes, but not necessarily physical processes.

It is useful to view the reading of electronic text in similar terms, only more so, or at least, more variously so. Readers of on-screen text interact physically with the text. Through the mouse, the cursor, the touch screen, or voice activation, the text becomes a dynamic object, capable of being physically manipulated and transformed. The presence of the text is heightened through the virtual reality of the screen world: readers become participants, control outcomes, and shape the text itself.

Reading text on screen tends to be much a more behaviorally interactive process than reading text on paper (Duchastel). The parallel activities of reading and writing create the interaction. Screen readers are actively engaged with screen text, as they key in information, or capture texts from one file and move it somewhere else, or annotate or add to existing information in a file. A similar interactivity is sometimes sought in books, as writers try to engage the reader in solving problems, considering scenarios, or attempting various learning activities while reading the text. However, writers of print material cannot *force* the interaction, they can only *invite* it; readers can play along or skim past the problem sets, brain teasers, or tutorial activities. Writers of on-screen texts can *force* interaction, making it necessary for the reader to do something physical in order to get to the next step.

The contrast in interactivity distinguishes other genres as well. Consider printed novels and their screen counterparts: text-based "novels" or adventure games. Readers of novels are constrained by the linearity of the text. While there are fundamental differences in how readers respond to a text, the book presents the same face to each reader, and the choices of approach are very limited. One might choose to read the ending first or to peek at various chapters, but these are fairly impoverished choices. A reader of a text-based electronic novel or adventure game, in contrast, has to make constant decisions about where to go, what to do, who to follow or question. In doing so, the reader is forced to construct not just a mental representation of the work, but a physical representation as well (the succession of screens), through concrete manipulations of the text. Out of many possible physical constructions of the text, the reader creates one, a particular chronological and experiential ordering of the text, a reading that belongs to no other reader.

I am not holding up increased interactivity as a goal of print and I am not suggesting that, for example, electronic novels are richer or more satisfying than print novels. Such is clearly not the case. However, we should not underestimate the developing genre of electronic novel: writers are discovering new forms of literary textuality and engaging in some very interesting experiments. (See, for example, the special issue of *Writing on the Edge*, with its accompanying hypernovellas on disk.) In these experiments, authors engage readers in new forms of interaction, encourage readers to take control over the text, and blur the lines separating author and reader. We need to be alert to interactivity as a deeply interwoven feature of electronic texts, one we are just beginning to exploit.

FUNCTIONALLY MAPPED TEXT

Text, whether on page or on screen, performs a function of some sort: informing, directing, questioning, or posing situations contrary to fact. Such functional variation is often expressed linguistically through the grammatical systems of mood (indicative, imperative, interrogative, subjunctive). Readers can also usually make some rhetorical determination as to what a chunk of text is doing—whether it is making a generalization, committing a vow, stating a fact, offering an example or definition, offering metacommentary on the text itself, or some other text act. In many printed texts, such functional variation is mapped semantically—one interprets the functional roles of various chunks of text by inferring purpose from the meaning of the words or phrases. Often, semantic or rhetorical function shifts are mapped by cohesive devices, phrases like "for example," or "to consider my next point." When text shifts from one function to another, the rhetorical tension at the boundary tends to demand some kind of signal, and the language is rich in such signal systems (Bernhardt, "Reader").

Both sorts of text—print and screen-based—also use visual cues of layout and typography to signal functional shifts. The visual system maps function onto text, signaling to the reader how the text is to be read and acted upon. Thus tutorial writers (print or online) might use a numbered list of action steps, with explanations indented below each action, or they might use a double-column playscript format, with actions on the left and results or explanations on the right. Boldface or other typographical signals might highlight actions, while parentheses or italics might signal incidental commentary. The visual structuring that functionally differentiates text is reinforced by syntactic cues that highlight the action being performed—imperative or declarative grammatical structures, sequence cues like *next* or enumeratives, and explanatory phrases like "to complete the installation" or "pressing the return key enters the value."

When language is on-screen, readers must be able to distinguish different functions:

- Some language cues interaction with the system: how to manage files, execute commands, or control the display.
- Other language cues navigation: where one is, how to move around, or how to get help.
- Still other language offers system messages, showing that errors have occurred or that the system is currently processing some command. Some language simply reminds readers of the system status or default settings.
- And some language is informative/ideational.

The tight interworking of text and action leads to frequent system requests for action that the reader must interpret and respond to correctly (from the system's point of view). These functional discriminations are not unique to electronic text, but they tend to be much more important to efficient reading, and they tend to demand highly planned and carefully structured formatting decisions on the part of the writer.

Not all areas of the screen are equal, and functional mapping tends to be richest on the borders—in the peripheral areas a biologist would call *ecotonic*. There is always a rich diversity and abundance of life on the edges of systems. On screens, the language is the richest, there is the most going on, there is the greatest range of things to do around the edges, on the perimeters. It is on the edge that we recognize where we are, what we can do, where we can go, or how we can get out.

Increasingly, various programs are adapting consistent functional mapping of options. With pull-down menus, for example, available options often appear in a regular or bolded black font, while program options that are not currently available are shown in a shadowy, gray font. A simple, efficient cue such as this can greatly help readers use the functional mapping of programs they have never seen before. As readers become increasingly sophisticated and as interfaces coalesce around predictable design strategies, readers will develop their skills to the point where they efficiently and correctly recognize text-as-information versus text-as-signal-that-something-can-be-done-with-it.

Unlike paper texts, screens offer a dynamic medium for mapping text in highly functional ways. Relying largely on visual cuing, readers acquire knowledge of how to do things with words and images. The traditional cues of paper texts—margins, indents, paragraphs, page numbers—appear impoverished next to the rapidly expanding set of cues that facilitate functional writing and reading on screen.

MODULAR TEXT

Most texts reflect some modularity of structure: a text is composed of other texts. Books have chapters or individual articles; magazines have articles, sidebars, letters-to-the-editors, advertisements, tables of contents, and so on. Many forms of print are in some way or another compositions, pieces of text positioned with other pieces of text, and often the individual modules are very different in type or function. A newspaper, for example, with its large pages, allows many modules of several sorts to be composed on the same page, and readers can efficiently scan large amounts of information. And the direction in popular newspapers, such as *USA Today*, is toward modularization, with pages composed of short, self-contained, highly visual exposition. An encyclopedia, too, is composed of many individual modules, each of which constitutes a text that can stand on its own.

The movement of text from paper to computer screen encourages further modularization of text structure. The screen is a window on a text base—only so much can be seen at one time. Just as an 8 ½" by 11" sheet of paper to some extent determines the shape of printed text (titles, headings, white space, line length, indentations), the size and shape of the screen constrains the shape of electronic text. The screen, or a window contained within the screen, becomes the structural unit of prose, with text composed in screen-size chunks, no matter what their subject or function. In such systems, texts is highly localized. Reader attention is arrested at the level of idea grouping—the single topic that is represented on a single screen.

Because text is fragmented and localized, on-screen text has problems with local cohesion. Closely related ideas must frequently be separated by screen boundaries. Even lists of strictly parallel, coordinate information must often bridge screen divisions, and the break from one screen to the next presents a larger gap than that from one page to the next. Consider that in a book, even when chunks of information must be broken at page boundaries, there is a 50% chance that the boundary will be at facing pages. And print layout can be manipulated to keep related information on one page. The problem is more difficult with small screen dimensions and strictly modular text fragments. Each module must, to some extent, stand on its own, interpretable without close logical cohesion with other screens. The writer must assume that a reader can arrive at a given screen from practically anywhere, so there can be no assumption that the reader has built up a model of the logical relations of the text from processing pages in a linear order.

It might be argued that since screen text can easily be scrolled, text need not be fragmented into screen-size modules. While it is true that most windows allow scrolling of text that is longer than a screen, scrolling is inherently unsatisfactory. When text must be scrolled to be viewed, readers hesitate, not knowing whether to scroll down or skip the text. And while the reader can quickly skim a stack of information if each card is completely contained within the window, it is time-consuming and ultimately wasteful to have to scroll to see if text should be read.

Also, when text is not composed in screen-size bites, readers tend to lose their places and become disoriented. An example of this occurs with the ERIC CD-ROM indexes on Silver Platter.

The *Page Up* and *Page Down* keys are used to move through lists of references, but these commands take the reader across the boundaries of individual entries. Readers (at least this reader) constantly lose track of whether entries have been read or not, since top-of-screen is not also top-of-page. Information that identifies titles or authors is frequently separated visually from other important text (such as abstracts or keywords), and a given type of information (such as title or author) is never in the same place on the screen. The whole system feels jumpy and erratic, and a general sense of disorientation prevails. The problem is alleviated to some extent through the use of *Control-Page Down*, which takes the browser to the top of each entry. But then text is missed that does not fit on a screen. Thus, with document databases containing huge

```
SilverPlatter 2.01        ERIC 1982 - September 1991        F10=Commands  F1=Help

  implementation, and evaluation.  The final chapter includes a summary and
  recommendations.  The 12 appendixes, which constitute more than half of the
  report, include the pre- and posttests, personal data and summative reaction
  questionnaires, the task analysis, a skills and interests questionnaire,
  project PERT and Gantt charts, screen design prototypes, and a user's guide.
  Tables and figures appear throughout.  Field test data are also included.  (26
  references) (GL)
                                                                        19 of 29
  AN: ED308829
  AU: Morrison, -Gary-R.: And-Others
  TI: Reconsidering the Research on CBI Screen Design.
  PY: 1989
  NT: 20 p.; In: Proceedings of Selected Research Papers presented at the
  Annual Meeting of the Association for Educational Communications and
  Technology (Dallas, TX, February 1-5, 1989).  For the complete proceedings,
  see IR 013 865.
  PR: EDRS Price - MF01/PC01 Plus Postage.
  AB: Two variables that designers should consider when developing

MENU: Mark Record    Select Search Term    Options    Find    Print    Download
Press ENTER to Mark records for PRINT or DOWNLOAD
```

Figure 1. Screen from ERIC CD-ROM Silver Platter. Notice how the scrolling interfaces with reading, since top of screen is not top of module. The text also has a homogeneity problem, since the lack of typographic cues leads to a homogeneous visual surface and attendant reading difficulties.

Screen from ERIC (Educational Resources Clearinghouse) CD-ROM Silver Platter. Boston: Silver Platter, 1986–91.

numbers of entries that must be browsed quickly, avoiding scrolling text modules is preferable.

Modular text does have its advantages. One distinct advantage is that the same text base can serve multiple audiences and multiple purposes for reading (Walker). When texts are composed in screen size chunks, the same modular text fragments can be used to build different documents or different paths through a document. Novice and expert tracks, for example, can be structured out of the same set of information. Texts of various sorts can be *compiled* instead of *written*, constructed out of interchangeable parts. Both the WordPerfect and the Microsoft Word manuals are examples of highly modular writing, with topics arranged alphabetically under headings, so the manuals work simultaneously as alphabetically-organized tutorials and reference volumes. Each page has predictable information in predictable slots. Such a book can be written in any order, and modules can be revised as needed without much effect on the other modules. And a modular approach can work well in both paper and electronic media. (Although, of course, a printed manual needs careful adaptation to work well as an online manual.)

Whatever we may wish for, modular text is definitely the shape of text to come. For many pragmatic uses of screen-based text (such as online help), highly localized, non-sequential, fragmented pieces of text work fine. Such modularization leads to tremendous economy—a single piece of text can be written once, but read and used many times, by various writers and readers, for various purposes. It is well suited to mass storage on CD-ROM disks and to search-and-retrieve operations using keywords or browsers.

We might speculate on the effects of modularity. Will readers become less tolerant of extended arguments and reasoning? Will all texts disintegrate into fragments—a chopped up hash of language—with texts of 75-words-or-less dominating the presentation of information? Will we stop thinking of reading as an extended, engrossing transaction with a text and its author and think of reading, instead, as gleaning or grazing across a range of *textbits?* Yes and no. Some of us will continue to engage with extended, lengthy, integrated text for certain purposes under certain conditions. And all of us will be exposed to increasing quantities of textbits—bits that are skimmed and scanned, compiles and compositioned, presented through various text databases that help us organize and exploit the information explosion.

HIERARCHICAL, LAYERED, EMBEDDED TEXT

To a limited extent, printed text can achieve a special sort of modularization through layered or embedded effects. Within passages of text, semantic cues signal that information is peripheral, or supportive, or explanatory, or defining. Parentheses, footnotes, asides, and facsimile or boxed text all allow writers to escape the immediately present text, to move down or across a level in the text hierarchy, to assign a different status to information, to put it next to or below the predominant text level. In longer printed texts, writers can assemble glossaries, indexes, information on authors, prefaces, notes on the edition, or notes to specific groups of readers. These devices give print some texture of hierarchy, indicating that not all information is on the same level. Readers can pursue the mainline text, but they can also read peripheral or supporting information that has a status other than mainline. Texts digress.

Books, however, are imperfectly suited to hierarchical or embedded text. They essentially are a flat medium, meant to be read in linear fashion. Readers can escape linearity; they can jump around or use different sections of a text in different ways. (For a provocative presentation of ways that texts can escape two dimensions, see Tufte). The programmed textbooks that reflected behavioristic models of learning, such as Joseph C. Blumenthal's *English 2600*, were one attempt to escape the linearity of print. These books took learners on various tracks through the text. Short quizzes over material would assess learner knowledge and then send the learner to appropriate pages for explanation and practice. Advanced learners would speed along on the advanced track. Such books were always a little odd. The habits of approaching text in linear fashion were too ingrained on learners. As one worked through such texts, one wondered what was being skipped and whether learning was being accurately evaluated.

Unlike books, computers *are* well suited to nonlinear text. *Nonlinear text* is, in fact, probably the best definition of the kind of text generated by the rapidly expanding technologies of hypertext. Hypertext programs allow texts of various sorts to be combined into large text bases, allowing readers to move freely across various sorts of information in nonlinear ways.

Though two-dimensional, screens offer the compelling illusion of depth. In a windowing environment, active files and various kinds of text can be displayed and stacked up on the screen. Somewhere behind an active file, help can exist, to be called with a simple command or click of the mouse. Glossaries can exist behind words, levels of explanation and example can exist below the surface of the text. Text can be put on clipboards or pushed out to the side of the work area. Information can be present without being visible except through subtle reminders: a shaded term suggesting a connection to another text, a dog-eared page icon pointing toward a personal annotation on a file, a pull-down bar offering access to other texts. The reader can be in two (or more) places at once, with a definition popped up alongside an unknown text or with a palette of shading patterns placed alongside a graphic. The desktop can be stacked with open files—multiple applications running simultaneously—each with its own text in its own screen areas.

Paper text must embed signals of hierarchy within the linear text itself or in some remote location, such as a table of contents. But electronic text can actually be hierarchically or loosely structured, and it can show its structure schematically or in full detail. A screen-based technical manual, for example, can have a cascading design, with top-level screens offering statements of purpose, scope, and audience definition. At a next level of detail, overviews of steps in a process can be offered. Each step can be exploded to show detailed procedures, and behind the detailed procedures other sorts of information can reside—troubleshooting advice, specifications, or code (Herrstrom and Massey). Such hypertext features essentially allow text to escape linearity—there need not be a Chapter One because there need not be a declared linear order of information. Text can be loosely structured, built by association, linked in networks or multidimensional matrices.

Linguists have long noted that syntax is deeply recursive. Sentences can contain sentences, clauses can contain multiple other clauses, and phrases can themselves contain clauses, so that, in effect, lower-level units within a hierarchy can contain higher-level ones. With electronic text, what is true at the syntactic level—the recursion that gives language extreme structural flexibility—is true at the discourse level. Like Chinese boxes, text can be nested within text, and huge texts can reside within tiny fragments. With the combination of both hierarchical subordination and lateral links from any point to any point, hypertext offers greatly expanded possibilities for new structures characterized by layering and flexibility.

NAVIGABLE TEXT

Readers of all text must navigate; they must find their ways through sometimes large or diffuse collections of information. And they have developed navigational strategies for print—using signposts such as tables of contents, indexes, headings, headers, pagination, and so on. Print readers can flip around in a text, scan very quickly, size up the whole, and generally learn from physical and directional cues where they are in the text and where information they need is likely to be.

Imagine your own strategies for reading a newspaper: how it is you decide what to read and how much of it, how your eyes work the page, how quickly and efficiently you take in information. There are highly developed skills operating here, and it shouldn't be too surprising that the early forms of teletext news, presented as a simple scrolling panel of information, did not enjoy much acceptance since they did not allow readers to exercise existing, efficient strategies for using print. People do not want to read extended text on screen, especially when the machine controls the content and the pace. Readers want control.

Books are highly evolved forms: what they do, they do well. A reader can come to a book with highly evolved strategies for getting information from print, but users of computer systems are often handicapped by not having useful, productive strategies for approaching computer-based text. They are often frustrated when they apply learned strategies from print or from other software, only to find that one system doesn't work the same way another one does. Because the screen lacks the total physical presence of a printed text, screen readers have difficulty sizing up the whole, getting a full sense of how much information is present and how much has been viewed. One knows immediately where one is in a book, but it is often difficult to maintain the same intelligence in screen-based text. And so readers of on-screen text have a difficult time navigating. They must read through a window onto a text, and that window limits what the reader sees at any one time. The window is a flat, two-dimensional space, and it is notoriously difficult to know exactly where one is, where one has been, or where one is going. And when an on-screen text is complicated by multiple windows and multiple active files, levels of embedded texts, or a hypermedia environment, navigation poses significant threats to coherence.

A critical threat to the usefulness of on-screen text is the *homogeneity problem* (Nielsen 299). Text on a computer screen tends to be uniform; because of consistent display fonts, spacing, margins, color, design, and size of text modules, it all starts to look the same. Contrast a book with a newspaper or a shopping list to get a sense of the variation in surface that print presents and it becomes clear why on-screen readers are frequently lost in textual space. The challenge of designing text on screens rests in large part on overcoming the machine's tendency toward a homogeneous surface.

Many initial attempts to provide navigation aids for screen-based text are analogically borrowed from paper text. Menus are something like tables of contents, except that when one makes a decision about where to go for information, the page turning is automatic. Indexes look similar in both media and work equally well if designed well. Still borrowing on paper cues, screen headers and footers—as well as titles on menus, pop-up windows, or text modules—can tell readers of screen text where they are, much as one can tell in many books what chapter or what article one is reading by looking to a title or a page header. Screens can be paginated (borrowing even the term *page* for *screen*), often in the form *Page 3 of 6*, but also iconically as in PageMaker documents, with sets of tiny, numbered pages on the lower left of the screen that can be clicked on to move through the document. Readers need a sense of how much they have viewed and what is left in the set of related screens they are scanning.

While some navigation aids are borrowed from print, other navigation options work best only within electronic media. Graphical browsers (looking like cluster diagrams) can offer readers a visualization of the structure of information, so that one can see at a glance the scope and nature of large collections of information. The information contained in a large text base is mapped onto a network representation—with key terms constituting nodes and lines showing relationships among the nodes. Each node represents a group of related information. Like electronic menus and indexes, such browsers offer more than a cue to structure; they facilitate interaction with the text base. Readers can point and click their way from one node to another, explode a node to explore sub-nodes, and so build mental models of the structure of information in interactive, highly intuitive ways. Books might provide similar browsers, such as timelines or the *Encyclopedia Britannica's* Topicon, but these devices simply do not have the fluid or interactive qualities of electronic browsers.

Ties, or links, or buttons—hot spots in the text that link one screen or term with other screens or terms in the text base—work much better in screen-based text than in paper texts. Some books achieve a limited level of such linking through, for example, endnotes or references to appendices, but the general mechanism is much better suited to electronic text. (Students of mine have read and used Joseph Williams's *Style* for months without recognizing that boldfaced terms are defined in the glossary.) The links in screen-based text announce themselves by their typography or visual character; clicking on a link takes one immediately to some related text. The links can be visually distinguished by function—links from an index to relevant text, from a term to a glossary definition, from a menu to a chosen activity, from an overview to more detailed information, and so on. Links serve as *anchors* to a given screen; one is anchored to the screen icon while going off for an exploratory cruise. For navigating large text bases, the single device of links with anchors in a present screen provides a powerful control over text that cannot be approached within paper texts.

Standard navigation devices are quickly emerging, so that screen readers can bring learned strategies to new interfaces and new texts. In many programs, the perimeter of the screen is defined as a wayfinding area, containing cues about where one currently is (as in the title on the screen) and about where one can currently go (as represented primarily in the choices of active icons). Having worked with a few programs that use similar devices, readers come to expect the icons to be active—to respond to a point-and-click. They realize, too, that cues will generally allow them to determine where they are and where they can go. They relate to the home menu—the familiar, top-level screen that offers a breakdown of wayfinding options at the broadest level. Such screens constitute *landmarks* to the navigator—familiar, easily recognizable locations. Readers come to expect to be able to do certain things, and well-designed systems use the navigational knowledge readers have naturally acquired through interaction with other programs, just as book designers offer readers an index, or a page number, or a chapter title.

SPACIOUS TEXT

Print is constrained by sheer physical bulk. Consider the constraint of bulk on the compact *Oxford English Dictionary*, with its print compressed to the point of practical illegibility to the naked eye, crammed onto pages full of abbreviations and omitted information. Or consider the sheer bulk of paper documentation necessary to run a complex piece of machinery—an aircraft carrier or an airplane. The sheer weight of paper makes a strong argument for online information. The tons of paper documentation that burn the precious fuel supply of a submarine have a negligible weight in electronic form. The same physicality that makes books easy to use—portable, handy, laptop—makes them impossible to use as systems grow larger and more complex, and as the need for documentation increases proportionately (or geometrically).

No similar physical constraint shapes electronic text. The result is a spaciousness in both the amount of information that can be recorded and in the design of information display. Steven Jobs can include the *Oxford English Dictionary* and Shakespeare's plays in the NEXT computer's memory—no problem. The price of memory has been decreasing quickly while new technologies increase storage limits. Large stacks of information can be duplicated for the price of a disk; huge quantities of information can reside on a single compact disk. A CD-ROM disk might hold 550,000 pages of text with 1,000 characters per page. But it weighs only ounces, fits into your pocket, and will soon be replaced by more compact storage media.

Writers of paper text are always contained by length (as I am here!): writing is a process of selection, cutting, paring away at what is non-essential or redundant. Paper text forces absurdities upon writers—squeezing text into narrow margins and using smaller fonts to keep the overall page count within limits. But screens introduce the luxury of open space. There is no demand to run unrelated text together in the interest of saving page space. If a writer hasn't much to say about something, space can be left blank without worrying about cost. The effect on prose is liberating, freeing it from the economic constraints of inscription.

GRAPHICALLY RICH TEXT

Print is a graphic medium; it displays its meanings in the spread of ink on page (Bernhardt, "Seeing the Text"). Writers of printed text have many options at their disposal to make texts visually informative: white space, font sizes, line spacing, icons, non-alphabetic characters like bullets and daggers, margins, and the whole range of pictorial displays—graphs, charts, drawings, etc. The use of computers for word processing has heightened our awareness of the graphic component of meaning. Both student writers and experts have at their command a wide range of graphic tools and an expanding base of research and aesthetic insight to guide the design of text on page (see Barton and Barton, "Trends"). Wholly new products—desktop publishing and graphics software—give authorial control over text/graphic integration. More than ever before, writers are page designers; they are composition specialists.

Electronic text extends visual composition by offering a surface with more graphic potential and greatly augmented options for text/graphic display and integration. Some of these display options are shared by print and screen-based text. White space (though often not white), space breaks, and margins actively signal divisions within a text, showing what goes with what and where the boundaries are. Bullets and numbered lists cue sequences of information. Font sizes and varieties, headings, color, boldface and italics show hierarchies within a text, cuing subordinate and superordinate relations. Headings, text shape, and callouts in the margins can provide filters for readers, tracking them toward or through various information paths so that each reader is guided to appropriate text for the task at hand.

But screen-based text goes beyond print in its visual effects. Readers can zoom in and out on screen text, editing graphics at the pixel level or looking at facing pages in page-preview mode, with Greeked text downplaying verbal meanings in favor of a visual gestalt that allows writers to evaluate design. Sequences can be animated, procedures can be demonstrated. Text can flash or take on spot color or be outlined or presented in inverse video. With CD-ROM integration, video, voice, or musical sequences can be part of a text, achieving effects that print can only struggle to suggest. Exploded diagrams, so important to technical writing, can actually explode, and readers can view technical illustrations at varying levels of detail, with high resolution on close-up shots of delicate mechanisms (Jong). Readers can travel in virtual space, examining an object such as a building or an automobile from various perspectives, moving around the object in three-dimensional CAD space.

Screen-based text takes information in iconic, visually metaphoric directions. We know people learn about complicated systems best when they have organizing metaphors. Electronic information allows us to exploit metaphors, so that the screen is a *window* onto a *desktop* and information is kept in *files*. We use *control panels*, complete with *gauges*, *switches*, *bells*, and *alarm clocks*. We relate easily to the icons of control, throwing text into the *garbage can* or moving icons for pages (representing files) from one location to another.

Of course, the phosphor glow of screen text causes its share of problems. We are subjected to flicker, glare, and electronic interference. The screen image suffers, and so we do, from non-optimal light conditions. Our eyes complain of fatigue from attempting to maintain focus on a curved screen. We are hampered by screen size and resolution. But that same phosphor offers a fluid, dynamic medium, with many more options than print has for displaying information and exploiting visual intelligence.

CUSTOMIZABLE, PUBLISHABLE TEXT

Little can be done by the reader of paper text to customize the text itself. The few customizing devices are well exploited: turning down the corner of a page or leaving a bookmark or a self-stick note, writing notes in the margin, or highlighting and underlining passages. Such modest adaptations of the static text to the uses of an individual reader make the book more valuable to the owner but less valuable to other readers.

Electronic text, in contrast, benefits from being infinitely more fluid, expansive, and adaptable to individual uses. Readers can annotate without the boundaries of hard copy. Text on screens can be changed—that is one of its essential properties. Lines can be written between the lines, notes can be appended to the text itself or as pop-up annotations behind the screen.

The display of the text itself can be customized. Readers can reduce screen clutter by suppressing the display of rulers, spaces, returns, mark-up language, stylebars, borders, and menus. Or readers can show properties of screen text—with every space and return signaled in a fashion that has no print equivalent. Readers can decide the level of on-screen prompting they want, with menus and help cues displayed or suppressed. Background colors, screen borders, audio messaging, cursor speed—any number of features of the display can be set to individual parameters. Individual user profiles can be stored that automatically adjust the parameters based on predetermined settings. This means the "same" text can display itself differently, depending on the preferences of individual readers.

Screen-based text has the potential to adapt to individual users automatically by keeping histories on users and responding in intelligent ways to likely scenarios based on what a particular user has done in the past. Individualized glossaries, dictionaries, macros, indexes, authoring levels, search procedures, bookmarks, and stylesheets all give the readers of screen-based text real ownership of their texts. Readers own the text because they can do what they want with it; they can make it their own, unlike any other reader's texts.

The controls over fluid, customizable text are shared by the system designer and the user. Systems can be made sensitive to user context, providing help based on best guesses about where the user is in the program and what sort of help might be needed. Shortcuts that allow individual control can be built into the system. For example, many programs offer novice and expert paths, with menus and on-screen prompts for new users. Such prompting speeds learning for new users. But power users want menus and cues suppressed—they know what they need to do and want to do it in the fewest number of keystrokes. Good design allows both types of user to coexist.

The control over the shape of texts that microcomputers grant users leads inevitably toward not just customizable but publishable text. Just as the printing press eventually puts books into everyone's hands, desktop publishing systems put the printing press into everyone's hands. Anyone can now design, display, and print work that is potentially indistinguishable from professionally printed work.

Traditionally, much of the cost of print has been in the production stage—the human and machine costs of typesetting, paper, binding, and distribution. Longer length or fancier graphics meant higher prices. The high production cost per unit for books and magazines made copies fairly expensive, but highly portable and accessible to anyone who could read. With screen-based text, however, much of the cost of production is shifted from the printer to the author and the audience. It is cheap and easy to duplicate disks. And disks (whether floppy or hard) hold immense quantities of information in a small format, so issues of length are no longer so important to overall cost. A disk can hold graphics and animated sequences, color diagrams and fancy fontography, interactive tutorials and reference materials. Once the information is coded to the disk, reproduction is a simple, inexpensive matter.

But the more complicated the on-screen text, the higher the overhead demands on authors and readers. Instead of printers needing high-priced equipment and expensive materials to produce fancy texts, writers need high-priced equipment to author texts, and readers now need high-priced equipment to run the disk. And whereas there never were compatibility problems between readers and books, there are now multiple and vexing problems of matching hardware and software.

Once printed, paper text is fairly static. It presents the same face to all readers, so that my copy of a book looks just like yours. In contrast to the static quality of paper text, on-screen text is fluid and customizable, updatable and expandable. These qualities lead to multiple versions, to individually adapted texts, and give an elasticity to electronic text that changes the nature of publication. And with the advent of desktop publishing, the movement from screen-based text to paper is eased, so even print loses its static quality. A writer can produce papers or books in multiple versions, easily redesigned and updated. Print is no longer permanent, because the cost and effort of updating editions is negligible. The fluidity of the screen has begun to overcome the static inertia of print.

THE SHAPE OF TEXTS TO COME

The shape of text changes as it moves from paper to screen. On-screen text is eminently interactive, closely embedded in ongoing action in real-time settings. It borrows heavily on the evolved strategies readers possess for interacting with printed texts, but provides a more fluid, changeable medium, so that the text itself becomes an object for manipulation and change.

As texts change, we will develop new strategies for reading and writing. Text bases will grow, becoming huge compilations of information stored on disk with no corresponding printed versions. It will feel natural to move through large pools of information, and we will rely on learned strategies for knowing where we are, where we want to go, and what we want to do when we get there. We will develop new sorts of reading skills, ones based around text that is modular, layered, hierarchical, and loosely associative. We will demand control over text—over its display, its structure, and its publication.

We are now at a point of transition of the sort described by Ong, similar to transitions from orality to literacy or from handwritten manuscripts to printed. The computer is becoming increasingly dominant as a primary medium for presenting and working with texts. As we take control of computer-based texts, the existing lines between reading and writing will tend to blur into a single notion of use (Slatin). Texts will have multiple authors and grow incrementally as readers individualize and structure text for their own uses. The presence of screens will become increasingly common, a part of our daily lives, close at h and in a variety of situations.

As with the interrelation of spoken and written media, so between paper and screen-based text: we will see crossbreeding, with the uses and forms of one medium shaping the uses and forms of the other, so that as the predominance of and our familiarity with screen-based text increases, the dimensions of variation discussed here will have a greater and greater shaping influence on paper text. But the real potential for full exploitation of these dimensions of variation lies in text on screens. It is the dynamic, fluid, graphic nature of computer-based texts that will allow full play of these variables in shaping the texture of print on screens.

Works Cited

Barton, Ben F., and Marthalee S. Barton. "Simplicity in Visual Representation: A Semiotic Approach." *Journal of Business and Technical Communication* 1 (1987): 9–26.

—. "Trends in Visual Representation." *Technical and Business Communication: Bibliography Essays for Teachers and Corporate Trainers*. Ed. Charles Sides. Urbana: NCTE, 1989. 95–135.

Barrett, Edward, ed. *The Society of Text: Hypertext, Hypermedia, and the Social Construction of Knowledge*. Cambridge: MIT P, 1989.

Bernhardt, Stephen A. "The Reader, the Writer, and the Scientific Text." *Journal of Technical Writing and Communication* 15.2 (1985): 163–74.

—. "Seeing the Text." *College Composition and Communication* 37 (Feb. 1986): 66–78.

Bolter, Jay David. *Writing Space: The Computer, Hypertext, and the History of Writing*. Hillsdale: Erlbaum, 1991.

Brockmann, R. John. *Writing Better Computer User Documentation: From Paper to Hypertext*. Version 2. New York: Wiley, 1990.

Chignell, Marc H., and Richard M. Lacy. "Project Jefferson: Integrating Research and Instruction." *Academic Computing* 3 (Oct. 1988): 12–17, 40–45.

Duchastel, Philippe C. "Display and Interaction Features of Instructional Texts and Computers." *British Journal of Educational Technology* 19.1 (1988): 58–65.

Harward, V. Judson. "From Museum to Monitor: The Visual Exploration of the Ancient World." *Academic Computing* 2 (May/June 1988): 16–19, 69–71.

Herrstrom, David S., and David G. Massey. "Hypertext in Context." Barrett 45–58.

Horton, William. *Designing and Writing Online Documentation: Help Files to Hypertext*. New York: Wiley, 1990.

Jong, Steven. "The Challenge of Hypertext." *Proceedings of the 35th International Technical Communication Conference*. Washington: Society for Technical Communication, 1988. 30–32.

Kostelnick, Charles. "Visual Rhetoric: A Reader-Oriented Approach to Graphics and design." *Technical Writing Teacher* 16 (Winter 1989): 77–88.

Lynch, Anne. "Project Jefferson and the Development of Research Skills." *Reference Services Review* (Fall 1989): 91–96.

Merrill, Paul F. "Displaying Text on Microcomputers." *The Technology of Text*. Ed. David Jonassen. Vol. 2. Englewood Cliffs: Educational Technology Publications, 1982. 401–13.

Nielsen, Jakob. "The Art of Navigating through Hypertext." *Communications of the ACM* 33 (March 1990): 296–310.

Ong, Walter J. *Orality and Literacy: The Technologizing of the Word*. London: Methuen, 1982.

Redish, Janice. "Writing in Organizations." *Writing in the Business Professions*. Ed. Myra Kogen. Urbana: NCTE, 1989. 97–124.

Rubens, Philip M. "Online Information, Hypermedia, and the Idea of Literacy." Barrett 3–20.

—. "A Reader's View of Text and Graphics: Implications for Transactional Text." *Journal of Technical Writing and Communication* 16.1/2 (1986): 73–86.

Rubens, Philip M., and Robert Krull. "Application of Research on Document Design to Online Displays." *Technical Communication* (4th Quarter 1985): 29–34.

Slatin, John M. "Reading Hypertext: Order and Coherence in a New Medium." *College English* 52 (Dec. 1990): 870–83.

Stricht, Thomas. "Understanding Readers and Their Uses of Text." *Designing Usable Texts*. Ed. Thomas M. Duffy and Robert M. Waller. Orlando: Academic Press, 1985. 315–40.

Tufte, Edward R. *Envisioning Information*. Cheshire, CT: Graphics Press, 1990.

Visible Language. Special issue on adaptation to new display technologies. 18.1 (Winter 1984).

Walker, Janet H. "Authoring Tools for Complex Document Sets." Barrett 132–47.

Writing on the Edge. Special issue on hypertext. 2.2 (Spring 1991).

DEVELOPING YOUR UNDERSTANDING

1. Summarize the differences between linear and nonlinear texts, illustrating your summary with concrete examples.
2. Explain how an increased demand for situationally embedded text might affect readers' expectations, and also the work of professional writers.
3. Bernhardt clearly explains how computers change the ways we interact with texts. Summarize these effects. Then, shift focus: instead of focusing on how people interact with texts, explore how electronic texts affect the ways people interact with one another, individually and/or in groups.
4. Discussions about the effects of computer technologies on writing can quickly become outdated, and Bernhardt's 1993 article (probably written in 1991 or 1992) is no exception. Assess where Bernhardt's arguments about the differences between print and on-screen texts might now be outdated by advances in either on-screen or print technologies. Then, assess what such rapidity of change suggests for writers in the workplace.

FOCUSING ON KEY TERMS AND CONCEPTS

Focus on the following terms and concepts while you read through this selection. Understanding these will not only increase your understanding of the selection that follows, but you will find that, because most of these terms or concepts are commonly used in professional writing and rhetoric, understanding them helps you get a better sense of the field itself.

1. limners
2. Stationers' Company
3. fair use
4. public domain
5. site license

WHO "OWNS" ELECTRONIC TEXTS?

THARON W. HOWARD
Clemson University

New information technologies make it increasingly difficult for authors and corporations to claim that ideas and information are property which can be sold. To understand the problems of authorship in electronic environments, this chapter examines the historical development of U.S. copyright and three historically distinct theories of ownership upon which it is based. The author ultimately argues that a revised social constructionist perspective best addresses the challenges of ownership created by new technologies.

> ***The Congress shall have the power . . . to Promote the Progress of Science and useful Arts, by securing for limited Times to Authors and Inventors the exclusive Right to their respective Writings and Discoveries.***
>
> **—U.S. Constitution, Art. 1, Sec. 8**

> ***Notwithstanding the provisions of section 106, the fair use of a copyrighted work, including such use by reproduction in copies or phonorecords or by any other means specified in that section, for purposes such as criticism, comment, news reporting, teaching (including multiple copies for classroom use), scholarship, or research, is not an infringement of copyright.***
>
> **—17 U.S. Code, Sec. 107**

For most people, including a large number of practicing professional writers and professional writing teachers, the issue of intellectual property isn't something they usually consider particularly problematic. Most writers today, particularly those of us

Source: Howard, Tharon W. "Who 'Owns' Electronic Texts?" *Electronic Literacies in the Workplace: Technologies of Writing.* 1996, pp. 177–198.

who spent a lot of our academic careers in and around English departments, tend to subscribe to the view that authors "own" their texts. We tend to believe (as is implied by the excerpt from the Constitution, above) that we have the right to expect remuneration for "our" writing, and furthermore, that we ought to be able to have some control over how our texts will be used.

However, recent trends toward more collaborative writing projects in the workplace, along with the use of online computer conferences, electronic discussion groups, hypertexts, multimedia presentations, groupware, and other computer technologies aimed at enhancing and promoting collaboration, are all seriously challenging the popular, romantic view that an author owns his or her text. More and more frequently, professional writers are finding themselves confronted with intellectual property and copyright issues which result from the increased reliance on computers in the workplace, and, in many cases, writers are finding themselves unprepared to deal with these issues. Consider, for example, the following scenarios:

Scenario 1

You work in the document design department of a large corporation, and, traditionally, your department has made it a point of pride to produce dramatic covers for the company's annual report. One of your co-workers finds a reproduction of a famous photograph in a popular magazine, and the image would be perfect for the theme of this year's annual report with some cutting, pasting, and a few other modifications.

Since the photograph is famous, since you're going to use only part of the image, and since you're going to modify the image in order to produce something which is essentially a new image, should you go ahead and scan it? Or do you first have to have permission from the magazine which first reproduced it, the publishing house which sells reproductions of it, or the photographer who originally took the photograph?

Scenario 2

You've just been hired to do some desktop-publishing work for a large consulting firm. The office manager bought you a new computer system to use, but the system came with a new software package that is incompatible with the old version of the software used by the rest of the office. As a result, you can't share files with co-workers and do your job effectively. Fortunately, however, the office still has the installation disks for the old version of the software, and the office manager tells you that, since these disks were purchased by the company, you can install the old software on your system.

Should you go ahead and copy the software since the office has already paid for it?

Scenario 3

You're doing research on an article about usability testing for *Technical Communication*, and, as part of your research, you join an electronic discussion group on the Internet, where people doing human-factors research exchange e-mail messages about their works-in-progress. As you're writing your article, someone posts an e-mail message to the group describing the results of her unpublished research project. These results are central to your article's thesis and force you to completely revise your thinking about the subject. Since these results haven't been published elsewhere, you wish to quote from the e-mail message in your article.

Can you legally and ethically quote from an e-mail message? Indeed, are you obligated to cite the message since it has had such a profound impact on your own thinking? If so, does anyone own the copyright on the message? Do you need to seek the author's permission? Or, since the message was electronically "published" by an electronic discussion group, do you need to have the permission of the person(s) who created and operate the discussion group or the university or company which owns the computer that hosts the group?

Scenario 4

You work for a large corporation in which e-mail is the primary means of communication. Instead of using informal notes, memos, short reports, or phone conversations to contact each other, people in your company use e-mail. In keeping with this "paperless office" milieu, you have maintained an electronic correspondence with a co-worker in another department for some time. You and your co-worker (who happens to be of the opposite sex) are careful to keep your electronic interaction limited to your breaks and lunch periods so that it does not interfere with your work. Your supervisor knows what you're doing and has said that she actually prefers that you correspond via e-mail on your breaks since that way, you're not tying up the office telephone. However, one afternoon, you discover that your electronic exchanges are being monitored and even shared as jokes among people in the computer operations department. You're furious at this violation of your right to control how your texts will be used, but your supervisor tells you that the company owns the computer and, therefore, has the right to monitor its use.

Can you stop this monitoring of your e-mail? Who actually "owns" the messages you've been sending? Do you, as the author, own the messages? Does the addressee who received them? Or does the owner of the system on which the messages were produced? Furthermore, what rights does ownership of the messages entail?

Scenario 5

You're a faculty member in a professional writing program at a large university, and one of your responsibilities is to serve as the placement director for the program. In order to help your graduates find information about companies which routinely hire writers, you decide to create a Hypercard stack which will allow students to click on a map of the United States. Then, depending on the state students select, students would receive information about specific companies located in that state. You construct your stacks on the university's computers, and from a book which provides an alphabetical list of national corporations, you select data on companies which you think might routinely hire technical writers. The resulting hypertext is so popular among your students that several publishers learn of it and are interested in publishing it.

Can you publish your hypertext? Have you infringed on any copyrights by providing your students with your hypertext in the first place? If you can publish your text, are you legally obligated to pay any royalties to your university or to the publisher or author of the book from which you selected your data?

As these scenarios illustrate, the new electronic environment in which professional writers must now function makes intellectual property and copyright issues more and more a part of their everyday experience in the workplace. Indeed, these sorts of issues are becoming so commonplace that we may well wish to make an understanding of intellectual property in an electronic environment a criterion of "electronic" or "computer literacy." As I will show when we return to these scenarios at the end of this essay, even a relatively clear understanding of the principles of copyright law may not allow writers to answer the questions posed in these scenarios.

However, not only do professional writers need to have a better understanding of copyright issues because they are more likely to encounter them than ever before, but they also need to better understand questions about intellectual property in an electronic environment because new information technologies are forcing us to reshape traditional notions about authorship and ownership. In a world where, for example, a software package can reorganize and rewrite the information in databases (thereby "virtually" creating or authoring texts without human intervention), colloquial ideas about authorship and ownership may no longer be enough. In fact, these sorts of technological challenges to traditional ideas of ownership are particularly troublesome to writers in the workplace because (a) they may diminish writers' claims to remuneration for their work, and (b) they may strip writers of the right to control how their texts will be used.

In order to better understand the problems of ownership in the electronic workplace, I will offer a brief historical examination of the origins of U.S. copyright law since it is through copyright laws that the rights of individual authors and corporations have come to be defined. Furthermore, by examining the evolution of current copyright law, I will explore why electronic publishing, electronic discussion groups, computer conferences, and other new information technologies represent such a challenge to current copyright law. A historical examination of another new publishing technology, i.e., the printing press, will show that then, as now, the introduction of new technologies challenged existing systems for owning and controlling texts. Furthermore, this examination will show that, although many people are not aware of it, current copyright law reflects an interesting struggle among at least three historically distinct and competing theories of textual ownership. First, there is, of course, the romantic and commonplace notion that authors have a "natural right" to the fruits of their intellectual labors. Second, there is the assertion that the public has a right to all knowledge since "Laws of Nature" and absolute truths cannot be the property of any one individual. And third, there is the view that all knowledge is socially constructed, that a text is a product of the community the writer inhabits, and that the text must therefore be communal, rather than individual, property. These three theories have tended to compete when the question has been whether a copyright is a natural right of private property or whether a copyright is a privilege granted to individuals by the public's representatives.

A HISTORICAL OVERVIEW

When I've taught my professional writing students about the history of copyright laws, and even when I've discussed the subject among some of my faculty colleagues, one of the things that always seems to surprise them is the fact that the original impetus to develop copyright laws did not come about through a desire to protect the "natural property rights" of authors. Indeed, most people I've encountered tend to have the same misimpressions about copyright issues that they have about driving their cars. Most people tend to think that it's their "right" to operate any motor vehicle they care to purchase. Similarly, they tend to believe that, since they also own the texts they write, they ought to be able to control how those texts will be used and ought to be able to profit from that use. And, of course, in actual practice, there are few things in our day-to-day experiences to challenge these notions. Today, the use of an automobile is so pervasive in our society that we just expect everyone to have access to them.

And yet, those unfortunates who either fail to receive or somehow lose their driver's license serve to remind us that operating an automobile is not a right we can expect; rather, it is a privilege we are granted by the government under certain specific circumstances. Similarly, as legal historians such as Joseph Beard are quick to point out, a copyright or (literally speaking) the right to reproduce copies of a particular text was not and, indeed, is not a "natural unlimited property right." Instead, it was and is a "limited privilege granted by the state" (Beard, 1974, p. 382). As with a driver's license, the government gives writers license to "operate" texts in the public domain. What's more, it makes the license so easy to obtain that we seem to forget that we're dealing with an issue of privilege rather than of natural right. Yet, if we consider the origins of English and American copyright laws in the sixteenth century, we can quickly see that protecting an author's natural right was never really an issue then either.

During the fifteenth and sixteenth centuries, the great new technological development was the printing press, and, just as today's "computer revolution" is stimulating the growth of new industries in information technology and electronic publishing, the printing press was producing tremendous growth in the book-publishing industry. Prior to the introduction of printing-press technology, the book trade depended on an excruciatingly slow and tremendously expensive publishing technology. Scribes, illustrators (then called limners), and book binders worked laboriously to produce each single copy of every book. Because of the enormous expense involved in this technology, most book-publishing efforts required the funding of either the Church or the Crown, a situation which made it easy for those in power to control the kinds of texts which would be produced and consumed.

Of course, the printing press changed all this. The radical reduction in production costs meant that texts could be produced and marketed cheaply and easily; yet, with a limited number of popular and lucrative texts available for publication, there was a dramatic increase in competition among book publishers. Two significant developments resulted from this increased competition. First, people involved in various aspects of the book-publishing industry (i.e., limners, book binders, printers, etc.) banded together into a cooperative organization which came to be known as the Stationers' Company. As Patterson and Lindberg (1991) point out, the Stationers were essentially a "group of businessmen who agreed to allow one of the[ir members] the exclusive right to publish a specific work in perpetuity" (p. 22). Thus, the Stationers created a voluntarily enforced form of copyright, which (though it did not carry the force of law and said nothing about the "natural rights" of authors) still offered book publishers limited protection against competition. In other words, the increased competition which brought about the development of the Stationers' Company clearly established the need to protect a publisher's (though not an author's) copyright. Indeed, as Martha Woodmansee (1984) points out, it wasn't until the eighteenth century that writers were able to realize any real profits from the competitive book trade in the form of royalties. In fact, even in the eighteenth century, copyrights were valuable properties:

> . . . a flat sum remained customary, upon receipt of which the writer forfeited his [*sic*] rights to any profits his work might bring. His work became the property of the publisher, who would realize as much profit from it as he could. (pp. 435–436)

The second result of this increased competition was that the Church and the Crown lost what had been their de facto control over the production and consumption of texts. Because of its new, more economical printing technology, the book-publishing industry no longer needed to depend on Church or State subsidies, and, consequently, publishers were free to produce texts which would not have received the economic sanction of the Church or Crown. Indeed, given that the public is always fascinated with controversial texts and is therefore going to purchase more of them, it seem likely that sixteenth-century publishers found new economic incentives to publish texts which, ironically, challenged the same religious and governmental authority which had been their chief means of support before the introduction of the printing press.

As a result of these two developments (i.e., the Stationers' desire to protect themselves from competition and the Crown's inability to control the publication of subversive books), in 1556 Mary Tudor and Philip of Spain granted the Stationers a royal charter, which stated in its preamble that it had been issued in order "To satisfy the desire of the Crown for an effective remedy against the publishing of seditious and heretical books" (Beard, 1974, p. 384). Furthermore, the Stationers' royal charter "limited most printing to members of that company and empowered the stationers to search out and destroy unlawful books" (Patterson & Lindberg, 1991, p. 23). As a result, modern copyright law finds its origins not in the recognition and protection of an author's natural property rights, but, rather, in the "ignoble desire for censorship" and in the greedy lust to "protect profit by prohibiting unlicensed competition" (Beard, 1974, p. 383). And yet, despite the disturbing motives behind this early form of copyright law, the Stationers' royal charter is significant because it firmly established the principle that a copyright is not the natural, absolute, or unlimited property of any individual or company. Instead, to the degree that a copyright can be considered a form of property at all, the Stationers' charter made it clear that to own a copyright is essentially to own a limited license or a privilege which the state grants in order to promote intellectual activities that are deemed to be in the best interests of the state and its citizens.

Although Mary Tudor, Parliament, and the U.S. Congress probably had very different views of the desirability of censorship and book burning, the same principle of privilege that Mary established in the Stationers' charter can be found in Parliament's 1709 passage of the Statute of Anne, the statute which in turn provided the basis for Article I, Section 8 of the Constitution. Unlike the Stationers' charter, both the statute of Anne and the Constitution recognize the rights of authors. In fact, Article I, Section 8 of the Constitution provides that authors shall have the "exclusive Right to their respective Writings and Discoveries," thereby offering writers the kind of protection which the Stationers' charter gave only to publishers. However, this provision does not assert that texts are the exclusive property of their authors; instead, what the Constitution does is to give Congress the legal authority to grant authors limited copyrights in order "To promote the Progress of Science and the useful Arts." In other words, as was the case with the Stationers' charter, copyright is still a privilege or license granted by the government for a limited period of time in order to promote not only the right of authors to profit from their labors, but also the enhancement of the public's collective welfare. Hence, just as the State of South Carolina makes laws which give me the right to profit from certain uses of my car for four years and under specific circumstances which are intended to protect and benefit my fellow citizens, the Constitution empowers Congress to make laws which give me the right to profit from certain uses of my texts for seventy-five years from their publication or for a hundred years from their creation (whichever is shorter) and under specific circumstances which are intended to promote the economic and intellectual well-being of the American public.

MAJOR PRINCIPLES OF U.S. COPYRIGHT LAW

Now the upshot of all this law-making, privilege-granting, condition-making legal-speak is that the Constitution has come to represent the delicate balance between the rights of an individual and the good of the public. It represents a sometimes uncomfortable compromise, "balancing an author's interest against the public interest in the dissemination of information affecting areas of universal concern, such as art, science, history, and business" (Van Bergen, 1992, p. 31). Copyright law in the United States recognizes that in order to encourage authors to produce the texts which will lead to the artistic, scientific, and technological discoveries that drive business and industry, it is essential that authors be allowed to realize a profit from their texts. Obviously, without the hope of profit, there is little incentive for a software developer to invest in the research required for the production of new computer applications, nor is there sufficient cause for a publishing house to pay large sums of money to photographers and writers in order to produce books which they cannot sell because the articles and photographs can be obtained more cheaply through some other means. In short then, U.S. copyright law is based on the simple principle that one has to spend money in the short term in order to make money in the long term; we have to pay for intellectual and economic progress by first investing in the mechanisms of research and development.

On the other hand, copyright law doesn't give authors and publishers the legal right to prevent the public from the "fair use" of texts. Indeed, I have already shown that individual authors are granted copyrights not because authors have a natural property right, but because such protection is in the public's best interests. Thus, as Pierre Leval (1990) notes, "Fair use is not a grudgingly tolerated exception to the copyright owner's rights of private property, but a fundamental policy of copyright law" (p. 1107).

The public's right to the fair use of texts is provided for in Statute 17, Section 107 of the U.S. Code, and essentially what it does is to place limitations on the "exclusive Right to their Writings and Discoveries" that authors and inventors received in the Constitution. Section 107 grants the public the right to copy a work "for purposes such as criticism, comment, news reporting, teaching (including multiple copies for classroom use), scholarship, or research." Thus, the doctrine of fair use allows the use of texts for noncommercial purposes which are in the public's best interests. However, this does not mean that, for example, teachers can freely make photocopies of entire textbooks for their classes or that a textbook publisher developing a multimedia presentation on the Vietnam War for high school history classes could freely use sequences from *Apocalypse Now* and *The Deer Hunter* in its stacks. Beyond granting the right to copy a work for educational purposes, the law further states that in determining whether the use made of a work in any particular case is a fair use, the factors to be considered shall include the following:

- the purpose and character of the use, including whether such use is of a commercial nature or is for nonprofit educational purposes;
- the nature of the copyrighted work;
- the amount and substantiality of the portion used in relation to the copyrighted work as a whole; and
- the effect of the use upon the potential market for or value of the copyrighted work. (17 U.S. Code, Sec. 107)

Thus, even though the two parties mentioned here (teacher and publisher) are using copyrighted texts for teaching purposes, they would both be considered guilty of copyright infringement because, in the first case, the teacher is copying the whole text and is interfering with the "potential market for" the textbook; and in the second case, the publisher of the multimedia presentation would be profiting from the commercial sale of its product to schools. Hence, in the doctrine of fair use, the balance between individual rights and public needs can once again be seen.

In addition to the "fair use" of texts, the copyright statute imposes other limitations on the exclusive rights of copyright holders, and one of the most important of these restrictions is on those features of texts which are copyrightable. According to the copyright statute in the U.S. Code, only the tangible expression of ideas belongs to the copyright holder. Ideas are not copy protected. This limitation is of particular interest because it gives perhaps the clearest articulation of the ways in which authors can be said to own their texts, and clearly this limitation undermines the commonplace and romantic notion that a "person's ideas are no less his property than his hogs and horses" (Woodmansee, 1984, p. 434). Instead, there are two principles of ownership being advanced here: first, that ideas are like universal laws of nature which, because they obtain for everyone, cannot be owned by any single person; and second, that a new discovery, even though it may be the product of one individual's intellectual labor, owes its origins to the realm of public knowledge and should therefore be considered communal property.

In terms of actual practice, current copyright law does grant authors the right to demand remuneration for their intellectual labors, and it does this by protecting the ways authors express ideas. However, it does not allow them to claim ownership of the ideas they express; authors cannot expect to have and maintain a monopoly on truth. According to copyright law, since truths are either universal absolutes or social constructions, they cannot be owned. Hence, if I write a piece of software that uses the mathematical equation $2 + 2 = 4$ as part of its code, I don't have to pay anyone for its use, nor can I expect to receive an honorarium every time someone in the United States calculates the sum of $2 + 2$, because mathematical principles and algorithms are thought to be universal truths. On the other hand, if people copy the way I used an algorithm in my software, if they borrow my code's structure or organization, then they are using my expression, and that expression is copyrighted. Consequently, while authors can't expect to profit from ideas and truths, they can expect to receive remuneration for the labor required to uncover and to formulate those ideas and truths. Although it's important to remember, as the doctrine of fair use makes clear, there are still certain public uses even of an author's form of expression for which the copyright holder cannot expect to be compensated.

Before reading further, answer Question 1 in "**Developing Your Understanding.**"

COPYRIGHTS IN THE ELECTRONIC ENVIRONMENT

This examination of the historical origins and principles which inform modern U.S. copyright law reveals that the commonly held belief that authors own the texts they produce does not accurately reflect the actual legal status of textual ownership. Although it is correct to say that authors do indeed own a copyright as soon as a text is produced and that they therefore enjoy the rights of copy protection, it is important that both the producers and consumers of texts understand that those rights exist in the form of a limited privilege granted by the State and that those rights obtain only under certain conditions specified by the State.

And yet, while professional communicators need to understand the general principles upon which current copyright law is based in order to function effectively in the electronic workplace, it's also important to understand that those same principles don't always yield clear answers when we have to deal with electronic texts. A professional writer may know that copyright is only a privilege, that the public has the right to certain "fair uses" of texts, and that only the form of expression is protected in a work; yet, this knowledge may still leave the writer unsure as to the exact copyright status of a particular electronic text in a particular situation. As Marilyn Van Bergen (1992) has noted, "there is good reason why the law is often symbolized by scales used as weighing instruments" (p. 31), and this is particularly true for copyright law since, as I have shown, it seeks to balance the rights of the individual against the needs of the public, since it represents a compromise among three competing theories of intellectual property, and since technological changes have, historically, represented challenges for existing forms of copy protection.

Still, in spite of the fact that the nature of copyright law makes it difficult to say for certain that a particular situation does or does not represent a copyright infringement, an understanding of copyright principles can still serve as a useful guide for professional communicators. To show how this is the case, I wish now to return to the scenarios with which I began this essay in order to illustrate how these principles can at least help writers either avoid litigation or recover the remuneration they are due.

Scenario 1

In this scenario, a member of a document-design team is planning to scan a famous photograph from a popular magazine in order to manipulate a portion of it for the cover of the company's annual report. The central questions here are (1) does such a reproduction fall under the doctrine of fair use and (2) who owns the copyright on the image?

As far as the question of fair use is concerned, it seems highly probable that this would not be considered a fair use of the original work. Since the reproduction is not being made for educational, news reporting, or critical purposes, its use is still copy protected. Furthermore, as Brad Bunnin (1990) points out in his extremely informative article "Copyrights and Wrongs," even though an image has been manipulated, it may still be "legally considered a derivation of an original work" (p. 77), and therefore its reproduction will require the permission of the copyright holder. In his article, Bunnin also reproduces an electronically manipulated version of Munch's famous painting *The Scream;* yet, in spite of the fact that *The Scream* is a famous painting in the public domain, and in spite of the fact that Bunnin's derivative reproduction is a new image, Bunnin received permission from the museum which owned the painting "and paid a $250 fee to manipulate it" (p. 77). Similarly, the member of the document-design team should receive permission before reproducing and manipulating the photograph.

As to the question of who should be contacted in order to receive permission to reproduce the image, the issue is a bit more complicated. Probably the safest course for the document designer is to purchase and receive permission to reproduce a copy of the photograph from the publishing house which owns the copyright on the original photograph rather than using the magazine's reproduction. The reason for this is that derivative works are also copy protected. As Nicholas Miller and Carol Blumenthal (1986) observe, "Copyright laws protect an author's rights in his own expression even when that expression makes use of nonoriginal information" (p. 229). For example, were I to copy Bunnin's manipulated version of *The Scream,* I would be responsible to Bunnin's publisher. Thus, in order to avoid infringing on what may be considered a derivative work in the magazine's reproduction of the photograph, the document designer should obtain a copy of the original photo from the original copyright holder.

Scenario 2

In this comparatively straightforward scenario, a writer is instructed by the office manager to copy desktop-publishing software which the company had previously purchased for use on another employee's computer. Here, the central question is whether purchasing a copyrighted work gives one the right to copy it.

There is an unfortunate, though common, misconception that, when individuals or companies own a copy of a book or a piece of software, they can use their property as they see fit. However, as the privilege principle makes clear, "owning" a text or even a copy of a text is not the same as having the right to copy a work. Typically, when I purchase a text, the only "property" that I "own" is the actual physical copy of the book, computer disks, photograph, painting, compact disk, etc. However, ownership of this physical property doesn't give me the right to copy the text. In order to copy the work, I also have to have purchased a license to copy it.

Today, most software publishers do, in fact, sell consumers limited licenses to copy their software. Usually, diskettes are sold in shrink-wrapped or sealed packages so that opening the package constitutes an acceptance of the conditions of the limited license to copy the software. Exactly which copyrights are granted in these licenses varies from software package to software package; however, the most common form of licensing agreement allows consumers only to make backup copies for protection and to install (i.e., copy) the software on one system for use by one individual. In terms of the scenario, if this is the sort of license which was purchased by the company, then it would be a violation of copyright law to install the desktop-publishing software on a second system.

However, before the writer in this scenario refuses the office manager's instructions to copy the software, it would be a very good idea to check the exact terms of the license agreement. It may well be that the company purchased a "site license" for the software, in which case the software might legally be copied onto the second system. Companies often purchase site licenses which allow them to copy software on several machines or to install software on their local-area networks so that the software can then be copied into the memories of a number of individual computers at the site, the exact number of copies possible being specified by the terms of the site's licensing agreement.

Scenario 3

In this scenario, a writer is preparing an article for the journal *Technical Communication* and wishes to quote a passage from an e-mail message that had been posted to an electronic discussion group. The central questions here are (1) whether such a use is protected by the fair use clause and (2) whether the author of the message, the owner of the discussion group, or the university which owns the host computer for the group is the copyright holder for the message.

Currently, it would probably be considered legal for the writer to quote a short passage from such an e-mail message. The doctrine of fair use allows the reproduction of short passages for the kinds of news reporting and critical purposes typical of articles found in *Technical Communication*. However, the situation here is clouded by the technology involved and the lack of specificity in the fair use clause. The fair use clause requires that, in addition to considering the purposes for and the amount of the work being copied, "the effect of the use upon the potential market for or value of the copyrighted work" should also be taken into consideration (17 U.S. Code, Sec. 107). Thus, the author of the e-mail message may feel that her copyright has been violated since she has not been given the opportunity to publish the work through more traditional means, where the potential for remuneration is greater. In other words, the author of the e-mail message may be able to argue that her right to report on the research has been "upstaged," and therefore its potential value has been diminished.

However, although there is some validity to this argument, it seems more likely that sending an e-mail message to a discussion group would be considered a form of publication, so the author of the e-mail message can't really argue that the work has been upstaged. Indeed, while the exact copyright status of texts sent to and distributed by electronic discussion groups is still unclear and can vary widely from group to group, more and more groups are operating as electronic publications. In fact, groups like PACS-L, PMC, and E-Journal have received ISSN numbers, giving them the same copyright status granted to more traditional print publications. In other groups where discussions are open and unmoderated, the groups' owners may explicitly state that the copyrights belong solely to the authors of the messages sent to the groups. And in yet another type of group, members of the group may have a more or less tacit agreement not to quote or cite each other's messages at all, making it unethical (though not necessarily illegal) to quote their messages. Thus, even though quoting from the e-mail message sent to the electronic discussion group would probably be considered fair use regardless of the type of group, the writer of the article in this scenario should first contact the discussion group's owner for more information since the owner operates as an agent for the university and would be able to describe the quoting practices of the group. If the group has an ISSN number, then quoting from the message is acceptable under the conditions specified by the doctrine of fair use. If the group does not have an ISSN number, then the safest and most ethical course is to attempt to secure the permission of the e-mail message's author before quoting from the message.

Scenario 4

In this scenario, an employee discovers that his personal e-mail messages to a fellow employee are being monitored and redistributed by managers in the company. The central question here is whether copyright laws offer the employee any protection against this use of his messages.

Although the employee may have some legal means of preventing the management from monitoring and redistributing his messages in this case, it is unlikely that this problem can be best solved through an appeal to copyright laws because the privilege principle upon which copyright law is based does not give authors a "natural unlimited property right" to their texts (Beard, 1974, p. 382). Because a company pays for an employee's time and provides the resources the employee uses to produce texts, the company has certain rights to the use of the texts created. Usually, in fact, the company is the sole copyright holder of the texts its employees produce while in the company's employ. However, in some cases (particularly in university settings), an institution may receive only a percentage of the remuneration due to the copyright holder since part of the work was accomplished with the institution's resources and part of the work was done on the writer's own time. In this scenario, since the company's resources were used to produce and distribute the e-mail messages, this does give the company some limited rights in the use of those messages. Consequently, the issue here is probably not one of copyright infringement; rather, it is one of privacy.

In a similar case at Epson America, an employee was allegedly fired because she questioned her supervisor's right to read employees' e-mail messages. The employee is currently suing Epson not for copyright infringement, but because monitoring and redistributing employees' private e-mail messages "violated a California law that makes it a crime for a person or company to eavesdrop or record confidential communication without the consent of both the sender and the receiver" (Branscum, 1991, p. 63). Similarly, in this particular scenario, the employee should probably seek appeal to either state or federal privacy laws rather than claiming copyright infringement.

Scenario 5

In this scenario, a university employee is attempting to publish a Hypercard stack which was produced on the university's computers and which reorganizes the job information compiled from a copyrighted source. The central questions here are (1) whether the university is entitled to some portion of the royalties received for the stack's publication and (2) whether using the data but not the organization or expression from another work constitutes a copyright infringement.

As was discussed in the previous scenario, since the university's resources were used to develop the stack, the university has the right to expect some remuneration for the use of its facilities. Thus, the faculty member should make arrangements to share a percentage of the profits with the university.

The question of whether the reorganization of data compiled in another source constitutes a copyright infringement is much more difficult, however. As was previously discussed, copyright law protects only an author's expression. Yet, in the case of reference materials and databases such as business lists, telephone directories, bibliographies, or indexes, virtually the only form of tangible expression is the way the data are organized. Furthermore, as Miller and Blumenthal (1986) have pointed out, with the recent developments in information technologies, "computer databases contain randomly stored information which can be retrieved by a computer program in a wide variety of ways. There is no 'organization' to protect" (p. 229). Consequently, two fundamental principles of copyright law come into conflict in this scenario. On the one hand, there is the principle dating all the way back to the Stationers' charter, which recognizes that publishers and authors must be able to expect a profit from their labors if they are going to continue to have the incentives required to produce valuable new texts. On the other hand, there is the principle that ideas and knowledge cannot be the property of any one individual and that only the expression of the ideas belongs to the author or copyright holder.

As is the case with most electronic texts today, it is not yet clear how Congress or the courts will decide to deal with these kinds of challenges to the fundamental principles of current copyright law. It may well be that, because hypertexts and electronic databases allow users rather than authors to determine the ultimate organization and shape of these electronic texts, future copyright laws will need to find radical new foundations. In fact, in a 1976 act, Congress did make a number of changes to the copyright statute in the U.S. Code precisely because of technological developments in the television, music, and computer industries. One of these changes was to Section 103, which now "provides that copyright may be had for compilations, but protection extends only to the material contributed by the author, not to preexisting material that is used in the compilation" (Patterson & Lindberg, 1991, p. 93; see also 17 U.S. Code, Sec. 103). In terms of the scenario here, then, this suggests that the faculty member's use of the information would not be a copyright infringement because the original compiler's expression has been avoided and also because Section 103 seems to reaffirm the notion that data are part of the public domain.

However, law courts are conservative institutions, and it seems likely that a scenario like this one will also be resolved according to precedents such as *Leon v. Pacific Telephone and Telegraph Co.* In this 1937 court case, the defendant essentially changed the alphabetical organization of a telephone directory to a numerical order based on telephone numbers, thereby using the data but not the plaintiff's mode of expression. Yet, in spite of the fact that the defendant did not encroach upon the plaintiff's expression, the court ruled that this was, nevertheless, a copyright infringement. As Miller and Blumenthal (1986) point out, the effect of this decision has been that "some of the recent cases which follow Leon have explicitly stated that the compiler's labor is what should be protected" (p. 229). In other words, when the courts have been required to choose between protecting a publisher's incentives to produce texts and consumers' rights to use a work's content but not its mode of expression, the courts appear to believe that protecting a producer's incentives is in the best long-term interests of the public. Thus, in this scenario, the safest and most conservative course would be to negotiate some kind of financial arrangement with the persons holding the copyright on the reference materials used in the stack.

CONCLUSION

As these scenarios have illustrated, the new electronic environment in which professional writers must now function makes intellectual property and copyright issues more and more a part of their everyday experience in the workplace. Today's professional communicators need to have a more thorough understanding of the principles upon which modern copyright laws are based than ever before. As the discussion of the scenarios has shown, an understanding of these principles may not allow a writer to predict with any degree of certainty how a court of law will rule in a particular case; however, I would argue that such an understanding can at least offer professional communicators some sense of how to avoid copyright infringements. And given the enormous cost of litigation, both in terms of actual dollars and potential damage to a career, I would argue that knowing how to navigate through the intellectual property minefield is a tremendously valuable skill.

References

Beard, J. (1974). The copyright issue. *Annual Review of Information Science and Technology*, 9, 381–411.

Branscum, D. (1991, March). Ethics, e-mail, and the law: When legal ain't necessarily right. *Macworld*, 63, 66–67, 70, 72, 83.

Bunnin, B. (1990, April). Copyrights and wrongs: How to keep your work on the right side of copyright law. *Publish*, 76–82.

Leval, P. (1990, March). Toward a fair use standard. *Harvard Law Review*, 1105–1136.

Miller, N., & Blumenthal, C. (1986). Intellectual property issues. In A.W. Branscomb (Ed.), *Toward a law of global communication networks* (pp. 227–237). New York: Longman.

Patterson, L.R., & Lindberg, S.W. (1991). *The nature of copyright: A law of users' rights*. Athens: University of Georgia Press.

Van Bergen, M. (1992, July/August). Copyright law, fair use, and multimedia. *EDUCOM Review*, *27*, 31–34.

Woodmansee, M. (1984). The genius and the copyright: Economic and legal conditions of the emergence of the 'author.' *Eighteenth-Century Studies*, *17*(4), 425–448.

DEVELOPING YOUR UNDERSTANDING

1. **[FOR THIS QUESTION, STOP READING BEFORE THE LAST TWO SECTIONS] Based on the general principle of copyright that the goal is to maintain "balance between the rights of an individual and the good of the public," argue what the professional writers should do in two of the opening scenarios. Assess how your recommended actions achieve this general principle of balance. Then, after reading the final two sections of Howard's article, be prepared to discuss in class the points, if any, where you and Howard disagree.**
2. **Referring to the arguments laid out by Howard, create a one-page copyright handout that professional writers could use as a quick reference and reminder. Your handout should have a clear definition of copyright and offer copyright principles that can help profes-**

sional writers think through issues of copyright within particular circumstances. The handout should also be carefully designed, so writers can use it as a quick reference (this is not an essay). Apart from these requirements, remaining content is up to you.

3. Referring to the five scenarios opening the article, Howard argues that "the new electronic environment in which professional writers must now function makes intellectual property and copyright issues more and more a part of their everyday experience in the workplace." For three of the five scenarios, identify and discuss the features that make these scenarios part of a "new electronic environment." Then, analyze how these copyright issues would be less possible, less frequent, or even impossible in an environment lacking the electronic features you identify.
4. What is legal may not necessarily be ethical; and under certain circumstances, what is ethical may not be legal. Still, professional writers are bound by both legal and ethical systems. Furthermore, technological contexts can complicate the balance between legal and ethical "goods." For instance, technologies have dramatically increased our capabilities of surveillance, making surveillance almost ubiquitous. Though it may be legal, for instance, for companies to "observe" what their employees are writing on e-mail, strictly policing e-mail use during breaks and lunches creates a relationship between employers and employees which may not reflect an ethic that is desired by either party.

   Choose one or two of the opening scenarios as examples to discuss how technological context(s) can make it difficult for professional writers to strike an appropriate balance between legal and ethical responsibilities.

Chapter 5
Projects

1. At the opening of his article, Bernhardt argues the following: "We need to constantly appraise the broad drifts in the shape of text—to anticipate what now constitutes and what will soon constitute a well-formed text. We need to think about how readers interact with text—what they do with it and how. We need to anticipate where text is going: the shape of text to come." Only recently have professional writers begun grappling with the changes in text triggered by the Web and other online media. Yet currently, and certainly by the time you read this, texts' shapes are becoming increasingly more dynamic and fluid with such technologies as Macromedia's Flash and Shockwave, as well as the increasingly more affordable high-speed computer processors and band-width connections that support them.

   Individually or in groups of three of four, develop three versions of one document: a print version, a basic hypertext version, and a dynamic hypertextual version (e.g., using Flash, steaming video, and/or at least some of the dynamic features, like roller-overs, now built into many Web authoring packages). When choosing your project, keep the audience and purpose stable; consider this a project that requires you reach the same set of readers for the same purpose but you need to offer your readers several media alternatives.

   Also, do not approach this project as one requiring you to simply translate a print version into HTML, or vice versa. Instead, work with each medium independently and make it your goal to take advantage of each medium's strengths. (To avoid the impulse of simply translating print texts into these other media, consider working first with hypertext.)

   Based on your experience developing these three versions, and supported by other forms of research (e.g., literature reviews, interviews, and/or observations), write a brief article length argument anticipating the shape(s) of texts to come. (The report should be co-authored if you have been working as a group.) Your reports can be submitted as presentations at professional or undergraduate research conferences, like the Society for Technical Communication (STC) and the National Council for Undergraduate Research (NCUR); submitted as articles for professional writing publications, like *Technical Communication;* or collected and published as a class research project on the Web.

2. "With automated print and Web design templates, boilerplating, and automated grammar checks," your boss argues, "there is really no reason to keep a publications department." She goes on to argue that "most routine writing tasks can be delegated to department secretaries and staff and the less routine writing tasks can be easily accomplished by department managers, especially now that the voice recognition word processing software is working well."

   This is not such an unlikely scenario. As Karen Schriver reports, "Although the rhetorical approach to writing dominates the academic writing scene, it has not yet made a large impact on writing in the workplace. Regrettably, some managers still think of writers as glorified secretaries" (*Dynamics in Document Design,* p. 75). As a result, Schriver goes on to argue, "writers are finding that they must be able to demonstrate how they 'add value' to organizations. . . . They must be good at explaining what they do—its complexity and sophistication—in ways that educate managers, colleagues, and clients" (77). Now is a good time to prepare for such situations, especially since in the workplace you rarely have the luxury of saying to your boss, "I've got reasons why what you're arguing doesn't make good sense, but let me do some research and get back to you on this later."

Individually or in groups of two or three, research the value that professional writers add to organizations. Then, develop an easily distributable print document or a Web page arguing the organizational value of professional writers, even in the face of such technological advances as those represented in the scenario above. Your document should serve one of two audiences and purposes. Either it should be a resource professional writers can use to help them argue their value, or it should be a document directed at managers with the aim of arguing the organizational value of professional writers.

3. Writers have always been responsible for helping readers easily navigate their way through texts. Different writing technologies have encouraged, and even demanded, different forms of navigation, and as writing technologies change, issues of textual navigation become increasingly important for writers and readers. This project asks you to focus on the navigational aspects of writing/reading.

   In pairs or groups of three, locate six to eight samples of four different types of documents (e.g., brochures, a Web site, and a technical manual). This should lead you to a collection of 24 to 32 different documents arranged in four categories. Closely analyze the various navigational strategies writers build into these documents and readers use to find their way around them. Based on your analysis of your document sampling, develop your own scheme for categorizing navigational strategies. Finally, prepare a 30-minute presentation and hands-on workshop designed to help writers expand their repertoire of navigational strategic knowledge. Potential audiences for your presentation/workshop include another group in your class, another advanced writing class on campus (e.g., a technical writing service course), a conference workshop session group, or a group of working writers meeting for a brown-bag session.

PART 3

Professional Writing as Productive Art

Chapters 6 and 7 include readings that examine more closely how professional writing and rhetoric is affected when writing is understood as a productive art—a way of producing or making something. As you have read elsewhere in this collection, professional writing and rhetoric can be understood as either a productive art or a form of practice, or even as both. *Professional Writing and Rhetoric* approaches it as both. Part 2 examined professional writing and rhetoric as a form of socially contextualized practice and action. Part 3 now looks at professional writing and rhetoric as an art of production.

Most people find it natural to think about writing as an art of making, what the classical rhetoricians called a technē, and find it rather difficult to reconsider writing as a form of social action. When people write, it seems common sense that something is produced or made. The something that writers produce is a text. Easy enough. But is that all writers produce? And is there anything that distinguishes the kinds of texts that professional writers produce?

Chapter 6 picks up on this latter question: Is there anything that distinguishes the kinds of texts that professional writers produce? *Professional Writing and Rhetoric* defines professional writing as organizationally situated authorship, which narrows the field to writing done within organizational contexts of some form. This does not necessarily imply industry and government contexts. Civic, church, and community groups also form organizational contexts. Even freelance writers, whose title suggests a freedom from organizational life, work typically for and/or with organizations of some sort.

In any of these organizational contexts where professional writers ply the arts of situational authorship, one feature distinguishes their writing from other sorts of writing: it is user-centered. In other words, professional writers produce texts that are to be used by readers. Such use can take various forms: readers can use texts to make decisions, to explore problems, to collect information, to guide action, to govern, to judge, etc., etc. No matter what a text will be used for, though, the art of professional writing and rhetoric is distinguished by its focus on producing user-centered texts. The readings in Chapter 6 explore this issue with some depth.

Chapter 7 picks up on the first question above: Do writers produce only texts? Certainly, the primary product of professional writing is text. That term should be understood quiet broadly. A text can fall under a number of genres—newsletters, brochures, memos, reports, presentations, advertisements, Web pages, etc.—as well as a number of media—print, video, CD, hypertext, etc. If we understand text in this broad sense, then professional writers certainly produce texts, and texts are certainly their primary product. But do professional writers produce anything other than texts?

As you have probably noticed in earlier readings and maybe from internship or job experience, professional writers are often involved in project management. In order to get texts to publication, writers have to work with a number of other people: graphic artists, researchers, marketing specialists, publishers, etc. In this managerial work, professional writers produce something other than a text: they produce workplace processes and relationships. Chapter 7 refers to such a product as "social space." That is, by managing the production of texts, professional writers are often active in producing the social spaces in which people work together. But social space is not only something external or tangential to texts. Texts themselves participate in producing different sorts of social space. Think of this simple example: If I am writing a report and want to make sure that certain people have read and signed off on the report, by adding a set of signature spaces on the front of my report, I have produced a text that produces a subscribed set of relationships between people and ideas. My text has participated in producing social space.

As you read Chapters 6 and 7 in Part 3, focus on the ways professional writers are and might be involved in the production—and reproduction—of social spaces.

CHAPTER 6

Professional Writers Produce User-Centered Documents

INTRODUCTION

When people think about what professional writers produce, the obvious first response has to do with texts of some sort. But do professional writers produce texts that carry any distinguishing characteristics? Or, perhaps professional writers' processes for producing texts distinguish their products from other written products? One of the distinguishing characteristics that is true across a broad range of professional writing contexts is that professional writers produce user-centered texts.

That said, two questions still loom: what makes a text user-centered? and how does a user-centered focus affect the writer's production processes and thinking? These, and the following, are questions you should consider as you read the selections in Chapter 6:

- What does it mean to be reader- or user-centered?
- How does user-centered writing affect what writers do?
- How does user-centered writing affect what writers must know?
- What can writers do to develop user-centered practices?

One of the key issues in producing user-centered documents is determining how writers understand their audiences/readers. Most introductory writing textbooks urge writers "to think about their audience." Being sensitive to one's audience has been a hallmark of rhetoric since its inception. Aristotle spent great care categorizing and analyzing possible audiences, and such audience analyses can be traced throughout the rhetorical tradition. Even though rhetoricians have long focused on analyzing one's audience, their methods for doing so have shifted greatly.

Unlike many writers who simply imagine possible audiences and assess their possible reactions to their texts as they write in order to draft and revise in an audience-sensitive way, professional writers have developed more interactive methods to help them produce user-centered documents. Part of the reason professional writers have done so is that many of the documents they write can be more clearly assessed to be usable or not. It's hard to tell if a critical essay has been usable by its readers; it's much easier to tell whether or not a set of instructions, for instance, has been usable by its readers. If a critical essay is not usable or effective, readers will not read it, won't finish reading it, or will not change the ways they think or act because of it. The impact of being unusable or ineffective is not that great on the reader/user. The same does not hold true for a set of instructions. If instructions are not usable, readers/users cannot properly and completely use the products they have purchased. When they can't, there are a wide range of possible impacts: the product can get a bad reputation, the product could have low sales, the producer of the product could be taken to court, or the reader/user could be injured.

Instructions are just one genre that professional writers might produce. As you read the selections in Chapter 6, think about how other genres might be used by readers and focus on how you as a professional writer can increase the usability of the documents you might eventually produce.

One way professional writers increase the usability of documents is by paying careful attention to the visual aspect of texts. Kostelnick, in the first selection of the chapter, develops a systematic approach to designing rhetorically effective visual texts. Johnson, in the following selection, focuses our attention on the ways we think about being user-centered and the processes we practice to produce user-centered documents.

FOCUSING ON KEY TERMS AND CONCEPTS

Focus on the following terms and concepts while you read through this selection. Understanding these will not only increase your understanding of the selection that follows, but you will find that, because most of these terms or concepts are commonly used in professional writing and rhetoric, understanding them helps you get a better sense of the field itself.

1. low visual intensity
2. high visual intensity
3. visual signs
4. alphanumeric mode
5. spatial mode
6. graphic mode
7. intra-textual level
8. inter-textual level
9. extra-textual level
10. supra-textual level

A Systematic Approach to Visual Language in Business Communication

CHARLES KOSTELNICK
Iowa State University

Although business communication relies heavily on the visual, current approaches to graphics and text design are prescriptive and unsystematic. A 12-cell schema of visual coding modes and levels provides a model for describing and evaluating business documents as flexible systems of visual language. Emphasizing clarity and objectivity, the "information design" movement has generated guidelines for creating functional visual displays. However, visual language in business communication is seldom rhetorically "neutral" and requires adaptation to the contextual variables of each document, a goal the writer can achieve by combining visual and verbal planning in the same holistic process.

Writing does not communicate until it is seen, until it becomes an artifact for visual inspection (Gelb, 1980). From the primitive making of lists, business communicators have created permanent records by transcribing symbols onto two-dimensional surfaces. With the proliferation of print technology formerly reserved for specialists, contemporary business communicators have unprecedented control over visual design. However, participation in a visually intensive "print culture" (Ong, 1982) does not guarantee visual sensibility. While the "technologizing of the word" opens new avenues for visual expression—everything from page design to bar charts to scanned images—the thing-like quality of texts and the perceptual reliance on the eye for information processing are largely unacknowledged and unexamined. In theory and application, visual language in business communication remains rudimentary and prescriptive, and confined to a superficial or decorative rather than a functional or rhetorical role in the communication process.

To compensate, we need a systematic approach for analyzing how visual elements affect the readability and the rhetoric of business documents, and even for describing the visual elements themselves. Terms such as "layout," "format," and "graphics" lack precision and connote low level skills rather than serious visual thinking. The visual language of business communication is a flexible system of symbols, marks, and spatial variations that operates on several levels within and outside the text. Used effectively, visual language enables readers to process information—in letters, brochures, reports, newsletters—clearly and efficiently. The "information design" movement has generated guidelines for creating functional, user-oriented documents. However, far from being generic or rhetorically "neutral" (Kinross, 1985), visual choices are bound to a perceptual and rhetorical context. Each document embodies an autonomous sys-

Source: Kostelnick, Charles. "A Systemic Approach to Visual Language in Business Communication." *The Journal of Business Communication,* Volume 25, Issue 3 (Summer 1988), pp. 29–48. Reprinted with permission.

tem requiring adaptation to a particular audience and purpose. Visual planning and invention, therefore, must be part of the global strategy of the document and thoroughly integrated into the writing process.

THE SPECTRUM OF VISUAL INTENSITY

Writing and reading are intensely visual acts in which writer and reader negotiate meaning on a two-dimensional field. Since this field can be varied and manipulated into infinite configurations, all documents are not visual in the same way or to the same degree (Bernhardt, 1986). Thus we could devise a spectrum of visual intensity ranging from very low to very high:

LOW VISUAL INTENSITY ⟷ HIGH VISUAL INTENSITY

What, then, are the limits of such a continuum? What characteristics determine whether a document has a high or low level of visual intensity?

At the low end we might place a conventional essay or report, which submerges explanations and arguments in continuous text, or a novel with its symbols, metaphors, and complex narrative strands that readers must extract from seamless blocks of text (Bernhardt, 1986; Herrstrom, 1984). Still, even a novel relies on spatial segmentation—books, chapters, paragraphs—to signal transitions in place, time, and voice (Sterne even includes his own diagrams). The low end of the continuum calls for something far less visual: ideally, a string of unembellished signs flowing in a single direction, unrestricted by margins or pages. Rendering the text entirely in upper case and eliminating punctuation would further reduce visual dependency. The reader would then have no sense of spatial direction other than the syntactic relations of linguistic signals. Like a message moving across a video screen in an airport, the text would approximate the perceptual impermanence of an utterance.

At the high end of the visual continuum, designing and seeing assume a far greater role in shaping and receiving the message. As text is segmented into chunks, chunks into lists, and lists into matrices, charts, and diagrams, words become "thing-like" (Ong, 1982) entities transfixed spatially and graphically (through line work, arrows, geometric forms) on the page. Even more thing-like and reliant on visual inspection are representative images—for example, a picture in a proposal of key employees seated around a conference table. Still, visual intensity does not necessarily depend on realism because messages can be purely visual but abstract. For example, to show the enrollment patterns of a college, I can use pictures of each class or even of each student (achieving optimum realism), but a bar chart or a line graph would display the trends more clearly and concisely and with greater visual immediacy.

Thus on one end of the spectrum we might place a purely (or as pure as possible, given that it must be seen) linguistic message, on the other end a purely visual message which defies simple translation into words. However, the conditions for the extremes rarely obtain: most business documents are visual hybrids. When transferred to the page, the linear message flashing across the screen undergoes a visual metamorphosis: words are shaped by a type style, underlined, thickened, segmented into lines, distributed vertically and horizontally across the page, and circumscribed by graphic marks. The picture of the employees at the conference table, on the other hand, can be reduced to a line drawing, a plan view of the office, a corporate logo, or a series of abstract marks, each of which (including a color photo) may depend on linguistic coding to make any sense. And so inside the extremes of the continuum occurs a vast intermingling of visual signs, and of the visual and the verbal. Like a chemical reaction, the signs are interdependent: different combinations alter the visual effects of individual signs, which collectively alter the meaning of the message. Because these combinations are bound perceptually and rhetorically to context, each document contains an idiosyncratic *system* of visual language. The visual intensity of each system depends on the interplay of several levels and coding modes.

A SCHEMA OF VISUAL LANGUAGE

What, then, are the constituent elements of visual language? First, we need to define the coding modes that comprise the raw materials of visual language; and second, the levels at which the modes operate within actual documents.

Visual language can be encoded in three modes:

- *Alphanumeric/symbolic:* textual particles, including letters that comprise linguistic units; numbers; and symbols such as parentheses, dollar signs, and ampersands.
- *Graphic:* marks, lines, tones, and colors which encode geometric shapes (squares, circles, bars), realistic images (a face, a building), or abstract forms (a corporate logo).
- *Spatial:* the distribution of alphanumeric/symbolic and graphic signs across a plane, creating visual syntax among textual or graphic particles.

These coding modes are integrated on four distinct levels: intra-, inter-, extra-, and supra-textual. Together these four levels and three modes comprise a 12-cell schema of visual language.[1]

The *Intra-textual* level (cells 1–3) is coded primarily in the alphanumeric/symbolic mode and controls the local form, size, posture, and embellishment of textual elements. Consider the following sample of text:

> During the *past month*, our clientele in each region has increased as follows: East Region, 31 clients; South Region, 40 clients; and West Region, 25 clients.

In the alphanumeric/symbolic mode (cell 1), I can select a typeface with or without serifs, change the thickness (bold, light) or posture (e.g., italics) of the typeface, choose upper or lower case, and use symbols (&, $, #) in place of words; in the spatial mode (cell 2), enlarge or shrink the type size, choose the linear spacing between characters and words (12 pitch; 10 CPI; tight or loose spacing, kerning); and in the graphic mode (cell 3) use marks to regulate the linear flow of text (commas, periods, colons) or to emphasize key ideas (underlining), or transform letters into expressive

A 12-Cell Schema of Visual Communication: Four Levels and Three Coding Modes

	Alphanumeric/Symbolic	Spatial	Graphic
Intra	1 micro-level textual form: style, size, weight, and posture of letters, numbers, and symbols	2 local spacing between characters and textual particles: CPI, picas, kerning	3 marks: punctuation, underscoring; iconicity of letters and words
Inter	4 serial and segmenting devices: headings, letters, numbers; typestyle variations showing textual structure	5 vertical/horizontal arrangement of text: line endings, indentation; lists, matrices, trees, flow diagrams	6 cueing devices: bullets, icons; syntactic devices: line work and arrows on tables, charts, diagrams
Extra	7 decoding devices: legends, captions, labels, numerical description of data	8 configuration of schematic and pictorial sign systems; plotting of data on X-Y axes, circles, other forms; depth of pictorial images	9 schematic: lines, tones, colors, and textures on data displays; pictorial: resolution of details on images, abstract to realistic
Supra	10 macro-level serial and segmenting devices: section titles, numbers; page headers, pagination	11 cohesion of entire document over several planes: page breaks, size; location of extra-textuals within text	12 coding marks unifying pages or sections of text: line work, color, icons, logos, tabs

icons. Intra-textual choices, therefore, regulate thousands of points on a plane: given only the common variations in each coding mode cited above, the possible permutations of our sample easily number in the thousands! Intra-textual choices can be local (a word in boldface) or global (a type size for an entire text), but are essentially a one-dimensional, point-by-point regulation of text.

The *Inter-textual* level (cells 4–6) is coded primarily in the spatial mode, generating visual cues that enable readers to search for and retrieve information. Distributed horizontally and vertically on an invisible two-dimensional grid, a seamless text can be transformed into a highly variegated system "surfacing" (Herrstrom, 1984, p. 229) textual structure.[2] In our sample, of course, some inter-textual decisions had already been made: the movement of the text linearly to the right, the line break and continuation below and to the left, the distance between lines. We can introduce numerous other inter-textual elements to reshape the text:

5. During the *past month*, our clientele in each region has increased as follows:
   - East Region, 31 clients
   - South Region, 40 clients
   - West Region, 25 clients

In the alphanumeric/symbolic mode (cell 4), "5." cues us that the statement is part of a series; headings, roman numerals, initial letters, and symbols are also serial devices. Spatially (cell 5), the second half of the text is segmented into units and aligned vertically as a list. Graphically (cell 6), bullets identify each item in the secondary series. In more dynamic inter-textual configurations—tables, matrices, flow charts, decision trees—spatial coding (the positioning of words and figures) creates a visual syntax, and graphic cues (arrows, lines, geometric shapes) form a system of "macro-punctuation" (Waller, 1980), among textual particles.

The division between the inter- and extra-textual levels is subtle but perceptually and semantically critical. The two examples below can illustrate the distinction:

Except for the spatial positioning of "PLAN," both messages use similar visual language. But their meanings and visual codes differ radically. On the left, the textual particle "PLAN" (cell 5) is circumscribed graphically (cell 6), indicating one of perhaps several functions in a decision-making process. Framed and embellished, the word is the thing-like center of attention, like the word STOP on a road sign. The same applies to textual particles in tables, matrices, flow charts, decision trees, and so on: all are inter-textual arrangements, simply at a higher level of intensity than a seamless essay. In the example on the right, however, the textual particle is merely a tag (cell 7) identifying an image (a plan of an office, a desk, a post card) coded spatially (cell 8: top view, proportional to the real thing) and graphically (cell 9: outlining the perimeter). Alternatively, the image may be nonrealistic, representing a quantity, say the dollar amount or time in hours of generating a plan (as opposed to other costs). In either case, textual elements assist the decoding of the message (legends, labels, captions), but information is processed primarily through seeing rather than reading. In the right-hand example, "PLAN" is useful for decoding the message but secondary to the visual image, which stands outside the text.

The *Extra-textual* level (cells 7–9), therefore, may rely on the alphanumeric/symbolic mode for interpretation, but coding is primarily spatial and graphic. Sign systems range in level of abstraction from pictorial (low level of abstraction—photographs, renderings) to schematic (high level of abstraction—logos, graphs, diagrams; see Twyman 1979). Below, the data in our sample message are displayed schematically:

During the *past month,* our clientele in each region has increased as follows:

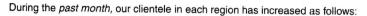

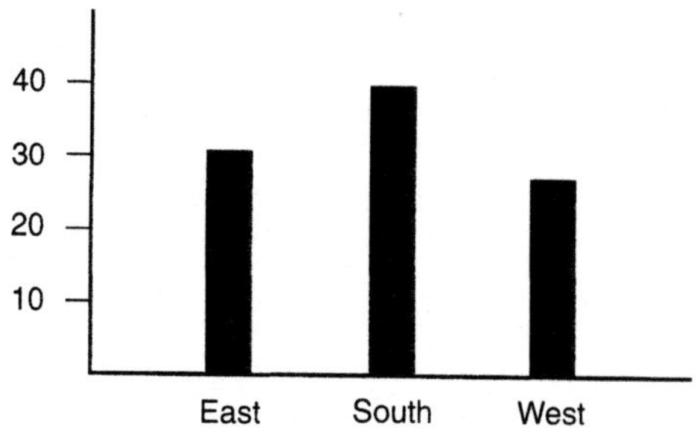

Alphanumeric signs (cell 7) encode the X axis with geographic locations, the Y axis with figures; spatially (cell 8), plotting the data on the grid determines the vertical distance between figures and the horizontal distance between regions; graphically (cell 9), shaded bars encode the data, lines locate the axes, and tick marks gauge the scale. All of the coding modes are open to variation. "East" can be simplified to "E," and figures on the Y axis replaced by exact values above each bar. Spatially, the distances between values and between regions can be increased or decreased, or the axis shifted 90 degrees so the bars run horizontally. Graphically, the bars can be cross-hatched, colored, or changed to simple lines, and the tick marks replaced by a light grid.

If the extra-textual were pictorial rather than schematic, the same variables would come into play: spatially, the image of the employees at a conference table can be close or distant or rotated at different angles from the real thing; graphically, it can be coded at a high level of realism (with plenty of surface details, down to the wood grain on the table) or reduced to an abstract stick drawing or a stylized logo. The coding choices for either schematics or pictorials are interdependent: the graphic coding of a logo may flatten an image spatially; the spatial coding of data on a bar chart may restrict the range of graphic choices. Like the intra- and inter-textual levels, visual language at the extra-textual level depends on the contextual variables of the system.

The *Supra-textual* level (cells 10–12), like the inter-textual, operates chiefly in the spatial mode, though on a more global scale, arranging extra-textuals (graphs, pie charts, pictures) within the text and regulating the flow from one page or section (paragraph breaks at the bottom of the page; breaks between parts of a report) to the next to create a coherent document. For example, in the last version of our sample, the spatial mode (cell 11) locates the extra-textual relative to the text ("During the past month. . ."): centered directly beneath the text, or to the left or to the right; or isolated on a separate page in close proximity or in an appendix. In the alphanumeric mode (cell 10), supra-textual elements include segmenting and serial devices (page headers, section titles, pagination); for example, I could label the bar graph in the sample "Figure 6," indicating a supra-textual relation with the other extra-textuals in the document. In the graphic mode cohesive devices—line work, page color, logos—unify sections of a document, providing visual markers for readers. In our example, a line border around the bar chart would further demarcate the transition from text to extra-text. By mapping out the terrain, the supra-textual level controls the three-dimensional configuration of the document, enabling readers to comprehend its logic and hierarchy and to retrieve essential information.

Each document blends the coding modes and levels into an idiosyncratic visual system incorporating most of the cells on the matrix. Consider the visual choices for a simple analytical memo report. At the intra-textual level, typefaces, point sizes (cell 1), spacing (cell 2), and graphic cues (cell 3), selected from a menu of computer-generated options, create interest and emphasis and regulate the intra-linear flow of signs. At the inter-textual level, headings (cell 4), indentation, and vertical spacing (cell 5) structure the text, and graphic coding around textual particles (cell 6) forms tables and matrices. At the extra-textual level, spatial coding distributes data (monthly sales figures, production costs) on a grid (cell 8), graphic coding records data with shaded bars (cell 9), and alphanumeric coding (cell 7) explains each variable. At the supra-textual level, pages are numbered and labeled with subject, date, author, and number (cell 10), composed spatially to locate the bar charts within the text and to ensure continuous flow from page to page (cell 11), and coded with icons, colors, tabs, or separate pages (blank or titled) to differentiate sections of the report (cell 12).

Individually, none of these choices may be novel, but when combined in the same document they compose a system of visual language dependent, as all visual images are, on context for their perceptual qualities (Arnheim, 1969). Hence, the levels and coding modes are interdependent. The cell 1 selection of a type size, posture, and style expands or contracts the document and therefore has supra-textual consequences. Conversely, a cell 11 decision to write a "one-page resume" impinges on intra- and inter-textual choices. The cell 4 decision to deploy a variety of headings may require cell 1 assistance to establish a hierarchy. The spatial and graphic coding of a bar chart (cells 8, 9) influence supra-textual page design. Choices for each coding mode and level, therefore, are contingent upon other choices.

The 12-cell matrix opens up these choices to public purview, enabling us to describe how visual elements coalesce to form coherent systems. Just as each writing situation engenders a different combination of linguistic choices, so too with visual language: each document requires conscious planning so that the modes and levels work harmoniously to achieve the same goals. To the extent that the cells on the matrix compose an integrated system that contributes to the purpose of the message, the document obtains a certain level of visual sophistication or intensity.

INFORMATION DESIGN: TOWARD A FUNCTIONAL VISUAL LANGUAGE

Of course not all visual systems succeed in accommodating the reader. What makes visual language reader-oriented? In a fast-paced business environment, effective visual display simplifies the readers' information processing. As Hartley (1985) puts it, "The printed page should provide a reliable frame of reference from within which the readers can move about, leave and return without confusion" (p. 18). Above all else, visual language should be functional. And on an idiomatic and conventional level, it usually is: words ordinarily flow horizontally across the page; letters have signature blocks; bars on graphs stand for measured units. These generic conventions enable the reader to place visual codes into the slots of mental schemata (see Schumacher, 1981). A one-page document that follows the conventions of a letter and arrives in an envelope with a booklet "fits" nicely into the slot of "cover letter," a persuasive message, widely distributed, that entices the reader to examine an accompanying document. On the other hand, a letter set of boustrophedon (as the ox plows, back and forth), or horizontally across the length of the page, poses an obvious "misfit" to most readers. By activating schemata, generic conventions foster reader expectations about the whole message.

Visual conventions, however, define only general patterns, change over time and from one discourse community or culture to another, and may not provide the optimum visual patterns for information processing. To compensate for these shortcomings, the "information design" movement has tried to establish more stable, culturally neutral, and empirically verifiable guidelines. Combining theory, practice, and research, information design draws upon several disciplines—graphic design, perception, educational and cognitive psychology, and semiotics—to discover reader-oriented visual patterns for many of the levels and coding modes on the matrix.

From the outset, the information design movement sought to create a modern, rational, ideologically free system of signs accessible to a mass audience. Its intellectual roots can be traced to the functional modernism of the Bauhaus in the 20's (Kinross, 1985) and indirectly to the visual dynamism of Futurist and Constructivist text displays. Each of these movements experimented with radically new modes of visual design, emphasizing clean, nontraditional forms and typefaces compatible with Machine Age technology (see Crouwel, 1979). Together these fresh approaches to textual display and image-making provided an aesthetic and functional springboard for early information designers. For example, Neurath's "isotype" (international system of picture education) symbols epitomized the quest for a rational, cross-cultural language of extra-textual signs for displaying quantitative data. After World War II the search for simple, universal, and technically precise forms continued in the invention of typeface systems, for example the serifless Univers style (Kinross, pp. 23 & 27). On a supra-textual level, the appeal to the universal was reinforced by the laws of gestalt psychology (similarity, good figure, equilibrium), which transcended style and culture by codifying perceptual experiences detached from language or ideology.

The information design movement, therefore, has evolved into a modern philosophy of visual display emphasizing objectivity, clarity, and the integration of text with simple forms. The movement has had a wide impact on advertising, government documents and forms, educational texts, and scientific and technical publications. Today theorists and practitioners continue to develop and test methods for functional designs. On the inter-textual level, Hartley (1985), Wright (1977), and several others have devised methods of employing headings, vertical spacing, and graphic cues to surface discourse visually, while Horn (1982) has codified these elements into a comprehensive system of "structured writing." On the extra-textual level, Tufte (1983) and Bertin (1981) apply functionalist principles to the visual display of quantitative data. Both advocate economy through density, arguing that graphs and charts should display the maximum information in the smallest space with the least ink. Bertin invokes semiotic theory to define visual clarity: graphic sign systems should be "monosemic" (pp. 177–179), the relations among signs conveying a single, unambiguous meaning. Consistent with its functionalist tradition, contemporary information design theory values economy, objectivity, and semantic and structural transparency.

Further supporting the functionalist approach are a wealth of empirical studies aimed at identifying optimum visual displays. At the intra-textual level, Tinker (1963) found that type sizes between 9 and 12 point maximize reading efficiency, lower case letters are more readable than upper case, and excessive italic print reduces legibility. On the inter-textual level, Hartley (1984) has studied, with mixed results, the effects of horizontal and vertical structuring of text on readability, while Hartley and Trueman (1985) have shown, through extensive experiments, that headings improve reader performance. Wright (1968) experimented with the spatial configurations of tables, finding that users of currency conversion tables perform better with redundant lists than with matrices. On the extra-textual level, Macdonald-Ross (1977) summarized over five decades of empirical research on the graphic display of quantitative data, concluding for example that horizontal bars are superior to unkeyed circles. Psychologists have studied the effects of spatial depth and degree of graphic detail in the processing of representative images (Perkins, 1980). They found that readers need a modest amount of acculturation, should be given only selected information pertinent to the task, and perceive depth on a two-dimensional surface differently from the "real thing." At the supra-textual level, guidelines can be derived from hemispheric brain research (see, for example, Welford, 1984). Because readers process visual information more effectively in the right brain (the left field of vision), pictures and schematics are generally better placed to the left, explanatory text to the right.[3]

Many of the empirical findings and theoretical positions cited above presume a universal reader performing generic, noncontextual tasks. Wright (1977) sums up the empiricist's dilemma: "Researchers often consider themselves free to ask if format A is better than format B, without needing to specify better for *what?*" (p. 112). Some of the findings—in particular, those of Hartley and Wright—are user-centered and tied to specific audiences and contexts. Wright's experiments with tabular displays are bound to the readers' experience with visual arrangements of figures, the situation in which the display is used, and the readers' familiarity with the subject. Many case studies of public documents are contextually sensitive to the special needs posed by audience and situation. Such studies entail pretesting and interviewing during the design process to assure quality control and to convert visual language to a "plain English" style intelligible to the nonspecialist (Redish, Felker, and Rose, 1981; Wright, 1979).

The functional guidelines of the information design movement, therefore, specify which configurations afford the most efficient processing of information, enabling us to engineer clear visual conduits. The emphasis on information privileges function over decoration, just as the Bauhaus did. A well-designed text replicates on the page the sleek steel and glass of a modern high-rise: functional, aesthetically austere, and culturally neutral. By melding text and form, the information design movement revolutionized the visual language of texts and made design a bona fide part of the creative process, from planning to production.

THE MYTH OF RHETORICAL NEUTRALITY

Ideally, information design deploys visual language to ensure an unadulterated channel between sender and receiver. However, visual elements themselves—especially in business messages—obtain meaning beyond the simple transfer of information. "'Pure' information," asserts Bonsiepe (1965), "exists for the designer only in arid abstraction. As soon as he begins to give it concrete shape, to bring it within the range of experience, the process of rhetorical infiltration begins" (p. 30). Simply put, a "rhetoric of neutrality" is a "pipe-dream," for even the mundane visual language of a train timetable has a certain rhetorical force, intended or not, that affects the reader's judgment (Kinross).

This rhetorical "infiltration" stems partly from taste and cultural values. Futurist and Constructivist text displays, the forerunners of information design, were considered iconoclastic when they first appeared; today they seem modern but passé. Then as now, they relinquish any clams to neutrality. The Machine Age modernism of the information design movement also constitutes a style—streamlined, simple, and low-key—but also subject to the vicissitudes of intellectual and aesthetic favor, and hence anything but "objective."[4] The invention and evolution of styles continually transform taste and perception, in art and in business communication. An early corporate logo, letterhead, or trademark juxtaposed to its contemporary version demonstrates the visual distance between past and present, and presages future recreation.

Technology also influences the rhetoric of visual language. At the outset of the century most texts were handwritten in one of four styles: business, civil service, text, or legal. The appearance of the typewriter transformed "micro-level" (Waller, 1980) visual choices, forcing characters into uniform, equally spaced units on a grid and initiating a whole new set of intra- and inter-textual conventions (Walker, 1984). Ninety years ago a letter done in business script would have signaled the nature of the message but otherwise attracted little attention. Today it would totally alter the rhetoric of the communication: the writer would suffer (an entry-level application letter) or benefit (a letter from a CEO to a consumer) from the unorthodox visual coding.

Cultural and technological changes are long-term, evolutionary tendencies that gradually modify the shared perception of large discourse communities. The choices business communicators make everyday are highly contextual and vary from one audience and visual system to another. Information design can weed out perceptually poor choices—illegible type styles, line spacing, and data displays—but cannot necessarily prescribe rhetorically appropriate choices governing the arrangement, style, or tone of a particular document.

For example, Tinker found that many common typefaces were equally legible but that readers preferred some faces over others. The "atmosphere value" or "semantic quality" of typefaces (Bartram, 1982) opens the door to subjectivity and rhetorical judgment.[5] What "pleases" readers in one message may not be pleasing or rhetorically consistent in another. Visual choices demand adaptation to the particular audience, purpose, and circumstances of the document. A simple intra-textual (cell 1) choice of a typeface can have a major impact on the rhetoric of a message. For an informal message to a co-worker, a handwritten note or a dot-matrix printout reinforces the familiar tone and transience of the document. However, a manager (one that values job security) submitting an annual report to a board of directors or a proposal to a government agency will specify a typeface that projects credibility and permanence.

Although technical capability helps, visual rhetoric is not necessarily constrained by available technology. Direct mail sales letters use typewritten serif styles even though virtually any typeface could be used, given the bulk printing. Indeed, to make messages appear more spontaneous, senders frequently underline key points with insouciant strokes of the pen (in red ink) to personalize the visual style. Urgent messages (e.g., collections) are customarily typed entirely in upper case, sacrificing legibility for a more official, authoritarian tone. These choices may seem trivial and subjective, but they are made for rhetorically sound reasons. Add to these variations the ability to highlight text with boldface and multiple fonts and point sizes now commonly available in desk-top printers, the business communicator will experience far greater freedom to adapt intra-textual tone and style to the rhetorical task.

The same holds for information design on any level of the matrix. Inter-textual structuring—headings, vertical spacing, graphic cues—has rhetorical consequences, surfacing certain aspects of the message while embedding others. Opening up a text with spatial and graphic coding is equivalent to persuading visually, impelling readers to value selected pieces of information and to acquiesce to logical and hierarchical connections that make them cohere visually. In the second version of our sample text, the list with bullets establishes visual parallelism among the three regions. Perhaps the West Region is new and the writer wants to emphasize its status as a full and equal sector. Perhaps isolating these elements argues that the vice-president for sales is doing satisfactorily in all regions, or that by comparison the West Region is faltering, or that more employees should be hired in the South. Depending on the context, selective textual surfacing advances the writer's argument. In an extremely sensitive message (a performance appraisal), the writer may even eschew any sort of inter-textual variation to embed key information, sacrificing legibility to maintain rhetorical integrity.

The third version of our sample detaches the data from the text, reshapes them graphically, and tells the reader in a resonant voice: "Here are the facts, bare and irrefutable!" The visual style, bordering on hyperbole, can of course be modified. Spatially, the bars can be extended or contracted on the grid, the graphic coding ornamented or streamlined, the alphanumeric coding expanded—depending on the designer's argument and rhetorical stance. The differences among regions can be accentuated or diminished, the graph made to appear scientific or aesthetically pleasing, clumsy or self-serving. Any of these effects influences the reader's impression of the data as well as of the author's integrity and is far from neutral. Even acquiescing to someone else's configuration of the data is still to choose, and the reader to receive, the visual rhetoric of the message.

Together these choices coalesce to drive the visual rhetoric of the document. Consider the range of visual decisions that go into an annual report to stockholders: the company logo, charts and graphs to display sales and production figures; a plan and line drawing of a new plant; photographs introducing board members; and textual displays of expenses, revenues, and earnings. Along with these are numerous supra-textual decisions regarding the location of extra-textuals, the size and number of pages, and graphic cues such as color and line work intended to coordinate the whole document, add interest to complex data, instill confidence in investors, and accurately project the company's image (innovative, consumer oriented, high-tech). The visual system is both informational and persuasive, enabling investors to locate pertinent facts and, more importantly, enabling the company to market its performance. Although annual reports are more complex than everyday messages, maintaining full rhetorical control over any business document requires careful visual planning. In other words, the writer must attend to the visual system *during* the composing process as rhetorical strategies take shape.

COORDINATING VISUAL AND VERBAL RHETORIC: A HOLISTIC PROCESS

Planning an entire document is a "holistic process" (Barton & Barton, 1985) in which the visual and the verbal modify, complement, and compensate for one another to accomplish the rhetorical goals of the communication. Linguistic invention entails innumerable lexical choices; visual invention materializes from the wealth of elements defined by the matrix. Both are driven by the same purpose and by many of the same principles and concepts: style, tone, level of technicality, interest, benefits, arrangement, readability, and so on. During the planning stage of a document—say a proposal persuading a skeptical reader to act on a recommendation—the visual rhetorician considers several options:

- Stimulate *reader interest* through a logo, picture, or data display (cells 8, 9) and through highly variegated inter-textual structure (cells 4, 5, 6).
- Surface selected *arguments* and *benefits* through intra- and inter-textual variations (cell 1: boldfacing and all capitals; cell 3: underlining; cell 6: bullets to highlight items in a list).
- Adapt the *style* of data displays (cells 8–9) to a nonspecialist (cell 7 alphanumeric coding), and project a serious *tone* necessary to engage this reader (cell 1 choice of a typeface; cells 6, 9, 12 graphic coding).
- Map out the document with inter- and supra-textual cues (cells 4–6, 10–12) that guide the busy reader through the information, increasing *readability*.

Because these and many other strategies apply to visual as well as verbal problem-solving, early planning synchronizes the two from the start.

Early planning also exploits the special rhetorical capabilities of the visual and the verbal. For example, a bar graph can often argue with statistics more clearly and emphatically than inter-textual displays of figures; a logo can establish the identity of an organization or a project more succinctly and forcefully than an explanatory paragraph. On the other hand, in a sales letter a poetic description of the scenery at a resort may be more provocative than a high resolution photograph, and in a negative recommendation letter a seamless text more tactful than one that is visually transparent.

Because the visual and verbal are interdependent, holistic planning enables the writer-designer to monitor the interaction between the two. A highly variegated inter-textual system changes sentence structure and syntax, increasing the frequency of parallelism and fragments and eliminating the need for verbal transitions (Bernhardt, 1986). Converting text-bound figures to a data display (as in the third version of our example) contracts and reshapes the text. On the other hand, a complex line graph may require textual expansion to highlight and explain key data. Style and tone are also interactive: a set of informally written instructions on how to pot plants calls for simple line drawings, instead of photographs, to illustrate the task; the linguistic tone of a sales letter theme or slogan sets the visual tone for images on the envelope and enclosures.

Globally, supra-textual decisions can regulate the linguistic evolution of the text. Thing-like, the page is a tablet, the writer's unit of workspace, like an artist's canvas, an architect's story, or a sculptor's marble slab. The tablet as visual workspace both complicates and simplifies the writer's task. Some undergraduates struggle to find enough material to fill up a "one-page" résumé, while others try desperately to conflate several pages into one. As items are arranged on the tablet, the writer strikes a balance between the functional exigencies of information design (the user may give the document only a cursory glance) and an appropriate rhetorical stance—the need to appear trustworthy, ambitious, successful, and employable. To these ends, elements are moved about, shaped, cut, and recast linguistically (in phrases and fragments). Through the symbiotic growth of the visual and the verbal, the document evolves within the supra-textual boundaries of the tablet.

Visual design, like writing, is a process of discovery. This discovery process occurs within and outside the text. Articulating a text with intra- and inter-textual cues—typographical variations, paragraph breaks, headings, lists, tables—unfolds and structures ideas, enabling the writer to resee the logic of the message. Creating visual clarity demands clear thinking, and rethinking, about content (Hartley, 1984, p. 502). Extra-textuals also act as discovery tools (Bertin, 1981). The process of translating data into a visual sign system surfaces patterns and relationships submerged in a table of figures. This happens rudimentarily in the third version of our example: the designer must experiment with graphic and spatial coding to find a rhetorically suitable configuration. If we had additional figures—abut client increases in previous periods, the clients themselves (individuals, businesses, government agencies), or dollar volumes per client or region—permuting these data on the X-Y axes would enable us to find an insightful pattern. Supple and unpredictable, data displays are powerful discovery tools, much like freewriting, zero drafting, and outlining.

At the outset of the holistic process, therefore, visual thinking may be provisional: designing a document is an evolutionary process entailing invention, revision, and editing. Visual language is rich and flexible, allowing infinite syntactic variation of the constituent elements outlined on the matrix. However, if these elements are to be transformed into inventive patterns, they must be treated as bona fide rhetorical tools guided by the same strategies that govern the linguistic text. Asserting complete rhetorical control over a document, therefore, demands visual thinking from planning to production.

CONCLUSION

Business communication is intensely visual: as practitioners, researchers, and instructors we are inundated with an array of visible forms. Although we use, teach, and presumably value effective visual language, we have no system for describing, analyzing, or evaluating it. This is partially because design is viewed as a ready-made set of conventions or as a task performed by specialists—in either case extrinsic to the writing process. However, in the making of business documents, writing and designing, reading and seeing, are interdependent, complementing and impinging on each other to satisfy the goals of the communication. Combined in any business document, the three coding modes and four levels of visual language comprise an integrated system enabling the reader to process information efficiently from the rhetorical perspective the writer intends. Adapting the visual system to the message is essential in business communication because audience and purpose vary from one message to another and because writers have the autonomy to design their own documents.

Successful adaptation requires attention to design throughout the writing process. We have only rudimentary knowledge about how visual and verbal thinking interact during this process or how the contextual variables of visual language affect readers. What we do know, instinctively though perhaps unconsciously as members of a "print culture," is that visual thinking is intrinsic to the discovery and communication of meaning. To use visual language fluently, or to understand its structure and nuances, we must explore it explicitly and systematically, with the same rigorous scrutiny given the linguistic text.

Notes

1. Twyman (1979) has proposed a graphic schema containing four modes of symbolization and seven methods of configuration. His matrix contains twenty-eight "cells" and classifies primarily spatial variations at the inter- and extra-textual levels.

2. Some of Twyman's methods of configuration are particularly relevant here: Pure linear (a continuous flowing text; e.g., in circular form), Linear interrupted (conventional text with line breaks), List, Matrix (tables), and Linear branching (a tree, an organizational chart).

3. Design guidelines derived from hemispheric research are contingent upon at least two variables: the kind of information processing required by the task (part-by-part vs. holistic analysis) and the amount of verbal thinking the viewer is doing at the time.

4. See Kinross for a detailed account of the aesthetic and ideological evolution of the information design movement and its claims to rhetorical "neutrality."

5. The "atmosphere value" of typefaces has long been the subject of study and speculation. For example, Ovink experimented with several typefaces used in literature and advertising, grouping their subjective effects under categories such as "luxury/refinement," "economy/precision," and "strength." But the moods and cultural values associated with intra-textual styles can change dramatically: serifless letter forms, which are typically allied with modernism, were initially revived in the eighteenth century because they evoked "rugged antiquity" (Spencer, 1969, pp. 29–30).

References

Arnheim, Rudolf (1969). *Visual thinking*. Berkeley, CA: University of California Press.

Barton, Ben F. & Barton, Marthalee S. (1985). Toward a rhetoric of visuals for the computer era. *The Technical Writing Teacher*, *12*, 126–145.

Bartram, David (1982). The perception of semantic quality in type: Differences between designers and nondesigners. *Information Design Journal*, *3*, 38–50.

Bernhardt, Stephen A. (1986). Seeing the text. *College Composition and Communication, 37*, 66–78.

Bertin, Jacques (1981). *Graphics and graphic information-processing* (William J. Berg and Paul Scott, Trans.). New York: De Gruyter.

Bonsiepe, Gui (1965). Visual/verbal rhetoric. *Ulm, 14–16*, 23–40.

Crouwel, Wim (1979). Typography: A technique of making a text 'legible.' In Paul Kolers, Merald E. Wrolstad, and Herman Bouma (Eds.), *Processing of Visible Language*, (*1*, pp. 151–164). New York: Plenum.

Gelb, I. J. (1980). Principles of writing systems within the frame of visual communication. In Kolers, Wrolstad, and Bouma (2: pp. 7–24).

Hartley, James (1985). *Designing instructional text* (2nd ed.). New York: Nichols.

Hartley, James (1984). Space and structure in instructional text. In Ronald Easterby and Harm Zwaga (Eds.), *Information design* (pp. 497–515). New York: Wiley.

Hartely, James & Trueman, Mark (1985). A research strategy for text designers: The role of headings. *Instructional Science, 14*, 99–155.

Herrstrom, David Sten (1984). Technical writing as mapping description onto diagram: The graphic paradigms of explanation. *Journal of Technical Writing and Communication, 14*, 223–240.

Horn, Robert E. (1982). Structured writing and text design. In David H. Jonassen (Ed.), *The technology of text* (pp. 341–365). Englewood Cliffs, NJ: Educational Technology Publications.

Kinross, Robin (1985). The rhetoric of neutrality. *Design Issues, 2*(2), 18–30.

Macdonald-Ross, Michael (1977). How numbers are shown: A review of research on the presentation of quantitative data in texts. *AV Communication Review, 25*, 359–409.

Ong, Walter (1982). *Orality and literacy: The technologizing of the word*. New York: Methuen.

Perkins, D. N. (1980). Pictures and the real thing. In Kolers, Wrolstad, and Bouma (2, pp. 259–278).

Redish, Janice C., Felker, Daniel B., & Rose, Andrew M. (1981). Evaluating the effects of document design principles. *Information Design Journal, 2*, 236–243.

Schumacher, Gary M. (1981). Schemata in text processing and design. *Information Design Journal, 2*, 17-27.

Spencer, Herbert (1969). *The visible word* (2nd ed.). New York: Hastings.

Tinker, Miles (1963). *Legibility of print*. Ames, IA: Iowa State University Press.

Tufte, Edward (1983). *The visual display of quantitative information*. Cheshire, CT: Graphics Press.

Twyman, Michael (1979). A schema for the study of graphic language. In Kolers, Wrolstad, and Bouma (*1*, pp. 117–150).

Walker, Sue (1984). How typewriters changed correspondence: An analysis of prescription and practice. *Visible Language, 18*, 102–117.

Waller, Robert H. W. (1980). Graphic aspects of complex texts: Typography as macro-punctuation. In Kolers, Wrolstad, and Bouma (*2*, pp. 241–253).

Welford, A. T. (1984). Theory and application in visual displays. In Easterby and Zwaga (pp. 3–18).

Wright, Patricia (1979). The quality control of document design. *Information Design Journal*, *1*, 33–42.

Wright, Patricia (1977). Presenting technical information: A survey of research findings. *Instructional Science*, 6, 93–134.

Wright, Patricia (1968). Using tabulated information. *Ergonomics*, *11*, 331–343.

DEVELOPING YOUR UNDERSTANDING

1. Explain what Kostelnick means when he argues that visual communication has remained "confined to a superficial or decorative rather than a functional or rhetorical role in the communicative process." Support your explanation with examples you have located, and be prepared to discuss your response and share your examples in class (bring your examples in electronic, transparent, or print-copied versions, depending on the technologies available in your classroom).
2. Locate a one-page document and analyze its use of visual language using cells one thru nine of the twelve-cell schema. Be prepared to discuss in class your analysis of any of the nine cells (bring your document example to class in either electronic, transparent, or print-copied versions, depending on the technologies available in your classroom).
3. Kostelnick argues that visual language cannot be rhetorically neutral, and yet the information design movement retains some sense that visual language can be rhetorically neutral. Argue why you believe information design/visual language can or cannot be rhetorically neutral. Support your position with examples.
4. Towards the end of his article, Kostelnick begins using a hyphenated "writer-designer" designation to designate the professional writer. Define the writer-designer's work (e.g., work she must attend to, her writing process, who she works with, and her final product) by contrasting it with the non-hyphenated writer's and designer's work.

FOCUSING ON KEY TERMS AND CONCEPTS

Focus on the following terms and concepts while you read through this selection. Understanding these will not only increase your understanding of the selection that follows, but you will find that, because most of these terms or concepts are commonly used in professional writing and rhetoric, understanding them helps you get a better sense of the field itself.

1. text
2. artifact
3. system-centered approach
4. user-friendly approach
5. user-centered approach
6. readability
7. iterative design

When All Else Fails, Use the Instructions: Local Knowledge, Negotiation, and the Construction of User-Centered Computer Documentation

ROBERT J. JOHNSON

> *Consider the "stop print" function in some database programs. You make a mistake or you see that it's not printing the way you want and you want to kill it. So you quickly open the manual under "print" and look for "stop" and there is nothing. So you look under "k" for "kill"; you even look under "a" for "annihilate." And there is nothing in the index that tells you how to make the darned thing stop.*
>
> *There should be a law: No one who hasn't managed a database should be allowed to program one. I call this law: "No one should be allowed to make menus who hasn't had to eat off them."*
>
> —Ellen Bravo, *9 to 5*

Users of technologies, at one time or another, use some form of written instructions to help guide them through the "correct" procedure of whatever artifact they are currently using. Whether we are putting together a barbeque, repairing a bicycle, tuning a car, or following the directions for taking a medical prescription, we often are compelled to consult the instructions: the texts[1] that contain the knowledge of what it is that we want to do. Instructions come in many shapes and sizes, but they need not be complex or voluminous to cause confusion or frustration. Early studies of how people comprehend warning signs (see Chapanis 1965; Hartley 1985) indicated that even the most rudimentary information can be misleading or ambiguous. More recent investigations have demonstrated that the brief but crucial information included on drug prescriptions often is not followed correctly, thus leading to possible medical complications and potential lawsuits (Williams, et. al. 1995). Clearly, despite their ubiquity, written instructions have something less than a glowing reputation.

Nevertheless, as much as we may try to avoid them, we simply cannot escape the presence of instructional texts in our everyday lives. In some instances, such as those involving computer technology, we might not actively seek instructional assistance

---

[1] In this chapter's discussion of instructional documents, "texts" will be used as broadly as possible: printed pages, on-line screen information, medicine bottle labels, road signs, explanatory cartoons—anywhere instructional material might reside.

but it is offered to us anyway. For example, I unexpectedly received the following message while loading a new piece of software on my personal computer:

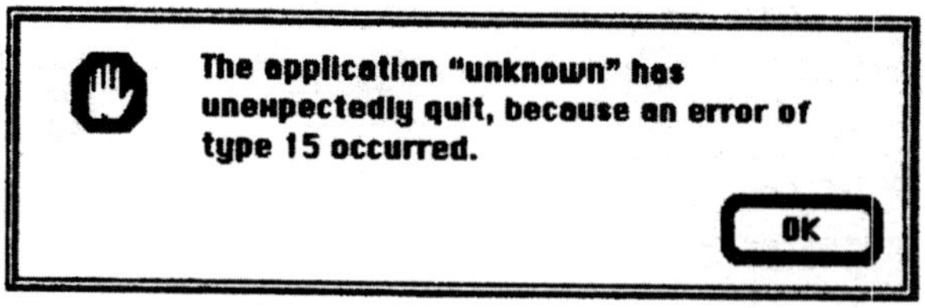
The application "unknown" has unexpectedly quit, because an error of type 15 occurred.
OK

Unless I am a super user of the system, or I happen to have an up-to-date definitive reference document explaining all of the error messages of the operating system, there is little useful information contained in this computer displayed "dialogue box."[2] It provides no visible knowledge of why the system stopped working or what the error might be. Was it my error or the system's error? Am I compelled to agree with the machine and just commit myself to agreeing that everything is "OK"? As the authoritative upraised hand might suggest, am I to fear that by choosing "OK" I might be harming the machine or the software? Above all else, if I attempt the procedure again (which I did), and I receive the same message again (which I did), what am I supposed to do?

Instructional materials have more far-reaching consequences than merely frustrating or confounding a user. James Paradis's study of the litigation that surrounded the operating manual for a direct-acting stud gun is but one example of the legal ramifications of written procedures (1991). Such discussions of the legal issues associated with instructions have given me pause as I have gleaned numerous documentation over the course of my research. Possibly the most frightening example was the procedure for cleaning a rifle. In the fourth step of the procedure a warning (all in upper case letters) stated:

4. NEVER WORK ON A LOADED GUN!

It appears that the authors of these instruction assumed the user would make it to step 4—a deadly assumption, given the context.

Curiously, with the vast number of texts dedicated to instructing humans *how to do*, the potential legal problems surrounding them, and the significant effort that several disciplines have devoted to the study of instructional materials, it is ironic that the reputation of these texts remains so dismal. Try to remember the last time you heard someone say, "Wow! Those were really fantastic instructions!" Chances are you can not recall such a statement, and if you can then it was probably pure sarcasm. Why have instructions received such a "bum rap?" Why are they used only when "all else fails"? Why do instructions seemingly fail to do what they are supposed to do—instruct humans on how to use technologies?

[2] The term *dialogue box* is curious in that it actually represents a monologue presented through the computer. Probably this terminology occurred because the computer has been characterized as an intelligent machine that communicates with people. At the same time, it also appears to be a reflection of a false sense of the reality of human–computer interaction, discussed in chapter 4.

Part of the difficulty with instructional materials is due to their deceptive simplicity. Instructions are meant to make the assembly or use of an artifact appear simple: they are a masking device for the complexity of systems and artifacts. Instructions are expected, as the renaissance courtier would have said, to possess *spretzaturra*—the ability to make the difficult appear mundane. This aim of instructions is primarily persuasive, as it can promote the commercial marketing of the technological products they accompany because it makes them appear easy to use. Ironically, the apparent ease of use that the texts display makes the instructional documents themselves look "easy." They look simple, so they must be simple to write (and simple to use, thus invoking the image of "idiot-proofing," discussed in chapter 3).

Further, instructional documents are what visibly reside between the user and the black box of technology. Therefore, they play the role of messenger in the transfer of the technology to the user and are more apt to be blamed for any breakdowns that occur during use because they are present at the moment of frustration or breakdown. As such, instructions are intruders into users' spaces. Humans have a proclivity to approach technologies without instructions in hand (literally). We have a strong desire to "get going" when we are using new artifacts, and instructional texts only intrude on our mission. Technologies are *means* to our ends, after all, and the activity of using instructions can be an additional barrier to achieving our goal. That is, when we use a technology we are too busy trying to figure out how the artifact can help us obtain our goal, and the instructional document that accompanies the artifact is perceived as a time-consuming nuisance.

Perhaps, as commonly suggested, the poor reputation of instructions can simply be blamed on bad writing. Many technical communications textbooks proclaim that the lack of clarity, brevity, or syntactic quality can be a major downfall of effective instructions. The evidence is too strong, however, to blame the monolith of poor writing as the essential problem of unusable instructions. As I mentioned at the opening of this chapter, readers can often comprehend the syntactic structure of the vocabulary used in instructional materials, but they cannot complete the intended actions that the instructions are meant to support.

Indeed, the aforementioned problems contribute to the difficulties we have with instructional documents. However, as accurate as these complaints may be, they do not actively address the fundamental *end* of usable documentation: the use of technology by users.[3] Instructions have been written with the artifact or system (or the marketing of the artifact or system) as the focus, and the *end* of written instructions has thus been to merely describe the "thingness" or sleekness of artifacts and systems instead of their situated use. *There is, in short, a deeply imbedded assumption that instructional materials are adequate merely because the information is there in either print or online form.* Never mind where or how the instructions will be *used*, this assumption dictates; the fact that users have a text in front of them is enough. Ironically, almost insidiously, this assumption places virtually the entire burden of comprehending instructional text on the user.

For instance, in a recent study of medical patients, it was concluded that there is something akin to a literacy "epidemic" in regard to the information with which peo-

---

[3] Here I am referring to the "end" of technology as being in the user—see chapter 2 for an elaboration on this concept.

ple are provided as they seek medical care. When patients at two large urban hospitals were studied concerning their ability to comprehend written materials provided by medical personnel, it was concluded that many patients suffered from "inadequate functional health literacy" (Williams, et. al., p. 1677). As the researchers explain, "Functional literacy is the ability to use reading, writing, and computational skills at a level adequate to meet the need of everyday situations. Functional literacy varies by context and setting; the skills of a patient may be adequate at home or work, but marginal or inadequate in the health care setting" (pp. 1677–78). In the study, they found that patients had difficulty comprehending, remembering, or acting upon medical information contained on such things as appointment slips, birth control instructions, prescription dosage information, and consent forms. With the exception of the consent forms, which the researchers admit are difficult to read due to legal concerns regarding malpractice (see p. 1681), the study contends that most of the documents used in the study are simple texts that are written at or below a sixth-grade comprehension level. In other words, the assumption here is that regardless of context (even though these authors seem to admit the problem of context in their aforementioned definition of functional literacy) the "medically illiterate" are just plain unable to read and write. I guess the same can be said of the "computer illiterate," who cannot understand simple texts like the error message shown on page 288. The information from within the medical black box is made visible, but those who cannot comprehend it are illiterate. Heaven forbid that the system should be changed to expose the black box! The knowledge of the system is assumed to be correct in the present arrangement. In addition, the users have no business taking part in the decision-making processes of the system, because they do not understand the "complexities" of it in the first place.

"User beware!" is the appropriate slogan. The user is relegated to the position of a one-way receiver who has little knowledge of the technology itself or how the technological system might be refigured through an active negotiation of designers, producers, and users. Instead, the situated activities associated with use are supplanted in favor of the static, *correct* description of technology, *ala* the knowledge of the "expert" who designed and developed the artifact. Thus instructional materials have, innocently or not, played a significant role in the continuation of the modern technology myth that the role of experts is to invent, while the role of novices is to await, with baited breath, the perfectly designed artifact. The complicity of instructional documents with this top-down, expert-to-novice ideology has a number of crucial ramifications that have been mentioned in previous chapters regarding the determinist tendencies of technology development, the perception of users as "idiots," and the use of the word "human" in technology development to give credence to technology-driven initiatives. Most obviously, a disproportionate binary is created that at once empowers the developers and disseminators of technology while it disempowers the users who receive the technologies at the end of the production cycle. In this chapter, I will argue that instructions are more often than not an afterthought of the technology development process. While this in itself is not a revolutionary claim, the consequences of this arrangement are substantial.

I have chosen computer documentation as an example of instructional text development for several reasons. First, the problems associated with instructional text have been magnified as a result of the personal computer. The computer has emerged into our culture through numerous contexts, many of which include people who have had no previous experience with computer technology. Even though these users may be highly educated, they received little, if any, formal instruction concerning computers in their traditional educational experience. Consequently, computer documentation has become the primary vehicle for educating this wide variety of users about this rapidly developing technology.

Second, computer documentation resides in more than one medium (print and on-line forms), and thus further complicates the challenge of user-centered theory. Such multiple media involvement opens doors to a number of fascinating and fundamental questions of reading, writing, text design, human comprehension, and the concept of self, to name only a few. In particular, the differences in media illuminate problems with traditional instructional genres. As we shall see later, traditional genres often do not address the needs of users in the ever-changing contexts of computer-mediated media.

Third, computer documentation writing is arguably the largest source of employment presently for technical communicators. This in itself offers more than one reason to seriously pursue computer documentation as it brings to the front questions regarding the practice of writing, the role of writers in industry, and even the position of power that technical communicators hold in institutional settings. For instance, technical and scientific writers have traditionally been defined by the discipline they work within—such as medicine, environmental science, industrial technology, and so on. The computer, however, complicates these divisions as many writers now perform functions that were once primarily within the jurisdiction of the computer documentation specialist. Most technical and scientific writers, regardless of their specialty, write computer-related instructional materials in print and on-line forms because their audiences are increasingly using the computer medium as a text.

Fourth, computer documentation is a marginalized text in the sphere of academic research. By that, I mean it has seldom received the serious theoretical, historical, or general scholarly attention that many other forms of texts have received.[4] Here I am not just including literary texts (like novels, poems, short stories, diaries, essays, etc.) as the favored sites of analysis, but I also include scientific articles, laboratory reports, legal documents, and grant proposals. These texts have been given rigorous attention at one time or another by scholars in a variety of disciplines.[5] Computer documentation is one of the most complicated forms of text we possess—both in terms of product and process. It simply deserves a serious analysis apart from the usual "applied" discussions that usually revolve around it.[6]

Finally, computer user documentation is a valuable lens, not only for the study of the texts themselves but also for studying the users who use them and the constituent

---

[4] The most notable exceptions are studies like Stephen Doheny–Farina's or Roger Grice's about documentation processes in industry settings.

[5] A short list of work done with these other forms of "nonliterary" texts would include Greg Myers' *Writing Biology*, Bruno Latour and Steve Woolgar's *Laboratory Life*, Bazerman's *Shaping Written Knowledge*, Yates' *Control Through Communication*, Dorothy Winsors' (and others) work on the Challenger disaster.

[6] Apart from the practical discussions of writing computer documentation, most research using computer documentation as a textual "site" has employed it in the service of explaining visual presentation of text (see Kalmbach or Koselnick), or as a model of products that exemplify usability methods (see Dumas and Redish).

cultures[7] that arise/evolve from the activities associated with computer technology. It is, in fact, the social arrangements and processes of computer documentation development that will provide much of the focus for the upcoming discussion.

In this section, three approaches to computer documentation development will be discussed: system-centered, user-friendly, and user-centered. These three terms have been borrowed from various places, but the use of them as a taxonomic device to describe computer documentation development processes is my own invention. Therefore, a definition of each term as it applies to the current discussion of computer documentation theory will be provided here.

ARTIFACTS, EXPERTS, AND IDIOTS: THE SYSTEM-CENTERED VIEW OF COMPUTER DOCUMENTATION

This view of computer system development has dominated the field of computer science for many years, and it has driven much of the technological development that computer systems have enjoyed in recent years. The reality that computer systems and applications need user documentation, however, has sprung from the recent phenomenon of the microcomputer and the proliferation of end users. System-centered design practices, in contrast, have evolved from an earlier time in computer technology where users were experts in the use of the systems, thus had little need for instructional texts in computer usage.

Documentation that was developed from these system-centered practices was aimed more at the static description of the computer's components and features than it was at any kind of end use of the computer. A now classic example of this is the huge library of documentation that accompanies UNIX, an operating system for mainframe computers. The documentation for UNIX (both its print library and its cryptic on-line commands and error messages) has been designed from a system-centered perspective that assumes more than a casual familiarity with UNIX. The major sections of Volume 1 of the UNIX documentation (see Figure 6.1) depict the system-centered focus on system features instead of user tasks. The documentation in the system-centered approach, as exemplified by the UNIX system, is a literal documenting of the static system: a description of the system's features removed from any context

1. Commands
2. System calls
3. Subroutines
4. Special files
5. File formats and conventions

Figure 6.1. Section headings of a UNIX manual.

[7] "Culture" is used broadly here to mean any community that might have common bonds due to context or practice: that is, workplace cultures, classroom cultures, and even electronic cultures such as those emerging from the use of the Internet or other computer networked arrangements.

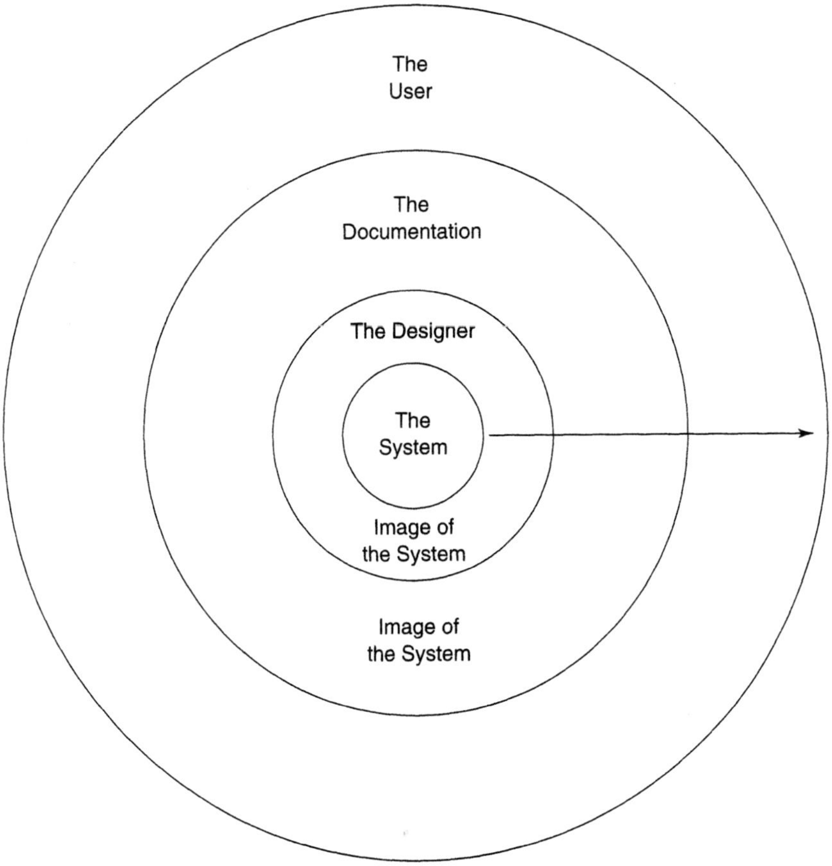

Figure 6.2. The system-centered view of computer documentation.

of use. The issue of where the computer will be used, who will use it, and for what purpose(s) it will be used is assumed in the system-centered view. A consequence of this approach for user documentation has been to place user documentation far from the center of system development (see Figure 6.2).

System-centered documentation places the needs of the technological system at the center and treats the system as the source of all knowledge pertaining to the development of documentation (as the arrow indicates). This view assumes that it is not only focal, but that the system exists before any documentation exists. There is no need for the user to have an impact on the central concerns of system development, this black box perspective of technology suggests, because the system is too complex and therefore should be designed and developed by experts who know what is most appropriate in the system design. The image that radiates from the system and eventually finds its way into the documentation, then, is an image that is envisioned by the system designer. Consequently, it is commonplace in system-centered computer designs to find cryptic error messages, long strings of difficult to memorize commands, and system components hidden from the interface that could prove potentially helpful to the uninitiated user.

From this designer's image follows the *documentation image of the system*. Here the documentation is written (often by the designers themselves, at least in draft form) to reflect what the designer views as the important components of the system. A common practice in this documentation development process is for writers to collect the "knowledge" of the system from the designers through notes written by the designers, or in interviews conducted by the writers. Consequently, the documentation is written to reflect the image of the system designer (refer to Figure 2.1 for an earlier description of the general system-centered process). The user, then, far removed from the central concerns of the system design, receives documentation that contains a system designer's image of the system.

I should explain that there are few published advocates of a system-centered approach to computer documentation, at least among professional writers.[8] Most technical communicators who write computer documentation are concerned with the needs of their audience—users—and therefore would not advocate a view of documentation development that places the user on the periphery. Unfortunately, technical communicators are often forced to be part of the system-centered approach because it is mandated (consciously or unconsciously) by the institution or organization they work within. Consequently, technical communicators write documentation at the end of the system development process and have little choice but to create system-centered documents.

Caught in a loop of technological determinism, system-centered documentation reflects the system designers' description of the system's features and its anticipated actions. The users' views of the system are irrelevant because they are not only figuratively but literally removed from the system development process. System-centered computer documentation has been viewed by developers as a necessary evil that must be created for the "idiots" who cannot figure out the system through their previous knowledge or trial and error. Put simply, system-centered documentation is the result of the deeply embedded dominance of system-centered philosophy in the world of computer technology development—a view that is far more technologically than humanly driven.

TEXTS, READERS, AND "REALITY": THE USER-FRIENDLY VIEW OF COMPUTER DOCUMENTATION

User-friendly is a term that has been applied to technology usage in general (see chapter 2). In terms of computer technology, it refers to a computer product that has been designed as an attempt to fit the product to the background and expectations of the end user. Familiar icons on the screen to replace strings of commands, the advent of computer screens that look like the actual printed page ("what-you-see-is-what-you-

[8] One published example of a system-centered approach to documentation is *Mastering Documentation* by Paula Bell and Charlotte Evans. Most system-centered documentation is produced in-house (and thus proprietary) with little or no published material explaining the process. The main evidence of the system-centered process remains the textual evidence it produces, like the UNIX manuals discussed previously.

get"), and the development of hand-operated pointer devices (i.e., the "mouse") to replace most actions done through a keyboard are a few examples of how computer interface designers have begun to make computers more accessible to a larger number of users.

The user-friendly approach to documentation development is characterized by an emphasis on the clarity of the verbal text, close attention to structured page design, copious use of visuals (often computer "screen shots"), and a warm, sometimes even excited tone that "invites" the user to enjoy learning the new computer system or software application. On a spectrum that places the system-centered computer documentation view on one end and user-centered views on the other, the user-friendly view would be in the middle, although in truth it is closer to the user-centered end. This view has more interest in the user than the system-centered view, and undoubtedly it has improved the quality of user documents to a great degree. The user-friendly view, however, still assumes that the design of computer systems is primarily the charge of the system designers and developers. That is, the system is assumed to be complete in the user-friendly approach, and user-friendly documentation is viewed as the vehicle for carrying the "reality" of the system image to the user. Therefore, user-friendly computer documentation has most often focused on the verbal and visual quality of texts that support the description of the system and the readability[9] of the text itself (see Figure 6.3).

The user-friendly approach is largely an outgrowth of the intensive work that has been carried out in recent years pertaining to document design and readability. During the past two decades, document design researchers have argued that documentation should be designed in accordance with a process model of document design (Felker, 1981), and that the functional nature of the texts should be accounted for through careful structuring of their verbal and visual elements. From this perspective, the user-friendly development process begins approximately at the same stage of the system development represented by the third ring of the system-centered view in Figure 6.2: the documentation, or text, image. Consequently, the user-friendly development process concentrates almost solely on text that has been written in accordance with the system designer's view of the system where there is little or no early user analysis before the documents are drafted. Most of these approaches also lean heavily on the traditional genres of user documentation, such as tutorials, user guides, reference guides, on-line help, computer-based training, and so on. This focus on text is clearly evident in most of the computer documentation textbooks currently on the market (Grimm 1984; Price 1993; Brockman 1991; Weiss 1991; Price and Korman 1993; Low, et. al. 1994).[10]

---

[9] I use the term *readability* in a strict sense, indicating the use of formal measures of determining the quality and comprehensibility of text, such as the Gunning Fog Index or Cloze tests.

[10] I do not mean to imply that these texts completely ignore the user. Nearly all of the prominent how-to books for computer documentation writing include chapters devoted to methods of audience analysis or to brief discussions of user testing. The audience analysis methods, however, are usually of the "audience invoked" variety (see Ede and Lunsford 1984), which calls for a writer to imagine the needs of the user. User testing is most often presented in a late chapter as a way to test already drafted documents.

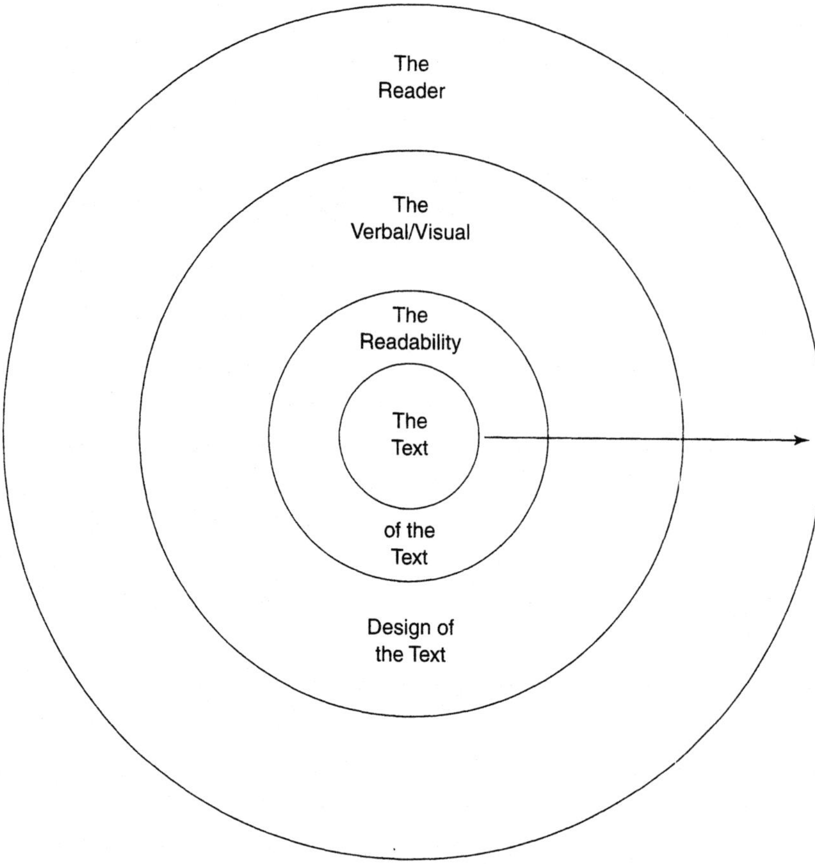

Figure 6.3. The user-friendly view of computer documentation.

Radiating from this central concern of the text lies the readability of the text. The user-friendly approach emphasizes the importance of reader comprehension and learning and often advocates either readability measures or the results of previous readability tests to measure the validity of the texts that they are developing (see Duffy 1985 or Duin 1989). Great importance is placed on the coordination between the verbal and visual elements of user documents (as seen in the third ring of the user-friendly view, Figure 6.3). The integration of text and graphic is certainly not new to technical communication, as some of the earliest general technical writing textbooks spent considerable time on this practice (see, for instance, Mills and Walter 1978; Lannon 1997), and it has been the source of much important scholarly work in recent years (see Barton and Barton 1989; Tufte 1983; Killingsworth and Sanders 1988).

Finally, the fourth ring of the user-friendly view attempts to account for the user as a reader who is situated in a particular occasion of text usage: readers who have particular needs that can be fulfilled by documents designed to meet that occasion.

Traditionally, the situation of the reader has been defined in terms of *learning* or *doing*,[11] and the texts that result from this interpretation of the user's situation fall into two categories: those that support learning (i.e., print genres of tutorials, user guides; on-line genres of computer-aided training or tours), and those that support doing (i.e., print genres of reference and quick reference; on-line genres like "Help").

The user-friendly approach has many advantages over the system-centered view, due to its focus on visual design and the needs of the user as reader: differences that have greatly improved the textual quality and user reception of computer user documents. The approach is a natural outcome of the historical movement toward creating more usable texts. It does, however, have at least two chief drawbacks when applied to computer documents. To begin, because this approach is based on reading, it focuses on how well readers comprehend and follow printed text. Readability research has become closely linked to the quality of textbooks, and thus focuses on the comprehensibility, organization, coherence, and audience appropriateness of the texts' verbal content (see Duin 1989). This limits the results of such research to primarily the verbal elements of the text. Some researchers have expanded this view to encompass the use of visuals in user documents (Redish 1988), but most have concentrated on the reader's understanding of the verbal components (Charney, Reder, and Wells 1988). Such research yields valuable insights into reader behavior, yet it should be questioned in terms of how easily these findings can be transferred to the use of other media—primarily the computer screen. It is well known that people do not read computer screens as they do printed texts. Eye strain, impatience, poor resolution, and so on all play a role in the difficulties of reading the computer screen. In addition, users are not readers in the sense of using text in a linear fashion (i.e., reading from sentence to sentence, page to page, and chapter to chapter). Instead, users browse, access, skim, and jump from screen to print and back to screen again (Wright 1983; Sullivan and Flower 1986; Hartley 1985).

Secondly, user-friendly research points toward the active engagement between a reader and a text (see Charney, Reder, and Wells 1988), but it usually does so outside of the context of a user's actual situation of use. Even though the user-friendly approach attempts to account for the reader's situation (see the fourth ring of Figure 6.3), such research is often laboratory controlled for the purpose of collecting data that can be analyzed under the pressure of statistical significance. Laboratory controlled conditions cannot adequately reflect the complexity of a user's actual situation (see Duffy 1985), as the social contingencies brought about by organizational or cultural forces are rarely a focus of this type of research.

Let me reiterate that this is not a criticism of user-friendly research per se, because such things as readability measures for the specific purposes of determining comprehension and memorization of texts can be useful. Instead, I am questioning the findings that reader-centered, user-friendly research can bring to the problems of designing documentation for *user* needs. Users often act differently than readers (Wright 1989), and a theory of user-centered design should focus on how these differences rearrange our view of documentation. This means, for example, that the tradi-

[11] See Sticht ("Understanding Readers") and Redish ("Reading To Learn To Do").

tional guidelines regarding the verbal and visual integration of text might be redefined to meet the needs of the multimedia of user documentation. The medium can cause differences in the use of visuals, the format of the written components, and the quantity of information to be contained on each page or screen (see Bernhardt 1993).

To build upon and then move beyond user-friendly conceptions of user documents—toward a user-centered approach—means, in part, to complicate the documentation production process. Instead of looking at users as being merely active readers of text, user-centered design must ask questions of the user's situation, the medium of the documentation, and the organizational and cultural constraints placed upon the user and the documents. As I have already said, the research aims of the user-friendly perspective are driven by traditional conceptions of reading, and, as the following discussion of the user-centered view will emphasize once again, users are not just involved with the act of reading. Instead, they are involved in a complex of discourse that incorporates reading as one of a host of communication, social, and technology interactions.

CONTEXT, NEGOTIATION, AND THE MEDIUM: THE USER-CENTERED VIEW OF COMPUTER DOCUMENTATION

The user-centered view is philosophically and practically at the opposite end of the spectrum from the system-centered view. The philosophy of user-centered computer documentation places the user at the center, thus arguing for the user as the driving force in the process (see Figure 6.4). By placing users at the center of the approach, neither the text nor the system are dominant features as they were in the other views. Instead, the focus of documentation development is placed on the user. In addition, this view envisions the user as situated in a particular time and place: the user is not using the documentation to learn software abstractly, but rather is learning the computer application for a specific purpose or purposes.[12]

The core of the user-centered view, then, is the *localized* situation within which the user resides. For example, a computer user may be using a word processor to complete his or her work. The act of using the word processor, however, is not a localized activity but rather a generalized description of what the user is doing. A localized description of the user's situation attempts to explain the specific work-related activity that the user is involved with, such as the role that the user holds in an organization or institution. Thus, the localized user of a word processor would be described as a "bookkeeper," a "teacher," "a nurse," "a documentation writer," and so on. The specific nature of users' work, then, drives documentation that is customized to the context of use instead of generically explained to a universal user who is the construct of

---

[12] I should emphasize here the approach that I am advocating is primarily suited to computer users who are conducting what Bodker would call "purposive human work" (*Through the Interface*, 1991). It also is akin to the philosophy of work-oriented design of computer artifacts as described by Ehn (1988) and others of the "Scandinavian Approach" to technology design (see chapter 4). Thus, the computer documentation I will focus on will primarily be for users involved in a work activity, not those using computers for entertainment purposes.

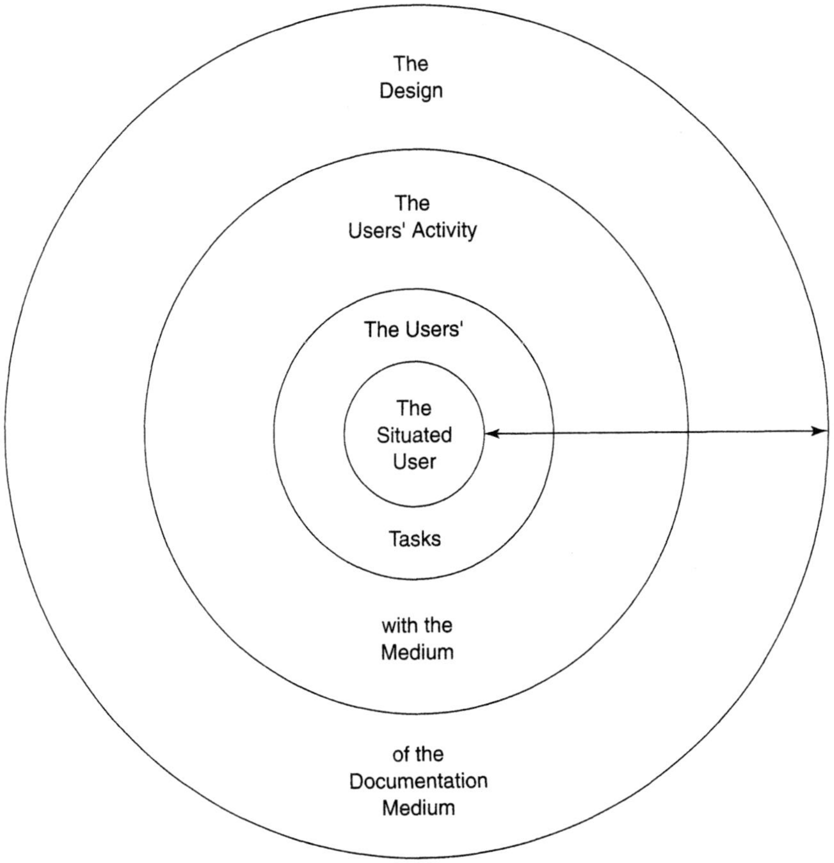

Figure 6.4. The user-centered view of computer documentation.

a writer's imagination. The focus on the users' situation also nudges the writer away from an immediate emphasis on the medium (which is still a very crucial component, as we shall see), and instead forces an analysis of the user as being central to the documentation process. This removes the writer from a premature preoccupation with the drafted document, as can be the case with readability or late-in-the-process usability tests where there is typically a relatively complete document in existence.

The user-centered view continues outward by taking into account the *tasks and actions* the user will be performing as a result of the users' situation (see the second ring of Figure 6.4). The focus on tasks within the user's context separates this approach from the system-centered approach, because the use of the computer is portrayed through user activity, as opposed to system features or system actions. A user-centered interpretation of a user's use of the word processor would be "writing a proposal," "laying out a page," or "producing a brochure." All of these activities could be considered "using a word processor," but the general activity depicted in the center of the user-centered scheme is too generic to accurately describe the actual activity of the user, especially concerning the documentation that will be written to help the user achieve his or her specific job. A more specific analysis of actual user tasks, as they are conducted within the context of use, is necessary.

I want to make clear that my concept of user tasks is considerably different than has been presented through *traditional task analysis* (see Bradford 1984; Weiss 1991; Barker 1998). In a traditional task analysis, a writer refigures the description of system actions into a vocabulary that is more relevant to the user's mental model of what the system is doing. Once a hierarchy of overall tasks has been outlined, writers delineate the subtasks that move the user from the start to the completion of the overall tasks. Traditional task analysis, which historically emerged about the same time as user-friendly approaches to computer documentation, was a significant improvement in documentation planning because it forced a user perspective into the system and software development processes.

The difficulty with traditional task analysis in terms of a user-centered perspective, however, is twofold. First, the user is relatively unsituated because, although traditional task analysis does move toward an understanding of user actions, it does so generically, not locally. Based upon a rational description of how the user *should* act, traditional task analysis merely reflects the anticipated actions of an idealized, logical user. The task analysis approach, in other words, is based on determining generalized knowledge that often ignores the local, domain knowledge of the actual, localized user.

Second, the tasks in traditional task analysis are still dictated by the system. The conception of traditional task analysis is embedded in a hyperrational approach to systems development: an approach that presents a completed system to the user with virtually no user input during any of the development phases. To be blunt, traditional task analysis is user-friendly, but certainly not user-centered. To avoid the rational replication of the system's tasks for the purpose of forcing the user to mimic them, the user-centered approach couches tasks within the situation of the user and seeks answers to questions like, What tasks will the user be performing within the given situation? Are these tasks truly user tasks, or are they couched within the terminology of system features? Are the tasks placed in an order appropriate to the situation of use?

In the user-centered approach, a close analysis of other situations and user tasks goes well beyond describing the rational system-dictated tasks expected of the user. Instead, the analysis attempts to understand the irrational or contingent occurrences that users experience within their local, everyday spaces. For instance, it is important in user-centered documentation to illuminate fundamental characteristics of users' situations to describe those *cunning* solutions that users have developed for dealing with technology. User knowledge of this sort was discussed in chapter 3 in terms of *metis*, the ancient Greek conception of the cunning intelligence that users display in everyday contexts. In a similar vein, feminist sociologists have termed this type of knowledge "articulation work," and as Susan Leigh Star explains, it is a type of knowledge that can only be "captured" through observation of users in their everyday environments, or direct discussion with them in that everyday experience.

> Articulation work is the work of fitting things together in real time in the workplace. . . . [For example, in a medical setting] Articulation work may mean tending to a patient, repairing a sentimental mistake in a conversation with a family, or finding a substitute implement in the middle of an emergency surgery. You cannot predict ahead of time what

> might be needed because, by definition, articulation work is contingent and occurs in "real time." Some people's daily work, such as that of secretaries, homeless people, and parents, seems especially rich in articulation. . . . (Epilogue, ST&HV, p. 504)

To reveal these moments of user knowledge is crucial to documentation. It "gets at" the essence of human involvement with technology in a situated way that describes the productive nature of user knowledge within context. These moments of *metis* or *articulation work* depict users producing knowledge, or at least displaying knowledge that they themselves have constructed/produced in the past and are now using to perform in the present situation. Such localized, domain knowledge is unaccounted for through most computer documentation development processes, and, subsequently, the localized cunning knowledge of the work environment fails to surface in the written texts themselves.

Beyond the task ring represented in Figure 6.4, before the medium of the documentation begins to play a central role in the design of the documentation, is the *users' activity with the medium*. One of the most complicated aspects of computer documentation specifically, and of instructional documents, generally, is that they are *used in conjunction with the act of using a technological artifact or system*. Consequently, it is important in user-centered design to determine which medium will best fit the particular user situation and tasks. For computers, this distinction has become especially relevant since the advent of on-line information and has been the source of numerous questions for documentation writers. Whether the documentation should be delivered in print or on-line is a question asked most often in discussions of documentation development. Will the user be better served by a print document, on-line document, or both? Will the user be using the medium in a linear or random manner? Is the medium capable of providing the user with a familiar model? For example, are the on-line system capabilities of the system too embedded within the interface, and thus users would be better served by the more familiar medium of print?

In addition to a choice of medium, the type of activity that the user is engaged in must be assessed. In the Discourse Complex of User-Centered Technology (see chapter 2) the three general categories of activity were described as *doing*, *learning*, and *producing*. In the context of computer documentation, *doing* describes activities where users are not reflecting[13] upon their actions for the sake of long-term retention. For instance, intermittent users of a certain piece of software may just want to be "refreshed" concerning the procedures of an action, but they will have little reason to retain that information because they rarely perform the task. Instructions that support the role of *doing*, then, usually will be less elaborative, more action-centered, and highly specific concerning the nature of the users' situation.

---

[13] The term *reflection* as used here is not meant in any way to indicate that users are unreflective, or that the ethics that might be involved with reflection on one's actions is absent from such a scenario. I merely mean that users are doing tasks that must be accomplished and that there is little time for memory retention. A good example of this are the pull-down menus of computer interfaces. People often cannot remember under which menu item a particular command exists, but once they are placed into a context of use (e.g., they are actually using a computer interface) they can immediately go to the correct menu choice. They have not taken the time to learn the actual placement of certain commands, but they nevertheless are knowledgeable users of the system.

The term *learning*, in the computer documentation realm, is revised to *learning through doing*. This change in terminology accounts for the problems associated with learning about computers while simultaneously using the computer: a paradoxical situation where you are compelled to learn (maybe because your job depends upon the new technology), but you are actually more interested in just completing the activity at hand.[14] This complicated paradox of learning and doing is, nevertheless, a real problem for users, especially those who are experiencing electronic technologies as they emerge into existing domains, such as the workplace or classroom. For example, if a medical technician is using a computer for the first time to analyze results of a blood test, then the technician is bringing expert knowledge of one domain (blood test analysis) to a new domain (computer technology) where the technician has little or no knowledge. For a technical writer who is determining what to include in the documentation for the blood test analysis software, it can be unclear which domain of knowledge is affecting any breakdowns that might occur.

More importantly, the user (the medical technician in the aforementioned case) is learning to carry out tasks *at the same time* he or she is doing the tasks. The scenario of this activity would be that the user is, at one moment, using the artifact (the computer), "reading"[15] the documentation (print or on-line), and attempting to extrapolate data that is meaningful in his or her knowledge domain. To account for this complex of activity, I have expanded the term learning to *learning through doing* to better represent the activity that users are engaged in while learning computer software. Documentation is compelled to manage this dichotomy for users through the careful design of print and on-line documents that can accommodate learning *and* doing simultaneously.

Producing describes two specific activities of users as they are involved with documentation development: activities that are not accounted for in either the system-centered or user-friendly views because neither approach invites actual users into the development process. First of all, users in a user-centered approach actually take part in the production of the documentation. They are involved in such things as initial design planning, iterative evaluation and testing of the documentation or software, and finally the decision making concerning the implementation of the documentation in their respective contexts of use. Second, users are producers in the sense of knowledge production (see chapter 3). What they know about the technology in question—its documentation, interface characteristics, and situations of use—are necessary prerequisites for involving users in technology development and decision making. The user's physical presence in technology development is not the only thing of importance; the knowledge of the user must be valued as well. In a

---

14 Mary Beth Rosson and John Carroll refer to this as "the paradox of the active user." Their observations of users showed that users want to learn, but often are more interested in just "getting on with it."

15 Reading is in quotation marks to designate the problems of reading discussed earlier in terms of the user-friendly view. I contend that computer users do not read in any conventional way, and thus the concept of reading is problematic. This is true if we are designing documentation that we anticipate as material to be read as if it were an "armchair" document or a textbook that one reads in a library carrel, removed from the context of technological use.

user-centered approach to documentation, the *producing* user is crucial because the process of user-centered documents is a collaborative, negotiated affair that is perpetually building the knowledge of users and their various contexts into the products being developed.

In itself, the social interaction of users, developers, writers, and marketing specialists should be adequate reason for adopting user-centered approaches to many different forms of technology development, including documentation. Unfortunately, the act of involving users so intimately in the development process is met with much resistance. One barrier is the already mentioned schism between experts and novices. Users, because of their role as the "unknowing subject," are deemed unworthy of participation in the highly specialized world of the technology developer. More likely, however, the reason for noninvolvement of users is the perception that it simply is too costly. Such arguments concerning profits margins have merit, but only within the context of short-term economic strategies. Certainly it is more expensive to commit time and resources to user involvement on a per project basis, especially if each project is seen as being unconnected to future versions of, or extensions of, the product. User input, though, is something that has value far beyond the present moment of testing a particular product. Capturing the knowledge of users and tracking such information over the long term could only be beneficial.

The long-term advantages of allowing users to play a role in the production process are potentially great, however, and some research has pointed to the positive effects for financial reward and customer satisfaction when actual users are allowed to work with the production teams (Ehn 1988; Nielsen 1993). Full effect of user involvement in production, though, has never been measured because of our culture's proclivity to base most production decisions on short-term motives and outdated cost accounting procedures that focus, in a very limited way, on month-to-month (and sometimes even day-to-day) analyses of the "bottom line" (see Johnson and Kaplan 1987). User-centered approaches to technology development are, in part, counter to the short-term mind-set of business planning because the benefit of user involvement can best be measured over the long term.

At the outer ring of the user-centered view is the *design of the documentation*. User-centered designing of the medium is now possible because the prerequisites of the user's situation, tasks to be performed, and purposes for using the chosen medium have been determined. The placing of the design or writing of the documentation at the outside of the user-centered view should not be taken pejoratively, however. Rather, it is the end of a production process that is consciously driven by the needs of the user. This does not diminish the role of documentation. On the contrary, it is an enhancement because it argues for user documentation that is in the image of the user's model, not the system model or system designer's model. A user-centered approach, in essence, is a thorough form of audience analysis that is aimed at designing documentation that fits what a user *actually* does, not necessarily what we *think* he or she should do.

A final feature in the user-centered view is represented by the arrows that indicate movement in both directions, instead of the singular inward to outward movement of the other two perspectives. In the system-centered and user-friendly views,

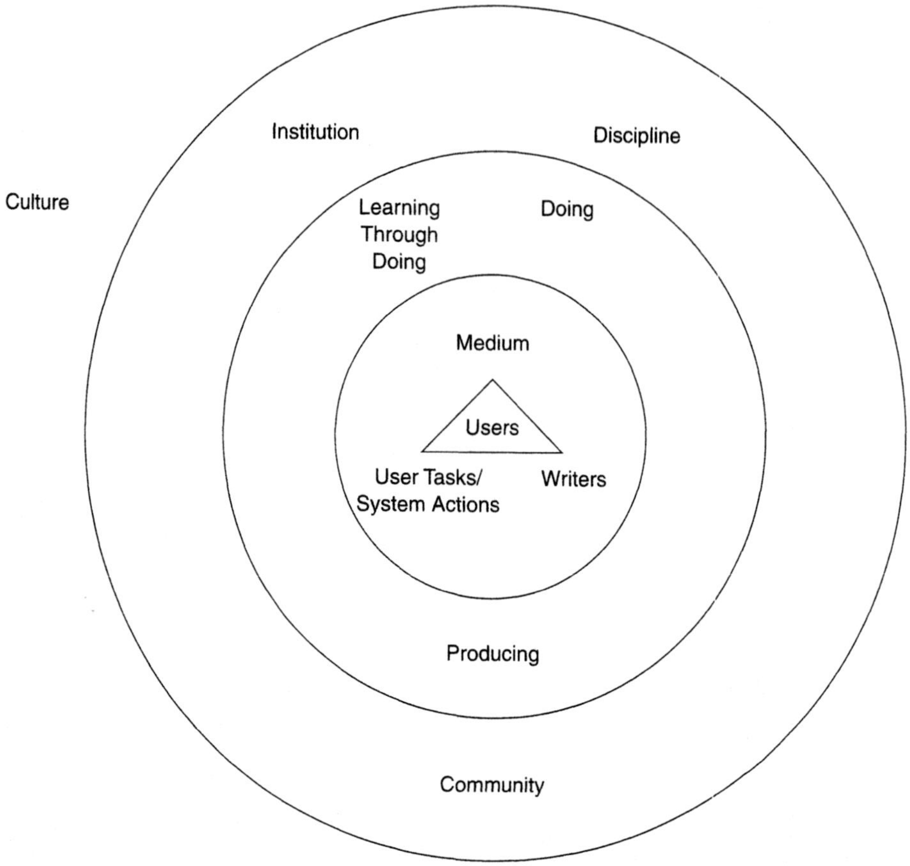

Figure 6.5. The user-centered rhetorical complex of computer documentation.

knowledge is put forth from the place of a singular authority (either the system or a text). These nonnegotiable views not only treat the user as a mere receiver, but they also exclude the user from an active involvement in the processes of production, as mentioned earlier. The two-way movement of the user-centered view depicts the recursive and negotiated nature of the user-centered approach to computer documentation, specifically, and technology, generally. The user-centered perspective invites involvement by the user throughout the process. The two-way movement attempts to disperse authority through a recursive process that is always in motion and always correcting itself, dependent upon situational contingencies. In an actual documentation production cycle, this recursive movement from user out to the medium and back again could occur numerous times—a truly iterative design technique.

To further elaborate the user-centered perspective, I will now turn to the user-centered rhetorical complex of technology (see chapter 2) for the purpose of revising it to meet the context of computer documentation (see Figure 6.5). I will begin by describing the major changes that alter the complex for the context of computer documentation.

THE RHETORICAL COMPLEX AND COMPUTER DOCUMENTATION

I do not intend to fully elaborate upon the entire complex here (for that you can turn to chapter 2). Instead, I will focus on either those elements of the complex that have been renamed for the purposes of defining user-centered computer documentation, or on those that need additional explanation because of their role in the processes of computer documentation production.

The artisans/designers of computer documentation are the *writers* who plan, design, and produce the user documents. There have been many discussions over the years pertaining to the label of "writers" as a meaningful designation for those individuals who create the instructional texts for computers and other complex technologies (see Haselkorn 1988). Companies have termed the producers of documents as information developers, document designers, communication specialists, or any number of other titles. The Council of Programs in Technical and Scientific Communication has carefully chosen the term *communicator,* in part to clarify that technical communicators do more than just work with written text. I do not wish to debate the relevance of accurately describing the activities that technical communicators do in their jobs: technical writers clearly do more than just write text. Nevertheless, given the potential shortcomings of the term, I have chosen *writers* to designate those who plan, design, and produce the various types of texts we call user documentation. Thus, writers become the artisans/designers in the complex in the context of documentation development.

The artifact/system is changed to *medium* to reflect the nature of the artifact that the writers work with. Documentation writers may be writing about computers, but the artifacts they create are the documents that reside in different media. Documentation, of course, comes in two primary media—print and on-line—but there are numerous variations of these media as a result of video, audio, animation, color, voice, telecommunications, and hypertext structures. Consequently, what was once the relatively "simple" and stable artifact of print is now much more complex, at least from a writer's point of view. Also, the advent of desktop publishing has made the role of the writer much more diverse than just a mere scribe or wordsmith who transfers the information about the system from the *system designers* to a text. Those writers who work with print technology are expected to produce camera-ready copy, a phenomenon that was rarely seen before the development of the laser printer and postscript technology. The medium, then, is not only an important factor for the user of technology but for writers who must meet the challenges of learning to "write" for different communication forms.

The terminology of the third point of the triangle, the *user tasks/system actions*, does not change literally, but the characteristics of this feature are of utmost importance in computer documentation and thus deserve some further explanation. As explained in chapter 2, the *user tasks* are the actions of use as perceived by the user. The *system actions*, on the other hand, are the actions that the computer system was programmed to accomplish by a system developer. The point at which these two aspects of technology meet is commonly called the *interface*, the place where the user meets the machine during the context of use.

The disjunction between user tasks and system actions is at the heart of the problem of user breakdown. Users often perceive their use of technology differently than the represented actions of the system. For instance, in fax machine documentation the action of sending a message to multiple recipients is commonly called "polling." While this terminology might be clear to the developer of the system, it is the rare user who perceives the action of sending to multiple recipients as "polling." Users would perceive this task as "sending to multiple recipients," or "distributing to more than one client." It is this visual and verbal description of the system (in print, online, or as a label on a control panel) that represents the use of the technology that is meant by the rhetorical element of user tasks/system actions. For writers, the crucial problem concerning user tasks and system actions is to understand the nature of these breakdowns and adjust the documentation (in the best of worlds, to fix the system itself) to represent the users' perception of tasks.

The aforementioned descriptions of user-centered concepts should offer some insight into the possibilities for application of the concepts to actual documentation writing practices. In any case, they offer a foundation for further discussion of what the impact of user-centered theory might be on documentation development practices. In the next section, I will use this foundation to discuss the issue of documentation genres—the types of documents that are written for users of computers—to investigate the nature of documentation within the rhetorical complex.

GENRES IN THE MAKING: THE CASE OF THE UBIQUITOUS TUTORIAL

Genres are taxonomic devices that provide order and meaning to many everyday artifacts. While genres can be used to categorize any number of artifacts, they have been used widely to categorize virtually all types of text.[16] When we pick up a newspaper, for example, we immediately have certain expectations about what we will be reading, how we will move through the text, and what we will do with the text once we have finished with it (wrap fish, of course). We know the category of text (journalistic) and what subcategory of journalistic text it is (newspaper). This tacit knowledge of text places it into groupings that aid us as we interpret it, and use it, in particular contexts.

Genres like newspapers have been around for quite some time: they have historically established context of use, and the expectations for these genres is generally understood by a wide audience. When traditional genres are resituated within a new

[16] For more on genre, see chapter 3 of JoAnne Yates' *Control through Communication*.

context, though, the revised use and purpose of the text can be the cause of a reformulation of the genre itself. This is currently happening with the genre of newspapers as they are put on-line over the WorldWide Web (although I hesitate to guess what the new genre might be called because the phenomenon is so new). The generic structure and purpose is fluid and being redefined as people learn how to use, and how they want to use, the new on-line medium of journalism. Thus, the situations we use texts within, and the technological medium of the text, have a great deal to do with how they are used and how they should be designed. In a discussion of technological artifacts as genres, John Seely Brown and Paul Duguid offer the following example of this phenomenon:

> No artifact is self-sufficient. The spot-lit, pristine artifact commanding the center of attention among the usual array of potted plants at a trade show reveals very little about whether, or how, or why it will or will not be used. But the artifact in the workplace, plastered with stick-ums or Scotch tape, modified or marginalized by practice, and embedded in social activities, can tell a rich, well-situated story. Consequently, in contemplating usability, designers cannot just consider isolated (and isolating) notions of functionality. Instead, they have to relate these to the socially and physically embedded practices. (p. 174)

Brown and Duguid go on to explain that in the case of genres, even though they may appear generally static, they are continually changing. "The underlying generic conventions, though well established and remarkably widespread, are not predetermined. They evolve locally and continually in practice. Change is prompted from many different directions" (p. 177). Practice, or use, is a determining factor in the development and recognition of genres. It is not the only determinant, but certainly the context of use frames genres in certain ways that make text recognizable to different users who are bound within a common situation, such as a community of users in a place of work.

Computer documentation genres are an interesting case to investigate concerning fluid, evolving genres. Documentation, either print or on-line, certainly is constrained within situations of use, and, as a genre, computer documentation should be perceived as mutable in terms of how it is designed and constructed. In addition, the user, as the primary agent of practice or use, also should be perceived as an agent in documentation design. In reality, however, the genres of computer documentation are rarely so flexible. Computer documentation historically has been designed to conform to predetermined notions of users and use: notions that come from a time when computers were designed for use by experts, or as we have seen in earlier chapters, for idiots. Thus, computer documentation is driven by preconceived ideas of what they should contain, even if those contents have little to do with the intended audience's needs.

As one peruses most general technical communication textbooks, or texts written specifically for the writing of computer documentation, it is evident that the genres of computer documentation (print or on-line) generally are broken into two overall categories—*texts for doing and texts for learning*. The *doing* texts usually are suggested for the advanced, or expert, user and include print documents such as reference guides, quick reference guides, and troubleshooting guides. The on-line forms of the *doing* texts are loosely placed in the large category known as on-line assistance or

"Help." Interestingly, the *doing* texts are found in a somewhat wide variety of shapes, sizes, and other design features. For instance, quick reference information has long been customized to meet the need of users ranging from computer users in an office setting to engineers using pocket-sized guides for information needed "in the field" (see Reitman 1988).

The *learning* texts, however, are not so flexible in structure. These documents—although subcategorized under labels like the user guides, guided tours, and computer-based training—are in most circumstances generically termed *tutorial*. The genre *tutorial* is supposedly well defined as a "use once, throw away" document—a no-deposit, no-return text that is supposed to get the user "up and running." Theoretically, the tutorial will be used in a linear, step-by-step manner as the user moves from one lesson to the next, each lesson presumably slightly more difficult or advanced than the previous one. Once the user has completed the assigned lessons in the tutorial, it is assumed that the user will abandon the tutorial and proceed either on his or her own volition, or with the occasional assistance of a reference guide.

In practice, however, tutorials are seldom used once and then abandoned. In the worst case, the tutorial is the computer world's epitome of the "when in doubt, use the instructions" attitude that began this chapter. In these cases, it is more common than not that the tutorial will remain in the protective shrink wrap, not to be used at all.

In the best case, the tutorial can become a long-term resource for the user. Because the user has become familiar with the structure, look, and language of the tutorial, he or she returns to it often to refresh the memory, or maybe to learn from a lesson that was never completed because either time ran short or the relevance of the lesson was unclear at that moment. Tutorials also can appear less intimidating because they often are shorter and less bulky than the large reference documents that portend to contain all of the knowledge of the system. Unfortunately, tutorials that are consciously designed to support long-term, user-centered goals are far less common than those designed for one-time use. In fact, the tutorials that are used in the long term probably were not consciously designed for that purpose. Instead they have been adapted serendipitously for long-term use by the user with techniques that use sticky notes, highlighter pens, and dog-eared page corners.

There is no reason, however, that tutorial documents could not perform a broader array of functions for users. First, however, the genre "tutorial" will have to be redefined. The concept of tutorials as *learning through doing* documents offers one perspective on how this can be accomplished. If learning documents such as the ubiquitous tutorial are re-thought through the problems of synchronous learning and doing—learning through doing—then they can become a more customized text in terms of design. The process of creating such documents, of course, entails analyses in accordance with the user-centered view (see Figure 6.4): the user's particular situation, specific tasks, and proclivity for and access to print and on-line media would be necessary. While I will not carry out such a process in detail here, I will use an example from a user-centered document that in many ways overcomes some of the difficulties of the use once, throwaway tutorial (See Figure 6.6).

Figure 6.6 is from a manual intended to introduce a new user to a database software (File Maker Pro® 2.1). The specific user of this document is a secretary who has the responsibility of keeping the records of applicants and current students in a graduate program at Miami University. The secretary's computer experience had for many years consisted of using an early Wang® system that had no graphical capabilities and limited word processing capabilities (in other words, it was not a what-you-see-is-what-you-get system). This particular secretary, however, was highly experienced with the context of this office setting, as she had held the position for nearly a decade before the new Macintosh® system was introduced.

As you can see from the two pages excerpted from the manual, the user's situation is central as the document has been customized to focus on just the tasks associated with applicants to, or current students of, the graduate programs. Thus, specific instances of entering information appropriate to this context are used in the examples of the manual. On the first page of Figure 6.6, the italicized information immediately to the left of the computer screen "shot" indicates how to size a box of "Degree Sought" information. This direct link to the user's specific context reduces ambiguity pertaining to the uses of particular fields in the database; it also creates a familiar context within which the user can immediately learn an action while actually accomplishing a necessary task related to her work.

Also notice that the information of "how-to" is explained in more than one way. At the top of the second page in Figure 6.6, for instance, the user is given three reasons for changing the layout of the screen. These three suggestions are then followed by an explanatory note that further justifies some of the choices for layout arrangements. It could be argued that such choices might interfere with a user's notions of completing a given task because the choices could lead to confusion on the user's part. In the case of this manual, however, the writers observed that she specifically wanted information that could be used more than once: she needed the document to be an asset in her different stages of learning the database.

Another user-oriented characteristic of this manual is the use of the two icons at the bottom of each page—one depicting a computer screen, the other a printed text. During observations of the secretary carrying out her work, and through interviews with her in the design process, the writers discovered that she never turned to the database's on-line "Help" for assistance. The old Wang® system did not have on-line Help, and consequently she had never witnessed on-line assistance in any of her previous experience with computers.

Through further analyses of the secretary and her use of on-line Help, the writers defined two problems related to the use of different media. First, they observed through user testing that when she was prompted by them to use Help she appreciated, to some extent, that the information was "at her fingertips." Despite this appreciation for the expedient nature of the on-line documentation, her proclivity was to use print documents when she needed assistance. To provide assistance to her that would account for this proclivity, the writers devised the print and on-line options at the bottom of the pages. They additionally felt that the icons would be highly visible and that, eventually, she might begin to use the on-line Help information with greater frequency if she was reminded of its presence through the icons on the printed pages of the manual.

The icons at the bottom of the pages also serve the purpose of providing the user with immediate feedback concerning the whereabouts of further instructional elaboration. As you might expect, this customized, situation-specific manual would not be

CREATING AND EDITING LAYOUTS

Changing the Format of a Layout

Each arrangement of Graduate Program Applicants information is called a layout. You can easily create a new arrangement of information or change the way information is displayed or treated in fields by creating a new layout or editing an existing one.

To create or edit layouts, change to the *Layout* mode.

Getting to Layout Mode

Get to *Layout* mode by

- choosing it under the **Select** menu

  or

- clicking the mode button to the left of the bottom scroll bar to bring up a pop-up menu. Select LAYOUT from this menu.

Layout mode looks like *Browse* mode, except you get rulers, section breaks, and editing tools for changing the size, appearance, and placement of fields.

The square corners around the "Degree Sought" field indicate that this field has just been selected. It may now be moved or resized with the cursor, or its appearance may be changed by selecting options from the pull-down menus.

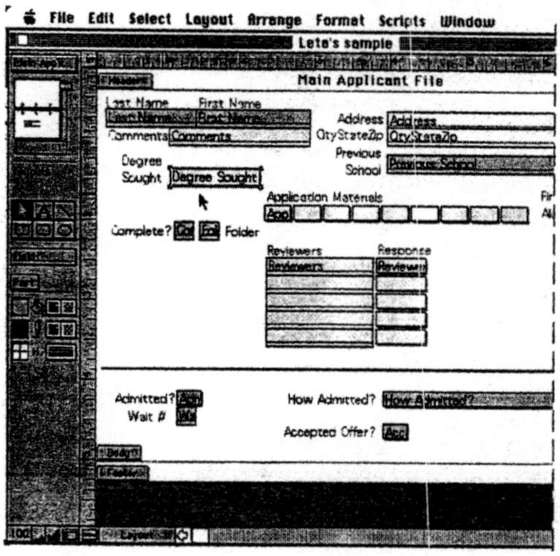

For On-Line help see

Designing Layouts

For help from manual see

Working with Layout Parts and Fields 3-27 to 3-59; Adding, Editing, and Formatting Layout Text 3-60; Moving Objects 3-82

Figure 6.6. Two pages from the database document.

APPLICANT DATABASE

Changing the Arrangement of a Layout

You may want to change a layout for several reasons. You may want to adjust it so it's easier to read on the screen, you may need to add or delete a field, or you may need to set up a layout for printing different versions of a report.

NOTE: It seems to work best to have separate layouts for viewing and printing, since features like colored boxes and field borders make reading information on the screen easier, but significantly clutter up a printout. Also, some fonts and font sizes will be more readable on the screen than others, but they may be too big or less readable when printed. Separate layouts for viewing and printing allow you to control the appearance and of your text with formats you set up once and use over and over. (For an example of a good print layout, see the section on compiling a list of applicants by program, p. 12.)

In *Layout* mode, items in a layout may be moved around simply by clicking and dragging on them with the pointer tool (the arrow). Use rulers, and grids to help guide placement.

Changing the Appearance of Text

To change the appearance of text in the layout

- Use the text tool (the A box) to select or highlight the text you want to change. Select the new appearance from the options in the **Format** menu.

- Use the pointer tool (the arrow) to select the fields you wish to change. Each selected field will be highlighted by small boxes in the corner. Select the new appearance from the options in the **Format** menu.
  (Holding down the shift key while selecting fields allows you to change more than one field at a time. You may also SELECT ALL fields on the **Edit** menu.)

For On-Line help see

Designing Layouts

For help from manual see

Adding, Editing, and Formatting Layout Text 3-60; Changing the Shape and Size of Objects 3-82;

Figure 6.6. (continued) Two pages from the database document.

expected to contain the extensive "complete" elaboration found in the commercial documentation accompanying the software package. To overcome this potential deficiency, page number references and topic identifiers were added adjacent to the icons. Through the page number references and topic identifiers taken from the industry-produced publications, the writers provided access to greater amounts and varieties of information: a feature that produces something of a "hyper" quality of the documentation as the user can move in random, yet linked, ways through the entire library of documentation for the software. Thus customized reference by specific page number and topic allows for integration of the entire set of user documentation for this database software—an integration guided by the actual situation of use.

The writers also found some interesting vocabulary problems that may very well have been the result of an expert-driven task analysis procedure, as discussed earlier in this chapter regarding user-friendly documentation. While investigating the problem of integrating the two mediums of documentation, the writers uncovered inconsistencies between the respective vocabularies of the existing print and on-line documentation for the database software. As they collected the task vocabulary for both the print and on-line information, they encountered quite different terminology for the different tasks. For instance, at the bottom of the first page of Figure 6.6, the on-line terminology is "Designing Layouts," but the print documents refer to this task as "Working with Layout Parts and Fields," or "Moving Objects." Instead of using one term for each task, the writers decided to offer the various renditions of the task vocabulary as they appeared in the different pieces of the documentation set. This feature offered a mini-index on the same page as the how-to "learning" information, making the "tutorial" a quasi-quick reference manual too.

It is somewhat easy to get involved with the analysis of texts and forget that there might be other issues at stake when we think about instructional documents. The title of this chapter, I believe, should be a constant reminder to technical writers that users have little affinity for instructional assistance: they will use it only when all else fails. Put another way, users of most technologies, and most pointedly users of computer technologies, often see documentation as something that can get in the way. The very presence of documents, whether print or on-line, presents the danger of users becoming disengaged from the learning/doing processes, because the documents can draw the attention away from the activity at hand.

An obvious consequence of this propensity by users to avoid instructions is that documentation, and those who write it, will remain invisible in the technology development and implementation processes. To a great extent this is already the case, as the perceived need for on-line texts is somewhat dubious. For example, I would argue that the tendency for some software manufacturers to put all of their products' documentation into on-line formats is driven more by economic concerns than by concerns for the users of the documents. Some on-line documentation is nothing more than the print documentation reentered into on-line Help with little or no attempt to alter the information to fit the use of the new medium. Such lack of interest in the design and use of on-line documentation, although not a new phenomenon, certainly does not bode well for its future as an important, viable component of everyday user technologies.

The invisibility of writers also is a problem as they are hardly at the top of the technology development hierarchy. Too often the role of writers has been "to write it up" at the end of the development process: a problem that clearly is a source of the system-centered, functionality-based documentation and interface designs that have dominated the personal computer industry for most of its short but influential history. In brief, the issue of empowering writers within their workplace contexts looms large in the domain of instructional text. In the conclusion of this chapter I will take up these two issues of invisibility regarding technical writers and the texts that they produce.

TO DOCUMENT OR NOT TO DOCUMENT?

As we already have seen in several instances, users face a double bind when they are confronted with the problem of learning new technologies through instructional texts. To engage with a technical artifact *and* a text at the same moment is a complex and frustrating task that illuminates the paradox of learning through doing. In their research on the minimal manual, John Carroll and his associates characterize this paradox in a way that is most useful in clarifying this fundamental difficulty of technological use. Carroll et al. differentiate between two types of learning procedures: *learning by the book*, and *learning by doing*.

> Learning by the book models the kind of learning process for which self-instruction manuals as a genre are designed. Learning while doing attempts to model procedurally the task-oriented skipping and self-initiated approach. . . which we have taken to be characteristic of real learning. . . . (Carroll 1990, p. 176)

In the previous discussion of the secretary's database, we saw how learning by the book and learning by doing might be confronted through documentation that challenges traditional notions of genre. The lenses used to design the new genres, however, may be rose colored; it can be argued that the presence of documents in the context of active learning is itself futile. Will users even use these documents that have been carefully crafted for their contexts? If texts get in the way of active learning, this skeptical perspective might ask, why create user documents at all? Why not just make the interface so usable that documentation is never needed?[17]

HyperCard® for the Apple Macintosh® is an example of this perspective that advocates the dissolution of documentation. Especially in its early days, Hypercard® was touted as being so easy to use, due to the graphical interface of the Macintosh® and the icon-driven HyperCard® application, that the documentation for learning Hypercard® was merely created as adjunct material hardly needed once a user had a very brief tutorial introduction to the application. Anecdotal evidence of this atti-

[17] This certainly is not a novel claim, as the eradication of user manuals through better interfaces has long been the goal of interface designers like Donald Norman.

tude toward HyperCard® can be given, but the following statement from a HyperCard® User's Guide provides some evidence:

> This ends your tutorials on HyperCard's basics. You know enough now to get off to a great start. You can go on reading in this manual to learn more about HyperCard or you can go off on your own, browsing and editing as you like. . . . Happy wandering! (p. 59–60)

One of the concepts of HyperCard®, and hypertext in general, is that it is a user-controlled environment, and therefore exploration, play, and experimentation are activities users can engage in to learn. Such a concept can be good. If a user is enjoying himself or herself, then it is probable that learning will occur. The depth of the learning, and the relevance for that style of learning within a particular context, though, is problematic. Play and experimentation would be fine when there is little pressure on the user to perform, but in a workplace or educational context some guidance and strategies for use probably are more appropriate. In the case of HyperCard®, for example, other levels of the software are useful, even necessary, for users (i.e., scripting, importing and exporting graphics or video) that would be difficult or nearly impossible to learn without some documented advice or guidance.

It does not seem at all clear to me that computer documentation will disappear. But we can be assured that the role, look, and medium of documentation will shift. It may even be the case that computer documentation will have to blend into ill-defined contexts, like those of intense collaboration between users as they learn networked electronic technologies. To ensure that these shifts are user-centered will take a concerted effort by writers to assure that it is not a determined shift: a shift dictated by the technological imperative that we hear so often in phrases like, "It just cannot be helped. This is the world we live in now!" In short, we should be ready to help define any shifting before the shifting pressures us into documentation practices that neglect user-centered goals.

BEYOND THE TEXT: WRITERS, WRITERS, EVERYWHERE

As the discussion of shifting genres demonstrates, the user-centered argument concerning computer documentation is not without roadblocks. There is one issue, however, that may be the most problematic of all—creating customized documentation for specific, localized contexts and the corresponding number of writers needed to design and develop the documents. Arguments for situating documents within contexts are not entirely new. Literacy advocates, for instance, have argued that literacy training is much more effective if situated within contexts relevant to the learner (Sticht 1985). This makes sense. Many people are more motivated if they can see a connection between what they are learning and the intended application of their new knowledge. Unfortunately, putting such localized ventures into practice is more easily said than done.

A most obvious difficulty is that writers who are knowledgeable in user-centered design must be hired to create these documents. This not only means that software development companies might hire these writers, but more specifically, it argues for institutions using the software to employ writers who can, on an ongoing basis, create documentation for the constantly fluctuating audience of users in the institution's employ. At present, such a scenario—where writers would be hired to create numerous custom documents—might appear absurd. How would institutions that are facing downsizing and other cost-cutting pressures make room for writers? There is no easy answer to this. But let me attempt to explain.

The usability evaluation and testing that must accompany user-centered design is costly, but, as I have already mentioned, only in terms of short-term management. These costs can be recovered over time through better design decisions that hopefully make the products more accessible to a wider body of users. This argument has already been made in several quarters in terms of "discount" usability (see Nielsen 1993). Most of these solutions, however, do not include technical writers. To include writers will mean efforts to recast writers. Writers will not only need to change the moniker that denotes them, but they will need to find avenues into other domains of technological design and development. For example, I have observed institutions that regularly increase the number of system administrators as computer usage increases within the organization. These administrators, in turn, take on various roles that include user liaison or software support. Writers could easily fulfill many of the functions of these system-trained administrators, particularly those tasks that involve the learning of systems. Companies and institutions that recast writers in new roles such as these could also benefit from the development of the print and on-line texts the writers would produce. The information about how to use the software would no longer be in oral (word of mouth) format, but instead would be formally documented, thus making documentation updates and more widespread, consistent training possible.

In addition, if institutions learn to value the worth of training materials more highly, especially materials customized to fit specific user situations, then writers will have more central roles in the maintenance of user systems. Surely many institutions have users who regularly underuse the computers in their domain because they are either unaware of the potential or are too strapped for the time to learn new features that would aid them in their jobs. Consistently updated custom documentation would be a great asset in such situations.

As documentation increasingly goes on-line, the role of writers should increase as well. Writers will certainly be called upon to enter the old print information into on-line formats, and this should be done, of course, through user-centered methods if there is any intention of creating usable products. In addition, writers should increasingly take on the role of interface designer. The interface, after all, is a "text" that resides between the user and the black box of the system—a multiple media text that includes documentation. Writers already are interface designers, in other words, but recognition of this technical communicator skill is rare outside the realm of technical communication.

A recasting of the technical writer, though, will ultimately call for an increased role in the decision-making processes of technological development. This case has already been made strongly for the user, both in this book and elsewhere (see Bravo 1993). Users are low on the proverbial totem pole and are rarely allowed access to the decision-making circle. Technical writers, I fear, are equally invisible in these circles where the design, development, and potential impact of technologies are (at least to some extent) determined. To employ writers in decision making is not that far-fetched; it is at least not any more far-fetched than the aforementioned argument to rethink writers as interface designers. Consider the following scenario:

The Washington Metropolitan Transit Authority was experiencing a problem concerning the opening of their newly refurbished subway system. The system is a wonderfully designed, efficient technology ready to whisk people across the Washington metro area. An especially important aspect of the new system is the automated ticketing system. To use the system is, on a surface level, relatively easy. You choose your destination, locate the cost for the ride on a chart, deposit dollar bills in the automated ticket machine, and then you receive a ticket with the appropriate amount for your destination. A glitch, however, was noticed in the new automated ticketing system. Maybe "glitch" isn't the right word, because the problem wasn't in the mechanics of the system—this worked perfectly well. The problem was in the anticipated confusion that new users to the subway would have learning how to use the ticketing machines.

Imagine that you are one of several thousand commuters using the new system on your way to work. You have never used the ticketing machine, and as you stand in line waiting to purchase the ticket, you read the instructions posted beside the machine. As your turn to use the machine draws closer, you realize there are actually a number of tasks to carry out in order to be certain you have purchased the correct ticket for the correct price. Befuddled, you are jostled by others in the line and eventually lose your place in line. Finally, you figure out the procedure, but now you must wait for the next train.

During design evaluations of the system, the designers saw the possibility of this learning problem. How were they to be sure that thousands of new users of the system, in a hurry on their way to work, would use the new ticketing system (paid for, in part, through taxes levied on the new users)? The answer was ingenious. The transit authority hired a number of people to act as "experts" during the early days of operation of the new system. These "experts" were trained to use the ticketing machines, and when the system opened on its first day, they were there to act as expert users so the new users could observe. These "expert" ticket buyers merely went though the lines time after time so others could watch them. They were not in uniforms or any other official clothing: they merely looked and acted like users who knew what they were doing and who the "novices" could watch.

The decision to solve the problem of the ticketing machines in this way could easily have been from the minds of technical communicators: people who have an affinity for and are advocates of users. I am sure technical communicators had something to do with the writing of the instructions that sit beside the ticketing machines, but I am less sure that writers would have been involved with the decision to create "expert" users who serve as models. Such decisions are traditionally left to the designers, engineers, and developers: a tradition that portrays the developers as decision makers and writers as scribes. Once again, we see the issue of knowledge rearing its head. Who has it? How is it produced? In the final chapter we will conclude by taking this problem of knowledge into yet another context—the academic sphere. Here we will see the possibilities, and the inevitable limitations, that exist in the realm of the technical rhetorician.

DEVELOPING YOUR UNDERSTANDING

1. **Explain the differences between the following two statements: the end of documentation is explaining X technology vs. the end of documentation is the use of X technology by Y users. As part of your discussion, explain how these two different conceptions of the ends of documentation affect the work of instructional writers/designers, such as computer documentation writers.**

2. Define the distinction between "readers" and "users" and summarize how the distinction affects professional writing.
3. Johnson claims that "instructions are more often than not an afterthought of the technology development process." Explain what he means, and describe how the resulting position of instructional writing within the technology development process affects what instructional/documentation writers do. Finally, describe how the writer's work might change if writing instructions were work integrated throughout the technology development process.
4. Explain what Johnson means when he says writers must be willing to "recast" themselves in order to overcome the limitations (e.g., increased costs) of user-centered writing. Assess what writers and workplaces gain and lose as writers recast themselves in order to increase user-centered writing.
5. Johnson, in the chapter's conclusion, argues that professional writers are "advocates for users." Explain what he means, comparing and contrasting this definition of the professional writer with whom professional writers are in both the system-centered and user-friendly models of writing.

Chapter 6

Projects

1. On your campus, the Web, or around town, locate a document (e.g., a brochure, product instructions, Web site splash page, or drug advertisement back page) that requires whole-document level revisions. Develop a revision plan based on your assessment of the document and produce a mock-up of the revised document. Do this work individually.

   Then in groups of three or four, share your documents, discuss your assessments of each document, share with one another your revision plans and mock-ups, and solicit feedback to improve revision plans/mock-ups. Based on your group feedback, work individually to produce a second mock-up for your document. Then, using the technologies available to you, produce as complete a revision as possible.

   Be prepared to discuss your original and revised document in class (bring both documents in electronic, transparent, or print-copied versions, depending on the technologies available in your classroom).

2. Locate a brief document, like a brochure, restaurant menu, flier, or set of simple computer software task instructions. Produce two revisions of the document you have chosen, one revision that employs less reader-oriented visual language than the original and one that employs more.

   In class, form groups of three or four. Share the three documents each member has found/produced, and assess what makes visual language more or less reader-oriented. Be prepared to present your findings and refer to your examples in class (come with electronic, transparent, or print-copied versions, depending on the technologies available in your classroom).

3. In groups of three or four, identify a helpful task in a software program that your classmates might like to employ. Adopting a user-centered approach, develop documentation that incorporates what you have learned in this chapter about good design.

4. As a class, choose two generally unfamiliar technologies for which you would like to learn and develop instructions. Divide the class into six groups: two system-centered design groups, one for each technology; two user-friendly design groups, one for each technology; and two user-centered design groups, one for each technology. Each group should develop a set of instructions for their technology and write clear project plans based on the design approach they represent. (All groups should have a shared due date.)

   After the groups have completed their instructions, they should pair-up so that one group represents one technology and the other group represents the other (it does not matter if groups are or are not paired according to design approaches). Now, each group is paired with another group that can serve as a test group for their instructions. Test the usability of each group's instructions. As a class, discuss what was learned about different design processes and how those design processes may or may not have affected usability.

CHAPTER 7

Professional Writers Produce Social Space

INTRODUCTION

Do professional writers produce anything more than user-centered texts? One answer to this question, which is the focus of Chapter 7, is that professional writers also produce social space.

Texts are the obvious products of professional writers' work, just as the obvious product of a shoemaker is a shoe. What is less obvious—even invisible—in both of these cases, though, is that writers and shoemakers also produce what can be called social space. In fact, all of us are constantly participating in producing social space.

Let's think about an athletic shoe shoemaker. One form of social space produced by a large athletic shoemaker is located in the social space of labor. In order to produce a shoe, the large shoemaker must employ laborers who can assemble the shoes and run the factories. The shoemaker, thus, produces the social space of daily work for these laborers. The shoemaker is also involved with other shoemakers in producing a range of social positions for the wearers of their shoes. One style within a brand symbolizes an elite social position, while another style within a brand symbolizes a less elite position. In these two ways, the shoemaker produces much more than a shoe; the shoemaker produces social space.

The same can be said for professional writers. While reading the following selections, keep these questions about social space in mind:

- In what ways might professional writers participate in producing social space?
- In what ways might professional writers be constrained by the social spaces within which they work?
- How can professional writers work to produce social spaces that enhance effective communication?
- How can professional writers work to produce social spaces that serve their users?
- What forms of social space might professional writers participate in producing?

Social space comes in a variety of forms. In general, social space refers to the relationships produced (or made impossible or difficult to form) between people and ideas. As already implied in the shoemaker example, social space can refer to the relationships we experience on a daily basis at work. This kind of space is quite concrete. Even more concrete social spaces refer to the material structures that surround us, like buildings and roadways. But social space can also refer to abstract spaces, like the elitism produced by shoes in the preceding example. As you read, think about the forms of social space, both concrete and abstract, that professional writers might participate in producing.

What you may be sensing at this point is the tight relationship between products and social practices. In many ways, the things we produce become, or are themselves, actions. If I produce a shoe, I am not simply producing a shoe. I am also acting on my world and others within it. At the very least, I am saying, "Hey, you should wear shoes!" This is the same sort of tight relationship professional writers must become aware of in their work. When writers produce texts, they are also producing (or making impossible or difficult) certain relationships between people and ideas. In this productive process, writers are also very much acting on their worlds. As you read the selections in Chapter 7, think about not only the social spaces that professional writers produce but also how these products become social actions in and of themselves.

In the first selection by Killingsworth and Jones, attention is drawn to the ways professional writers' increasing roles in project management produce workplace relations and how these relationships affect the production of texts. Johnson-Eilola and Selber discuss how hypertexts participate in constructing and re-producing certain kinds of organizational structures. In this selection, they draw attention to the ways that texts themselves are social acts. Grabill and Simmons examine, in the final selection of the chapter, how professional writers partcipate in producing social understanding of "risk," as well as the citizens faced with risk. In their article, they show us that writers, through their practices, produce much more than texts.

FOCUSING ON KEY TERMS AND CONCEPTS

Focus on the following terms and concepts while you read through this selection. Understanding these will not only increase your understanding of the selection that follows, but you will find that, because most of these terms or concepts are commonly used in professional writing and rhetoric, understanding them helps you get a better sense of the field itself.

1. division of labor model
2. integrated teams model
3. Taylorism
4. linear process
5. recursive process
6. self-directed team

Division of Labor or Integrated Teams: A Crux in the Management of Technical Communication?

M. JIMMIE KILLINGSWORTH
BETSY G. JONES

Two models of collaboration predominate in the process by which technical documents are produced. We call these the *division of labor model* and the *integrated team model*. Division of labor has tended to prevail since the 1950s, when "technical writer" first made its appearance as a job title. Writing, editing, and illustrating—not to mention communication management—had formerly been integrated into other activities of document production. Engineers and scientists wrote and edited their own reports, for example. The addition of technical communication support personnel no doubt has improved document readability and aesthetics. But the writing process has become steadily more fragmented and linear. Certain people are in charge of planning, others do the research and writing, and still others are responsible for editing, illustrating, and typing. The proliferation of specialists among communications professionals also creates a need for managers who find their place in the division of labor as reviewers of the documents and perhaps as planners.

This fragmentation is further abetted by the development of specialized "discourse communities"—groups such as nuclear physicists, electrical engineers, systems programmers, and, now, technical communicators, who are, in turn, specializing as manual writers, newsletter and brochure managers, medical editors, and so on. These groups distinguish themselves from others by specific language practices. Technical terminology, jargon, and shoptalk reinforce distinctions based on job titles or job descriptions.

Yet an alternative to this divided practice has begun to emerge in industry, an approach that could involve full integration of all contributors to a given document in every stage of the process—planning, research, writing drafts, revising, and even (given full access to automated systems for word processing and desktop publishing) typing and illustrating. This alternative to the division of labor—the integrated team model—demands a new socializing of individuals from disparate discourse communities and thus represents a challenge for managers. Their main task in this approach is to oil the machinery of collaboration

DIVISION OF LABOR VS. INTEGRATED TEAMS: BACKGROUND

The purpose of this section is to ground our research in the general history and theory of organizational management and to report briefly on recent literature in technical communication that is relevant to our topic.

Source: Killingsworth, M. Jimmie, and Betsy G. Jones. "Division of Labor or Integrated Teams: A Crux in the Management of Technical Communication?" *Technical Communication* (Third Quarter, 1989), pp. 210–221. Reprinted with permission from Technical Communication, The Journal of the Society for Technical Communication, Arlington, VA, U.S.A.

Historical and Theoretical Perspective

The two organizational models have a long history in managerial theory, the essentials of which are effectively summarized in *Managing Communication in Organizations: An Introduction* by Cummings, Long, and Lewis [1]. The division of labor approach, with its origins in the classic economics of Adam Smith, was first described as an aspect of modern bureaucratic rationality by Max Weber in 1909. The Weberian analysis of bureaucratic organization recognized the following principles:

- Hierarchical distribution of power and control
- Standardized rules and procedures
- Specialization and division of labor according to tasks and subtasks
- Employment based on technical competence
- Detailed job descriptions
- Prescriptive and rigid information flow
- Subordination of individual needs to organizational goals.

Although Weber was ambiguous, even cynical, about the overall effects of this type of organization upon human society [2], he was quite clear about the advantages he observed in bureaucratic rationality:

> The decisive reason for the advance of bureaucratic organization has always been its purely technical superiority over any other form of organization. The fully developed bureaucratic apparatus compares with other organizations exactly as does the machine with the non-mechanical modes of production. Precision, speed, unambiguity, knowledge of the files, continuity, discretion, unity, strict subordination, reduction of friction and of material and personal costs—these are raised to the optimum point in the strictly bureaucratic administration [3, 973].

The Weberian description of efficient division of labor was supplemented by Frederick Taylor, the founder of "scientific management," also known as Taylorism. The first "efficiency expert," Taylor contributed four new principles to the model of bureaucratic rationality:

- Work should be selected scientifically through rigorous testing.
- Effectiveness should be objectively measured.
- Managers are planners.
- Workers are doers [1, 36].

In 1916, Henri Fayol added to this theory of managerial control the five now classic principles of management: *planning, organizing, commanding, coordinating, and controlling* (review and evaluation) [1, 36].

As early as 1927, the beginnings of the integrated team concept began to emerge as a dialectical challenge to what was by then the well-established division of labor model. The founder of industrial psychology, Elton Mayo, developed the so-called "human relations approach," which offered alternatives to the system of bureaucratic management, including the following:

- People-oriented rather than production-oriented management
- Use of informal work groups

- Emphasis on cooperation rather than competition, building of community among workers and managers
- Inclusion of workers in planning and decision-making
- Increased concern over worker satisfaction [**1**, 38–39; **4**].

Like the bureaucratic model, the human relations approach claimed to have found the key to the manager's foremost concern—productivity. The difference is that the emphasis in Taylorism and its descendants has fallen upon time and cost efficiency, whereas the emphasis in human relations management falls on worker motivation, which, it is claimed, assures the highest quality products and which is neglected and even suppressed in bureaucratic management despite any advantage that is realized in terms of time and cost.

Three recent developments within the human relations tradition support the emergence of integrated teams: the concepts of *meaningfulness of work*, *self-directed teams*, and *informating versus automating*.

Meaningfulness of work. Hackman and Oldham claim that this essential ingredient in a worker's motivation—and thereby the key to increased productivity—depends upon three factors:

- Skill variety: the chance to use and develop a number of skills in a task
- Task identity: the degree to which the worker identifies with the task, usually an effect of being involved from beginning to end
- Task significance: the worker's recognition of the task's importance in the lives of other people within the organization and in the world of human relations in general [**1**, 261; **5**].

Self-Directed Groups. Businesses have begun to experiment with project groups that work on a task from beginning to end and have a high degree of decision-making power in their project [**6**, **7**]. This practice appears to have been influenced by the trend among Japanese managers to favor employee participation and the use of small, informally structured groups [**8**]. But the American system has a motive different from the tradition-sensitive managers of Japanese industry: "Given rising educational levels, demands for greater individualism, and alienation from traditional authority, it is becoming increasingly important to consider human needs" [**9**, 1188]. In one company designed to accommodate these worker needs, "members of autonomous work groups rotate jobs, select their own members, decide on assignments, monitor their own performance, provide training for each other, and are paid for the number of tasks they know how to perform" [**9**, 1189].

Rossabeth Kanter, an advocate of this approach, observes, "The organizational chart with its hierarchy of reporting relationships and accountabilities reflects one reality; the 'other structure,' not generally shown on the charts, is an overlay of flexible, ad hoc problem-solving teams . . ." [**10**, 488]. Kanter stresses that the purpose of such groups is to increase the fertility of the setting in which ideas are generated and implemented by drawing team members from a diversity of sources and backgrounds: "It is not the 'caution of committees' that is sought—reducing risk by spreading responsibility—but the better idea that comes from a clash and an integration of perspectives" [**10**, 489].

Table 1. Summary of Two Organizational Models

Division of Labor	Integrated Teams
1. Managers hire and assign highly specialized employees to discrete roles in a closely analyzed production process.	1. Project teams composed of individuals from various specialties share in a holistically determined effort to complete a task.
2. Tasks are accomplished (i.e., documents are produced) in a step-wise, linear flow.	2. Tasks are accomplished in an informal, often recursive manner that varies according to team decisions and capabilities.
3. Process of document production is managed and reviewed "from outside" by professional managers.	3. Process is managed "democratically" by members of the team or "representatively" by a leader within the group who serves as a liaison with management.

Automating versus Informating. In an important new book, *In the Age of the Smart Machine*, Shoshana Zuboff argues that the computerization of the industrial environment creates two possible managerial options which stand in direct opposition to one another and which run parallel to the two models we have been outlining [11]. According to Zuboff, managers can use a computer networking either to *automate* their organization, increasing the mechanical control and impersonality of Taylorism, or to *informate* their organizations, using computer networking as a means of providing widespread access to information and of opening lines of communication formerly closed in the rigid bureaucratic hierarchy [see also 12]. The tendency to automate falls within the division of labor tradition of bureaucratic management, while the tendency to informate creates a way of empowering self-directed, integrated teams.

We are now in a position to summarize the two models against the background of these dialectically opposed traditions. Table 1 provides a brief comparison of the two models.

Review of Relevant Literature in Technical Communication

The literature on the management of technical communication often takes division of labor for granted. The implicit adoption of the values of bureaucratic and "scientific" management is suggested by the predominant concern with such issues as increased specialization, automating, and efficiency in matters of time and cost. Thus, the editors of the special issue of *Technical Communication* dealing with the topic echo the classic managerial values of Fayol in their claim that "information . . . is a productive resource that can be evaluated and managed for the purposes of planning, controlling, and decision-making," though they stop short of giving a mechanistic or Tayloristic definition for input (number of hours spent on a project) and output (number of pages produced) [13, 216].

Likewise, while reporting on managerial innovations in technical manual production, Killingsworth and Eiland nevertheless rely on the Fayolian description of the manager's tasks—*planning, organizing, commanding, coordinating, and controlling*—without recognizing the possibility that management can be participatory [**14**].

In a case study of automated management, Shirley Anderson describes a "Job Tracking System" that exemplifies quite well the use of computers for the kind of authoritarian automating described by Zuboff and that thereby finds its implicit justification in the orientation and values of the division of labor model. With the system, the communication managers are able to monitor—

- Daily status of individual inputs and projects
- Unit productivity, product cost, and vendor performance
- Employee time entered by task [**15**, 119].

If the classical managerial rationales do prevail in the field, how are the tasks of communicative labor divided? At least two articles have drawn on the division of the writing process into prewriting (planning and researching a topic), drafting, revising, and editing—much as it is treated in composition textbooks. Krull and Hurford consider the effect on productivity enabled by computer assistance in each stage of the process [**16**]. Manyak encourages managerial intervention at the stages of pre-writing, revision, and editing—leaving drafting (or "scribing," to use the term of Krull and Hurford) to the individual communicator [**17**].

No articles have as yet explicitly considered the *concept* of integrated teams, but a number of authors provide case studies and anecdotal evidence that such practices are emerging in technical communication. Dressel, Euler, Bagby, and Dell present a system of managing proposal production that involves all members of an integrated team—proposal manager, text coordinator, writers, contract officer, text-processing leader, and graphics leader—in every stage of the writing process: outlining, drafting, editing, reviewing, and producing the document in its final form; moreover, this system employs informing technologies closely resembling the ideal of Zuboff [**18**].

Proietti and Thomas report on an integrated and democratically managed group in a project to develop an interactive video course; the group includes the project manager, course developer, programmer, editor, instructional designer, consultant, funder, and subject matter experts [**19**].

Dilbeck and Golowich describe a company reorganization that effected a fuller integration of software developers, trainers, and technical writers, with the result that all the participating employees achieved a higher level of job satisfaction: the writers attained a better sense of "the big picture," the trainers were grateful for a reduced writing load, and the developers gained a new respect for the writers and trainers through a better understanding of their functions; they also attained a higher level of job performance: training manuals were better written, and user manuals were more technically astute [**20**].

The need to develop effective user manuals has given rise to a number of innovative efforts at collaboration and integration in the computer industry. Wendy Milner, in asserting that technical communicators can make a contribution to online documentation, seeks an expansion of the writer's role into the area of product design [**21**]. Barstow and Jaynes make essentially the same point in insisting on effective integration of on-line and hard-copy instructions for users [**22**]. Mark Smallwood agrees that documentation can be improved if writers are included in the initial development of software products, though he admits that communicators crossing boundaries in this way may be greeted with hostility by designers; writers can win credibility, he suggests, only by achieving a degree of technical proficiency and role flexibility [23]. In the same vein, Fowler and Roeger advise close collaboration between writers and programmers [24]. Chew, Jandel, and Martinich also insist on the "enhanced role of writers in the product development cycle," especially in human factors engineering of user interfaces in computer products [25; see also 26, 27].

Just as the need for good user manuals has spurred new collaborations among technical experts (programmers and designers), communicators (writers, editors, and artists), and managers in the computer industry, the need to produce a quality product in a hurry has prompted proposal teams to seek innovative techniques of integration. The literature on proposal writing for some years now has reported on the success of storyboarding and other graphical techniques of document planning as a way of achieving maximum input from team members at various stages of the project [28, 29, 30]. No longer considered merely a way of illustrating written points, visuals are used early in the writing process as a means of encouraging collaboration of various team members. This practice breaks the traditional linear path of document development (from planning to drafting to revising to editing to printing) and allows for overlap and recursion in the stages of the writing process.

Table 2. Summary of Two Organizational Models Applied to Technical Communication

Division of Labor	Integrated Teams
1. Only managers observe and participate in the whole process by which documents are planned and produced.	1. All members of a document team participate in all stages of production.
2. Individuals are responsible for a clearly defined task which they are trained as specialists to perform.	2. Team members are assigned roles based on their training as specialists but share and exchange roles within the team if this will improve the final product.
3. The process of writing is accomplished in a step-wise, linear fashion, with managers and technical staff planning the document, technical staff and technical writers drafting it, writers and editors revising and editing it, artists illustrating it, and word processors and printers producing final copy.	3. The stages of the writing process are overlapping and recursive, with all team members offering advice and content at every phase: writers, editors and artists, for example, may offer outlines and graphics for story-boarding during planning; and technical experts may participate in producing final copy through the use of desktop publishing software.
4. Computer technology is used to automate (monitor and manage time and costs).	4. Computer technology is used to informate (allow team members full and fast access to company files).

Against this background, we can now refine our description of the division of labor and integrated team models to show how they work in technical communication. Table 2 gives a summary.

A SURVEY OF TECHNICAL COMMUNICATORS

To get a better sense of how deeply entrenched the division of labor is in today's industry and to chart the inroads cut by the alternative of the integrated teams model, we conducted a survey of technical writers in various organizations.

Research Questions and Hypotheses

In general, our research questions were these:

- How does a document progress through the planning, drafting, and revision phases; is the process linear, overlapping, or recursive?
- What kinds of personnel participate at each stage of the document preparation process?
- How does the training of the personnel affect their participation in the process?
- Is the process different for different kinds of documents?
- To what extent is the process automated or informated?
- To what extent is the process assisted by graphical techniques?
- What are the employees' attitudes toward their place in the process of document production?

We begin with the hypothesis that the production process in industry is most frequently modelled according to the division of labor along a linear path from planning to drafting to revising and editing. We also expected that variations in the process would depend upon the types of individuals (managers, technical experts, marketing people, writers, editors, artists) involved at different stages in the process, the training of the writers and editors, the degree to which the process was automated, how the process was managed, and the rhetorical and graphical techniques employed. Finally, we expected, on the basis of our reading of the literature, that integrated teams would be used most frequently in organizations that produced contract proposals and computer manuals for users.

Methods: Data Collection and Analysis

We developed a questionnaire to elicit information about the document production process and the individuals involved (Figure 1). We asked the participants to answer questions about the participation of various individuals in different stages of document production, about the relation of the different stages of the process, about automation of the process, and about the structure and management of the process. We realized that, in labeling various stages of the process ("planning," "drafting," and "editing") or in providing descriptive terms for the flow of information development ("linear," "recursive," "overlapping"), we ran the risk of biasing our data. In all our questions, therefore, we left space for comments, hoping the rel-

Questionnaire on the Process of Developing Technical Documents

Return as soon as possible (no later than March 15) to:

Professors Jimmie Killingsworth and Betsy Jones
Department of English
Texas Tech University
Box 4530
Lubbock, TX 79409

________Check here if you would like a copy of the results of this questionnaire.

1. Please give your name, title, and the name and address of your company, division, and/or organization.

2. Which form(s) of technical communication does your office often produce?
   Proposals________ *Comment:*
   Manuals (including computer documentation)________
   Reports________
   Other______________________________
3. What kind of personnel participates in each stage of document production (for example, technical writers, engineers/technical staff, editors, graphic artists, marketing staff, management)?

   Planning?

   Drafting (that is, early stages of writing)?

   Revising/Editing?

Comment:

4. Which stages involve teams of authors/editors/artists, and which are normally done by individuals?

	individuals	teams
planning	________	________
drafting	________	________
revising	________	________

Comment:

5. How automated is the process; that is, which stages are handled "on line" or in some way make use of word processors and other computerized equipment?

   planning________ *Comment:*
   drafting________
   revising________

6. In which stages are graphics used as devices for organizing, illustrating, or replacing written text (in some organizations, for example, flow charts and other graphic devices play a significant role even in the early stages of planning)?

   planning________ *Comment:*
   drafting________
   revising________

7. Which of the following term(s) best describe the writing process at your company or division?
   ________*linear* (The process flows directly from planning to drafting, then back again, for example.)
   ________*recursive* (Drafting begins while planning is still in progress, then back again, for example.)
   ________*incremental* (All operations are completed on one section of a document before the next section is written.)
   ________*other* (Please give term and definition):
8. Who oversees and approves publication design?
9. Who oversees, manages and/or controls the overall process of planning, drafting, revising, and publishing documents?
10. How does the process vary as the type of length of document varies?
11. What kind of formal training in technical communication have you and your staff had? (Include degrees in technical writing, workshops, seminars, graduate courses, etc.)
12. Please use the rest of this page (and, if necessary, the back of the page) to comment on any features that you think are unique in the way your company handles the process of document production.

Figure 1. Questionnaire used for data collection.

atively simple check-the-blank questions would ignite an attitudinal response in the comment section which would prove more valuable than a simple count of how many organizations used division of labor versus how many used integrated teams at the time of the survey.

We sent the questionnaire to 120 technical writers whose names were taken from the Society for Technical Communication's directory of members. Sixty-five, a little over 50 percent, responded, most with ample comments about their practice and their attitudes. The moderate return is most likely due to the demanding nature of the three-page questionnaire with its many requests for comments. One respondent complained vigorously about the design of the questionnaire (while nevertheless supplying us with extremely useful responses on her attitudes about her work).

To analyze the data we collected, we tallied the responses and attempted to discern patterns within and among them. Some of the data could be quantified (see the *Results* section below). But we were most lucky to receive many long responses in the comments sections. These gave us insights not only into the structure of document production, but also into the respondents' perceptions and feelings about that structure. Such responses yielded less to quantification than to careful scrutiny and dialectical interpretation in light of similar or vastly different responses. The "numbers" elicited by the questionnaire often do little more than reinforce our common sense about the profession. They are useful primarily in providing a context for the comments and a means of locating and identifying the authors of the comments. We ask, therefore, that our readers bear with us as we report this data, all the while realizing that our real interest is the reporting and analysis of the rich comments our subjects generously provided.

RESULTS

Of the 65 respondents, 16 (25%) were from companies involved in the computer or electronics industries, and 24 (37%) worked for other private sector companies involved in such fields as petroleum, agriculture, and consumer products. Eighteen (28%) represented military or government organizations, including defense contractors. Four responses (6%) came from university research or publications departments, and 3 (4%) were returned anonymously—with responses but without names, titles, or addresses.

Though the sample hardly represents all technical writing fields or indicates the breakdown of types of industries employing technical writers, our responses do provide a representative cross-section of organizations and companies. Strongly technical and scientific, those organizations further suggest active bureaucracies—multiple layers of departments and individuals that might participate in the document-production process.

Manuals and reports are the most commonly produced documents (45 and 39 of the 65, respectively), according to the respondents. A sizable number (23) reported that they prepare proposals as well. The "other" forms mentioned show a wide array of writing activity: brochures, marketing literature, journal articles, conference papers, newsletters, even speeches, maps, and videos.

Team Involvement at Different Stages of the Production Process

We found that collaboration (specifically, team involvement) is the normal procedure at many companies, especially at the planning stage. The following shows the answers to the question, *Which stages involve teams of authors/editors/artists, and which are normally done by individuals?*

	Plan	Draft	Revise
Individuals	30	47	40
Teams	30	13	

Clearly, individuals are entrusted with significant duties. That drafting, in particular, is an individual's task demonstrates to some extent the independent nature of this stage of the writing process, even in complex and bureaucratic companies. In contrast, planning relies on such social activities as brainstorming, setting multiple objectives, and problem solving, just as revising and editing require satisfying multiple agendas.

Participation of Different Discourse Communities at Different Stages

The personnel involved in production, also a function of organizational nomenclature, suggests significant participation from a variety of people and groups. We asked: *What kind of personnel participates in each stage of document production (for example, technical writers, engineers/technical staff, editors, graphic artists, marketing staff, management)?* The results:

	P	D	R
Writers	37	40	40
Tech. Staff	36	39	38
Editors	13	9	26
Artists	9	4	7
Mgmt	27	8	17
Other	17	8	13

P (Planning), D (Drafting), R (Revising)

These results suggest that it has become common for technical communicators to be involved throughout the production process and to interact with diverse groups and individuals. And the specific discourse communities begin to emerge as well: the engineering, scientific, and technical staff, artists, managers, and "others"—marketing staff, auditors, product specialists, reviewers, quality assurance personnel, and users. Technical staff, specialists in various scientific and technological fields, join technical writers in participating in the full process more frequently than such other groups as managers, editors, and artists, few of whom create drafts.

We found that of the three stages in the writing process, drafting and revising take most advantage of word processors and other computerized equipment. In fact, the involvement of personnel may be a function of the level of automation; for processes that individuals rather than teams dominate, computers are used.

Not surprisingly, the responses from computer companies show significant use of word processors and other computer equipment. One respondent commented, "This place looks like hi-tech heaven to me. I haven't seen a typewriter since I got here—everyone who has a place to sit has a computer terminal, connected to a small printer [and] a line printer, and has access to a QMS printer."

Still other responses suggest that the automation varies even among divisions within an organization and is currently undergoing change. One respondent from a large government laboratory indicated that only recently have a few editors begun to edit on-line. In such organizations, where scientific authors have strong jurisdiction rights, attempts to edit soft copy, even when a tracking system is used, can represent a serious encroachment on territorial boundaries.

Graphics

In the answers to our question, *In which stages are graphics used as devices for organizing, illustrating, or replacing written text?*, we learned that graphics are integrated during planning, drafting, and revising, but especially the latter two. Yet the approach to graphics varies significantly across and among individuals and organizations. A sampling of responses:

From a private company with significant military contracts: "Especially in 'new look' manuals prepared for the U.S. Army, pictures replace text wherever possible. User is less able to read than, say, a typical Air Force or Navy technician, so pictures of *every step* are needed." The implication, though not altogether clear, is that written text no longer assumes a priority over graphics—in the process or in the finished product.

From a consumer products firm: "Wherever an illustration will increase clarity and conciseness, it is used. Most frequently, editors determine where text might be augmented or replaced." Typical of many responses, this suggests a late introduction of graphics into the writing process.

From a free-lancer who produces manuals and marketing literature: "Graphics always come first in the material I produce. They are sketched, and constantly revised during the generation of text until they evolve into a source of final artwork." This is one of the few responses that unambiguously showed a full integration of graphical techniques in all stages of production.

The Writing Process Described

Our request to describe the company or division's writing process yielded an interesting response. We suggested four possible descriptive terms—*linear*, *recursive*, *overlapping*, and *incremental* (plus *other*)—with the following definitions provided:

> *Linear:* The process flows directly from planning to drafting to editing, with perhaps a separate person or group in charge of each stage.
>
> *Recursive:* The process flows from planning to drafting, then back again, for example.
>
> *Overlapping:* Drafting begins while planning is still in progress, and so on.
>
> *Incremental:* All operations are completed on one section of a document before the next section is written.

We also left open the possibility that more than one term might apply. Only three respondents selected *incremental*; none suggested *other* definitions. But the remaining terms were identified in almost equal measure: linear 27, recursive 25, and overlapping 23.

Supervision of Publication Design and the Publication Process

Our question designed to yield information about the supervision, management, and approval of publication design as well as the overall process we left open-ended, and thus we received a variety of responses. We expected the responses to reflect the array of titles used among organizations, and indeed they did. Most indicated that a project manager or supervisor was responsible for overseeing the production process, but many listed editors or writers as well.

We also expected most to indicate a single person or title, and there we were surprised with the results: Even where management and approval are involved, individuals such as a manager and editor or writer frequently share responsibility. Assignment of responsibility is also determined in some cases by the type of document under production. Interestingly, though, whereas engineers and technical staff were mentioned often when we asked who participated at which stage, few respondents indicated engineers' participation in publication design or management of the entire process.

Thus, collaboration does seem common in the management phase of document production, but the discourse community changes to include primarily writers, editors, and project managers, the latter of whom seem to have writing, editing, and design responsibilities. This finding supports the assertion of Killingsworth and Eiland that more and more often communicators rather than technical personnel are assuming broad managerial duties [**14**].

Formal Training in Technical Communication

As a corollary to our questions about personnel who participate in the document production process, we asked, *What kind of formal training in technical communication have you and your staff had? (Include degrees in technical writing, workshops, seminars, graduate courses, etc.)* The growing role of the communication professional, who must interact with members of many discourse communities, as well as the proliferation of college-level technical writing programs, piqued our interest in the writers' training, if any. We hypothesized that staff members who had received formal writing training would identify more strongly with their organization "niche."

The answers to the question varied enormously, as did the organizations, of course. Some respondents carefully listed each staff member and his or her training; others gave general comments. A sampling of responses:

- In general, we have had little formal training—less than 1/3 of our staff has ever completed a [technical writing] class, except for in-house training, which I conduct. All our writers have either a technical degree (EE or EET), or else have equivalent experience in military schools or industry. All writers and editors require at least a bachelor's degree.
- Peers' educations vary—most were teachers and English majors—one is a lawyer. I believe the science background gives an edge by giving you more credibility in the eyes of the programmers, etc.

- All 18 of us have at least one degree (BA) in English or a related field.

The following indicates the results (the number of staff members in the organization with such training), to the extent that they can be quantified:

English/Technical Communication degrees	36
Technical/Engineering degrees	16
Journalism degrees	7
Art degrees	2
Business degrees	1
Continuing education seminars & workshops	31
Military or on-the-job experience	10
Technical writing or graduate courses	8

Thus, formal training—degrees in technical communication or related fields—is common, and several respondents indicated that staff members hold master's and Ph.D. degrees in technical communication, English (in at least two cases, Ph.D.s in literature), or other fields. Yet seminars and workshops (no doubt including those sponsored by STC and ITCC) are also viewed as important vehicles for staff development.

These results indicate increasing specialization among communicators in industry. The occasional defensiveness about qualifications demonstrates a perceived *need* for specialized credentials, even if that need is self-imposed. And an increased need for specialization by task goes hand-in-hand with intensified division of labor. But technical expertise received in formal technical communication training might, on the other hand, be viewed as an effective means toward enhanced collaboration and team integration [23; 25].

"Uniqueness" of the Document Production Process

Finally, we requested, *Please use the rest of this page (and, if necessary, the back of the page) to comment on any features that you think are unique in the way your company handles the process of document production*. Granted, such an open-ended request (and the last one on a three-page document, at that), is likely to be ignored. More than 40% of the respondents declined to list any features unique to their companies.

From the 60% who responded, however, some significant patterns emerged. Five respondents suggested that a feature unique to their company is the important role of the individual writer in shaping the entire document in consultation with other members of a project team. In those cases, the writer is assigned or assumes responsibility for a project and follows that project through all phases from planning to publication. The result is a wide range of style and quality. A respondent from a large company noted:

> Perhaps there is some uniqueness in the non-uniformity of style in our finished product. If you were to review a year's output of tech manuals, you would be seeing the writing of about 25 different people. You would certainly note great differences in style, and be aware that there is a very wide range of writing abilities evident in the material we send out the door.
>
> . . . The reason we don't take greater pains to ensure that the style of every manual is indistinguishable from every other is that our management and our customers do not

> place great value on a uniform "company" style. The books we produce are essentially tools useful in operating and repair of specialized equipment.

Three other respondents mentioned a similar but even more significant role for the writer on the product design team. These writers, all of whom indicated a high level of specialized training in technical communication, nevertheless demonstrated a significant degree of satisfaction in their contribution to product engineering and other tasks "outside" their chosen place in the division of labor. We will return to their comments in the discussion section below.

In contrast, five other respondents suggested that the centralization of technical communication in their companies is unique. In most of these cases, the technical communication staff is responsible for almost all of the documents produced by the organization, an although they work with engineers and other writers and editors through the process, their departments carry a document from planning to publication.

Several who responded from small or developing companies, frequently in industries experiencing change (such as computer-related firms), noted that the document production process is itself in a state of flux that matches the company's own evolution. As those companies and industries mature, so do their procedures for preparing manuals, reports, and other documentation. One respondent wrote: "When I started at this company, we did not have a formalized process. The process is evolving as we mature. . . . *Planning, planning, planning* is the key to good documentation. The writing itself is the easy part."

Nevertheless, several other respondents—these from more established and bureaucratic companies—report extremely specific and unchanging processes. Three of them attached flow charts showing the paths from document inception to publication, as well as the individual responsible for each phase along the way.

Some who answered the final question used it to vent frustration with their companies' policies or procedures. Reasons varied, however, including unhappiness over the scope of job responsibilities to the role of the writer amidst the technical staff and management. One (from a military setting) noted, "Powerful people within the bureaucracy impose changes in a document's content that are harmful to the intended purpose. Frequently obfuscation is preferred over clarity. It takes both courage and diplomacy to create a good document. Usually the writer-editor is the only one sincerely interested in the poor reader." Another said, "I can't identify a formal 'flow' because I don't work with other professional communicators. I work with engineers and analysts, who draft reports, then ask me for help. I don't think they plan; I think they modify boilerplate."

DISCUSSION

This last question proved to be a key to the respondents' feelings about the way the production process is handled in their companies. "Unique" became the word that the employees attached to something about their companies that they took pride in. On the negative side, the denial of "uniqueness" served to indicate dissatisfaction with the process. Following a scathing review of the inefficiency in his company's operation, one respondent wrote: "Typical defense contractor publication operation!"

This respondent marked "linear" as the best description of the writing process at his organization.

We might suggest that the three most frequently marked descriptive terms—*linear, recursive*, and *overlapping*—are indexes of which model predominates in the various organizations. A linear process indicates a strong division of labor, a recursive process implies a high degree of integration and team interaction, and overlapping stages in the process suggest an organization in transition between the two models or an organization that draws upon both to some extent. Significantly, the strongest expressions of job satisfaction were associated with processes described as recursive and overlapping, especially the former. Since the responses were very nearly equally divided among the three descriptions, it is likely that an even distribution of the three organizational patterns (division of labor, integrated teams, and transitional or mixed) exists among the participants in our survey.

Generally, however, we were surprised by the high degree of integration in the process of document production. And the attitudinal findings related to this integration seem quite clear. When technical communicators are permitted entrance into all stages of the process and are given the chance to approve or make publication decisions and even to contribute to product design, they feel they are working at their fullest creative potential. When they are closed out of one of the phases of the process, they tend to be critical of their organization's practice. For one respondent, a member of a centralized technical communication operation, being out of touch with design decisions meant that no such decisions occur: "There is no such thing as publication design at our company that I am aware of. Publications are managed by a small technical publication section that is only interested in adherence to paragraph numbering and meeting schedules."

Perhaps the key issue in the evolution of integrated production processes involves the decision about whether to centralize the technical communication function. In the next section, we consider this problem at some length.

Centralized or Decentralized Technical Communication?

Centralized departments for technical communication usually support division of labor by separating the revision, editing, and printing processes from the processes of planning, researching, and writing of early drafts. Only one of our respondents suggested that a centralized technical communication department was responsible for all stages of the process, though some indicated that this might be the case for a few kinds of limited publications—brochures, newsletters, and the like.

Several comments indicate a recent move away from centralization in companies that have tried the increased division of labor and have found it unsatisfactory. Thus one respondent remarks: "This company recently underwent a reorganization. It decentralized *all* support functions, including technical communication. One result was to eliminate the editing staff and give writers much more responsibility." Typical of the respondents in decentralized organizations, this writer described the flow of the production process as "recursive."

Similarly, another respondent writes—

> Our company is unique in that the technical writers are on the software development staff, rather than in a separate department. . . . [The] programming staff is cooperative

> in teaching new systems and enhancements to technical writers, and in reviewing the writer's work for technical accuracy. Also, the development staff calls on the writing staff for help in organizing menu displays, editing on-line help, editing prompts and comments, and presenting effective screen formats. . . . We feel that our documentation procedures result in manuals which are both technically accurate and easily accessible to users.

Decentralization, says one technical writer, with obvious pride, requires "self-starting traits and confidence in your own skills."

In the fullest statement we received on the integrated and recursive approach to the process of document production, one technical writer from a software development company said: "In our company the distinctions concerning who does what and when are not concrete. This situation may change from one document to the next." In response to the question on uniqueness, she remarked—

> Though overall production plans are determined by management, the actual document is heavily shaped by the writer assigned to the project. After talking with a programmer, I provide an outline according to the organization that I feel best captures the material. I then can direct the amount and types of information I need.

Having received the structured information, the writer drafts the document and continues consultation with the technical staff.

Other respondents indicate that editors as well as technical writers have begun to move out of the communication department, with the result that conflicts among discourse communities are transformed into cooperative dialogue, a giving and taking of advice that can affect the document before it is "set in concrete." One respondent writes—

> The technical editor becomes a part of the design team very early in the project (more than a year prior to release). As the product develops, the editor understands it and the writing of documentation is a team effort. This eliminates most of the Engineer/Tech Writer conflicts. It works well for us.

Some organizations have taken a step half-way toward decentralization by retaining their technical publications department, but also making it possible for other departments and projects to claim an editor for their own. In one such organization,

> Editors have principal responsibility for overseeing the publication. This includes all interaction with the authors and establishing priorities and responsibilities for support groups (word processing, graphics, printing, etc.). The editors also have lead responsibility for working with various departments; therefore, if, for example, the engineering department has publication questions, they will consistently deal with the same editor. The approach increases the editor's role, while developing a higher level of visibility for the person actually doing the publications work.

One editor who has worked in an organization like this provides a clear perspective on her preferences:

> I don't work with the central publications organization [any longer]. I prefer to be closely affiliated with a project so I know the scientists and become acquainted with their field. . . . The editors in the central organization . . . can't get familiar enough with the subject to do substantive editing.

We thought that smaller companies would more likely use an integrated approach, but our findings show that size is not a major factor in this determination. One writer from a very small operation complained about the rigidity of the process and of role-assignments. And representatives from some of the largest organizations in the country indicated substantial degrees of integration.

Two facts demonstrate the high level of satisfaction among technical communicators involved in organizations with an integrated process of document production. First, none of the respondents from such organizations reported dissatisfaction. Second, the length of these respondents' answers to the uniqueness question is a further index of their pride in their work. They consistently wrote the longest descriptions of their companies' practices. Most respondents from organizations with a heavy division of labor wrote little or nothing in response to the question on uniqueness, and of the three who did respond at length, two used the space to complain about their organization's structuring of the process.

Despite the increasing use of the integrated process model in some companies, and despite its potential for employee satisfaction, the entrenchment of the division of labor deepens in many companies, as its most apparent symptom—increasing specialization—indicates. The most reliable index of this trend among our findings is the proliferation of job titles, some of which are specialized enough to require new words—like "formattist." Many of these new specializations arise from technological demands, as one of our respondents reports: "The severe challenge of [an elaborate text/graphics software system] has given rise to a new job classification—software support engineer."

The Limited Use of Informating Technologies and Graphics

It is ironic that a system for integrating text and graphics should produce greater division of labor, for in theory such systems could provide high degrees of integration; they should be, in Zuboff's language, "informating technologies"; they should promote new methods of collaboration and interaction. But the respondent who reports on the need for a "software support engineer" laments staff members' "virtual enslavement to the . . . computer system."

Indeed, our most surprising finding concerned the relatively minor, and occasionally negative, effect that graphics and computerization have on the shape of the process by which documents are produced. Only one of our respondents, for example, mentioned the use of storyboarding as a means of opening up the process of planning and drafting documents, even though the technique is widely advocated in the literature on technical writing. Our respondent definitely recognized the strong potential for this graphical procedure: "We've done best on proposals for which we used the storyboard technique, which calls for graphics to be planned at the outset." More typical, however, is the respondent who reports that graphics are not considered until late in the process: "Whenever an illustration will increase clarity and conciseness it is used. Most frequently editors [in a process marked "linear"] determine which text might be augmented or replaced [by graphics]."

Only one respondent mentioned a serious transition toward the use of computers to integrate the process of producing documents: "Some groups are beginning to handle whole processes on line (PCs) from draft to printing final copy."

Many respondents, especially those in computer companies, may have taken this procedure for granted; for, as we hypothesized, organizations that produce user manuals demonstrated the highest levels of team integration. With the advent of desktop publishing, however, we expected far more numerous comments on how computers have empowered groups of writers and researchers by giving them more control over the whole process of document production.

CONCLUSION

Though division of labor still dominates the practice of many companies, there is a perceivable trend in the management of technical communication toward what has become known in management circles as the "self-directed team." The chief advantage of this arrangement, from the perspective of overall efficiency, is enhanced motivation and job satisfaction. Our findings suggest the reality of this effect, though whether higher quality products emerge as a result of higher motivation is a question we are not yet prepared to answer.

Our findings also suggest the need for further research about the integrated teams approach in the types of organizations that, from the outlines provided by our survey, seem to have adapted it most fully—private sector companies, especially in the computer industry. An important question which we are not yet in the position to answer concerns the general applicability of the model; that is, beyond the question of whether it works, we must ask whether it works best in very particular settings and whether it can be applied elsewhere.

We can suggest that, among the organizations represented in our survey and in the literature, the integrated process model of document production, in addition to enhancing motivation, may also increase desirable integration in three other areas:

1. Integration of employees from various discourse communities, which will allow effective interchange and new opportunities for education and training. Our survey indicates that many organizations are currently realizing this benefit.
2. Integration of a document with the product or service it describes. In a few of the organizations we surveyed, this benefit has begun to accrue as, for example, the writers of user manuals begin to contribute to the development of online documentation and other human interface functions.
3. Integration of the different modes and styles of a particular document, the most obvious example being the balance of written text with graphics. A surprisingly few of our respondents seemed aware of the potential of informating technologies and graphics to achieve the objective of a fully integrated communication.

Managers of communication in various companies will no doubt be better able than we are to see the disadvantages of the integrated process model. From their perspective, in a field where a high rate of turnover is the norm, one drawback will be obvious. As classic economic theory suggests, division of labor is effective because it allows for easy replacement of "interchangeable parts." In short, it is much easier to replace a "formattist" than it is to replace a member of a "self-directed team." From the perspective of the worker in a technical communication, however, this can only be seen as an advantage.

References

1. H. Wayland Cummings, Larry W. Long, and Michael L. Lewis, *Managing Communication in Organizations: An Introduction*, 2nd Edition (Scottsdale, AZ: Gorusch Scarisbrick, 1987).
2. In his famous historical work, *The Protestant Ethic and the Spirit of Capitalism* (Trans. Talcott Parsons, New York: Charles Scribner's Sons, 1958:182), Weber writes despairingly of the evolution of modern bureaucracy: "For of the last stage of this critical development, it might well be truly said, 'Specialists without spirit, sensualists without heart; this nullity imagines that it has attained a level of civilization never before attained.'"
3. Max Weber, *Economy and Society: An Outline of Interpretive Sociology* (Berkeley: University of California Press, 1978).
4. Ernest Mayo, *The Human Problems of an Industrial Civilization* (New York: Macmillan, 1933).
5. J. R. Hackman, and G. R. Oldham, "Motivation through the Design of Work: Test of a Theory," *Organizational Behavior and Human Performance* 16 (1976): 260–279.
6. J. R. Hackman, "The Design of Self-Managing Work Groups," in *Managerial Control and Organizational Democracy*, eds. B. King, S. Streufert, and F. E. Fredler (New York: Wiley, 1978), 61–89.
7. E. J. Poza and M. L. Markus, "Success Story: The Team Approach to Work Restructuring," *Organizational Dynamics* 25 (1980): 3–25.
8. Robert Cole, "Work Reform and Quality Circles in Japanese Industry," in *Critical Studies in Organization and Bureaucracy*, eds. Frank Fischer and Carmen Sirianni (Philadelphia: Temple University Press, 1984), 421–452.
9. William Pasmore, Carole Francis, Jeffrey Haldeman, and Abraham Shani, "Sociotechnical Systems: A North American Reflection on Empirical Studies of the Seventies," *Human Relations* 35(1982): 1179–1204.
10. Rossabeth Moss Kanter, "Empowerment," in *The Leader-Manager*, ed. John N. Williamson (New York: Wiley, 1986), 479–504.
11. Shoshana Zuboff, *In The Age of the Smart Machine: The Future of Work and Power* (New York: Basic, 1987).
12. Mike Cooley, *Architect or Bee?: The Human/Technology Relationship* (Boston: South End Press, 1980).
13. Thomas E. Pinelli, Thomas E. Pearsall, and Roger A. Grice, "Introduction to Special Issue on Productivity Management and Enhancement in Technical Communication," *Technical Communication* 34 (1987): 216–218.
14. M. J. Killingsworth and K. Eiland, "Managing the Production of Technical Manuals: Recent Trends," *IEEE Transactions on Professional Communication* 29 (June 1986): 23–26.
15. Shirley A. Anderson, "Managing through Automation," in *Proceedings of the 33rd International Technical Communication Conference* (Washington: STC, 1986): 118–120.
16. Robert Krull and Jeanne M. Hurford, "Can Computers Increase Writing Productivity?" *Technical Communication* 34 (1987): 243–249.

17. Terrell G. Manyak, "The Management of Business Writing," *Journal of Technical Writing and Communication* 16 (1986): 355–361.
18. Susan Dressel, J. S. Euler, S. A. Bagby, and S. A. Dell, "ASAPP: Automated Systems Approach to Proposal Production," *IEEE Transactions on Professional Communication* 26 (June 1983): 63–68.
19. Hetty L. Proietti and Joyce L. Thomas, "Multifunctional Team Dynamics: A Success Story," in *Proceedings of the 35th International Technical Communication Conference* (Washington: STC, 1988): MPD 58-MPD 60.
20. Alice M. Dilbeck and Laura A. Golowich, "Developing the Integrated Documentation Team," *Proceedings of the 35th International Technical Communication Conference* (Washington: STC, 1988): MPD 31-MPD 33.
21. Wendy Milner, "The Technical Writer's Role in On-Line Documentation," in *Proceedings of the 33rd International Technical Communication Conference* (Washington: STC, 1986): 61–64.
22. Thomas R. Barstow and Joseph T. Jaynes, "Integrating Online Documentation into the Technical Publishing Process," *IEEE Transactions on Professional Communication* 29(Dec 1986): 37–41.
23. Mark S. Smallwood, "Including Writers on the Software Development Team," in *Proceedings of the 33rd International Technical Communication Conference* (Washington: STC, 1986): 387–390.
24. Susan L. Fowler and David Roeger, "Programmer and Writer Collaboration: Making User Manuals that Work," *IEEE Transactions on Professional Communication* 29(Dec 1986): 21–25.
25. Joe Chew, Julie Jandel, and Anthony Martinich, "The New Communicator: Engineering the Human Factor," in *Proceedings of the 33rd International Technical Communication Conference* (Washington: STC, 1986): 422–423.
26. Lee W. Wimberly, "The Technical Writer as a Member of the Software Development Team: A Model for Contributing More Professionally," in *Proceedings of the 30th International Technical Communication Conference* (Washington: STC, 1983): GEP 36-GEP 38.
27. Tim Bosch and Larry Levine, "Technical Communication and the Technical Design Process," *Proceedings of the 32nd International Technical Communication Conference* (Washington: STC, 1985): MPD 28-MPD 31.
28. David Englebert, "Storyboarding—A Better Way of Planning and Writing Proposals," *IEEE Transactions on Professional Communication* 15(Dec 1972): 115–118.
29. James R. Tracey, "The Theory and Lessons of STOP Discourse," *IEEE Transactions on Professional Communication* 26 (June 1983): 68–78.
30. R. Green, "The Graphic-Oriented (GO) Proposal Primer," in *Proceedings of the 32nd International Technical Communication Conference* (Washington: STC, 1985): VC 29-VC 30.

DEVELOPING YOUR UNDERSTANDING

1. **Summarize the advantages and disadvantages of each model: division of labor and integrated teams.**
2. **Compare and contrast the kinds of social space created by each model: division of labor and integrated teams. In particular, describe how the professional writer's work and organizational role differ between the two models.**

3. Killingsworth and Jones claim that "more and more often communicators rather than technical personnel are assuming broad managerial duties." This claim suggests that professional writers, in more and more instances, have greater control in designing/shaping the social space of document project work, at least. Describe the kind of social space you would, in general, work to design, and outline what steps you would/could take to develop such a space. (Consider Killingsworth and Jones' topic headings, like training, group/team formation, and writing process, as topics for creating possible "steps.")
4. Most people studying to be professional writers do not imagine themselves as also preparing to be managers, yet Killingsworth and Jones (as well as other articles within this collection) strongly suggest that a good part of a professional writer's work is managerial. Identify some of the managerial work for which professional writers might be responsible. Then, analyze how such managerial work involves professional writers in the production of social space.

FOCUSING ON KEY TERMS AND CONCEPTS

Focus on the following terms and concepts while you read through this selection. Understanding these will not only increase your understanding of the selection that follows, but you will find that, because most of these terms or concepts are commonly used in professional writing and rhetoric, understanding them helps you get a better sense of the field itself.

1. hypertext
2. automating
3. informating
4. contraction
5. expansion
6. magic bullet theory
7. articulation

After Automation: Hypertext and Corporate Structures

JOHNDAN JOHNSON-EILOLA
Purdue University
STUART A. SELBER
Clarkson University

Early claims for hypertext reveal some of the medium's social and intellectual and revolutionary potential, but specific hypertexts often merely support and deepen status quo, relatively hierarchical social and textual relationships. Because these texts

Source: Johnson-Eilola, Johndan, and Stuart A. Selber. "After Automation: Hypertext and Corporate Structures." *Electronic Literacies in the Workplace: Technologies of Writing,* 1996, pp. 115–141.

are seen as ways of automating existing patterns of work and control, they often act only to contract vital processes of communication. By thinking of hypertext as having the potential to expand communication processes, we might encourage a broad-based, positive shift involving not only new emphasis on the roles of the reader and writer, but reconsideration of the social situation and technology itself.

> ***If a skilled typist could consistently turn out sixty words per minute, why waste her time on filing or answering the telephone? A skilled typist was likely to be kept in her job for as long as her employer could keep her there. . . .***
>
> ***It should be emphasized that there was nothing inherent to the typewriter which compelled such an organization of clerical work. The typewriter, in fact, can be quite useful for people who operate it sporadically. . . . The organization of work is largely determined by the efforts of businessmen and scientific office managers to organize their clerical labor as profitably as possible, and not to make the "inefficient" error of having a typist do work that a lower-paid file clerk could just as easily do.***
>
> **—Margerie W. Davies (1988, p. 34)**

In this chapter, we offer a general framework for complicating the relations between various types of hypertext, corporate structures, and technical communication. We argue that commercial hypertexts, as they are currently constructed by technical writers, frequently tend toward automating and conserving traditional, hierarchical corporate structures and contracting the scope and importance of communication.[1] Although such forms of automation often constitute valuable improvements over old ways of work, an overreliance on the automation of communication activities often disempowers both users and technical communicators. The majority of our discussion centers on readers, rather than writers, of hypertexts in technical communication settings. We are convinced that the low value placed on the act of *reading* and *using* technical documents in the "automating" view of hypertext bears much of the burden for the parallel low status of the writers of such automatic texts. (Similarly, see Dautermann's claim in this volume that users who underutilize computers "may devalue writing in general.")

We begin with a brief sketch of the current state of hypertext in business and industry. Comparing historical conceptions of hypertext to the medium's most popular current uses, we argue that some important and powerful aspects of hypertext have been left largely undeveloped or restricted to specialized sites and users. A large degree of this uneven development is due not to the isolated technology itself, but rather, to emphases on efficiency and short- over long-term profit and productivity in some versions of corporate and industrial cultures (see Wieringa et al., this volume). Such dynamics are not in themselves repressive or disempowering, but often become so dominant that they override other concerns and spaces of action. In the final sections of this essay, we critique and attempt to extend distinctions Zuboff articulated between "automating" and "informating" technologies, highlighting important social relationships and tendencies that influence the shape of communica-

---

[1] For discussions, see Burke and Devlin (1991); Horton (1990); and Barrett's edited collections on hypertext (1988; 1989; 1992).

tion and communication processes. We offer a potentially profitable rethinking—and necessary complication—of the relations among work environment, technology, writer, and reader, as those concepts are embodied in hypertexts being produced and used in corporate settings.

WAITING FOR THE REVOLUTION: A HISTORY AND SURVEY OF HYPERTEXT

New technologies are commonly integrated into cultures in conservative ways, strengthening rather than defying existing relations of social and political force (Sproull & Kiesler, 1991; Marvin, 1988; Kanter, 1989). But the contemporary state of hypertext contrasts sharply with the revolutionary potential prophesied by some of its originators. Although hypertext has gained popularity in the last ten years, its history goes back to at least the 1940s. In the pages of both the *Atlantic Monthly* and *Life* in 1945, Vannevar Bush wrote enthusiastically about his design for the proto-hypertext "memex," a machine the size and shape of an office desk.[2] To Bush, the memex represented a powerful tool for drastically improving human communication and, therefore, society in general: "Presumably man's [sic] spirit should be elevated if he can better review his shady past and analyze more completely and objectively his present problems" (p. 1/54).

Like Bush, later hypertext pioneers such as Ted Nelson and Douglas Englebart sensed the failure of traditional print media to accommodate the ever-increasing and ever-diverging tide of interrelated information, as well as the restrictions print media placed on research and scholarships. Although Ted Nelson's Xanadu remains more of a conceptual than actual product, the influence of Nelson's vision of hypertext remains strong: a worldwide "docuverse" holding the interconnected web of all the world's literature—a category in which Nelson includes not only traditional works of high culture but also popular literature, scientific work, and informal communication; the system is designed to encompass any type of text (Nelson, 1982). In many ways, the growing World Wide Web begins to approximate some of the functions features of such a docuverse. Nelson (1990) envisions a simple royalty system for writers and publishers (p. 2/33); because readers have the same authoring privileges as writers, "publishers" can mean any users of the system interested in placing their own text in the network (p. 2/42–43). And Englebart, speaking of the NLS/Augment system he designed in the 1960s, characterized his version of hypertext as "the biggest revolution you had ever seen for humanity, in the sense of people being able to connect their brain machinery to the world's problems. And it was going to go on for many, many decades" (cited in Englebart & Hooper, 1988, p. 27).

What is most notable about these original versions of hypertext are the ways in which the medium pointed toward a revolution in not only ways of writing and reading, but also—and more important—profound *social* shifts. Although there are certainly numerous theorists and designers today who remain committed to exploring the use of hypertext in revolutionary ways, such uses are largely relegated to computer-oriented research and development facilities (e.g., NoteCards at Xerox

---

[2] For a fuller discussion of Bush's career, see Nyce and Kahn's edited collection (1991).

PARC; gIBIS and rIBIS at MCC) or in some educational sites (e.g., the use of Storyspace, Intermedia, and some Hypercard stacks). In general, hypertext has been designed—and perceived—as a tool for increasing the simple, technical efficiency of existing print-based tasks rather than as a forum for transforming tasks in a broader, social sense—not only making a task easier or faster, but reconstructing communication and work environments as well.

One obvious presupposition of such a conservative approach is that hypertext versions of paper-based documents should be little more than faster, electronic versions of original source text. Thus, these systems tend to encourage, in terms of design, hierarchical indexing of topics mapped politically and cognitively to book technologies. In terms of use, for example, these systems encourage browsing of indexes and existing connections between author-generated links (see, for example, online help in Microsoft Word or PageMaker), as opposed to less traditional, but perhaps more valuable, user-generated associative trails. Certainly, such traditional use of hypertext provides a valuable addition to the growing repertoire of technological aids that modern workers draw upon: more efficient retrieval of information represents an important investment for business and industry, especially as workers and users contend with increasingly large amounts of available information. But the capability of hypertext to virtually emulate other literacy technologies and dominant cultural forms can mold specific instances of this technology into well-worn channels of hierarchical control.

Shoshana Zuboff's critique of computerization in industry offers one important perspective in examining the reasons for technical communication's concentration on making efficient, rather than transforming, traditional work practices. Zuboff (1988) defines two main methods of computerization in corporate sites: automating and informating. Automating technologies act to speed up the pace of work by translating repetitive, predictable human activities (such as turning pages or locating cross-referenced material) into machine instructions. According to Ritchie (1991), such computerization strategies may tend to reinforce the traditional "logic of bureaucracy." Informating technologies, in contrast, produce new information based on automated tasks. For example, a hypertext-based procedures manual for equipment maintenance might *automate* a maintenance person's navigation of an online text. Henderson (this volume) provides the useful distinction between improving on *processes* and on *by-products:* the types of automation frequently found in hypertext improve on the by-products but do not encourage users to rethink the fundamental processes in question. In this case, the user's task has not changed in a substantial way. Informating texts provide users with new possibilities, but they do not (at least in theory) require specific uses of the new information they provide. Informating texts oscillate between cycles of automation and user control. So, in the hypothetical maintenance manual, the text could not only automate communication processes but also *informate* by offering the user additional information—suggesting alternate procedures for maintenance, allowing users to communicate with other users in similar situations, providing historical tracking of the performance of the equipment in question compared with similar technologies or contexts, etc.

A more specific example of informating technology comes from the aerospace industry, which is currently converting many of the paper-based technical manuals associated with its attack helicopters into electronic technical manuals (ETMs).

Although strictly automated paper manuals are, arguably, less useful during military operations than paper-based manuals—they represent additional electronic equipment that is not easily deployed or maintained in combat environments (Schnell, 1992)—informating versions of such manuals might prove useful. The new Comanche attack helicopters (the RAH-66) will contain ETMs that provide various access paths for maintenance or operational tasks. For example, "if a soldier's request is in the form of a trouble code from the onboard diagnostics, the actual corrective measures will appear on the screen along with the related logistics procedures to update the aircraft logbook and requisition parts" (Schnell, 1992, p. 25). As this example illustrates, such an ETM could be considered informating in that it highlights, for users, related information and suggested tasks. Notably, the system merely poses secondary activities instead of completely automating the process and removing control away from the user.[3]

Despite the seeming usefulness hypertexts, current workplace structures tend to mitigate against the dispersal of control encouraged by such technologies (Schrage, 1990). As Zuboff (1988) notes, informating technologies seem to threaten traditional hierarchical organizations because they encourage decision-making capabilities and skills to move outward from centralized control; in other words, they foster networked, rather than hierarchical, relationships (see also Kanter 1989; Drucker, 1988; Reich, 1991; Hansen, this volume). Although scholars and researchers have proved numerous examples of and arguments for informating-class hypertexts (VanLehn, 1985; Johnson-Lenz & Johnson-Lenz, 1992; Selfe et al., 1992; Johnson-Eilola, 1992; Selber et al., 1996), such systems are largely relegated to corporate research and development sites and educational institutions. Even World Wide Web browsers often act only to automate activities such as looking up entries in a library card catalog. For the most part, developers and users seem comfortable automating traditional tasks in order to realize immediate and easily distinguishable increases in simple, technical efficiency. Such tendencies—which are understandable but should not be unquestioningly accepted—have channeled the development of hypertext along relatively limiting and limited paths. To more fully understand the reasons behind this uneven development, we can examine the parallels between the hypertext and other technologies that were shaped by (and exerted shaping forces upon) the social environments from which they emerged.

CRITIQUING CATEGORIES: PROBLEMS WITH AUTOMATE VERSUS INFORMATE

Some of the difficulty in classifying hypertexts as either automating or informating stems from the fact that our distinctions between these two types appear, at this stage, to be a feature of the technology disconnected from social use. Such a flaw hints at a vague sort of technological determinism: the technology seems to determine its own

[3] The notion of agency remains problematic—but still productive—here because in one sense every user's decision takes place in a specific context and in a specific historical sequence. We do not want to assert that any of the forms of text we discuss here ever offer a mythical high ground divorced from ideology. We are interested in locating and complicating some of the forces that silence other concerns.

use regardless of how a person uses the technology. To a degree, this is true: a hammer embodies certain possibilities, significantly different possibilities than does a screwdriver. For each technology there exists a range of possible uses. These possibilities are not completely determined by the technology but are mutually constructed in the nexus of both the technology and social situation (which are, themselves, complex and often contradictory constructions). A simple technology such as a mirror encourages one type of use in dressing rooms and quite another in high-energy optics. But in neither situation is the user able to freely substitute any other technology. The computer offers a particularly ambivalent technology, a "virtual" machine that can be easily molded to emulate a wide variety of mechanisms (Feenberg, 1991; Bolter, 1991). Zuboff (1988), in discussing the ways in which computers can either empower workers or alienate them, depending on how the specific types of uses are constructed in differing environments, notes that

> In many cases, organizational functions, events, and processes have been so extensively informated—converted into and displayed as information—that the technology can be said to have "textualized" the organizational environment. In this context, the electronic text becomes a new medium in which events are both observed and enacted. As an automatic technology, computerization can intensify the clerk's exile from the coordinative sphere of the managerial process. As an informating technology, on the other hand, it can provide the occasion for a reinvigoration of the knowledge demands associated with the middle-management function. (p. 126)

The case appears more complex for technologies such as hypertext. The purely automating features of hypertexts are apparently the simple substitution of one technology (the computer) for another (the book). But, as we have already illustrated, some uses and designs of computer technology begin to include an informating component differently from that of most books. The nature and shape of this transformation—technological and social—is not frequently reflected upon in the use of hypertext as an automating device. As Zuboff (1988) warns, "It is quite possible to proceed with automation without reference to how it will contribute to the technology's informating potential" (p. 11). Likewise, Carolyn Marvin (1988), in her historical analysis of electronic communication, argues that "[e]arly uses of technological innovations are essentially conservative because their capacity to create social disequilibrium is intuitively recognized amidst declarations of progress and enthusiasm for the new" (p. 235). Frequently, especially in hypertext, both automating and informating aspects coexist. However, because the informating aspects are often not reflected upon or articulated, their shape and function can become absorbed by the current social situation. That is, the automating features are touted and discussed, while the informating features are ignored or dismissed. The inertia of automation restricts the informating capacities to an invisible development along lines of preexisting forces. In addition, even in cases where hypertext informates work processes, the use of that new information may be restricted (used, for example, by management to track worker productivity and learning).

The importance of context becomes apparent when we attempt to classify technologies such as style-analysis or grammar-checking programs commonly included with word processors. For some users and contexts, these programs informate by analyzing text and offering numerous possibilities that users can act on as they wish. Other users in different contexts, however, may not possess the required skills, confidence, or motivation to do anything but accept the program's advice as correct. As many writing teachers have found, users who are not already knowledgeable about mechanical, stylistic, and rhetorical issues in a variety of discourses may use such programs in automating ways. In the same technology there exists the possibility for both informating and automating uses. Similarly, the procedures software tool (PST) described by Wieringa et al. (this volume), for example, would both automate and informate the process of writing procedures.

The possibility for both types of use with this technology does not mean that the technology is itself neutral, only that there are multiple forces involved in the construction of specific uses.

From the standpoint of management, a primary difficulty that stems from informating aspects of a technology is the degree to which that technology might require or even encourage skilled decisions on the part of workers. An automating technology commonly represents an attempt to remove not only "drudge work" but also the skill located in any one worker. For example, consider Harley Shaiken's (1986) discussion about the introduction of numerical-control technology in machine shops and an automatic turret punch press designed to stamp sheet metal parts based on computer-tape instructions. The manufacturer of the punch press offered free training courses to shop staff. Despite the interest in these courses expressed by the machine crew, shop management allowed only "an engineer, a foreman, [and] an electrical shop supervisor" to attend the free classes (p. 115). "As one worker commented later, 'The work program is of great concern because it is being used as a basis for justifying a removal of work. . . away from the sheet metal shop and into the hands of draftsmen and engineers'" (p. 115). Even when the work crew surreptitiously trained themselves in programming the machine—a capability resulting in higher-quality work—management, sensing the encroachment of worker control in the technology, installed an override switch on the machine that prevented the workers from entering or modifying machine instructions (p. 116).

On the surface, cases of automating hypertext seem very much different from the numerical-control machines discussed above; hypertext does not appear to de-skill workers in any substantial way.[4] What has been automated are tasks such as turning pages, retrieving manuals from bookshelves or distant sites, and discussions with colleagues sometimes necessary in troubleshooting or learning new procedures. In this view, hypertext has improved the efficiency of day-to-day tasks in ways that most workers would applaud. What has happened with common introductions of hypertext, however, is a general limitation of the informating aspects that are also possible with this technology. In general, hypertext has been framed in strictly automating terms, without reminders that the technology might also be articulated in ways that can support technical communication of a different kind, working to expand rather than contract processes of communication.

---

[4] Admittedly, one might complain that hypertext users may lose (or never gain) important skills related to using printed texts, but the loss of aptitudes such as dog-earing pages is of dubious importance.

REARTICULATING INFLUENCES: CONTRACTIONS AND EXPANSIONS OF COMMUNICATION

At this point, it might be useful to rethink Zuboff's terminology in a way that allows us to get at not merely the functional characteristics of an isolated technology, but also at the social and political context in which technologies are developed, used, maintained, and reconstituted. Instead of categorizing hypertexts as either one type or another on the basis of only concrete technological determinants, we need to broaden our scope and take apart the technology as it is used in order to look at the relations among the various elements. Although the automating/informating distinction offers a useful starting point, what becomes primary (from our perspective) is not the specific characteristics of any one technology but how those characteristics are taken up, channeled, defined, and defied by people. Because most hypertext applications possess at least some degree of informating capacity, our point is not that a certain type of hypertext generates information while another merely automates processes. For those technologies that informate, what is done with that information becomes central. In other words, *does a specific hypertext primarily contract or expand communication processes?* Framed this way, we can rethink how this technology is used in social situations, noting the influences that traditional corporate structures can exert over such uses. First, however, it may be useful to more fully define our terms.

The distinction we want to make is based on two opposing views of writing and reading (activities that, in hypertext, sometimes begin to resemble one another). Technical communication theorists frequently construct similar categories (Dobrin, 1989; Katz, 1992; Slack, Miller & Doak, 1993). In the first view, something we will call *contraction*, technical communication is a process of information transfer from sender to receiver based on the classic Shannon and Weaver model of communication (1949). Communication in a contractive technology shrinks in conceptual/visual size so that the link between sender and receiver is constructed as a frictionless, noise-free, and if possible, completely invisible wire. In this model, communication technologies are designed to increase accuracy of information reception and the raw speed at which information moves—in discrete, ideally unambiguous chunks—from writer to (relatively passive) reader.

At the opposite end of the spectrum lies the *expansion* view, in which writing and reading are themselves modes of thinking, less information transfer than a continual process of constructing and deconstructing multiple, often contradictory meanings. Communication in this view is a social and political process rather than a mechanistic transfer of information packets.

The contraction/expansion view represents a broad range of social and technological possibilities rather than easy pegs on which to place specific hypertexts. An important difference between Zuboff's automate/informate and our own contract/expand is the idea that the production of information is *continual* in most work contexts (even if not apparently emanating from a specific piece of computer technology). We are attempting to widen the sphere of concern to include not only the discrete technology (e.g., a specific database) but also the social construction of that information—a construction that, in part, determines how information can be used by specific workers. Thus, "information" is not only object delivered to an end-user, but also recursive process taking in user, designer, technology, and context. In a different but related context (that of groupware), Johnson-Lenz and Johnson-Lenz (1992) have observed the polarization that often develops between "mechanistic" and "open" systems of computer-based communication. As they warn, "[I]t is tempting to grasp for easy answers—either tighter mechanisms of social control or its polar opposite—refusal to make responsible choices. . . .[T]he way forward reveals itself as a dynamic balance. . . " (p. 291).

At the most contracted extreme, a text cuts off discussion and reflective thinking—the text offers, perhaps instantly, one, and only one, "correct" chunk of information. At the most expanded extreme, the text offers no unqualified answers and a huge number of navigational choices (but few hard and fast rules for users to distinguish which choice to make); an expansion medium encourages users to play out and construct possibilities. The distinction here lies, however, not merely in the quantity of data produced, but primarily in the *social* options opened for the circulation and reconstruction of that data by user-workers. We cannot equate the number of explicitly available paths through a text space with the degree of "freedom" people have in using the text. The actual experience of reading the hypertext may still be relatively contracted if a user's current social situation requires or even strongly suggests a specific path. Consider, for example, someone using a phone book to look up a single, predefined name and corresponding number: the phone book itself, as a concrete technology, offers thousands or even tens of thousands of differing paths, but the current user's context contracts those numerous possibilities to a single one. (This is not to say that phone books are oppressive, only that if someone claimed they represented a general technological breakthrough with profound social and political implications, we should be skeptical.) It is important to note that this contraction happens *prior* to reading experiences rather than during the moment-by-moment process of navigation that must occur in a temporal stream of reading and thinking. Thus, the distinction between contraction and expansion lies in the dynamic convergence of both social and technological forces.

Currently, the more open type of text is most frequently found in experimental fiction such as Michael Joyce's *Afternoon* or Carolyn Guyer and Martha Petry's *Izme Pass*, texts that continually challenge readers to navigate and reconcile the postmodern territories of collapsing subjectivity, indeterminacy, and complicity (see, e.g., Moulthrop, 1989; Landow, 1992; Douglas, 1991; Johnson-Eilola, 1994). We also see examples of hypertext-based collaborative writing environments designed to support both developmental and design work (Selber et al., 1996) and in some areas of the World Wide Web (although, at least currently, the Web encourages browsing rather than authoring for users in most contexts). Although few nonfictional texts reach the extreme of experimental fiction, one might consider Jay David Bolter's *Writing Space* hypertext an example that tends toward expansion. Bolter's text, based on an accompanying print text, integrates both a hierarchical structure and abundant extra-hierarchical material. On the one hand, readers are encouraged to expand on the reading patterns suggested by print texts, moving in and out of the structure in order to gain a fuller (but never completely unified) perspective on the implications of Bolter's arguments. On the other hand, the text contains some elements that encourage an automated use—the hierarchy, for example, connotes hypertext as an automation of the book. More important, although many theorists claim that hypertext readers should always be able to also become writers, Bolter's text is relatively closed, presenting a situation that may signify to readers that the information is traveling in one direction, from Bolter to reader (Johnson-Eilola, 1992; Amato, 1991; Tuman, 1992). *Writing Space* offers a mixture of contracting and expanding capacities that emerge differently, depending on specific actions of users. The importance of such oscillations suggest that automation can and should be a crucial—but not sole—element of communication.

As we have highlighted throughout, contracting/automating texts are the most popular applications of hypertext in technical communication. A naive explanation would claim that expanding texts are inefficient and offer little value to corporate users. But in conceptually and functionally contracting the processes of writing and reading—decreasing time spent in these activities as well as diminishing a sense of personal responsibility for the construction of meaning in both activities—corporate users face difficulties immediately evident on the surface. As we have already mentioned, the contraction of writing and reading purchases much of its foundation from the outdated "conduit" theory of communication: that information passes, in packets, from sender to receiver. In this view, more efficient media—implicitly, "better" media—are those that transfer the information packets with as little "noise" and as much speed as possible. Although this theory has been replaced by a host of more complex communication theories, the commonsense view of the conduit model continues to exert great influence over day-to-day operations in technical communication environments: the popularity of hypertext that emulates the book (and also attempts to construct virtual books as a transparent medium) testifies to its continuing survival.

From the perspective of readers, the drawbacks to the conduit model seem significant. For example, little responsibility is given to the reader's role in constructing meaning. Because the information in this view is actually constructed by the author, only "carried" by the medium and "received" (passively) by the reader, only the author can bear responsibility for the effectiveness of a message—a situation highlighted in the term often used for this model, "the magic bullet theory." The writer constructs a bullet and shoots it at the reader; if the bullet misses its mark, it is because the writer constructed an ineffective bullet, chose a poor weapon, or aimed sloppily or at the wrong target. The violent nature of this model aside, readers are implicitly discouraged from assuming any real responsibility or credit for their readings in a contracting medium. This perspective is supported by the current Society for Technical Communication's Code for Communicators, where writers are mandated to make meaning of texts for readers.

In contrast to this view is perhaps the most unique potential of hypertext as it was defined by early thinkers: it increases the power of users because they actively make navigational decisions in the act of reading. This tension is resolved, however, in automatic uses of hypertext when readers internalize another system of control, that of their work. Although information is being produced, the social context of the information production constructs a very contracted range of possibilities. In an expanding medium, however, readers might be encouraged to consider such matters as the possibility of multiple (even conflicting) interpretations or views in the text, the accuracy of the information, or the ideological agendas of the technology, the author, or the task.

In addition, the contraction inherent in the information transfer model is distributed unequally across job classes. As Harry Braverman (1974) argues,

> The recording of everything in mechanical form, and the movement of everything in a mechanical way, is. . . the ideal of the office manager. But this conversion of the office flow into a high-speed industrial process requires the conversion of the great mass of office workers into more or less helpless attendants of that process. . . . The number of people who can operate the system, instead of being operated by it, declines precipitously. In this sense, the modern office becomes a machine which at best functions only within its routine limits, and functions badly when it is called upon to meet special requirements. (p. 348)

Historically, automation has displaced work at the lowest levels of responsibility and social class: the forms-processing tasks normally completed by office clerks (Zuboff, 1988, pp. 133-159; Machung, 1988); the sewing and mending performed by women (Kramarae, 1988); the hands-on skills of the pulp-mill operator (Zuboff, 1988; Hirshhorn, 1984); and the traditionally female task of housework (Leto, 1988). Frequently, when workers are not completely displaced from their work in such situations, they find that they are now both isolated from co-workers and also expected to substantially increase their output (normally without corresponding increases in pay). The association of automating or contracting tasks with decreased responsibility and increased stress is not guaranteed, but often constructed by hierarchical, efficiency-driven structures of many workplaces. Thus, hypertext takes its place in this history of technological efficiency as a way of stream-lining the processes of reading and writing, contracting the activities to the point of disappearance or, at best, low significance.

Our critique to this point may appear overly pessimistic: currently, hypertext does not seem to portend the massive automation and de-skilling that accompanied technologies such as the automotive assembly line. But as hypertext becomes more popularly conceived of as a technology that transfers information and contracts communication, it becomes less likely that hypertext will be developed along expansive lines, especially for those classes of workers (such as the clams clerks discussed by Zuboff or many writers in corporations) whose tasks often fall under the totalizing goal of easy efficiency.[5]

REASSERTING RESPONSIBILITY: OPPORTUNITIES FOR TECHNICAL COMMUNICATORS

Despite frequent claims for hypertext in general, the popular online documentation form of this technology (often the epitome of the contracting class) does not necessarily move control from author to reader, but may in fact begin moving control away from both parties and into the machine/technology itself. In this way, hypertext fol-

[5] We would argue, somewhat egotistically, that academic and especially humanistic uses of hypertext frequently break free of these constraints because of the community's relatively greater emphasis on increasing personal development and acceptance of postmodern positions on unity, truth, and subjectivity. As we describe elsewhere (Selber et al., 1996), theoretical and pedagogical positions from fields such as composition offer an important perspective on the use of writing and reading technologies in corporate sites.

lows in the footsteps of other de-skilling technologies that were used by managers to translate operator knowledge into computers in order to exercise more complete control over information processes and products (Shaiken, 1986; Zuboff, 1988; Hirschhorn, 1984). Paradoxically, then, from the perspective of technical communicators, the activity of writing is not the construction of meaning but often the attempt to render their own positions transparent—functional texts should ideally transmit meaning directly from technology to user. Even more complex views of communication, such as the translation theory of communication, in which writers translate technical material into lay terms, tend to ignore the fundamentally political nature of the balance of power inherent in the relations among writer, reader, technology, and work environment (Slack, Miller, & Doak, 1993).

In recognition of this diminution of importance, we call on technical communicators to begin reconceiving the broadly political influence exerted through their role in the process of communication. Although assuming such a responsibility is never simple or expedient—and often, given the traditionally low status of technical writers in industry relative to scientists and engineers, such an assumption is potentially dangerous—technical communicators must begin to slowly, but purposefully, recognize both contracting and expanding forms of hypertext. Technical communicators might, for instance, begin offering end-users the capability to not only "receive" information from the hypertext, but also to become full-fledged authors capable of adding their own links and nodes to texts. (Tentative forms of these capabilities can be found in the bookmarking facilities of recent online help documents and in World Wide Web browsers such as Netscape and Mosaic.) Such a facility would not only increase the importance of both writers' and readers' interactions with the text, but, perhaps more important, can be connected to management's concerns for efficiency and flexibility. Users can customize systems so that they are more effective (in the broad sense of the term) in their specific situations.

Other approaches to thinking about text and meaning provide technical communicators with possibilities for increasing their own responsibilities in the communication process. Articulation theory, a movement generally attributed to the cultural criticism of Stuart Hall (1985), represents an attempt to explain and act upon the complex political/power relationships between language and culture. In this perspective, meaning is constructed in varying ways, depending on both object and social situation; differing environments engender different articulations. The concept of articulation offers technical communicators a powerful method for reconstituting the shape and relevance of communication in corporate and industrial sites, giving new importance and responsibilities to the roles and activities involved in writing and reading (Slack, Miller, & Doak, 1993). As a brief example, consider the case of a technical writer who might define her position of "writer" as "the creator of documents that make technology easier for novices to use." In this articulation, the term *writer* is connected or articulated to a number of culturally powerful concepts that affect and partially construct the meaning of the term: "writer" can be easily seen as articulated to terms such as "transparency," "information/knowledge transfer," "efficiency," and "clarity."

Dominant articulations such as these frequently organize or reinforce social relations as hierarchical structures; those at the top of these structures are frequently able, through various coercive and ideological means, to enforce articulations that more deeply entrench unequal power relations. As such, participants continually struggle over the dominant and subordinate articulations (Hall, 1985; Grossberg, 1986; Grossberg & Slack, 1985; Hebdige, 1988). As our foregoing discussion has shown, the automating or contracting orientation of many functional hypertexts tends to articulate both "reader" and "writer" to positions of low power—readers being passive receivers of information, and writers, at best, being possessed with the knack for allowing some "true" meaning of a technological problem to flow through themselves, into computer memory, from there to the reader's brain. The process of communication contracts, in this articulation, to become a mere function, a component necessary for technological use but not of great importance (except when it fails). From this perspective, engineers and scientists—those who actually created the concrete technology—have the most power; technical communicators, as well as readers and users, are somewhat like acolytes.

But, as Jennifer Slack, David Miller, and Jeff Doak (1993) argue, "technical communicator" can also be articulated, with some effort, as "author"—a person of relatively greater prestige, responsibility, creativity, and power. The approach of the articulation view itself also suggests a much stronger articulation for reader—no longer a box into which meaning is put, but now the person who constructs meaning (albeit under some strong constraints). This transformation in many ways parallels the distinctions we have made between contracting and expanding forms of hypertext: by articulating hypertext to this new, empowered relationship, both writers and readers might be able to resist efficiency as an *overriding* articulation (although probably it will always be one of many articulations exerting force). Hypertext can be articulated in an expansive way to embrace the active construction of meaning in communication by a whole environment of interconnected agencies: "From sender through channels and receivers, each individual, each technology, each medium *contributes* in the ongoing process of articulating and rearticulating meaning" (Slack, Miller, & Doak, 1993).

The difficulty for writers and readers of hypertext in corporate settings lies in the tenacity of the dominant articulation, efficiency (which is itself complexly articulated in terms of simple, technical/mechanical expediency). As we have shown, it is more likely that upper and middle management will integrate hypertext in a conservative way, one that not only resists change, but, in fact, deepens the articulation of communication as contraction by speeding up the frames of access and the ease with which information can be "received" instead of "constructed." But the opportunity exists for a different articulation, one of mutual activity and influence, especially if technical communicators can discover ways in which expansive hypertexts might construct a different meaning for their own positions, perhaps one of increased responsibility and power within an organization.

FINDING COMMON GROUND: CONVERGENT GOALS FOR TECHNICAL COMMUNICATORS AND CORPORATIONS

Even though corporations and their employees sometimes work at cross-purposes, expanding hypertexts can be productively articulated to each group. Technical communicators might find that this form of hypertext grants them a higher degree of responsibility and prestige, but these increases do not necessarily come at the expense of the corporation. The major impediment to getting corporations to understand the benefits (of what is obviously a more complex approach to text and communication) is in convincing managers and administrators to recognize the need for new corporate structures and strategies. Robert Reich (1991), for example, notes the ways in which organizations are shifting from high-volume enterprises (prioritizing hierarchical managerial structures and automation) to high-value enterprise (encouraging networked social structures and flexibility and mobility). Overly rigid hierarchical structures, one-way communication, and simple efficiency may be detrimental to deeper concerns such as innovation, market growth, and long-term financial (if not structural) stability. According to Peter Drucker (1988) and many others, old ways of doing business are no longer adequate within today's and tomorrow's global market.

In this light, contracting hypertext represent what might be a limiting holdover from print technologies, automated factories, and rigid corporate structures. Designers who construct hypertexts that primarily automate or contract existing paper-based activities presuppose that users want an online text that operates identically to its print-based counterpart, only faster; what is missing is the idea that such a text might offer a completely new range of textual possibilities. Hypertext as informating or expanding—the original vision of Bush, Nelson, and Englebart—may become more common as the social and political situations in which it is used and developed begin to change. As work processes and products become increasingly integrated and collaborative, hypertexts may begin to resemble Nelson's Xanadu or Englebart's oN-Line System (NLS).

The need for such an expanding construction of hypertext is being expressed in projects such as the Virtual Notebook System (VNS) at Baylor College of Medicine. In the VNS project, the architecture of the system is specifically designed to "enhance information sharing among scientists" (Burger et al., 1991, p. 395). And for engineers and systems designers at Boeing, the need arises as workers collapse the common distinctions that programs such as Hypercard make between writers and readers (or authors and browsers) into interdisciplinary teams. Collapsing such categories is important at Boeing because "[e]ach engineer contributes a unique perspective to the design of a product and its processes. The information they create is interrelated and these interrelationships must be represented in the data" (Malcolm et al., 1991, p. 14).

Hypertexts that primarily contract communication processes often mirror conservative corporate structures and practices and reinforce hierarchical business operations; those that mainly expand (and there are fewer of these) hold the potential to make qualitative changes within a corporation, down to the level of individual worker. We might assume, therefore, that as corporations begin adopting new structures and strategies for competing in the global marketplace, parallel shifts will occur in their use of technologies such as hypertext. The move from "data" to "information" projected by Drucker (1988, p. 46) parallels what Kanter sees in "postentrepreneurial" corporations, corporations responding to competitive pressures by implementing the following five changes in how they organize and conduct business (1989, p. 88):

1. A greater number and variety of channels for workers are available to take action and exert influence.
2. Relations of power shift from vertical to horizontal (from chains of command to peer networks).
3. Distinctions between managers and those managed diminishes.
4. External relations are increasingly important sources of internal power and influence.
5. As a result of the first four changes, career development of workers is less intelligible but also less circumscribed.

These changes within modern organizations may encourage more expanding, as opposed to contracting, uses of hypertext. In addition, these changes may allow technical writers to take more active roles in corporate communication processes, as proposed by the articulation model previously discussed.

One difficulty with much of the writing on postentrepreneurial approaches is the way in which they frame causality: Frequently, corporations such as Boeing come to recognize the value of new approaches to technology because the corporate structure and strategies have already shifted and now require new ways of communicating. We argue that such transformations can occur in parallel, new visions of the technology driving corporate development and vice versa. However, management is not often situated in a position that encourages seemingly radical changes. From a conservative standpoint, it appears there is little to gain. But technical communicators have a higher stake in this potential transformation: expanding hypertext not only places more value on the role of writer and reader, but also might help drive shifts in corporate structure at large that also increase the perceived value of new, expanding methods of communication, not merely in documentation but in all phases of corporate life. As Henrietta Shirk (1988) argues, even the shift from paper to online documentation, which arose from efficiency-based needs,

> requires changes in how technical communicators function in organizations that produce successful online documentation and perhaps even in how these organizations are internally structured. These changes in turn raise important issues about the professional preparation and development of online documentors. (p. 321)

The next stage of hypertext may afford technical communicators with opportunities to address these issues. Certainly, no single technical communicator can initiate sweeping changes in a multinational conglomerate. But possibilities exist, particularly in corporations that might already be considering change but are not completely convinced. By articulating hypertext as supporting both contractions and expansions in hypertext use, technical communicators can construct all aspects of communication as constructive, social activities

References

Amato, J. (1991). Review of writing space: The computer, hypertext, and the history of writing. *Computers and Composition: A Journal for Writing Teachers*, 9(1), 111–117.

Barrett, E. (Ed.). (1988). *Text, context, and hypertext: Writing with and for the computer.* Cambridge, MA: MIT Press.

Barrett, E. (Ed.). (1989). *The society of text: Hypertext, hypermedia, and the social construction of information*. Cambridge, MA: MIT Press.

Barrett, E. (Ed.). (1992). *Sociomedia: Multimedia, hypermedia, and the social construction of knowledge*. Cambridge: MIT Press.

Bernstein, M. (1988). The bookmark and the compass: Orientation tools for hypertext users. *ACM SIGOIS Bulletin*, 9, 34–45.

Bolter, J.D. (1991). *Writing space: the computer, hypertext, and the history of writing.* Hillsdale, NJ: Erlbaum.

Brand, S. (1987). *The media lab: Inventing the future at MIT.* New York: Viking/Penguin.

Braverman, H. (1974). *Labor and monopoly capital: The degradation of work in the twentieth century*. New York: Monthly Review Press.

Brown, P.J. (1987). Turning ideas into products: The guide system. In *Hypertext '87 proceedings* (pp. 33–40). Chapel Hill: University of North Carolina, Association for Computing Machinery.

Burger, A.M., Meyer, B.D., Jung, C.P. & Long, K.B. (1991). The virtual notebook system. In *Third ACM Conference on Hypertext proceedings* (pp. 395–401). San Antonio, TX: Association for Computing Machinery.

Burke, E., & Devlin, J. (1991). *Hypertext/hypermedia handbook*. New York: McGraw-Hill.

Bush, V. (1945/1987). As we may think. In T. Nelson (Ed.), *Literary machines* (pp. 1/39–1/54). South Bend, IN: The Distributors.

Carlson, P.A. (1988). Hypertext: A way of incorporating user feedback into online documentation. In E. Barrett (Ed.), *Text, context, and hypertext: Writing with and for the computer* (pp. 93–110). Cambridge, MA: MIT Press.

Charney, D. (1994). The impact of hypertext in processes of reading and writing. In C.L. Selfe & S.J. Hilligoss (Eds.), *Literacy and computers: The complications of teaching and learning with technology* (pp. 238–263). New York: Modern Language Association of America.

Cohen, N.E. (1991). Problems of form in software documentation. In T.T. Barker (Ed.), *Perspectives on software documentation: Inquiries and innovations* (pp. 123–136). Amityville, NY: Baywood.

Conklin, J. (1987, September). Hypertext: An introduction and survey. *IEEE Computer*, 17–41.

Davies, M.W. (1988). Women clerical workers and the typewriter. The writing machine. In C. Kramarae (Ed.), *Technology and women's voices: Keeping in touch* (pp. 29–40). New York: Routledge.

Dobrin, D.N. (1989). *Writing and technique*. Urbana, IL: National Council of Teachers of English.

Douglas, J.Y. (1991). Understanding the act of reading: The WOE beginner's guide to dissection. *Writing on the Edge*, *2*(2), 112–125.

Drucker, P.F. (1988, January/February). The coming of the organization. *Harvard Business Review*, 45–53.

Eisenstein, E. (1983). *The printing revolution in early modern Europe*. New York: Cambridge University Press.

Englebart, D., & Hooper, K. (1988). The Augmentation System Framework. In S. Ambron & K. Hooper (Eds.), *Interactive multimedia* (pp. 15–31). Redmond, WA: Microsoft.

Feenberg, A. (1991). *Critical theory of technology*. New York: Oxford University Press.

Grossberg, L. (1986). History, politics and postmodernism: Stuart Hall and cultural studies, *Journal of Communication Inquiry*, *10*(2), 61–77.

Grossberg, L., & Slack, J.D. (1985). An introduction to Stuart Hall's essay [Essay introduction to Hall, "Signification."]. *Critical Studies in Mass Communication*, *2*, 87–90.

Hall, S. (1985). Signification, representation, ideology: Althusser and the post-structuralist debates. *Critical Studies in Mass Communication*, *2*, 91–114.

Hebdige, D. (1988). *Hiding in the light. On images and things*. New York: Routledge.

Hirschhorn, L. (1984). *Beyond mechanization: Work and technology in a postindustrial age*. Cambridge, MA: MIT Press.

Horton, W.K. (1990). *Designing and writing online documentation: Help files to hypertext*. New York: Wiley.

Johnson-Eilola, J. (1992). Structure & text: Writing space and STORYSPACE. *Computers and Composition: A Journal for Teachers of Writing*, 9(2), 95–129.

Johnson-Eilola, J. (1994). An overview of reading and writing in hypertext: vertigo and euphoria. In C.L. Selfe & S.J. Hilligoss (Eds.), *Computers and literacy: The complications of teaching and learning with technology* (pp. 119–219). New York: Modern Language Association of America.

Johnson-Lenz, P., & Johnson-Lenz, T. (1992). Postmechanistic groupware primitives: Rhythms, boundaries, and containers. In S. Greenburg (Ed.), *Computer-supported cooperative work and groupware* (pp. 271–293). San Diego: Harcourt Brace Jovanovich.

Kanter, R.M. (1989, November/December). The new managerial work. *Harvard Business Review*, 85–92.

Katz, S.B. (1992). The ethic of expediency: Classical rhetoric, technology; and the holocaust. *College English* 54(3), 255–275.

Kramarae, C. (Ed.). (1988). *Technology and women's voices: Keeping in touch*. New York: Routledge.

Landow, G.P. (1992). *Hypertext: The convergence of contemporary critical theory and technology*. Baltimore: Johns Hopkins University Press.

Leto, V. (1988). 'Washing, seems it's all we do': Washing technology and women's communication. In C. Kramarae (Ed.), *Technology and women's voices: Keeping in touch* (pp. 161–179). New York: Routledge.

Machung, A. (1988). 'Who needs a personality to talk to a machine?': Communication in the automated office. In C. Kramarae (Ed.), *Technology and women's voices: Keeping in touch* (pp. 62–81). New York: Routledge.

Malcolm, K.C., Poltrock, S.E., & Shuler, D. (1991). Industrial-strength hypermedia: Requirements for a large engineering enterprise. In *Third ACM Conference on Hypertext proceedings* (pp. 13–24). San Antonio, TX: Association for Computing Machinery.

Marvin, C. (1988). *When old technologies were new: Thinking about electronic communication in the late nineteenth century*. New York: Oxford University Press.

Moulthrop, S. (1989). In the zones: Hypertext and the politics of interpretation. *Writing on the Edge*, *1*(1), 18–27.

Nelson, T.H. (1982). A new home for the mind. *Dafamation*, *28*, 168–180.

Nelson, T.H. (1987). *Computer lib/dream machines*. 2nd ed. Redmond, WA: Microsoft.

Nelson, T.H. (1990). *Literary machines 90.1*. Sausalito: Mindful Press.

Nielsen, J. (1990). *Hypertext and hypermedia*. Boston: Academic Press.

Nyce, J.M., & Kahn, P. (1991). *From memex to hypertext: Vannevar Bush and the mind's machine*. New York: Academic Press.

Parsaye, K., Chignell, M, Khoshafian, S., & Wong, H. (1989). *Intelligent databases*. New York: Wiley.

Raymond, D.R., & Tompa, F.W. (1987). Hypertext and the new *Oxford English Dictionary*. In *Hypertext '87 proceedings* (pp. 143–153). Chapel Hill: University of North Carolina, Association for Computing Machinery.

Reich, R.B. (1991). *The work of nations: Preparing ourselves for 21st-century capitalism*. New York: Knopf.

Ritchie, L.D. (1991). Another turn of the information revolution. *Communication Research*, *18*(3), 412–427.

Schnell, W.J. (1992). Automated technical manuals. *Army aviation*, *41*(2), 24–25.

Schrage, M. (1990). *Shared minds: The new technologies of collaboration*. New York: Random House.

Selber, S.A., McGavin, D., Klein, W., & Johnson-Eilola, J. (1996). Issues in hypertext-supported collaborative writing. In A.H. Duin & C.J. Hansen (Eds.), *Nonacademic writing: Social theory and technology* (pp. 257–280). Hillsdale, NJ: Erlbaum.

Selfe, R.J., et al. (1992). Online help: Exploring static information or constructing personal and collaborative solutions using hypertext. In *SigDoc '92 Conference proceedings* (pp. 97–101). Ottawa: Association for Computing Machinery.

Shaiken, H. (1986). *Work transformed: Automation and labor in the computer age*. Lexington, KY: Lexington Books.

Shannon, C.E., & Weaver, W. (1949). *The mathematical theory of communication*. Urbana: University of Illinois Press.

Shirk, H.N. (1988). Technical writers as computer scientists: The challenges of online documentation. In Edward Barrett (Ed.), *Text, context, and hypertext: Writing with and for the computer* (pp. 311–327). Cambridge, MA: MIT Press.

Slack, J.D., Miller D.J., & Doak, J. (1993). The technical communicator as author: Meaning, power, authority. *Journal of Business and Technical Communication*, *7*(1), 12–36.

Sproull, L., & Kiesler, S. (1991). *Connections: New ways of working in the networked organization*. Cambridge, MA: MIT Press.

Tuman, M. (1992). Rev. of Bolter, "Writing space." *College Composition and Communication*, *43*(2), 261–263.

VanLehn, K. (1985). *Theory reform caused by an argumentation tool.* Report ISL-11. Palo Alto: Xerox PARC.

Walker, J.H. (1987). Document examiner: Delivery interface for hypertext documents. In *Hypertext '87 proceedings* (pp. 307–323). Chapel Hill: University of North Carolina, Association for Computing Machinery.

Zuboff, S. (1988). *In the age of the smart machine: The future of work and power.* New York: Basic Books.

DEVELOPING YOUR UNDERSTANDING

1. Summarize how hypertext can be designed to (re)produce either (a) hierarchical or (b) flattened, networked social space.
2. Describe an automated hypertext with which you are familiar. Then, explain how the hypertext could be redesigned to informate.
3. Johnson-Eilola and Selber call on professional writers "to slowly, but purposefully, recognize both contracting and expanding forms of hypertext," and yet they warn that "given the traditionally low status of technical writers in industry relative to scientists and engineers," accepting such a call is "potentially dangerous." Analyze the organizational constraints and related "dangers" that you, as a professional writer, might need to be aware of if you choose to accept their call.
4. Johnson-Eilola and Selber suggest that professional writers can alter social space through their writing: "expanding hypertext not only places more value on the roles of writer and reader, but also might help drive shifts in corporate structure at large." Summarize how changing writing—for instance, changing mediums from print to hypertext—might cause changes in social space.

FOCUSING ON KEY TERMS AND CONCEPTS

Focus on the following terms and concepts while you read through this selection. Understanding these will not only increase your understanding of the selection that follows, but you will find that, because most of these terms or concepts are commonly used in professional writing and rhetoric, understanding them helps you get a better sense of the field itself.

1. technocratic approach
2. negotiated approach
3. Shannon-Weaver model
4. asymmetrical power
5. dominance
6. oppression
7. participatory democracy
8. symbolic analyst

Toward a Critical Rhetoric of Risk Communication: Producing Citizens and the Role of Technical Communicators

JEFFREY T. GRABILL
Georgia State University
W. MICHELE SIMMONS
Purdue University

In this article, we build on arguments in risk communication that the predominant linear risk communication models are problematic for their failure to consider audience and additional contextual issues. The "failure" of these risk communication models has led, some scholars argue, to a number of ethical and communicative problems. We seek to extend the critique, arguing that "risk" is socially constructed. The claim for the social construction of risk has significant implications for both risk communication and the roles of technical communicators in risk situations. We frame these implications as a "critical rhetoric" of risk communication that (1) dissolves the separation of risk assessment from risk communication to locate epistemology within communicative processes; (2) foregrounds power in risk communication as a way to frame ethical audience involvement; (3) argues for the technical communicator as one possessing the research and writing skills necessary for the complex processes of constructing and communicating risk.

Since the 1960s there has been an increasing concern for the environment in the United States. A number of risk communication scholars credit the 1962 publication of Rachel Carson's *Silent Spring* with launching the contemporary environmental movement (Belsten; Rubin and Sachs). In *Silent Spring*, Carson detailed the dangers of pesticides that accumulate in water, soil, and food, and as a result, generated a public awareness of environmental hazards. This awareness was heightened in 1969 when a series of environmental dangers plagued the nation, including the Santa Barbara oil spill, the seizure of eleven tons of salmon in Wisconsin and Minnesota due to excessive DDT concentrations, the Cuyahoga River in Cleveland, and the smog alert days in Los Angeles when health officials suggested that children not play outside (Belsten 30; Rubin and Sachs 54). The year 1970 brought the first Earth Day, and a "call for new initiatives to resolve environmental problems" (Belsten 30). The public was no longer content to leave the fate of human health and the environment to government and scientific experts. Communities demanded an explanation of what risks were present, and what was going to be done about those risks. When the U.S. congress began passing environmental regulations that provided for public involvement, such as the Comprehensive Environmental Response, Compensation,

Source: *Technical Communication Quarterly 7.4* (Fall 1998), "Toward a Critical Rhetoric of Risk Communication: Producing Citizens and the Role of Technical Communicators," Jeffrey T. Grabill, and W. Michele Simmons, pp. 415–441. Reprinted by permission of The Association of Teachers of Technical Writing.

and Liability Act (CERCLA) in 1980, the public legally became an important component in the decisions of environmental management. Risk communication evolved out of the need for risk managers to gain public acceptance for policies grounded in risk assessment methodologies. However, conflicts began to arise between the quantitative approach to risk assessment—the characterization of the potential adverse health effects based on an evaluation of results of epidemiological, toxicological, and environmental research—and the public's perceptions of risk (Plough and Krimsky). As a result, experts in risk assessment and management worked to design models for effectively and efficiently explaining risk to the public—risk communication.

Although Craig Waddell argues that communicating the risk of both natural hazards and consumer products is "an increasingly important aspect of the work of both technical experts and professional communicators" (*Risk Communication* 1), risk communication has been investigated primarily by communication, cognitive psychology, and risk assessment scholars. Too many of these approaches to risk communication have been arhetorical—typically, decontextualizing risk and failing to consider social factors that influence public perception of risk. Such a characterization of risk communication has been used to explain why communication problems arise—either audiences fail to understand risk and/or they reject what they are hearing (see, for example, Belsten). In this article, we build in arguments in risk communication that the predominant linear risk communication models are flawed for their failure to consider audience and additional contextual issues. We argue, following scholars like Craig Waddell and Barbara Mirel, that risk is socially constructed, and the failure to see risk as socially constructed leads to an artificial separation of risk assessment from risk communications. This separation can lead to unethical and oppressive risk communication practices because the public is separated from fundamental risk decision making processes. The claim for the social construction of risk has significant implications for both risk communication and the roles of technical communicators in risk situations. We frame these implications as a "critical rhetoric" of risk communication that (1) dissolves the separation of risk assessment from risk communication to locate epistemology within communicative processes; (2) foregrounds power in risk communication as a way to frame ethical audience involvement; and (3) argues for the technical communicator as one possessing the research and writing skills necessary for the complex processes of constructing and communicating risk.

INSTITUTIONS OF TRUTH: THE PRODUCTION OF RISK

Three disciplines contribute to much of the risk communication literature: risk assessment, cognitive psychology, and communication. While each has made significant contributions to the field of risk communication, each has to some degree contributed to the positivist positioning of most risk communication studies. We argue that it is important to examine these disciplines and related institutions in an effort to understand how hierarchies of power are established and exercised through each. Foucault asserts that institutions exercise power by regulating and constraining knowledge-making, production, and consumption through a system of rules and practices. Foucault also argues that by understanding the ways in which power is exercised, and looking for gaps in this system, we can work toward resisting, even altering these unequal power relations. An examination of the institutions currently contributing to risk communication will reveal that each still grants asymmetrical power in decision making processes to experts, and that this asymmetry serves to mask the complex ways of knowledge about risk is socially produced. Our purpose here is to foreground the institutional and disciplinary production of knowledge about risk in order to look later for gaps or "space" within these systems for change. (The concept of "the institution" is important, and we explain our use of the term in more detail later in the article. Here we use the concept of "institution" in two senses, as a discipline [or regular, shared ways of producing and distributing knowledge] and as an organization [or bureaucracies with policy and decision making power]. Our concern at this point in the article is with the interrelationships between disciplines and the organizations that mark the institutionalization of knowledge production and policy making about risk.)

The first and still prominent contributor to knowledge about risk communication is risk assessment. Scholars of risk assessment work to identify and quantify risk to human health and the environment to determine the probability of accidents or the spread of disease. Risk assessment develops scientific methods for generating an assessment of risk. Regulatory agencies—in the decision making processes of risk management—evaluate and compare remediation options in order to select an appropriate regulatory response to a potential hazard (National Research Council). Risk assessment is commonly located within private and governmental institutions that create risks or regulate them. Because risk assessment is a function of scientific and technological disciplines and experts, the access of others to these institutions is necessarily limited. When environmental regulations such as CERCLA began mandating that the public be involved in approving the risk policy, risk assessors began developing models of risk communication (Hance, Chess, and Sandman; Slovic). While the risk assessment literature promotes public inclusion in decision making about handling a hazardous waste site or risk situation, it asserts that the solution lies in educating the public and bringing public perception into conformity with scientific rationality. For example, an early and still dominant definition of risk communication adopted by the Environmental Protection Agency (EPA) asserts that risk communication is "the act of conveying or transmitting information between interested parties about levels of health or environmental risks; the significance or meanings of such risks; or decisions, actions or policies aimed at managing or controlling such risks" (Corvello, Sandman, and Slovic 112). Scientists and government officials who advocate this definition believe that making the public understand the risk will bring citizen approval. Risk is determined by experts and communication is the transfer of information from those who produce knowledge to those who consume it.

When researchers in risk assessment began to realize that the public rarely perceives risk the way risk assessors do, they began working with researchers in cognitive psychology to explain the discrepancies between "expert" assessment and public perceptions of risk. As Barbara Mirel explains,

> Because psychological theorists of risk see that an individual's outrage about risk is generated by cognitive reactions to social and ethical interests, they argue that the goal of risk

> communication must not be to educate citizens in expert "facts" to change their opinions but rather to evoke dialogue through a focus on the sources of a particular audience's outrages and fears. (45)

Together, scholars in risk assessment and cognitive psychology adapted psychometric scales to predict how public audiences would react to specific risks. Paul Slovic explains that psychometric scales

> ask people to judge the current riskiness (or safety) of diverse sets of hazardous activities, substances, and technologies, and to indicate their desires for risk reduction and regulation of these hazards. These global judgments have then been related to judgments about the hazard's status on various qualitative characteristics of risk. (408)

Risk communicators currently use psychometric risk factors to determine how best to adapt their initial message to the public and how to negotiate the decision making process (Slovic). However, these factors focus primarily on the risk itself, not the public or on a range of other contextual issues. For example, a person might be asked to rate the hazard of living near a chemical company that produces toxic chemicals on a scale for nine different attributes. Attributes include how well known the hazard is to the person, how harmful the effects of the hazard would be, how frightening those effects are, how easily those effects can be controlled, how easily the hazard itself can be avoided, and so on. Risk is then characterized with reference to these attributes. Health assessors consider these risk types static and universal perceptions. As a result, they often approach risk similarly in all communities, often leading to inappropriate policies and hostile reactions from involved citizens (see, for example, Ross 176).

Slovic argues that one generalization that can be drawn from psychometric studies is that "Perceived risk is quantifiable and predictable. Psychometric techniques seem well suited for identifying similarities and differences among groups with regard to risk perception and attitudes" (408). We argue, however, that these generalizations pose a problem for risk communication because these labels, which try to provide objective estimates for the public's "irrationalities," often take on ontological status and do not account for differences across communities. People's risk perceptions are determined by real and localized situations, not hypothetical, decontextualized questions on psychometric scales. Barbara Mirel argues likewise:

> [S]ome researchers claim that such psychological theories are too limited for the needs of risk communication. They argue that by using the individual as the unit of analysis, psychological theories fail to capture critical social and cultural influences, influences that will not be adequately addressed by communications simply oriented to outrage factors. These critics see social and cultural structures and relationships as the units of analysis, contending that perceptions of risk are constructed and perhaps even determined by membership in certain social and cultural groups. (45)

When some of these factors are disregarded in the decision making process, the conflicts that result are conflicts over the very "truth" about risk. As Mirel argues, the "'real debates' going on in risk controversies. . . are over the institutions that different groups set up as 'decision processors'. . . questions about the distribution of power, the credibility of authority, and the legitimacy of decision making practices and procedures" (47). Methodologies drawn from cognitive psychology, like psychometric scales, do include public perception in the decision making process, but they deny citizens any real power in determining what factors should shape the risk policy. Here again, the institutional location of knowledge production is important. There are clear lines between knowledge producers and consumers, with citizens playing only limited roles in the construction of risk.

In addition to risk assessment and cognitive psychology, communication studies has contributed significantly to our knowledge of risk communication. Typical approaches in communication are encouraging in that they see risk communication as a two-way process—asserting that all involved parties have valuable information to contribute in decision making processes. However, with some exceptions (Juanillo and Scherer; Rowan "What Risk," "Goals," and "The Technical"), many studies attempt to establish rules and canons for effective communication that fail to acknowledge that social factors play a significant role in the public perception of risk and therefore should be considered in decision making processes (Adams; Heath and Nathan; Castelli). Work in communication, then, makes a significant start toward relocating aspects of risk assessment. However, little research has been done on these social factors, and therefore, little work has focused on how public perceptions and different interests, values, emotions, and rationalities can be incorporated into current institutional locations and practices of knowledge production about risk. Currently, risk assessment is separated from risk communication, and both stand in a more powerful (and removed) relation to the public. Understanding the exclusion of citizens from meaningful participation in the construction of risk itself is an important step toward understanding why many in risk communication feel it fails too often (e.g., Belsten). These institutional sites of knowledge-making, we will argue, constitute a location for changing risk assessment and communication.

CURRENT APPROACHES OF RISK ASSESSMENT AND COMMUNICATION

The institutional positioning of risk assessment, management, and communication is important for understanding how issues like risk, theories of communication, and audiences are perceived. The positions of risk assessors and communicators within industry and government often give a "practical" edge to published discussions of past risk situations or new models and tactics for future success. And it is in the practice of risk communication that most observers see systematic problems. Laura Belsten, for example, writes that the practice of environmental risk communication has "failed miserably," largely because agencies and firms dealing with risk decision making have excluded the public (31). She writes that most government agencies and private firms adopt some version of a "decide-announce-defend" policy in which decision making (risk management) happens behind closed doors with risk communicators then charged with defending the decisions to a sometimes hostile public. In this process, soliciting public comment happens only after the real decisions have been made. Belsten writes that the public's only recourse is to challenge decisions made for (and to) them in court or through public pressure on politicians, resulting in costly, time-consuming, and ineffective public policy processes (31-32).

The ineffective practice that Belsten describes is closely connected to the models of communication that inform practice. There are a number of models for risk assessment and communication to be found, and we think these models—to the extent that they are connected to practice—are descriptive of work in risk assessment and communication. Our review of the literature turns up four distinct models linked to risk communication, but here we have collapsed them into two categories. The first category we call "technocratic" approaches, a term we take from Waddell ("Saving") to describe positivistic, linear (one-way) approaches to risk communication. The second category we call "negotiated," a term we think describes a set of approaches developed as a critique of technocratic positions. Negotiated theories tend to work on an explicitly "democratic" model, yet as we will show, are still largely linear (and therefore limited) in their view of communication.

Technocratic Approaches

The technocratic approach is generally understood as a one-way flow of technical information from the "experts" to the public (Fiorino; Rowan "Goals"). In the technocratic approach, risk communicators to educate/influence the public to think about risk the way experts do (Plough and Krimsky 304). Technical aspects of risk, not the values, concerns, fears, and opinions of each local community are considered during decision making processes. This is a vision of communication with a long history. Before CERCLA, health assessors determined the quantitative risk of a hazard and dictated to the government or responsible industry how to alleviate that risk. The public, their concerns, questions, and opinions were excluded from the decision making process. Even after CERCLA, health assessors have had a difficult time understanding how to involved the public in the decision making process. The difficulty rests in the problems of accounting for a range of social, economic, cultural, political, and psychological factors in the largely quantitative decision making models used in risk assessment. For example, Steven Katz and Carolyn Miller, in their examination of a waste siting controversy in North Carolina, assert that in decision making contexts, "risk communication developed as an attempt to overcome these differences by 'correcting' the public's 'risk perceptions' so that they would better match the 'risk analyses' made by the experts" (116). Therefore, when there are conflicts between the determined risk of a situation and the public's perception(s) of that risk, the problems that result are constructed as problems of communication (e.g., lack of information). For example, Stratman et al. cit Milton Russell, an EPA administrator for policy, planning, and evaluation as characterizing the risk communication process in terms of a metaconduit model:

> Let's imagine risk reduction as a consumer-driven production and distribution process. Scientists, who assess the severity of the risks, are the manufacturers. Government regulators, who make risk management decisions, are the wholesalers. And professional communicators—network and newspaper journalists—are the retailers. We government regulatory wholesalers use risk characterizations from the scientists to explain the reasons for our decision. Then journalistic retailers pick up our product on the loading dock. . . [and]

> they present the news of the day. Based on those presentations, consumers of the news decide to buy the news or not, use it or misuse it, and change their behavior or demand that public officials change theirs. . . . If citizens misjudge risk, their orders will still come through, and the government machine still delivers, but the results don't necessarily leave citizens better off. (qtd. In Stratman et al. 10)

The model of communication here is strikingly similar to Shannon and Weaver's model. Knowledge is constructed prior to communication, and miscommunication is attributed to "noise" (or irrationality) along a one-way, linear channel. The "technocratic" approach of communication, then, sees risk as determined by experts prior to communication. "Effective" risk communication is the result either of transferring information to a public that understands and accepts it, or in some formulations, persuading the public to accept a given risk (see Sandman). In either case, knowledge is "scientifically" produced prior to communication, communication itself is largely linear, and audiences are seen as needing education and/or persuasion—management—and not as participants in the rhetorical construction of risk (see Porter, *Audience*, Chapter 3, for a discussion of audience management).

Negotiated Approaches

What we call "negotiated" approaches are actually a set of models that were developed as a critique and alternative to technocratic risk communication. They begin by questioning the technocratic assumption that risk assessment can be determined based solely on a defined set of principles and scientific norms independent of cultural values. Alonzo Plough and Sheldon Krimsky characterize the technocratic approach as one where "perceived responses to risk are important only in understanding the extent to which ordinary people's ideas deviate from the truth. . . from the perspective of technical rationality, risk can be studied independently of context" (305). Thus, scholars like Plough and Krimsky seek to solve communication problems through a more negotiated, two-way approach to risk.

Belsten's solution for poor risk communication is to offer a theory of "community collaboration," and we think her work is representative of useful negotiated approaches. Belsten's work rests on theories of public collaboration in public policy decision making (36). She argues, quoting others, that "collaboration occurs when a group of autonomous stakeholders of a problem domain [are] engaged in an interactive process, using shared rules, norms, and structures, to act or decide on issues related to that domain" (37). Any one who is affected by a given risk is considered a stakeholder, and community collaboration only works when a high degree of participation is included in public decision making about risk. In addition, stakeholders are included early in the process and have the option to say "no"—the acceptance of a given risk is voluntary (37-39). Negotiated approaches are important for their recognition of audience/the public as important participants in decision making processes. Furthermore, such approaches are useful in their recognition that participation must be wide-spread and take place early in decision making processes. Some scholars (e.g., Rowan "The Technical"; Ross) ground their work in a powerful theory of communicative ethics, usually Habermas. At its best, work relying on Habermas argues that decisions made currently within technical realms (e.g., administrative bureaucracies) should be deliberated publicly (see Blyler). The turn to Habermas makes sense; it is an attempt to argue questions of civic concern within a framework of discourse ethics that seeks to prevent coercion. But we see limitations with Habermas, and by extension, with solutions of many negotiated approaches. Habermas argues that only when a decision emerges from an argumentative discourse situation that is in accordance with the pragmatic rules of discourse is the result ("norm") justified (71). His rule-bound procedure ("practical discourse") "insures that all concerned in principle take part, freely and equally, in a cooperative search for the truth" (198). But Habermas's system of argumentation is idealized; participants in a "real" risk situation are not free and equal, and despite the Habermasian rules for what "should" structure ethical communication, risk situations rarely, if ever, approximate his ideal. In short, negotiated approaches are problematic for their failure to include a notion of power, and therefore, we are suspicious that they are capable of changing risk communication practice. As we argue for risk as socially constructed, we are looking toward an approach to participatory decision making that links a theory of power (and powerlessness) to the exercises of power involved in knowledge production (risk assessment/communication).

RISK AS SOCIALLY CONSTRUCTED

Our claim is that the risk of a given situation is socially constructed by a number of interests and factors. Indeed, when there are disputes and communication "problems" in a risk situation, we suggest that what is happening is not problematic or an impediment to be overcome. Rather, these problems are a public contestation over the meaning of risk—the "truth" about risk is actually a product of such disputes. Our concern with the social construction of risk, however, extends beyond epistemology. As disputes about risk are characterized by interactions between interests more and less powerful, the failure to account for power in decision making about the meaning of risk—a failure of both technocratic and negotiated approaches—can lead to the "oppression" of (typically citizen) audiences. Conceptualizing risk as socially constructed is important because (1) it locates knowledge-making within communication processes, and (2) it considers how power is differentially exercised in such processes.

As we have suggested, one of the problems associated with linear approaches to risk communication is the artificial separation of risk assessment from risk communication. This separation prevents a view of risk as socially constructed because knowledge production is not a function of communication processes. Rather than fostering an exchange and collaborative generation of knowledge that contributes to public policy, the technocratic approach, for example, sees the audience as consumers of information to be considered after decisions have been made. This audience-as-consumer stance is reflected in current CERCLA regulations on public participation and in Belsten's descriptions of typical practices, where public approval must be gained, but only after policy decisions have been made. The CERCLA section on public participation states that "before adoption of any plan for remedial action to be undertaken" the appropriate party must take both of the following actions: "(1) Publish a notice and brief analysis of the proposed plan and make such plan available to the public. (2) Provide a reasonable opportunity for submission of written and oral comments and an opportunity for a public meeting at or near the facility at issue regarding the proposed plan and regarding any proposed findings" (42 U.S.C. section 9617 CERCLA section 117). Further, a transcript of the meeting must be kept and made available to the public. While CERCLA section 117 (a) states that the analysis of the proposed plan must include "sufficient information as may be necessary to provide a reasonable explanation of the proposed plan and alternative proposals considered," public opinion does not have to be solicited or incorporated into the initial decision making process (42 U.S.C. section 9617 CERCLA section 117). This separation of assessment and communication denies the public the ability to actively participate in the production of public policy, too often resulting in public objections to policy and resistance to implementation (see Katz and Miller; Rowan, "What Risk"; and especially Stratman et al.).

The positivist view that science discovers an objective Truth through a rational, linear process has been called into question by many (e.g., Kuhn; Latour) who argue that knowledge is not an accumulation of facts that progress toward the Truth but is rather a collection of perceptions that are agreed upon by a discourse community. Rules for how science is conducted or how theory choices are made are negotiated by the practitioners of a community who hold to a shared set of methods and beliefs. The progression toward increased public involvement in the decision making about risk is contingent upon the concept of scientific knowledge shifting toward a negotiated understanding, a shift contingent upon creating a space within "the community" of risk assessors for "others." As a result, what some see as the problems of public outcry against risk assessments or policies can also be seen as the construction of "risk" itself. In this manner, risk communication becomes a complex web of issues and participants that work together to construct a risk policy. Similarly, Craig Waddell argues that

> risk communication is not a process whereby values, beliefs, and emotions are communicated only from the public and technical information is communicated only from technical experts. Instead, it is an interactive exchange of information during which all participants also communicate, appeal to, and engage values, beliefs, and emotions. Through this process, public policy decision are socially constructed. ("Saving" 142)

Rather than a linear flow of technical information from the risk assessors to the public, risk communication becomes a web, a network, an interactive process of exchanging information, opinions, and values among all involved parties. In contrast to all linear models, this approach flattens the hierarchy between the "expert" and "non-expert" and believes risk assessment must incorporate technical information about a risk within a broader framework, including social, political, and economic factors. Recently, similar socially constructed views of risk communication have been promoted by scholars such as Rowan ("Goals"); Juanillo and Scherer; and Plough and Krimsky. Additionally, Katz and Miller see this approach to risk communication as fostering participatory democracy, emphasizing "process more than results, with participating citizens gaining not only results but satisfaction and investment from their engagement in decision making" (133–34).

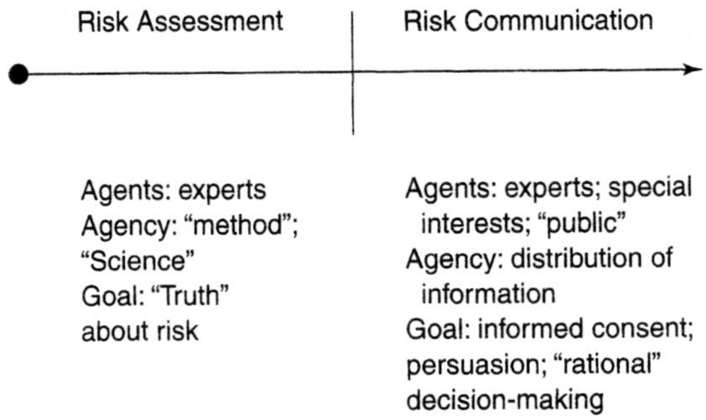

Figure 1. Traditional risk assessment and communication.

If, as we have argued, risk is socially constructed, then the separation between expert/public and assessment/communication cannot hold. We have constructed the separation of assessment and communication as a linear process of research and dissemination (see Figure 1).

According to this representation, the processes of risk assessment and communication are linear and sequential, and the risk assessors are typically "experts": scientists, statisticians, actuaries, economists, public health officials. Our characterization is not meant to dismiss or demean the expertise, insight, and inquires of intellectuals involved in risk assessment. Their work and expertise is crucial to the process. We are not arguing that risk assessment be thrown out, but rather than additional knowledge be added to the mix. Because, as we will argue, citizens are capable of contributing valuable knowledge to the decision making process, we do want to call into question the exclusive domain accorded risk assessors and the power of the expertise granted them. (As Winner argues, as groups contest public decision making about technologies, each often produces a set of experts who can differ widely on a given "scientific" question. Such a situation results in "[1] futile rituals of expert advice and [2] interminable disagreements about which choices are morally justified" [75]. The result is continual use of conflicting "expert" advice, tangled ethical and moral discussions about the design and use of technologies, and a range of intractable problems. Winner concludes that "political disputes about technology are seldom if ever settled by calling upon the advice of experts" [76]. In this way, Winner *dispenses* with expert opinion, at least in terms of the status typically given to experts.) Such a situation allows an arbitrary line to be drawn between assessment and communication activities, and such a line serves as a false border with epistemology linked to science on one side, leaving an impoverished rhetoric of "arrangement" and "style" on the other. The effect of this separation between assessment and communication is to frame communication in ways we have discussed earlier: risk communicators are given the task of disseminating information (the truth about risk) to various public audiences. The resulting rhetoric of risk communication—stripped of its epistemological possibilities—has at its goal the creation of consent, either via belief in the truth of the information or a range of persuasive strategies (see Katz and Miller). The problems risk communicators face often stem from the fact that the public resists their separation from the processes, and their resistance takes the only form available—rejection of risk communication and communicators (see Rowan, "What Risk," for examples of such rejection). (Stephen Doheny-Farina sees a similar problem in the area of technology transfer. He focuses on the notion of "uncertainty" in the largely linear communication models involved in technology transfer [uncertainty=difference between information needed and information available]. He writes that the theory of communication that underlies most practice separates knowledge [and knowledge production] from communication and sees knowledge as produced by experts in a realm separate from audiences and non-experts. In this way, knowledge becomes a commodity [8–9]. When communication is successful, there is little uncertainty and the "truth" is well-received. When there are problems, "the concept of uncertainty indicates that the communication problem is not one of meaning but one of the availability of the correct information" [9]. Doheny-Farina's description of the status of knowledge and models of communication in technology transfer describes how communication "problems" are constructed in technocratic approaches to risk assessment and communication.) Seeing risk as socially constructed situates knowledge making and audiences within communication processes.

The notion of risk as socially constructed also asks us to foreground power. While often alluded to in risk communication scholarship, "power" strikes us as the most undertheorized issue in the risk communication literature. In our view, technocratic approaches of risk communication can easily (and perhaps necessarily) oppress audiences. And even "negotiated" approaches are problematic because they tend to assume equal power relations within processes of negotiation. We take our notion of power and oppression from Iris Marion Young because it is directly connected to our central concern with participation in decision making. Domination and oppression are key terms in Young's work because they allow her to name certain practices "unjust" outside a distributive framework. For Young, domination occurs when people are systematically excluded from "participating in determining t heir actions or the conditions of their actions" (31). In many ways, most people are dominated in some aspect of their lives, usually at work or school—institutional systems that often do not allow people fundamental access to decision making. Young argues that the powerful draw on the resources of an institution through the everyday practices of the institution, thereby maintaining and extending their power. This is precisely what happens when experts, committees, or closed agencies control the decision making processes involved with risk assessment/communication (see Katz and Miller; Waddell "Saving" and "Defining"). To act powerfully within an institution, then, requires *access* to everyday processes and practices. James Porter writers that access can mean three things, infrastructural access (access to resources), literacy (education), and community acceptance (*RHETORICAL ETHICS*). In the sense in which access is important for decision making about risk, infrastructural access means access to the processes of decision making within an institution, literacy means the discursive means to participate effectively, and acceptance refers to a "listening stance," or a commitment to collaborative decision making. Access means not only "a place at the table," it means the rhetorical ability to participate effectively and the structured requirement that others listen.

For Young, oppression is a qualitatively different experience than domination. An oppressed group need not have an oppressing group; oppression is structural and relational as well as material and often the result of "humane" practices and intentions. In Young's schema, oppression has "five faces," yet we believe one is particularly relevant to risk communications—powerlessness. According to Young, powerlessness is an oppression that is the result of a lack of participation and a reliance on hierarchy. Many people have some power in relation to others. Young argues, however, that

> The powerless are those. . . over whom power is exercised without their exercising it; the powerless are situated so that they must take orders and rarely have the right to give them. . . [a] social position that allows persons little opportunity to develop and exercise skills. The powerless have little or no work autonomy, exercise little creativity or judgment in their work, have no technical expertise or authority, express themselves awkwardly, especially in public or bureaucratic settings, and do not command respect. (56–57)

The powerless, in other words, lack all three forms of access and therefore have little chance to change their position.

We believe that in order to be ethical and nonoppressive, decision making processes in risk assessment/communication must consider not only the scientific assessment of the risk posed, but also the values, emotions, and concerns of *all* involved parties, with preference given to the input of the less powerful (e.g., citizens, particularly those typically underrepresented). This interactive exchange of information among parties is a complex process that a linear model of risk assessment and then communication cannot accommodate. Rather, we envision risk communication as an intricate web of issues and participants that socially constructs policy. What we envision, namely, is a critical rhetoric of risk communication that sees risk as constructed socially and the *processes* of construction as the focus of concern.

RESEARCH AND THE TECHNICAL COMMUNICATOR: TOWARD A CRITICAL RHETORIC FOR RISK COMMUNICATION

A critical rhetoric for risk communication would be based on the following principles. First, it sees risk as socially constructed and rhetorical—an epistemic rhetoric that focuses on the construction and communication of risk. The meaning and value of risk in a given situation is a function of multiple and sometimes competing discourses. In this way, controversy about risk is reframed not as a problem or a negotiation between two parties (the risk maker and the audience) but as a complex web of stakeholders and positions that contribute to the meaning of risk in a given situation. Second, a critical rhetoric focuses on the processes of decision making, seeing these processes as the key institutional locations for knowledge making (e.g., within government agencies or legislation). In particular, a critical rhetoric focuses on the relations of power within decision making processes, asking questions about who participates and in whose interests decisions are made. And third, a critical rhetoric seeks to contextualize and localize risk situations and processes, a function both of its rhetorical approach and its concern with local participation. By contextualizing risk, a critical rhetoric counters the tendency to develop "scientific" and universal models of assessment and communication that treat risk algorithmically and audiences as universal, rational, and therefore silent. Viewing "the public" as universal and reasonable, traditional risk communication theory privileges logos and assumes all audiences think alike—that all rational individuals will be swayed by the same evidence. (We have a problem with the concept of "the public," and it is a problem we fail to solve here. In fact, quite the opposite is true. We have recourse to a [false] concept of "the public" as a rhetorical counter to "experts," but we believe the concept of "the public" to be largely fictional. The subject position of "the public" as coherent, unified, and identifiable strikes us as an unhelpful decontextualization of the people affected by a given risk and therefore important participants in the construction of that risk. Our hope is that the critical rhetoric we outline here necessarily asks risk assessors and communicators to help identify various "publics" and to take seriously the differences of a given town, neighborhood, or community.) In addition, by contextualizing risk, assessors and communicators must consider a range of issues related to the relationship of individuals to the social institutions involved in a given risk situation. Such a consideration would necessarily examine concepts of race, class, gender, and other issues important to a given community in the assessment and communication of risk, issues too little considered in the risk communication literature but which are nonetheless deeply relevant. An investigation of these factors hopefully would work to inform a more ethical approach to risk communication, uncover the suppressed voices in risk communication approaches, and be a starting point for improving the present condition of how certain groups are excluded from the decision making process in risk communication.

In effect, by changing the process we are looking toward more "democratic" and *effective* assessment and communication. However, we must be cautious in our use of the term democratic. In one sense, the technocratic approach is democratic because the public is involved in the last stage of the communication process—this is the participation of liberal democracy and interest-group politics (see Patterson and Lee). We are arguing for an approach that fosters *participatory* democracy; one that involves the public in fundamental ways at the earliest stages of the decision making process. Rather than telling the public about a risk or a decision, we are arguing for an approach that allows the public to actively participate in producing the policy itself. An approach that fostered participatory democracy would grant the public—and their contributions to the policy—equal and sometimes preferential status with the technical experts and their contributions. If we accept that citizens should be able to participate in decisions that affect themselves and their communities, and that they, themselves, are the best judge of their own interests, then we see that the technocratic approach is incompatible with participatory democracy. And because negotiated approaches often fail to account for asymmetries of power and conceptualize a limited number of stakeholders, we also see them as useful but limited. A critical approach, in short, seeks to solve problems of omission (e.g., people, positions, ways of knowing and talking), domination (e.g., failure to acknowledge differences in power), and indeed oppression. We think a critical approach will lead to *better* risk communication.

Research and the Technical Communicator

In their critique of the "environmental rhetoric of 'balance,'" Robert Patterson and Ronald Lee argue that the appeal to "balance" in the regulatory and political discourse connected to the Kinglsey Dam relincensure "procedurally diminish[es] the public" (26). Patterson and Lee write that the processes of decision making in relicensure procedures are concerned with identifying interests and then gathering information from these interests. The problem is that "the public" does not participate in these decision making processes but rather is represented and seen as "consubstantial" with the organized interest groups capable of generating the expertise and access necessary to participate in decision making (28–29). In this respect, Patterson and Lee argue that "the subjective experience and moral reflection of ordinary citizens are discounted" (29).

We agree with their reading of this decision making process but feel that their characterization exposes an even deeper problem of representation and participation. The process they describe is democratic—citizens, through organized representation, can and do participate in public policy decision making. However, the very notion of "the public" is a representation. For us, the problem, then, is not representation, but what kind of representation and how that representation takes place. It is possible, in other words, to construct risk communication practices that allow citizens (various publics) to represent themselves in decision making processes and thus add more of a participatory element to our current (and limited) representative participation in decision making about risk? We think it is possible and look to the skills of the technical communicator to help articulate such a view.

As a way to conceptualize how a critical rhetoric of risk communication can be implemented, we want to reconceptualize risk communication as a type of technical communication. But beyond this, we are interested in a more research-driven, analytical role for the technical communicator, one that we hope can help break down the line between risk assessment and communication.

There seems little question that technical communication has been changing due to changes in communications technologies and the political economy. Dan Jones writes that new technologies are challenging the professional identify of technical communication, but he also asserts that changes driven by these new technologies do not take technical communication away from a central purpose and expertise as advocates for users. To this discourse of professional change, Johndan Johnson-Eilola adds more fundamental changes in the political economy that have altered the value and nature of production from industrial production to information exchange. Johnson-Eilola argues for the necessity of relocating the value of technical communication by reframing it as "symbolic-analytic" work that "mediates between the functional necessities of usability and efficiency while not losing sight of the larger rhetorical and social contexts in which users work and live" (246). Johnson-Eilola's contextualization of technical communication is an important move because it forces technical communicators to wrestle with tough intellectual, ethical, and political issues.

We are interested in his adaptation of Reich's symbolic-analytic worker because such a framework describes, we believe, the complexities involved in risk communication—a type of symbolic-analytic work—and therefore articulates a connection between the research burdens of risk communication and the technical communicator. In Johnson-Eilola's adaptation of the symbolic-analytic worker as technical communicator, he lays out a set of characteristics that we believe are important for seeing the technical communicator as a key player in risk assessment/communication situations.

Johnson-Eilola writes that the technical communicator as symbolic-analytic works with information and produces a wide range of documents in a variety of media—typical for most professional communicators. The symbolic-analyst, however, is also capable of "experimentation," "collaboration," "abstraction," and "systems thinking." In other words, the technical communicator as symbolic analyst can conduct research, work with others from multiple specialties, "discern patterns, relationships, and hierarchies in large masses of information" (Johnson-Eilola 260), and think systematically in ways that construct relationships between disciplines and within messy situations. Furthermore, the technical communicator as symbolic analyst retains an advocacy role. While many professionals function as hired advocates, we want to retain the historical advocacy for "users" that is common to the ethos of technical communicators and expand that notion of advocacy to include those who normally do not hire professionals (e.g., citizens or "the powerless"). The connection of some of the symbolic-analyst's characteristics to risk communication is obvious. But we are particularly interested in crossing the gap between assessment and communication, and we think the research capabilities of the symbolic-analyst allows the technical communicator to cross that gap, an issue that warrants further discussion.

Usability Research: Participation and Epistemology

While risk assessment may traditionally be done by biological and chemical scientists, actuaries, statisticians, and psychologists, the technical communicator as symbolic analyst can add research practices other specialists cannot provide. In particular, it is the technical communicator who can insert the audience/public/citizens directly into the risk assessment process through usability testing. We are not talking about usability strictly in terms of document production (although we are not excluding this). Rather, we are looking toward a wider range of research practices—a range of contextual interviewing and observation practices in particular—that necessitates researchers work *with* audiences in the construction of knowledge (e.g., risk). For example, Katherine Rowan ("Goals") asserts that in order to create awareness and concern about risk, messages from all parties must be detectable, decodable, and considerate to all. Usability testing could examine such messages as well as their delivery mechanism in order to facilitate a more effective negotiation and decision making process among involved parties.

Usability is powerful because we believe that "users"—citizens, the public, stakeholders—have important knowledge often excluded from decision making, and it is this "user knowledge" that usability testing can get to. The analogy we are making here is that the user of technologies and documents found in most usability work is like the "user" of technologies and documents in risk situations. In this case, members of a given community are "users" of the public spaces and environmental resources as well as the risk communication distributed to them. But more importantly in our view, users in all contexts are *potential* participants in decision making about technologies—from computer interfaces and documentation to waste incinerators and construction projects. We feel they should be *actual* participants, and in this regard, we want to use Bob Johnson's work to develop a notion of user knowledge for that participation.

Johnson writes that most user are perceived as "dumb," at the "bottom of the proverbial epistemological ladder," and therefore designers of technologies seek to "idiot-proof" technologies to keep users out of harm's way. In contrast to this, Johnson believes that users are productive and therefore "have a responsibility for the design and implementation of technology" (57). He continues,

> The reason for the absence of discussions of users and use runs deep in the history of western culture, and at the root is the question of ownership of knowledge—in short, the question of epistemology. Who creates knowledge? Is it created only by those who we generally equate with knowledge, like philosophers or scientists? Or is knowledge production also within the province of those generally associated with "the practical," such as technicians or users of technology? (58)

The answer is that users are productive, and Johnson glosses three types of user knowledge. (The connection between Johnson's notion of users and use producing knowledge is clearly, connected to Scandinavian design of technologies. Designers like Pelle Ehn have long argued for work-oriented participatory design, and as a part of that work, have often articulated a notion of epistemology linked to users and use.) Users as "practitioners" are capable of "cunning intelligence" (*metis*), or the ability to use technologies in new, effective, and context-dependent ways not envisioned by front-end designers. Users as "producers" are more clearly connected by Johnson to the design of technologies, but most importantly for our purposes here, the notion of users as "citizens" focuses on how users can become "responsible members of the technological community" (77). Johnson writes that the "spaces in which the knowledge of users and the cultural environment intersect are difficult to describe. . . because there is no terminology for discussing users in large, social contexts" (77). We feel that risk situations provide the terminology that Johnson is missing, and furthermore, we feel that a recognition of user knowledge is essential for changing the epistemological order in risk situations.

For instance, Johnson cites an important example in which users helped design traffic flow in Seattle. The situation was one common to many large cities—too many cars on the road during peak commuting hours. Taking a common approach, transportation officials measured traffic flows utilizing various counting and statistical methodologies in order to determine how to reroute traffic and which roads to expand. However nothing seemed to work until a team of technical communicators from the University of Washington explored the same problem from a different angel. Rather than studying the traffic, this team studied the driving preferences and habits of Seattle citizens through surveys, interviews, focus groups, and observations (81). This information provided a workable solution to the traffic problems. The technical communicators sought user knowledge about the problem, and in fact, users had the best solutions to this particular traffic problem and were able to solve problems that "experts" could not. About this example, Johnson writes "In terms of the user as citizen. . . the point is made most strongly: the users are represented as an important force in the design of the system. . . because they are asked to help determine the best solutions to the problem" (82). Citizens possess knowledge about how a technology is used in, or would affect, a particular community. This understanding is something "experts" may lack but need in order to design a usable technology. As a result, citizens can contribute valuable information to the design and decision making process.

Our point is that users are intelligent and productive—like experts, users create knowledge—and sometimes user productivity takes decision making in new and important directions. But our point is also that users cannot be productive if they are silent. The power of Johnson's example comes from the fact that it was technical communicators who thought to ask users about the problems of traffic flow and implemented the research practices necessary to help construct user knowledge.

In fact, technical communicators are perhaps uniquely capable of participating in the construction of user knowledge. But what types of research practices are we talking about? We are not talking about testing methods commonly located in usability labs, for they are most appropriate for working with discrete technologies. Like the researchers in Seattle, we are talking about methods common to some usability work and most qualitative research: surveys, interviews, and perhaps most importantly, field-based observation of how people use the spaces and/or technologies associated with a given risk. The type of research and multidisciplinary work we are framing here expands the scope of work for technical communicators. We believe that the technical communicator may be one of the few professional workers trained for both the multidisciplinary perspectives and user advocacy necessary to help dissolve the boundary between assessment and communication and thereby facilitate a more participatory construction of risk. In fact, as Figure 2 helps us illustrate, we believe the technical communicator as symbolic analyst allows us to move between two key binaries in risk communication scholarship, expert/non-expert and assessment/communication. In this figure, we see the binaries of continua in an attempt to express a range of practices between "straight" assessment and communication and the positions between expert and non-expert. In keeping with our argument, we see the technical communicator as able to occupy multiple positions and play a variety of roles.

On this map, the top left quadrant represents "traditional risk assessment," or risk assessment done solely by experts. The technical communicator's role here is fairly traditional—to help "translate" knowledge to audiences. The top right represents non-expert risk assessment, or a range of informal assessment practices that people use whenever confronted with risk—conversation, reading, formal and informal writing, and meetings. The technical communicator, given that he or she is capable of collaborative research (as a symbolic-analyst), can gain access to these processes and help give voice to these audiences. The bottom left represents technocratic (at the extreme) and other linear approaches to risk communication. The bottom right represents a wide range of "non-expert" risk communication, or discourse produced by interests without the institutional designation as authorities. Of course, as we used Langdon Winner to suggest earlier, real life risk situations often have expertise on both sides of a given issue, and so our map fails to capture the complexity of audiences in such situations. Nevertheless, the technical communicator can move between ranges and varieties of experts and non-experts. In short, this figure illustrates the positions we assert the technical communicator can occupy as well as the movement we believe is necessary between positions—in fact, we believe that a critical rhetoric moves, that research and rhetorical actions must move between and across positions as situations and problems dictate. (We are taking our mapping practices and our sense of critical research practices from the work of Sullivan and Porter. See their book for a discussion of postmodern mapping and a significantly more substan-

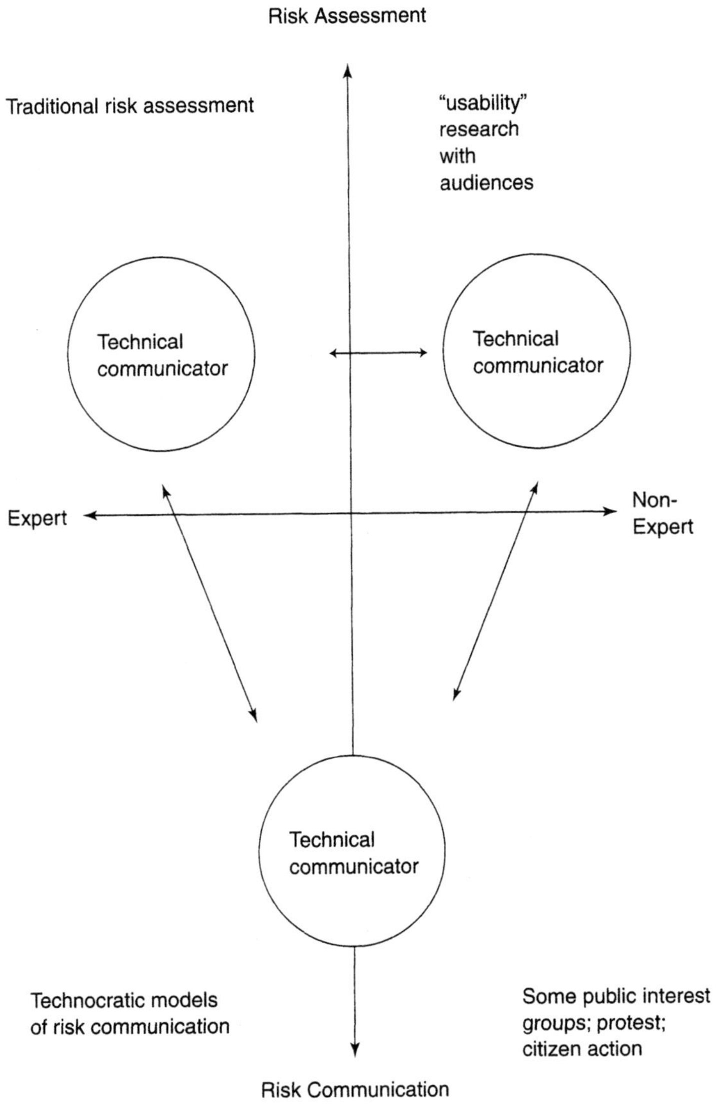

Figure 7.2. The multiple positions of the technical communicator.

tial discussion of critical research.) In so doing, we believe this figure captures the complexity of risk processes and is descriptive of the complex web that characterizes the social construction of risk. All of these practices, from expert to non-expert assessment to a wide range of communication practices happen, sometimes simultaneously, in a given risk situation. As Johnson-Eilola writes, "[b]ecause of the political, economic, and social aspects of all technologies, technical communication should not limit itself to simple functionalism, but must also include broader and more complex concerns" (259). Risk assessment/communication encompasses such broad and complex social, intellectual, and rhetorical concerns. Risk communication explicitly takes technical communication into the realm of civic discourse.

Usability Research: Participation and Power

The focus on research and user knowledge that we have been articulating as part of a critical rhetoric for risk communication has ethical as well as epistemological dimensions. Thus the issue with which we want to conclude is the same issue at the core of a critical rhetoric-power. If power and the abilities of professionals like technical communicators are located within institutions creating risks and used to manage not only risks but also stakeholders, then technical/risk communication and communicators are involved in a relation of power that can dominate and oppress. This is why we are interested in a critical rhetoric for risk/technical communication that can help prevent such exercises of power. (In our concern, we echo the work of others concerned with the uses to which "good" technical communication can be put and the potentially limiting institutions within which technical communicators work. See Katz; Savage; Sullivan; and Slack, Miller, and Doak.) We have argued that technical communicators as user advocates can bring about more participatory and ethical decision making processes. However, the implications of such a positioning create their own problems. We typically imagine clients for risk assessment/communication to be business, industry, and the government-risk makers. In our approach to risk communication, however, we suggest that "clients" can be citizens, the public, the poor, or the powerless. However, it is more likely that the technical communicator will be hired by the relatively more powerful rather than the powerless. Because technical communicators have an obligation to their client, whoever hires the technical communicator often has an advantage. As a result, it is possible for the technical communicator to be caught in the very power relations that we argue technical communicators can work to dismantle. This problem of clients and advocacy is why we believe the merger of risk assessment and communication practice is important. Such a merger-hopefully-inserts audiences and audience advocacy earlier in the decision making process, even when technical communicators are working for more powerful clients. Still, problems, of advocacy will not disappear.

Yet we believe that it is possible to change the processes of risk assessment/communication through research and writing practices in order to insert user knowledge and participation into decision making processes. To do so requires changing the institutions that make decisions about risk, and this is difficult. A common view of institutions monolithic and static bureaucracies makes institutional change an impossibility-how can we change government? business? However, Porter et al. have a view of institutions that sees then as rhetorical systems of decision making formed by the discourses that make them possible (e.g., legislation, business plans, policies, procedures, research protocols.) According to this view, institutions are "inherently dynamic and open to change through the very rhetorical practices by which they operate" (16). In short, institutions are written, and therefore, can be rewritten. In order to rewrite institutions, rhetorical (and sometimes material) space must be created within an institution; the creation of such space can sometimes have radical consequences. Our claim that users produce knowledge and that technical communicators are uniquely capable of the research necessary to enable users to participate in decision making is an attempt to frame how institutional space can be created. The institutional separation between assessment and communication is not only false (epistemologically) and unproductive, but it prevents the creation of "spaces" for users within decision making processes. We suggest that it is through research practices that institutional space for users/citizens within knowledge producing and decision making institutions can be constructed. It is one thing to talk about how decision makers *should* listen and *should* allow citizens to participate (they should!). It is an entirely different project to structure as part of the everyday practices of a given institution research designed to facilitate user/citizen participation as legitimate knowledge producers and decision makers.

How would the creation of institutional space for users/citizens solve problems or power and powerlessness? It wouldn't. At least not entirely. But we do think a critical rhetoric for risk communication would address problems of powerlessness because it enables users/citizens to speak, to develop autonomy within institutions-in short, to exercise power. The technical communicator as symbolic analyst can contribute to critical risk communication through research and communicative practices that can help provide a preferable space for those least powerful within assessment and communicative decision making practices. Such processes might indeed lead to a collective articulation of the good of a risk situation and subsequent implementation of that good. Not only might these processes lead to more ethical risk assessment/communication (the articulation of a good for a *we*), but we think such process would be *better* because they seek to avoid the problems of mistrust and conflict that mark traditional practices. Furthermore, given the inclusion of user/citizen knowledge, such processes might also be more "intelligent" by including multiple perspectives, different types of knowledge, and potentially a greater number of acceptable solutions. The risk communication literature is full of stories of corporate or governmental abuse, citizen resistance, and general "failure." We suggest that a critical rhetoric of risk communications and the role of the technical communicator within that rhetoric can help alleviate conflict and impasse through the facilitation of more effective assessment/communication processes. Our title refers to "producing citizens," and that phrase embodies a double meaning. Risk communication practices invariably produce citizens—the real or idealized audience who consumes communication. A much more "productive" view sees citizens as themselves producers-of knowledge, of values, of communities. A critical rhetoric locates its energy and hope in this sense of producing citizens by seeking to access user/citizen knowledge by creating the institutional space within which risk can be collectively constructed and more effectively communicated.

ACKNOWLEDGMENTS

The authors would like to thank Patricia Sullivan for her encouragement and guidance.

Works Cited

42 United States Code, Section 9617 (1995).

Adams, William C. "The Role of Media Relations in Risk Communication." *Public Relations Quarterly* 37.4 (1997): 28–31.

Belsten, Laura. "Environmental Risk Communication and Community Collaboration." *Earthtalk*. Ed. Star Muir and Thomas Veenendall. Westport, CT: Praeger, 1996. 27–42.

Blyler, Nancy Roundy. "Habermas, Empowerment, and Professional Discourse." *Technical Communication Quarterly* 3 (1994): 125–45.

Carson, Rachel. *Silent Spring*. Boston: Houghton Mifflin, 1962.

Castelli, J. "Welcome to the World of Risk Communication." *Safety and Health* 142 (1990): 68–71.

Corvello, Vincent, Peter Sandman, and Paul Slovic. *Risk Communication, Risk Statistics, and Risk Comparison: A Manual for Plant Managers*. Washington, DC: Chemical Manufacturers Association, 1988.

Doheny-Farina, Stephen. *Rhetoric, Innovation, Technology: Case Studies of Technical Communications in Technology Transfer*. Cambridge, MA: MIT P, 1992.

Ehn, Pelle. *Work-Oriented Design of Computer Artifacts*. Stockholm: Arbetslivscentrum, 1988.

Fiorino, Daniel. "Citizen Participation and Environmental Risk: A Survey of Institutional Machanisms." *Science, Technology, and Human Values* 15 (1990): 226–43.

Foucault, Michel. "The Subject and Power." *Michel Foucault: Beyond Structuralism and Hermeneuticsi*. Ed. Hubert L. Dreyfus and Paul Rabinow. Chicago: U of Chicago P, 1982. 208–26.

Habermas, Jurgen. *Moral Consciousness and Communicative Action*. Trans. Christion Lenhardt and Shierry Weber Nicholsen. Cambridge, MA: The MIT P, 1990.

Hance, B.J., Caron Chess, and Peter Sandman. *Industry Risk Communication Manual: Improving Dialogue with Communities*. Boca Raton, FL: Lewis, 1990.

Heath, Robert, and Kathy Nathan. "Public Relations' Role in Risk Communication: Information, Rhetoric, and Power." *Public Relations Quarterly* 35.4 (1990): 15–22.

Johnson, Robert R. *User Centered Technology: A Rhetorical Theory of Computers and Other Mundane Artifacts*. Albany: SUNY P, in press.

Johnson-Eilola, Johndan. "Relocating the Value of Work: Technical Communication in a Post-Industrial Age." *Technical Communication Quarterly* 5 (1996): 245–71.

Jones, Dan. "A Question of Identity." *Technical Communication* 42 (1995): 567–69.

Juanillo, Napoleon K., and Clifford W. Scherer. "Attaining a State of Informed Judgments: Toward a Dialectical Discourse on Risk." *Communication Yearbook*. Ed. Brant Burleson. New Brunswick, NJ: International Communication Association, 1994. 279–99.

Katz, Steven. "The Ethic of Expediency: Classical Rhetoric, Technology, and the Holocaust." *College English* 54 (1992): 255–75.

Katz, Steven, and Carolyn Miller. "The Low-Level Radioactive Waste Siting Controversy in North Carolina: Toward a Rhetorical Model of Risk Communication." *Green Culture: Environmental Rhetoric in Contemporary America*. Ed. Carl Herndl and Stuart Brown. Madison: U of Wisconsin P, 1996. 111–40.

Kuhn, Thomas. *The Structure of Scientific Revolutions*. Chicago: U of Chicago P, 1970.

Latour, Bruno. *Aramis or the Love of Technology*. Trans. Catherine Porter. Cambridge, MA: Harvard UP, 1996.

Mirel, Barbara. "Debating Nuclear Energy: Theories of Risk and Purposes of Communication." *Technical Communication Quarterly* 3 (1994): 41–65.

National Research Council. *Risk Assessment in the Federal Government: Managing the Process*. Washington, DC: National Academy P, 1983.

Patterson, Robert, and Ronald Lee. "The Environmental Rhetoric of 'Balance': A Case Study of Regulatory Discourse and the Colonization of the Public." *Technical Communication Quarterly* 6 (1997): 25–40.

Plough, Alonzo, and Sheldon Krimsky. *Environmental Hazards: Communicating Risks as a Social Process*. Dover, MA: Auburn House, 1988.

Porter, James E. *Audience and Rhetoric: An Archaeological Composition of the Discourse Community*. Engelwood Cliffs, NJ: Prentice Hall, 1992.

—. *Rhetorical Ethics and Internetworked Writing*. Greenwich, CT: Ablex and Computers and Composition, 1998.

Porter, James E., et. al. "(Re)writing Institutions: Spatial Analysis and Institutional Critique." Unpublished.

Ross, Susan M. "Two Rivers, Two Vessels: Environmental Problem Solving in an Intercultural Context." *Earthtalk*. Ed. Star Muir and Thomas Veenendall. Wesport, CT: Praeger, 1996. 171–90.

Rowan, Katherine. "Goals Obstacles, and Strategies in Risk Communication: A Problem-Solving Approach to Improving Communication About Risks." *Journal of Applied Communication Research* 19 (1991): 300–29.

—. "What Risk Communicators Need to Know: An Agenda for Research." *Communication Yearbook*. Ed. Brant Burleson. New Brunswick, NJ: International Communication Association, 1994. 300–19.

—. "The Technical and Democratic Approaches to Risk Situations: Their Appeal, Limitations, and Rhetorical Alternative." *Argumentation* 8 (1994): 391–409.

Rubin, D.M., and D.P. Sachs, eds. *Mass Media and the Environment*. New York: Praeger, 1973.

Russell, Milton. "Communicating Risk to a Concerned Public." *EPA Journal* 12.9 (1986): 19–21.

Sandman, Peter. "Getting to Maybe: Some Communications Aspects of Siting Hazardous Waste Facilities." *Readings in Risk*. Ed. Thomas Glickman and Michael Gough. Washington, DC: Resources for the Future, 1990. 223–31.

Savage, Gerald J. "Redefining the Responsibilities of Teachers and the Social Position of the Technical Communicator." *Technical Communication Quarterly* 5 (1996): 309–27.

Slack, Jennifer Daryl, David Miller, and Jeffrey Doak. "The Technical Communicator as Author: Meaning, Power, Authority." *Journal of Business and Technical Communication* 7 (1993): 12–36.

Slovic, Paul. "Informing and Educating the Public about Risk." *Risk Analysis* 6 (1986): 403–15.

Stratman, James E., et al. "Risk Communication, Metacommunication, and Rhetorical Stases in the Aspen-EPA Superfund Controversy." *Journal of Business and Technical Communication* 9 (1995): 5–41.

Sullivan, Patricia, and James E. Porter. *Opening Spaces: Writing Technologies and Critical Research Practices*. Greenwich, CT: Ablex and Computers and Composition, 1997.

Sullivan, Dale L. "Political-Ethical Implications of Defining Technical Communication as a Practice." *Journal of Advanced Composition* 10 (1990): 375–86.

Waddell, Craig. "Saving the Great Lakes: Public Participation in Environmental Policy." *Green Culture: Environmental Rhetoric in Contemporary America.* Ed. Carl Herndl and Stuart Brown. Madison: U of Wisconsin P, 1996. 141–65.

—. "Defining Sustainable Development: A Case Study in Environmental Communication." *Technical Communication Quarterly* 4 (1997): 201–16.

—. Risk Communication Course Syllabus. Michigan Technological University. Winter 1994–1995.

Winner, Langdon. "Citizen Virtues in a Technological Order."

Technology and the Politics of Knowledge. Ed. Andrew Feenberg and Alastair Hannay. Bloomington: Indiana UP, 1995. 65-84.

Young, Iris Marion. *Justice and the Politics of Difference.* Princeton, NJ: Princeton UP, 1990.

DEVELOPING YOUR UNDERSTANDING

1. Summarize what Grabill and Simmons mean when they assert that knowledge is made, and that knowledge-making is located within communication processes. Use an example to illustrate your discussion.
2. Summarize how professional writers/risk communicators participate in "producing risk."
3. Explain how linear models of risk assessment/risk communication constrain the social space of professional writing, reducing it to "an impoverished rhetoric of 'arrangement' and 'style.'"
4. Compare and contrast the social spaces produced by the technocratic, negotiated, and socially constructed approaches to risk. For each, describe what the professional writer does, and analyze how what she does contributes to producing a certain kind of social space.

CHAPTER 7

Projects

1. Without referring to the term "social space," interview a professional writer about the kind(s) of social space she has produced and would like to produce. To focus and concretize your interview, have the writer focus on two projects, one she has completed and one on which she is currently working. In addition to exploring the kind(s) of space she has and would to like to produce, find out *how* the writer participated in the production of this space (e.g., working collaboratively, drafting texts, and/or managing writers). Then, produce a report (oral and/or written) that summarizes the kind(s) of space she has produced and still wants to produce, that summarizes the strategies they used to produce such space, and that assess the limits organizational contexts have placed on what they could/can produce.

2. Grabill and Simmons argue that professional writers can "construct risk communication practices that allow citizens (various publics) to represent themselves in decision making processes" by inserting usability and qualitative research methods into the risk assessment/risk communication process. That is, professional writers can reconstruct the social space of risk so that citizens can more actively participate. However, Grabill and Simmons fail to acknowledge that in order to construct such social space, professional risk communicators, who typically work within and for governmental and industry organizations, must first reconstruct their own workplace social spaces in ways that make it possible for them to insert the research practices they suggest.

   In small research teams, study how professional writers, in general, have participated and can participate in reconstructing the social space of their immediate workplaces, so they can eventually have the space they need to insert new practices, like early qualitative research, into their work/writing processes. Construct a clear research plan, one that identifies the questions you would like to explore, the sources that you can use for each question, and the most effective research methods for exploring each question. Consider library research (for example, some of the sources in *Professional Writing and Rhetoric* and their bibliographies would lead you to helpful information), as well as field research (e.g., interviews with professional writers and disciplinary experts on your campus).

   Based on your research, develop a short report, written for professional writers, that lists and illustrates, perhaps through abbreviated cases/scenarios, how professional writers can participate in reconstructing the social space of their immediate workplace.

3. Several articles in this collection, just to identify a few in the field, suggest that professional writers reframe who they are either to better express the roles they are playing already or to help them carve out identities that better capture their expertise. For instance, Grabill and Simmons cite Johndan Johnson-Eilola's argument that professional writers recast themselves as "symbolic-analysts," and in Chapter 2, Slack et al. argue that professional writers "rearticulate" themselves as "authors." Redefining oneself, one's work, and one's field are ways people can produce new social space.

   As an act of producing the social space of professional writing, define what a professional writer is and does, and describe the kinds of workplace contexts—e.g., production processes, routine practices, structured social interactions, and technological access—that would best make it possible for professional writers to assume the role you define for them.

PART 4

Becoming a Professional Writer

No matter how long any of us has been a professional writer, whether we are taking our first class or have been studying and practicing professional writing for many years, we are all always still "becoming a professional writer." There are a number of other very good books that help professional writers with years of experience and education become better writers. *Professional Writing and Rhetoric*, however, is written for those of you who are in school and just starting into professional writing. That could mean it is your first class on the subject; it could mean you are soon to be entering an internship or the job market; or it could mean that you have had some background in writing but are just now exploring professional writing, especially as it is related to rhetoric. No matter the case, Chapter 8 addresses your stage of "becoming a professional writer."

The readings in Chapter 8 focus on the **transition** from school to the workplace, the stage you are or will soon be facing as a professional writer. As you will see in the readings, the transition from school to the workplace is not free of bumps and bruises. Because it isn't, many writers feel like they are leaving one "world" and moving into a completely different "world." With this kind of thinking, it is hard to form links between the two worlds, to see how they are different and yet to find ways that what one learns in one world is of value in the other. As a result, writers often feel like they are "starting over," and they also often lose contact with knowledge and practices they learned in school that could be of great value in the workplace.

Chapter 8 focuses on the transition between school and the workplace, so you can form better linkages between the two worlds. By focusing on these linkages, you do not have to "start over." Instead, you can see the transition itself as an important and educational stage in your "becoming a professional writer."

CHAPTER 8

Writing Yourself into Professional Writing and Rhetoric

INTRODUCTION

Using school as a place to study professional writing and rhetoric and to develop the arts and practices of writing is of great value. In the workplace, no one has the job of helping you learn how to analyze organizational contexts, or helping you act ethically, or helping you learn ways to work with readers to design documents that meet the needs of multiple audiences. In the workplace, you may find yourself feeling frustrated because a routine form does not give you space to include what you see as critical information. Or you may feel frustrated because you do not seem to be included early enough in the software product development processes to head-off what you see as "writing problems." Even though you may feel these kinds of frustrations, no one in the workplace is responsible for helping you see them as issues of social space. No one will help you see that forms are ways of producing what we consider to be legitimate organizational knowledge, and that revising the forms can change what counts as knowledge within that organization. No one will help you see that a division-of-labor model of product development often creates "bad" technical documents because writers are not included early enough in the "writing" process to intervene. These are the kinds of things school is designed to help you learn.

However, what school gives is not enough. In the workplace, you learn things that school is not prepared or designed to teach you. For instance, even though you may have the chance to work on projects in school that include multiple audiences, the complexity and impact on your writing of such a situation will be felt nowhere like in the workplace. School environments also have a very difficult time recreating the absolute *need* for documents that you can only fully experience, appreciate, and learn to deal with in workplace experiences. In school, people rarely need to have the documents you create, but in the workplace, your documents will take on an air of necessity that can be a bit overwhelming. In this sense, workplace experience, either through employment or internships, offers you learning experiences that cannot be offered to you through school.

Chapter 8 includes readings that help you explore the bridges between school and the workplace. As you read the selections in this chapter, keep the following questions in mind:

- what are some of the situations you might find yourself facing as you are becoming a practicing professional writer?
- how can you prepare for your transition from school to the workplace?
- what can you do once in the workplace to make your transition from school more effective?

All of these questions assume that you have also examined how school and the workplace are different. There are obvious differences between school and work, but these are not the differences you need to examine. One difference that often goes unexamined by writers is that what makes "good" writing in school is not necessarily what makes good writing in the workplace. As you read the selections in Chapter 8, think about how writing differs in terms, for instance, of purpose and function, audience, need, process, and collaboration.

But you should also consider which school practices and knowledge can be of advantage to you once you enter the workplace. As outlined in the opening of this introduction, school is designed to offer you knowledge and practices for which the workplace is not, or poorly designed. And as you consider the knowledge and practices you have gained from school, think about how these can help you make successful transitions into a practicing professional writer's work.

FOCUSING ON KEY TERMS AND CONCEPTS

Focus on the following terms and concepts while you read through this selection. Understanding these will not only increase your understanding of the selection that follows, but you will find that, because most of these terms or concepts are commonly used in professional writing and rhetoric, understanding them helps you get a better sense of the field itself.

1. lateral transfer
2. vertical transfer
3. alienated independence
4. resolution

This qualitative study examined the transitions that writers make when moving from academic to professional discourse communities. Subjects were six university seniors enrolled in a special "writing internship course" in which they discussed and analyzed the writing they were doing in 12-week professional internships at corporations, small businesses, and public service agencies in a major metropolitan area. Participant-observer and case-study data included drafts and final copies of all writing that the interns produced on the job (including texts and suggested revisions by other employees), an ethnographic log of data and speculations arising from the group discussions, written course journals from each intern, transcriptions of taped, discourse-based and general interviews with the interns, and a final 15-page retrospective analysis of each intern's writing on the job. Results showed a remarkably consistent pattern of expectation, frustration, and accommodation as the interns adjusted to their new writing communities. The results have important implications for the lateral and vertical transfer of writing skills across different communicative contexts.

Moving Beyond the Academic Community: Transitional Stages in Professional Writing

CHRIS M. ANSON,
L. LEE FORSBERG
University of Minnesota—Minneapolis/St. Paul

> *I guess I really screwed up on my last memos. The writing's so much different than I'm used to. One of my memos was too formal, the other was too touchy-feely. Most of my immediate supervisors are relaxed here but the memos I've seen are real formal and short and the language is relatively direct. But then there'll be something personal and witty. I don't know. . . I guess the thing to be understood is that lots of these writings are for upper management, and they don't have the time and energy to wade through a mess of expanded material or complicated prose. In school, it's always expand, expand. Here they want the information quick, and so we gotta give them. . . I don't know. I feel like I'm back at square one. I used to think I was a pretty good writer. You should see what my boss Chuck did to my last draft.*
>
> *—John, student intern at a large corporation*

Author's Note: The research reported in this study was made possible in part by a grant from the Office of Educational Development Programs, University of Minnesota—Twin Cities.

John is a 21-year-old college senior writing for the first time in a professional setting. He is an excellent academic writer—an honors student with a high grade point average and a history of "A" grades for his course papers and essay exams. As this excerpt from one of his taped monologues reveals, however, John is having considerable difficulty writing in the context of his student internship at a large company. Strategies that had worked well for him in college are no longer very effective, and he is unprepared to meet the demands of multiple readers whose status, power, and professional orientations vary considerably across departments and levels of the corporate hierarchy.

John's problems are also particular to his nonacademic setting—a high-power, "Fortune 500" organization with competent communicators. In a small company of

Source: Anson, Chris M., and L. Lee Forsberg, "Moving Beyond the Academic Community: Transitional Stages in Professional Writing," *Written Communication* 7.2 (April 1990), pp. 200–231, copyright © 1990 by Sage Publications, Inc. Reprinted by permission of Sage Publications, Inc.

low prestige, among professionals with less than effective writing abilities, John's abilities as an academic writer might place him in the opposite but equally difficult position—a young novice wanting to give advice to his superiors, who have many years of experience in the business world.

In this article, we examine the transitions that writers make when they move from an academic to a nonacademic setting and begin writing in a new and unfamiliar professional culture. While certain surface-level writing skills are "portable" across diverse contexts, such skills are less important to making a successful transition as a writer than coping with the unfamiliar epistemological, social, and organizational characteristics of a new context. A writer in such a context is in many ways "illiterate" until he or she begins to understand these characteristics and their manifestation in written texts. Our research shows that becoming a successful writer is much more a matter of developing strategies for social and intellectual adaptations to different professional communities than acquiring a set of generic skills, such as learning the difference between the passive and the active voice. By documenting a cycle of expectation, struggle, and accommodation typical of our interns as they began writing in new professional settings, we will support this developmental view and suggest some directions for further research on contextual adaptation and the transfer of writing ability.

BEYOND THE ACADEMY: WRITING IN NONACADEMIC CONTEXTS

Recently, research on writing has begun to look more closely at the influence of context on writers' composing processes, particularly the way in which the social dimensions of specific settings (e.g., audiences, purposes, assigned or imputed roles, institutional ethos, and so on) influence writers' rhetorical and linguistic decisions. As Odell and Goswami (1984) have reminded us, writing does not exist independently of the community in which it is immersed. Studying writing without considering the importance of its context is, to borrow the words of Beach and Bridwell (1986), "a little like studying animal life by visiting zoo cages" (p.6).

Belief in the importance of context to communication is, of course, not new (for an overview, see Harrison, 1987). Its rediscovery in the field of composition studies, however, represents the development of a more social view of the writing process. In contrast to literary or cognitive perspectives, social approaches to writing are interested in the relationship among writers, texts, and their surrounding context (Faigley, Cherry, Jolliffe, & Skinner, 1985). Writing is seen as a socially constituted act, shaped by the writer's "discourse community," members of an organizational or intellectual group who share specialized kinds of knowledge and textual competence (for recent discussions, see Anson, 1987; Bazerman, 1979; Bizzell, 1982; Cooper & Holzman, 1989; Faigley, 1985; Odell, 1985; Porter, 1986).

Interest in the cultural, noetic, and organizational features of discourse communities has led some researchers to turn away from school settings, whose "discourse communities [are] inevitably defined by the educational system" (Freed & Broadhead, 1987, p.156) and whose writing tasks are typically "assigned by a teacher or researcher in order to accomplish a pedagogical goal or provide data to answer a research question" (Odell & Goswami, 1982, p. 201). For these scholars, school writing is often rhetorically limited, used for what Britton, Burgess, Martin, McLeod, and Rosen (1972) have called "dummy-run" practice. Without rich, varied audiences and purposes, the consequences of such writing are substantially different from the writing found in nonacademic settings.

In exploring these differences, new studies of professional writing are showing some of the complex relationships between texts and contexts. Surveys across different vocations and job categories are revealing the quantities and varieties of writing required of professionals (for a review, see Anderson, 1985). Recent context-sensitive text analyses have pointed to relationships between the formal features of a text and its surrounding social and rhetorical situation (Bazerman, 1981; Colomb & Williams, 1985; Fahnestock, 1986; Miller & Selzer, 1985; Secor & Fahnestock, 1982). Analyses of entire organizations (Doheny-Farina, 1983; Fred & Broadhead, 1987; Knoblauch, 1979; Paradis, Dobrin, & Miller, 1985), as well as case studies of individual writers working within organizations or professions (Brown & Herndl. 1986; Gould, 1980; Odell, Goswami, & Herrington, 1983; Selzer, 1983), are further showing us how texts are shaped by what Odell (1985) has called the "internalized values, attitudes, knowledge, and ways of acting that are shared by other members of the organization" (p.250). Writing, in other words, cannot be seen apart from its functions in various social, administrative, and managerial processes. As Cooper (1986) has put it, the "characteristics of any individual writer or piece of writing both determine and are determined by the characteristics of all the other writers and writings in the [socially constituted] systems" (p.368).

The recognition that contexts, writers, and texts are in a dynamic state of evolution and mutual influence holds much promise for a fuller understanding of what Brandt (1986) has called the "public conditions" through which writers in professional settings must try to share meaning. An excellent way in which to understand the role of such conditions and their influence on the writing process is to study what happens when members of one discourse community begin to write in another, relatively unfamiliar community. Some preliminary work has examined differences between writers of similar age and experience in two different communities. Ordell, Goswami, and Quick (1983), for an example, studied two similar groups of writers working in different rhetorical domains: (a) five, young, recently-hired legislative analysts employed by a state legislature; and (b) five university undergraduates with academic preparation that might make them good candidates for similar positions in state or federal government. Comparisons of the writing of these two groups revealed fundamental differences in the tasks, audiences, purposes, information, and evaluative methods underlying each context, and these differences had important effects on the thinking and composing processes of each group. Similar findings have also been reported from research on the accommodations that students must make when writing in different academic disciplines, where assumptions about writing on a "professional" level vary considerably (see, for example, Faigley & Hanson, 1985; Herrington, 1985; McCarthy, 1987).

Virtually no research, however, has explored the *transitions* that writers make as they move into new and unfamiliar writing contexts. If, as Piazza (1987) has asserted, social norms, roles and relationships, status, and other social factors constrain the functions and uses of writing in a particular setting (p.113), then studying the transitions that writers make in a new setting may inform us further not only about the nature of those constraints but also about how one learns to accommodate them.

FROM ACADEMIA TO THE PROFESSIONS: THE WRITING INTERNSHIP PROGRAM

In this study, we chose to focus on an important transition that all writers make when they move beyond the academic community and begin working in professional settings. Our subjects were six college seniors at the University of Minnesota who were conducting professional internships in and around the twin cities of Minneapolis and St. Paul. Because most of these students had little or no experience writing in a professional setting but had become quite accustomed to writing at a major university, we felt that the differences in the two types of contexts would reveal the problems that writers face when moving between discourse communities and thus suggest more clearly some of the adaptive processes that writers undergo in new professional settings.

In the pilot phase of an educational development project at the University of Minnesota, we offered a special "writing internship" course. This course was designed for students whose internships required them to do a significant amount of writing. Most of the internships were full-time positions lasting from 10 to 15 weeks. Sponsors for the positions were quite varied, including major corporations, small businesses, public service agencies, local utilities, newspaper offices, and art museums. The kinds of writing demanded of the interns ranged within and across these settings from singly authored memos, press releases, news features, and informal surveys to collaboratively written reports, feasibility studies, grant proposals, and recommendations to upper management for funding various projects.

Instead of using a traditional pedagogy, that is, presenting principles of composition or asking the students to write practice essays, we taught the internship course as a workshop involving group analyses of the participants' work-related drafts and discussions of issues in their writing. One of us (Forsberg) played the rose of teacher/evaluator, coordinating the classes, leading the discussions, making assignments, and assessing the students' work at the end of the course. The other (Anson) played the role of participant, observer, attending the sessions, taking notes, joining in the discussions as appropriate, and reflecting on the issues raised by the group. In this way, we were able to gather and analyze data from two different perspectives: (a) helpful (though finally authoritative) facilitator, and (b) interested and engaged observer.

We wanted a research methodology that would blend productively and unobtrusively with the internship course. Our method was essentially qualitative, allowing us to collect and examine a greater variety of simultaneously gathered data than might have been possible using other methods (for representative discussions, see Green & Wallat, 1981). We collected everything that the interns were writing, in both draft and final form. We kept logs of data and speculations arising from the group discussions. We asked the interns to keep a course journal in which they explored their thoughts about (a) writing on the job, (b) the texts they were seeing there, and (c) whatever guidance or suggested revisions they were getting from their supervisors or fellow employees. We also conducted a series of general and "discourse-based" interviews (Odell, Goswami, & Herrington, 1983), which we tape-recorded, transcribed, and analyzed. At regular intervals during the course, we also asked the interns to turn on a tape recorder and reflect aloud on their experiences. For their final course paper, interns produced a 10- to 15-page retrospective analysis of their writing on the job.

Because the interns knew that we were interested in their experiences as part of our pilot project with the internship course, these data-gathering techniques neither intruded on the natural setting of the classroom nor placed the interns in an overtly "experimental" situation, which might have affected their behaviors or their reports to us about what they were thinking and doing (see Tomlinson, 1984). Instead, they were aware simply that they straddled two contexts—one academic and one professional—and that in the former, they would be analyzing and coming to a fuller understanding of what they were doing in the latter.

In selecting the six case-study subjects, we favored students who would be doing a considerable amount of writing on the job and who had performed well as academic writers. Because we wanted to examine the transitions of fairly competent writers, we limited our selection of majors to English and journalism, both of which provide students with a fairly strong background in writing and communication skills. Beyond these preferences, we aimed for as much diversity as possible in the subjects' age, prior experience, attitudes, and internship positions. The six subjects are identified here:

Jim, 20, a journalism major, worked for a nonprofit educational organization in Minneapolis, writing public relations materials. Jim had previously worked in telephone sales.

Louise, 21, an English major, worked for a nonprofit arts organization, researching and writing request letters and grants. Louise had previously worked for a short time as a secretary in a small firm.

Joan, 35, a returning student, was completing a journalism degree. She worked as an intern for a very small public relations agency, writing pitch letters and press releases. Joan had been a clerical worker for a large corporation.

Paula, 38, also a returning student, was completing an English degree. She worked as an unpaid feature writer for a weekly suburban newspaper. She had worked as an administrative assistant in a government office.

Rachel, 22, was an English major working as an intern for a two-person company that caters theme parties for corporations. She was mainly responsible for writing brochures and letters. Rachel had previously worked as a secretary in a bank.

Betsy, 21, an English major, worked as an intern in a large, nonprofit theater organization, writing public relations and fund-raising materials. She had previously worked for the Red Cross and a small theater group.

THE STUDY: STAGES OF TRANSITION

The wealth of data from journals, papers, logs, taped monologues, interviews, and workplace texts provided us with a number of issues to examine: how writers in a new context perceive and adapt to complex and unfamiliar audiences, how they learn to use the language of their workplace, how they make connections and distinctions between experiential and academic learning, how they describe the influence of the context on their texts, how they revise those texts, and what motivates their revisions.

In the early stages of our study, we expected to find that the interns all underwent unique transitions, unrelated to each other's experiences and therefore revealing only individual differences. As we collected and worked through the data from our six subjects, however, an unexpected but consistent pattern began to emerge. During the course of their internships, the students appeared to pass through several distinct stages as they adjusted to their writing contexts—stages marked by qualitative differences in the way they described their writing. This pattern became even more discernible when, focusing on our accumulating data, we saw that the same cycle of transition in all the accounts, regardless of the student's age, previous experience or academic writing skill, or specific work context. Within this cycle, the exact point at which a student experienced each stage, with what intensity, and for what length of time, varied considerably according to both the individual and the nature of the workplace, but that variance in itself suggests some new ways of thinking about the interplay between the individual writer and the writing context.

The relevance of this cycle of transition, we believe, is not just that it presents as interesting chart of a learning or developmental process, but that it reflects the relationship between the writer's new context and the way he or she begins to "read" it. In order to write at all, in order to produce texts that become transactionally real, writers must first be able to adopt a persona appropriate to their position in the workplace, acceptable to themselves, their superiors, and other eventual audiences of their writing. Although the word "audience" seems woefully inadequate here, it is a process of audience analysis that enables the adoption of such a persona, and there is no understanding of audience that more fully constitutes a text than the writer's perception of his or her immediate context, the workplace. The writer must first become a "reader," and the process of that reading must occur in a dynamic context of interpersonal action and adaptation (see Forsberg, 1987). The cycle of transition, then, impinges on and eventually informs writing decisions in the ongoing process of creating texts.

At the most general level, we identified from our data three stages of transition through which the interns passed as they moved from academic to nonacademic writing. Each of these may be further elaborated by some specific characteristics:

1. *Expectation*. In this preinternship stage, the writer builds a vision, that is, a social construct, of his- or herself working and writing in a new professional setting. Often, the picture is *idealized*, particularly if the student has been a reasonably successful writer in college. The image provides the impetus for often intense *motivation* to perform well. Occasionally, perhaps because the new context is perceived as different from the known academic context, the intern may express some apprehension about applying his or her knowledge to the new situation.
2. *Disorientation*. As the individual tries to determine a role in the organization, he or she becomes disoriented, and this in turn can lead to intense *frustration* and a sense of failure. The characteristics of disorientation, however, may take several forms. Some interns undergo a period of *alienated independence*, a sense of having to do things all on their own, of being expected to know already how to execute tasks and being apprehensive

about consulting others, and of not knowing how, or when it is appropriate, to ask for information. The frustration that results from the struggle to understand new tasks in a still unfamiliar setting often leads the writer to *evaluate* the nature and quality of knowledge in the workplace, and this process of evaluation in turn may lead to personal and professional *conflict*.

3. *Transition and Resolution*. As the writer begins to establish a role and forms new knowledge for adaptation, he or she may begin to take on greater *initiative*, understanding what is expected and forming new self-concepts. This initiative usually leads to *rewards and responsibility*, culminating in a *resolution* of previous frustration and often of conflict as well. The individual finally integrates experience and reflects on the intellectual changes afforded by writing in the new context.

For economy, we will illustrate each part of this cycle (and its more specific characteristics) by drawing on the data from the six interns. While all six interns showed a general process of transition involving the three stages that we have described, we do not mean to suggest that they did so in the same way or to the same degree. Organizing a human learning process into entirely discrete categories is impossible, except on a theoretical level. In reality, there is considerable concurrent development and "learning at the boundaries" of any of these stages. Learning and adaptation do not take place linearly; while one kind of knowledge may be developing very quickly, another may develop slowly or recursively. Attitudes may fluctuate, linger, or recur before being integrated. One rationale for identifying discrete stages of the adaptive process, however, is to specify those attitudinal patterns that can be considered typical. This heuristic can then provide a backdrop against which to analyze further data that appear to deviate from the norm.

Expectation

The students studies had various motivations for participating in an internship, ranging from a desire to explore something new and unfamiliar to a desire to take a first step on an already-established career path. Their feelings about the coming work experience also varied, but in each case, both motivations and expectations set the stage for the ways in which they would begin interpreting and adapting the their writing context.

Louise's expectations were characterized by a desire to explore and discover her own occupational future. As one of her early journal entries reflects, she was both excited and uncertain about writing in a new setting:

> I don't know exactly what I want to be doing five years from now. I'm very interested in the major I have chosen, and I like my classes—but what can I do with them? What do I *want* to do with them?. . . I came across a grants research internship. Up to now I had no knowledge of grant writing, and no interest in it. . . I became increasingly interested. . . I started thinking about maybe pursuing this as a possible career.

Joan had decided on a particular career and saw the internship as an opportunity to take an active role in selecting an experience. She voiced her decision in an interview:

> I knew I wanted to get into PR. . . When I approached the internship, I knew it was something I wanted to do, and I looked forward to it. It was something I sought, this particular job with this particular person in this particular environment. I had very methodically and logically looked for what I wanted and gone after it. Most of the things before had simply crossed my path, and I had said, OK.

A desire to be important in his new setting and to make a difference in the workplace was stronger for Jim than for any of our other interns. In an early log entry, one of us wrote that "Jim has a lot of expectations going into this internship. As he put it, he expects to get there and 'go gangbusters.'"

Rachel, another explorer, saw her internship as an opportunity to experience writing as a career. In her final paper, she wrote:

> I was excited about the prospect of having a job where what I did was write. . . I [imagined I] would come to work, spend my time writing in the back room, and sometimes even bring the work home with me. . . To have the desire to take any work home was amazing—this job was different. I was excited, thrilled with the prospect of being able to test my abilities.

As these excerpts suggest, there was a strong tendency for the interns to look positively ahead to their experiences as writers in nonacademic settings, and their anticipation often carried them through the first few days on the job. Most had done well in college and felt that the applications of their knowledge to a new setting would be challenging and interesting. What they were not prepared for, however, was finding themselves in a setting requiring not the "lateral" transfer of the skills they had learned in college, but the "vertical" transfer of situationally rooted knowledge—of what Teich (1987) has claimed is almost entirely a function of the context and content of a specific rhetorical situation (p. 198). Such knowledge allows one to make choices about how to think, how to organize experience, and how to express ideas coherently for solving problems in a particular discourse community (p. 197). For example, after remarking in her paper (quoted above) that she was thrilled with the prospect of being able to test her abilities, Rachel went on to write: "But what I found was an arena where all the rules had changed and no one could remember where the rulebook went." Much like being pushed suddenly into a pool, the interns' immersion in a new culture left them shocked and disoriented.

Disorientation

It is mundane to observe that life experiences seldom match expectations. Much of the disorientation expressed by the interns soon after they began writing on the job, however, originated not only from the disappointment of generally held expectations, but from the collision of what they saw in their new reality and what they had learned from previous experience in other discourse settings. For example, in her course paper, Rachel described the sudden shock of her new writing context:

> I was thrust into an environment [for which] I was less prepared than I thought. . . I was so used to professors basically telling you exactly what they want from you that I expected to be, if not taught, then told, what exactly it was that they wanted these brochures to accomplish. . . They have not taken the time to discuss it—they just put things on my desk with only a short note telling me when they needed it done. No directions or comments

> were included. I do not expect to be instructed, but I do expect some guidance.

This "shock," as the following excerpt from a teaching log about Jim suggests, may have its source in too grandiose an image during the earlier stage of expectation:

> Coming out of sales, Jim has a hard time dealing with the prevalent attitude at —, a non-profit organization. He doesn't understand why they don't have a more positive attitude, more of a "go get 'em" spirit.

Ironically, not being able to identify their role in the office often inhibited the interns' ability to find out what was expected of them. Jim's sanguine attitude at the start of his internship soon turned to frustration as he tried to balance his sense of self-esteem as a writer with the need for guidance:

> In other jobs I've had, like the office job I had this summer, if you had a question you would look through kind of a catalog situation to find the answer or maybe go talk to someone if you absolutely could not find it. Here if you have question, you automatically have to go to someone.

Likewise, a journal entry from Betsy revealed a struggle to find a professional role acceptable to both herself and her colleagues—a struggle which had profound effects on her persona as a writer:

> I am finding it difficult to find the point between complete shy quietness and rambunctiousness where I am comfortable. People have been commenting on how quiet I am and I fell uncomfortable being too quiet, but don't know anyone well enough yet to joke around. . . I realize that an internship is what you make it and you only get out what you put in and all that, but I am not sure how much is "putting in" and how much is being annoying and bothersome.

Similarly, Rachel described her small office as a place where the other three employees have "established their space. . . I can almost feel the invisible walls I'm about to walk through. They all have dominant personalities."

Customarily, independence on the job means having responsibility with a certain amount of freedom or autonomy. Our interns, however, felt neither free from supervision not, paradoxically, able to rely on it for help; they were expected to know already how to execute assigned tasks on their own. Reluctant to show how much they did not know by asking for information, they felt isolated, alone, and disconnected from the rest of the workplace. This perceived independence was sometimes overwhelming: too much responsibility, too much time on their own, too much expected of them—and too little guidance, as Joan remarked in her journal:

> I also found myself becoming quite frustrated when I was left on my own for long periods of time. . . As I am part-time, a lot of things are handled by [the supervisor]. The busier she gets, the less I know about what's going on.

From a teaching log about Jim:

> Jim's first memory of working at — was sitting at his desk chain-smoking, drinking Coke and staring at a blank piece of paper thinking, "My God, I've got to do this." He recalls being incredibly anxious, and said, "All that independence was really overwhelming."

From Rachel's journal:

> I'm not used to being given so much responsibility so soon. I wonder how many people walk into the job feeling so bold and sassy and then when given a task with no real definites along with it can breeze through, take the responsibility, and do a good job. The way *they* want it. I have to anticipate what result they want, what they want to accomplish, in a very short time. The bosses are extremely busy in this little office and apparently they don't have the time or maybe the inclination to sit down and tell me what they want. If I sound timid, that's exactly the way I'm feeling right now. This is my first real writing job and I'm more than a little scared.

And from an interview with Louise:

> I just remember from the beginning, feeling like I was being abandoned. Then later on I realized that it was probably better that [the supervisor] wasn't there . . . Because I knew if she was there, if I had a little question—I'd ask her. I'd want to ask her. And sometimes you have to ask a lot of questions. But sometimes it's better not to, because a lot of times it's things that you can figure out on your own if you just put a little more thought into it. So at first I was a little bit lonely, because I thought I'd have problems.

Of all the interns, Jim experienced the perceived isolation of his role in the office with the most intensity. He also held the highest expectations of his own performance and of his job. He began work feeling that he already knew how to write press releases because he had done so in class, unlike other interns who knew they would be confronting new writing tasks. Later, he found that "for the press releases, the correlation between what we did in class and what I actually had to do in the outside world wasn't close enough that learning it in class for J[ournalism]-school was that important, I guess. . . I've never written anything that short and been so worried about it before."

Jim also began work under a supervisor who, he wrote in his journal, "seemed to be a bad combination of highly critical and usually absent." She would hand pieces of writing back to him two or three times to be redone but (aside from some minor editing) with little guidance about how to revise. Like the other interns, he was anxious about the implication that he was "already supposed to know how to do whatever the job required."

In all the cases, the interns' perceived independence eventually gave way to a stage of sometimes intense frustration, particularly when they found it impossible to adjust quickly and easily to their new role, perform new tasks with little guidance, and communicate with their supervisor (either to understand what was expected of them or to confess the need for help). As an excerpt from Louise's journal suggests, the interns were still coping with a great deal of uncertainty:

> Right now it frustrates me that I can't write them [business letters] faster, but I have to remember that I'm new at this. One of my problems is patience—I expect to jump into a new job and know all about it and what to do right away. Realistically, this is impossible. It takes time to learn these things—especially when I'm on my own.

This sense of frustration and failure often originated, as Joan pointed out, from lack of knowledge about the conventions of the writing context:

> [The supervisor] needed 3 releases done and in 5 hours I didn't even finish one. I felt totally incompetent by the end of the day. [The supervisor] didn't seem to think it was a big

> deal. She said I shouldn't be so hard on myself—that it would come in time. This comment was no help at all. I've written press releases before and still have not been able to increase my speed. I think the problem centers around a formula and around my insecurity. What do they actually want? What is the most important part of the release? Can I call other people for more information? Am I doing this correctly?

Rachel, reflecting on learning a new writing task, expressed similar feelings:

> They want more hype. Oh perfect. I was having trouble with the hypersell writing so I compromised with a blend. I've never had any experience of this sort so I'm finding it difficult. . . These descriptions are difficult because I haven't been [at the events] and all I can work with are pictures and their old descriptions. I may go by mood.

Interestingly, in the early stages of frustration, the interns typically looked on their supervisors as caretakers and providers, and for this reason were willing to accept supervisors' criticisms and guidance—whenever offered—as a way of releasing uncertainty. For example, in an interview, Jim told us about

> a thick skin that I think I've developed—every time I'd write a little press release. . . and every single time they weren't satisfied with it. "Jim, can I speak to you for a minute?" I couldn't write one without [this process]. You just have to come to the conclusion that it's not a personal attack, it's just, you know. . . an attack on your writing. You just have to sit there and smile and nod and try to take their hints and their clues and what they really want.

Paula's journal also shows this mixed reaction to criticism:

> Initially, it was little intimidating. . . for the first interview in particular I was just told, why don't you do that article on –- that I had suggested. I said, O.K. Any pointers? And I was just told to be really accurate with my quotes.

Unable to resolve their frustration with limited guidance, however, many of the interns began noticing an adversarial quality in their relationships with their supervisors. At some point, for example, most of the interns' views of how things should be written or done began to conflict with those of their supervisors. Because the interns lacked security in their own roles, these conflicts seemed particularly threatening, and they reacted strongly. But this stage also seems closely associated with initiative, as the interns began to assert their own opinions against those presented in the workplace, to strengthen and define their own role by imitation or contrast. In her journal, Joan remarked on a growing conflict:

> [The supervisor] has also been getting a lot more testy about getting something done fast. I think that a problem is developing here. I feel that I need more research and she goes for the "quick and dirty" approach. I have a very hard time feeling comfortable with that.

Likewise, Paula's log showed the beginnings of a transition from insecurity to conflict:

> When I saw the. . . story I was furious. My opening paragraph had been changed again by the editor. It affected the whole lead-in that I had worked so hard on. The opening sounds boring to me now. . . What I think makes a story interesting she eliminates. I also think the humor should have been left in my bicycle lead-in. Maybe it's typical to resent any changes made to one's work—although some of the changes I do agree with so it's not entirely a case of sour grapes on my part.

And Rachel's summary described a conflict of knowledge and authority when she argued with her superiors about some grammatical details:

> You expect a professor to know grammar and syntax. . . [That you know] where to put the comma in a series [is] a given. [I had to discuss this] with one of my supervisors. I turned out to be right on this issue and that somehow made him very nervous. He still wanted me to write it his way on his correspondence because that was the way he had always done it and he doesn't want some "kid" to tell him the correct way of doing it. . . On one hand, I don't want them to think I am trying to overhaul the way they write, and on the other hand, I would like to send out material that I'm happy with, that is done correctly.

It is interesting to note in this excerpt how both Rachel's and her supervisor's sense of authority on the matter of surface correctness originates in their static view of the conventions of discourse—as "fixed" and unalterable. Many such conventions, as those of us working across disciplinary contexts well know, are much less universal than our educators prepared us for.

More subtle than the conflict produced by differing views of how to do something in the office is the internal conflict produced by an inability to receive guidance from the supervisor or to evaluate the supervisor as a role model and, by either means, gauge one's own performance. As we reflected on our interviews with Paula, we wrote the following in our field log:

> [Paula] also has conflicting attitudes toward her editor. She looks to her for guidance in her writing, as a role model for her as a journalist, and as a source of writing models. She receives little guidance from her, however, and finds that when she reads articles her editor has written, she can't read them easily and has a hard time extracting the information. Yet she can't analyze what's "wrong" and feels unqualified to criticize.

This difficulty finding an appropriate role for collaboration and critique is also reflected in an interview with Louise:

> [My academic] papers get a lot more writing on them. A lot more corrections. I'm not sure if [the supervisor] really did think that what I wrote was fine. . . or if she's not real particular about the way I said things.

Although the source of the supervisors' resistance to provide evaluative feedback may have come from the position of the intern as a "terminal employee," it had serious consequences for many of the writers' self-esteem, as Jim suggested in his journal:

> I did get input on my writing but that was basically it, and usually negative. As an example, the time cards had a place for [her] signature as my boss, and just above this were boxes to be checked if my work was satisfactory or unsatisfactory; I would be paid regardless. I waited patiently for my first time card to fill up so I could see what her basic opinion would be. I brought it into her office then left so she could think and check without having me over her shoulder. The time card came back unchecked! The only authentic job input I received, besides on my writing, was a thoroughly disgusted look and body movement after I told her I had missed paper deadlines, which I rightfully deserved.

The sense of conflict cannot be underestimated in this cycle of transition, in this process of learning to read the context. In the course of our work with the interns, we found quite often that supervisors gave very unilateral directives about writing ("don't ever use humor," for instance, as a general rule for journalistic writing) or gave students samples of their own or others' work as models that represented very poor writing.

Experienced journalists can weigh a new editor's advice against what they already know and evaluate it accordingly. Someone seasoned in a particular work context can understand that an employee can be a poor writer but an excellent fundraiser, for example, and so make balanced judgments about what writing techniques, formats, or conventions to "borrow." Lacking the knowledge to make such judgments, however, the interns reached eagerly for anything that sounded like a rule or looked like a model; it was something they could imitate to complete their work and fulfill their role. Although supervisors are not formally teachers in the workplace, new employees automatically look to them for rules, vesting them with a universal rectitude. As Jim put it in his journal: "I automatically assumed all persons, specifically my boss, operating within [this environment] were basically right because it was their home environment, and I was wrong. When the interns' own knowledge or education caused them to doubt or disagree, however, it threw them into a quandary. If they rejected one precept, they felt that they should reject them all because they had no basis for judging which advice to follow and which to discount. The dilemma hampered their ability to learn from their context. When their evaluative process clashed with the received wisdom, the interns doubted themselves.

For the interns, then, resolving conflict meant realizing not only that the rules are different in a new context but also that their role models may not be the best exemplars of those rules. The question, "This was the right way to do it in school, but is it the right way to do it here?" is more easily resolved than the question, "Is this person, his work, and his advice a good model for success in this arena?" The second question follows from the first, and is more difficult to answer; it is also more central to building a self-concept as a nonacademic writer.

The inability of the interns to answer the second question with certainty—indeed, their very resistance to asking it, particularly because years of schooling had militated against doubting the wisdom of authority—sometimes led to further internal conflict as they sought validation for their performance. Jim never really reached a balance in his evaluation of his first supervisor; she left before he could resolve his conflicts. She responded negatively to his writing, but the data—primarily samples of her own writing and suggestions that she made to Jim for his—reveal her to be a poor writer and model. Yet Jim continued to want some indication that she thought he was performing his job well.

Louise, who had one of the most positive experiences of the group and had been praised for her work, was still left wondering if she had, in fact, learned to write "well" by the standards of her workplace. For an evaluation to be meaningful, the evaluator must be respected; the grants and letters given to Louise as models were poorly written. Against what criteria, then, could Louise ever begin to rate herself as being a good writer? For Louise, the answer lay in taking action.

Transition and Resolution

Conflict and initiative seemed to be relatively concurrent stages in the cycle of transition for our interns. Initiative was a response to certain kinds of frustration and an eventual release from them. From what we observed, however, conflict may continue indefinitely—a tendency reflected in many business settings where well-adjusted, competent employees work with (and often thrive on) disagreements with their colleagues.

For our interns, personal frustrations centered on their role in the workplace and could not be resolved without some action. Most of our writers look initiative to gain experience in their internship and to make a contribution in the workplace. It seemed to be a point at which they could merge their own goals with those of the organization. Joan, for example, asked for opportunities to gain experience beyond the internship, and gave her supervisor a list of career objectives that she wanted to achieve. In her internship, Louise was primarily responsible for collecting research; grant writing was no more than a possibility. In an interview, she told us how she attended a grant writing seminar on her own before being assigned to write one:

> I have a feeling that if [the supervisor] didn't know that I was taking this class, she wouldn't have asked me to write [a grant]. But knowing. . . that it was a writing class, and that I needed exciting things to do rather than just look up foundations and write letters to them—I don't remember what I asked her. I think we were talking about this class and if I was getting enough information that I needed and. . . then she asked me if I wanted to write one.

At other times, the interns had to assert themselves to get information they needed to write. More at ease in confronting her supervisors and insisting on useful feedback, Rachel made an appointment to see her bosses to "test" their judgments:

> I [wanted to] pin them down on the simplest of issues. I tried to convince them to set the mood in the first part of the descriptions and to include pricing and options. Now I can proceed. . . How does it feel? It feels good. I almost think they were waiting to see if I had enough gumption to make them talk about it.

Jim, who earlier had struggled intensely in his internship, showed initiative by designing a project to get his agency some free time on a local television program. He thought the exposure would help raise enrollment in the agency's classes, but he met passive resistance and negative comments when he proposed it. His goals apparently did not merge with those of the organization, as reflected in an observation from our log:

> I asked Jim if he was sure they wanted to increase their class enrollments or if maybe they already has as much as they could comfortably handle. The expression on his face was wonderful when he said he didn't know—it had never occurred to him that increased class enrollments might not be desirable.

Transitional learning experiences are those points at which the intern develops an understanding of the context that facilitates his or her work—and writing. In the incident above, Jim had apparently misread the needs and goals of his organization. The failure of his attempt to take initiative brought another possibility to his attention, a factor that he can use to interpret other events. At this stage, therefore, foiled initiative does not seem to hurl one back into frustration and conflict; rather, it often leads to transitional learning experiences because mistakes test the context and force a more direct-communication of information. In her final project, Rachel offered us a concise overview of the process:

> I'm still learning how my supervisors think and operate and I think I have a better grip on how they want things done now. It has come to me through trial and error, though, but sometimes that is why it sticks in your head. It's a hard way to learn at times and I wonder if my supervisors see mistakes the way I do. I see them as building blocks and not as any indication that you aren't understanding what you're supposed to do.

Having had to rewrite her last feature story because it was too short, Paula's journal entry about her next assignment read:

> I was prepared to elaborate and make this review fairly lengthy. Another learning experience. The editor pointed out that since the readers could not go to see the performance (it was a one-time event) and in fact not many of them had seen it (about 20–25), I should keep it brief. I also had to control my own tendency to get carried away and use poetic metaphors to describe the presentation.

And, as reflected in her journal, Joan learned that a response letter can be a business opportunity:

> Sent out a letter to a lady [who] had requested literature on a new product one of our clients has. . . This letter again showed my naivete. I had assumed I was simply sending her a letter. [The supervisor] said no, this was a pitch. It only took me a half an hour to write this one!

Louise, who had become adept at tailoring request letters to the particular interests of a donor, gained an understanding of another aspect of the grant application process:

> Because of lack of time, we decided to send the same grant to all [four potential donors]. . . It might be nice to try and come up with a new one, but I must be realistic, the foundations all need to read the same information. . . These letters are already very concise. . . There is no room to go on and on explaining in great detail why this particular foundation is so wonderfully perfect for supporting [us]. This isn't that type of grant. And it's not as if they are all going to get together and compare grants, so why write up a different one for each?

As a direct result of action and initiatives, all our interns reported or intimated some level of rewards and/or extra responsibility toward the end of their experience. Some achieved rewards late in their internships simply because they had to be on the job a certain length of time before they were given responsibility. This was true for both Betsy and Louise. In her journal, Louise told about handing in the final draft of her proposal and "feeling pretty good about it":

> I worked long and hard on it, and feel confident about [the] outcome—but, unfortunately, I don't have the final say. I sat there trying to look busy while my supervisor read my 13-page draft. I have to admit I was pretty nervous. She was making a few marks on it here and there. She looked up and smiled and said, "Excellent job. I'll be proud to use this proposal again in the future." I'm sure I was just beaming.

In a similar vein, Betsy reported how her supervisor rewarded her with added responsibility as the internship came to a close:

> She wants me to do some research on the stories used for next season. . . [She] is trusting me to do a lot on my own. I'm not just doing grunt work—it's a lot of thinking and writing on my own.

We have no data on whether Jim achieved reward from his internship. In an interview, he reported that a potential opportunity presented itself just before the course ended:

> The new coordinator said now once the regular spring publicity is done that her husband [who] works video. . . would like me to work on a TV script. Which is kind of neat, I think.

Rachel did not finish her major writing project before the end of the course, but she was aware of its importance and potential:

> The promotional writing that I'm doing will have an impact, hopefully positive, on the future of this company. That is no small responsibility.

And Paula reported in her journal that she experienced an early success:

> Just found out my band director story was picked up by one of the other papers and printed on the front page. WHOOPEE!

Joan, however, seemed to realize only toward the end of the course, as she approached resolution, that the experiences she had had throughout her internship were rewarding and rare opportunities. As she wrote in her journal:

> I've been wondering if I would be able to have this diversity if I were working for another agency. I think not. I don't know of many places that would allow a novice to meet with their major clients; do marketing; work on an international seminar; and put together a seminar.

The interns we studied all seemed to reach, by virtue of their learning and the rewards it afforded, a final stage in the cycle which we have called *resolution*—the integration of self and role as the writer becomes more comfortable within the context and resolves previous conflicts to gain a more balanced view of the experience. At this point, the writers seemed able to assess and internalize the gains that they had made and to value them as growth and learning. We find it especially interesting, however, that resolution is dependent on earlier frustration, suggesting that instead of finding ways to *avoid* the difficulties inherent in learning, a writer does well to learn how to *embrace* these difficulties and make them productive (see Anson, 1986).

At the end of the quarter, Joan found herself resolving both the ambivalence she felt toward her supervisor and the anxiety that had given her difficulty with her writing. A late journal entry, for example, showed a remarkably positive attitude toward her supervisor which seems linked to a fuller understanding of the supervisor's responsibilities:

> After the presentation was over, I realized why I like working for [this supervisor] so much. She is a good businesswoman and knows how to sell. I may disagree with the

> amount of time she spends on research, but she usually is right in the end. . . I can now write much faster and without a crutch. I've decided that my main problem is needing approval. Now I write as if there will be no one to edit it. The final copy seems to come out much better.

Louise's journal also shows how she came to a relativistic view of the work on which she had expended so many hours:

> Now I can only hope that the foundations and companies I have contacted will respond positively. I would like to know that somebody out there is donating money in response to a grant that I have done. If not, that's life, and I'll try again another time. The experience was beneficial and rewarding.

And Paula made personal gains, as reflected in an interview:

> I think I've grown a lot in my self-confidence in my own writing ability. I've received a lot of compliments and have people looking for my stories, and that's really good strokes for the ego. I've realized that I probably have a facility that I assumed everybody else had and [that] may be not be the case. And I really want to improve my writing skills as a result.

Jim, perhaps the most anguished of the interns, found himself better equipped for his future career. As he told us during an interview:

> Now. . . I'd be able to work publicly for a place but also I'd be able to work independently in an office, which I think is important. Because on my other jobs I've worked in offices before, but never that independently.

Finally, we were surprised at how easily Rachel resolved the major conflict in her internship—the lack of time to write in the midst of all her office duties—and came to a very salient statement about working in context. In her final summary, she wrote:

> I said once in class that I didn't think I was learning much, but in the construction of this paper I have been forced to think differently about that statement. . . I was thinking about. . . a classroom setting. What I realize now is that I've been learning through experience and the actual doing, and gradually transforming my thoughts in ways that will complement the company. I haven't sat down with a textbook on how to operate within a professional setting and then taken a test on the subject. I've been gradually assimilating without really realizing it. There are many things that you think you don't know until you are forced to explain them.

REFLECTIONS AND PROSPECTS

Reviewing the beginning of her internship for her final project, Rachel wrote (as was cited earlier) that 'no one could remember where the rulebook went"; summarizing the conclusion of her internship, she realized that she had indeed learned much, without always being consciously aware of it.

In the time that transpired between these two statements, Rachel had found the rulebook through (a) trial and error, (b) learning to read the context in which she was writing, and (c) both unconscious and conscious assimilation. She too had come to know something of what the other people in the office knew—the tacit beliefs and conventions of the workplace. What her supervisors had was unspoken knowledge—something established that they no longer were aware of thinking about.

At the same time, Rachel also began to realize that the rulebook itself was not a rulebook at all, in the conventional sense, but was a strategic way of understanding and working in a continually evolving context. By watching the interns struggle with this concept, we began to see that there is no such thing as a "context" as a static entity, that is, a place or function whose rules are entirely visible and can be learned or known independently of the effect that the outsider creates by entering it. Communicative contexts, in other words, are subject to the Heisenberg Uncertainty Principle; our interns were not simply "filtering" a static context through their own development and consciousness, but in fact were writing on it and influencing its development.

Once we included this broader concept of transition in our study, we soon realized that at the center of the phenomena we were observing was an ongoing process of adapting to a social setting, involving not only the idiosyncratic textual features of a discourse community but a shifting array of political, managerial, and social influences as well. Just as our interns were developing their own new knowledge, so were they affecting the behaviors of those with whom they worked—particularly in the way their immediate supervisors perceived and communicated with them. The entrance of a novice member into the workplace tests the context, and the necessary interaction produces changes for all participants. At one point, Jim said that he was learning as much about offices as writing; we would suggest that the two types of knowledge are inseparable, and that their effects are mutual.

Our study also suggests that the writer must first become a "reader" of a context before he or she can be "literate" within it. This literacy does not seem restricted to mutual knowledge of some intellectual domain (cf. Hirsch, 1987) but includes highly situational knowledge that can be gained only from participating in the context, which itself is in a constant state of change. Although we did not measure it, we believe that, to some degree, the length of time it takes for such literacy to develop is dependent on the individual's ability for social adaptation. Recent research on the relationship between social cognition and writing ability has focused mainly on static, school-centered tasks (Beach & Anson, 1988; see Bonk, 1990, and Rubin, 1984, for reviews). In the present study, the greater the individual's ability to interpret the constituent elements of the environment in a positive, self-determining way, the more quickly the individual achieved resolution, both within him or herself and within the workplace. And with resolution came greater comfort and competence with writing tasks.

This is not to say, however, that the ability to achieve success in a new writing context rests solely with the *writer*. On the contrary, the context is a deeply enabling or disabling factor in the process. The small sample makes it difficult to draw definite conclusions; but it is revealing that, of our interns, the two (Louise and Joan) who had supervisors willing to act as mentors, even to a small degree, made easier adjustments than did interns who had less communicative supervisors. Jim, who worked with a very negative supervisor, in an office undergoing great change, fared worst of all. And while our accounts might locate the nexus of adaptation on the interaction between an employee and an immediate supervisor, some jobs do not provide for such a relationship. Betsy, for example, worked as an assistant to everyone in the theater administrative office. She made a relatively easy adjustment only because the entire staff was cooperative and communicative. Clearly, this complex relationship among writing, social adaptation, and the climate or culture of the community stands at the center of writers' experiences as they plan, compose, revise, and assess their texts.

In looking at the experiences related by our interns, we also tried to determine whether the degree to which the internships were related to the interns' career goals played any part in the nature or quality of their transition as writers. The work of William Perry (1970), for example, has suggested that students who reach an advanced stage of intellectual "commitment" might be shared. And recent studies of changes in students' intellectual growth during and following internship and other field-learning experiences would seem to support such a hypothesis (see Hursch & Borzak, 1979).

But we found no strong evidence of such a relationship. Betsy, Joan, and Jim were pursuing definite career choices; Betsy and Joan both had positive experiences, while Jim did not. Louise, Paul, and Rachel were all exploring. Louise felt that she had found a good career possibility, Paula was still ambivalent about her writing, and Rachel said that she had now eliminated one career possibility. Here again, we must look at the interplay between the individual and the workplace. Jim was the least skilled writer and worked in an apparently negative environment; Rachel worked in a very stressful entrepreneurial office. Paula worked outside an office environment, with little interaction. Louise, Betsy, and Joan all showed the greatest ability to interpret their experiences positively and to set goals and pursue them.

Our recognition of the interplays between the intern and the culture of the discourse community caused us to regret not having conducted a wider ethnography of one or two of the organizations where our subjects were working. Although we obtained clear glimpses of these settings through texts that our interns brought to us (both their own and others'), most of our information came to us already interpreted by our subjects. The relationship between what new employees see in their contexts and what might be seen by an ethnographer could further refine the stages of transition that we have documented.

It is also clear that our interns' experiences were somewhat different from what might be normally expected in moving from an academic community into a professional working situation. First, the internship course itself provided the students a unique opportunity to analyze and share their experiences with a group of sympathetic peers under the leadership of their teacher, an expert in business writing. Under normal circumstances, newly-hired employees are locked in single combat with their own composing difficulties, without a group of peers to either help out or lend support. Second, interns are usually perceived differently from "new hires"—the position is temporary, with fewer risks and professional demands than a permanent job—and this fact may change the intern's sense of status or security, relationship to superiors, desire to learn about the writing context, and development of long- and short-range goals. The process of bringing to the surface our students' otherwise tacitly developing knowledge and strategies, of helping them to gain a more conscious understanding of the phenomena that we (and they) were studying, no doubt affected their transitions as writers.

Finally, it is worth noting that all of these questions focus on the way in which one learns to write in a new professional setting. While the bulk of research on nonacademic writing is illuminating and extremely useful pedagogically, it has set out to understand writing in professional contexts through already proficient (if not expert) writers. It therefore explores only in a secondary way what it means to become such a writer. We know what sorts of knowledge professional writers have and what sorts of strategies they use, but much less is known about how these are acquired. Further research must begin to bridge the gap between academic and nonacademic writing by taking a more developmental perspective toward the factors that contribute to learning to write in professional settings (Anson, 1988). Sensitivity to issues of gender and class could also explore some of the cultural dimensions in adapting to a new professional context (it is interesting to note, for example, that the one male in our study had the most difficulty integrating himself into a new community).

Without further research validating and refining the stages of adaptation that we have documented, we hesitate to offer specific pedagogical implications from our study. However, the line of research we have presented begs for further development with an eye to improving our current teaching methods, both in business and in education. We do not know, for example, whether the structure of the internship and the concurrent course, with their definite starting and ending points, crystallized the stages and gave impetus to the individuals' achieving some form of closure before separating from the experience. Before developing ways for helping young writers to foresee an experience based on our scheme, we would want to know whether the stages would be extended or recur over a longer period of time. For example, given that the act of writing occurs within a dynamic context of ongoing change, can we expect this cycle of transition to recur throughout the course of an individual's work experience? Would it do so with less intensity or conscious awareness as the individual becomes more experienced? Over a longer work span, in what form (and with what results) does a writer set up new expectations, take initiative, undergo conflict and frustration, complete new projects, and receive rewards and responsibility? Would the entrance of new supervisors, new co-workers and new circumstances in the workplace—expansion, cut-backs, and layoffs—result in new frustration or conflict that then would require the need for new adaptation toward resolution?

In searching for answers to these questions, we may well discover some useful ways in which to extend our current pedagogy of professional writing beyond its primarily textual or "composing process" orientation and into the realms of territoriality, initiation and membership, ritual, and dialect—concepts that would seem to lie at the heart of writing as cultural adaptation.

References

Anderson, P.B. (1985). What survey research tells us about writing at work. In O. Odell & D.L. Goswami (Eds.), *Writing in nonacademic settings* (pp. 3–83). New York: Guilford.

Anson, C.M. (1986). Reading, writing, and intention. *Reader*, 16, 20–35.

Anson, C.M. (1987). The classroom and the "real world" as contexts. *Journal of the Midwest Modern Language Association*, 20, 1–16.

Anson, C.M. (1988). Toward a multidimensional model of writing in the academic disciplines. In D.A. Jolliffe (Ed.), *Advances in writing research: Vol. 2. Writing in academic disciplines* (pp. 1–33). Norwood, NJ: Ablex.

Bazerman, C. (1979). *Written language communities: Writing in the context of reading.* Unpublished manuscript. (ERIC Document Reproduction Service No. ED 232 159).

Bazerman, C. (1981). What written knowledge does: Three examples of academic discourse. *Philosophy of the Social Sciences, 11*, 361–367.

Beach, R., & Anson, C.M. (1988). The pragmatics of memo writing: Developmental differences in the use of rhetorical strategies. Written *Communication, 4*, 157–183.

Beach, R., & Bridwell, L.S. (Eds.). (1986). *New directions in composition research.* New York: Guilford.

Bizzell, P. (1982). Cognition, convention, and certainty: What we need to know about writing. *Pre/Text, 3*, 213–243.

Bonk, C.J. (1990). A synthesis of social cognition and writing research. *Written Communication, 7*(1), 136–163.

Brandt, D. (1986). Toward an understanding of context in composition. *Written Communication, 3*, 139–157.

Britton, J., Burgess, T., Martin, N., McLeod, A., & Rosen, H. (1972). *The development of writing abilities* (11–18). London: Macmillan.

Brown, R.L., Jr., & Herndl, C. (1986). An ethnographic study of corporate writing: Job status as reflected in written text. In B. Couture (Ed.), *Functional approaches to writing: Research perspectives* (pp. 11–28). New York: Ablex.

Colomb, G., & Williams, J.M. (1985). Perceiving structure in professional prose: A multiply determined experience. In L. Odell & D.L. Goswami (Eds.), *Writing in nonacademic settings* (pp. 87–128). New York: Guilford.

Cooper, M.M. (1986). The ecology of writing. *College English, 48*, 364–375.

Cooper, M.M., & Holzman, M. (1989). *Writing as social action.* Portsmouth, NH: Heinemann.

Doheny-Farina, S. (1983). Writing in an emerging organization: An ethnographic study. *Written Communication, 3*, 158–185.

Fahnestock, J. (1986). Accommodating science: The rhetorical life of scientific facts. *Written Communication, 3*, 275–296.

Faigley, L. (1985). Nonacademic writing. The social perspective. In L. Odell & D. Goswami (Eds.), *Writing in nonacademic settings* (pp. 231–280). New York: Guilford.

Faigley, L., Cherry, R.D., Jolliffe, D.A., & Skinner, A.M. (1985). *Assessing writers' knowledge and processes of composing.* Norwood, NJ: Ablex.

Faigley, L., & Hanson, K. (1985). Learning to write in the social sciences. College *Composition and Communication, 36*, 140–150.

Florio, S., & Clark, C.M. (1982). The functions of writing in an elementary classroom. *Research in the Teaching of English, 16*, 115–130.

Forsberg, L.L. (1987). Who's out there, anyway? Bringing awareness of multiple audiences into the business writing class. *Iowa State Journal of Business and Technical Communication, 1*, 45–69.

Freed, R.C., & Broadhead, G.J. (1987). Discourse communities, sacred texts, and institutional norms. *College Composition and Communication, 38*, 154–165.

Gould, J.D. (1980). Experiments on composing letters. In L.W. Gregg & E.R. Steinberg (Eds.), *Cognitive processes in writing* (pp. 97-128). Hillsdale, NJ: Lawrence Erlbaum.

Green, J.L., & Wallat, C. (Eds.). (1981). *Ethnography and language in educational settings*. Norwood, NJ: Ablex.

Harrison, T.M. (1987). Frameworks for the study of writing in organizational contexts. *Written Communication*, 9, 3–24.

Herrington, A. (1985). Writing in academic settings: A study of the context of writing in two college chemical engineering courses. *Research in the Teaching of English*, *19*, 331–362.

Hirsch, E.D., Jr. (1987). *Cultural literacy: What every American needs to know*. Boston: Houghton Mifflin.

Hursch, B., & Borzak, L. (1979). Toward cognitive development through field studies. *Journal of Higher Education*, 50, 63–78.

Knoblauch, C.H. (1979). Intentionality in the writing process: A case study. *College Composition and Communication*, *31*, 153–159.

McCarthy, L.P. (1987). A stranger in strange lands: A college student writing across the curriculum. *Research in the Teaching of English*, *21*, 233–265.

Miller, C.R., & Selzer, J. (1985). Special topics of argument in engineering reports. In L. Odell & D. Goswami (Eds.), *Writing in nonacademic settings* (pp. 309–342). New York: Guilford.

Odell, L. (1985). Beyond the test: Relations between writing and social context. In L. Odell & D. Goswami (Eds.), *Writing in nonacademic settings* (pp. 249–280). New York: Guilford.

Odell, L., & Goswami, D. (1982). Writing in a nonacademic setting. *Research in the Teaching of English*, *16*, 201–223.

Odell, L., & Goswami, D. (1984). Writing in nonacademic settings. In R. Beach & L.S. Bridwell (Eds.), *New directions in composition research* (pp. 233–258). New York: Guilford.

Odell, L. Goswami, D., & Herrington, A. (1983). The discourse-based interview: A procedure for exploring the tacit knowledge of writers in nonacademic settings. In P. Mosenthal, L. Tamor, & S.A. Walmsley (Eds.), *Research on writing: Principles and methods* (pp. 220–235). New York: Longman.

Odell, L., Goswami, D., & Quick, D. (1983). Writing outside the English composition class. In R.W. Bailey & R.M. Fosheim (Eds.), *Literacy for life* (pp. 175–194). New York: MLA.

Paradis, J., Dobrin, D., & Miller, R. (1985). Writing at Exxon ITD: Notes on the writing environment of an R&D organization. In L. Odell & D. Goswami (Eds.), *Writing in nonacademic settings*, (pp. 281–308). New York: Guilford.

Perry, W.G., Jr. (1970). *Forms of intellectual development in the college years: A scheme*. New York: Holt, Rinehart & Winston.

Piazza, C.L. (1987). Identifying context variables in research on writing: A review and suggested directions. *Written Communications*, 9, 107–138.

Porter, J.E. (1986). Intertextuality and the discourse community. *Rhetoric Review*, 5, 34–47.

Rubin, D.L. (1984). Social cognition and written communication. *Written Communication*, *1*, 211–245.

Secor, M., & Fahnestock, J. (1982). *The rhetoric of literary argument*. Paper presented at the Penn State Conference on Rhetoric and Composition, University Park.

Selzer, J. (1983). The composing process of an engineer. *College Composition and Communication*, *34*, 178–187.

Teich, N. (1987). Transfer of writing skills: Implications of the theory of lateral and vertical transfer. *Written Communication*, 9, 193–208.

Tomlinson, B. (1984). Talking about the composing process: The limitations of retrospective accounts. *Written Communications*, *1*, 429–445.

DEVELOPING YOUR UNDERSTANDING

1. Identify and briefly describe two or three of the main challenges faced by Anson and Forsberg's subjects. Then, locate some of the articles you have read in previous chapters that help professional writers prepare for these challenges. Explain how these articles help professional writers, especially ones entering new positions, understand and deal with the challenges you have chosen to discuss.
2. Anson and Forsberg argue that professional writers must first become readers. Summarize what they argue writers must read and why it is important to complete these readings. Then, evaluate the reading process of one of their subjects, as far as you can glean from the details they give you in the article.
3. If you were faced with Rachel's situation as an intern, where she was asked to make a brochure but given little to no instructions or guidance, what would you do? Refer to previous articles to discuss *what* you would need to know and do, *why* you would need to know and do these things, and *how* you could find what you would need to know and get done what you would need to do.
4. Respond to the following claim: Learning how to write is less about learning how to produce a certain kind of thing than it is learning how to analyze, strategize for, and respond to writing situations. Support your response with examples, both personal and drawn from the Anson and Forsberg article.

FOCUSING ON KEY TERMS AND CONCEPTS

Focus on the following terms and concepts while you read through this selection. Understanding these will not only increase your understanding of the selection that follows, but you will find that, because most of these terms or concepts are commonly used in professional writing and rhetoric, understanding them helps you get a better sense of the field itself.

1. macro-level writing processes
2. micro-level writing processes
3. low-level context
4. mid-level context
5. high-level context
6. social cognition
7. rhetorical development

Becoming a Rhetor: Developing Writing Ability in a Mature, Writing-Intensive Organization

JAMIE MacKINNON

While a good deal of research has been done on the development of writing ability in schoolchildren (e.g., Bereiter, 1980; Britton, Burgess, Martin, & Rosen, 1975), little is known about the development of writing ability in educated, working adults. Research indicates that on-the-job writers believe that they do develop significantly through on-the-job writing experience (Bataille, 1982; Brown, 1988) and that this experience is valued more highly than other factors such as writing courses or training in school (Paradis, Dobrin, & Bower, 1984). The nature of this development, however, has not been closely examined.

Indeed, research on the what, the how, and the why of development of on-the-job writing ability has hardly begun. As a result, we have an incomplete picture of writing development, one that implies that development is primarily school-based and thus conceived as "whatever the schools make it to be" (Bereiter, 1980, p. 88). The result is also a limited perspective on nonacademic writing, one that lacks an understanding of the intellectual abilities that nonacademic writing "demands and fosters" (Scribner & Cole, 1981, p. 76).

While interest has grown recently in writing as a "local" phenomenon, with context-dependent aims and characteristics (e.g., Bazerman, 1988), we are still far from understanding the relationship between the development of writing ability and the contexts in which it occurs. Britton was perhaps one of the first empirical researchers to underline the importance of context to development, especially to the significant audience—and function-related aspects of development (Britton et. Al., 1975). And Britton has warned repeatedly against "mistakenly treat[ing] writing as a single kind of ability, regardless of the reader for whom it is intended and the purpose it attempts to serve" (Britton, 1978, p. 13). Of course once the idea of development in context is accepted, the more difficult and problematic it is to conceive of a satisfactory *general* model of development, or even, perhaps, of "rhetorical maturity" (Miller, 1980), except in a local and contingent sense.

These very difficulties, however, point to the kind of theoretical contribution that might be made by research on the development of writing ability in various workplaces, especially in university-educated adults. In contrast to children, whose writing development intertwines with rapid growth in the mechanical and linguistic abilities needed to produce written text, educated adults who develop on-the-job writing ability should theoretically develop in a more purely rhetorical fashion, that is, in their ability to move the world through written language. And theoretically,

Source: MacKinnon, Jamie, "Becoming a Rhetor: Developing Writing Ability in a Mature, Writing-Intensive Organization." Originally published in the Work: *Writing in the Workplace: New Research Perspectives* edited by Rachel Spilka, pp. 41–55.

growth in the writing ability of educated adults should be determined largely by the specific characteristics and demands of the work context in which it occurs. A better understanding of how and why educated adults develop on-the-job writing ability would therefore extend development theory by suggesting elements of a theory for rhetorical development in context.

Such an understanding, even a provisional one, should also widen the perspective on nonacademic writing in two different ways. First, it would add a dynamic, diachronic dimension to the picture, a dimension that has been largely ignored. Writers move the world through writing, but they too move, even as they write. Second, if development is shaped by the demands of the context in which it occurs, knowledge of this growth should also help us understand some of the determining features of various workplaces as discourse communities.

These issues provided a theoretical context for a naturalistic, longitudinal investigation of writing-related change in ten newly employed, highly educated adult "knowledge workers." The investigation was an attempt to discover some early, provisional answers to two complex questions. First, what writing-related knowledge do university graduates acquire in their first one to two years on the job in a writing-intensive, mature organization? Second, how do they learn what they learn? These questions in turn suggested a number of lower-level questions about the role of attitudes and beliefs in developing on-the-job writing ability, the effect a growing understanding of the socio-organizational context might have on writing, and possible changes in the writing process and written products. The investigation was an attempt to provide a tentative answer to a third, higher-level question: How best to characterize writing-related knowledge acquired on-the-job?

SETTING

I am a writing instructor at the Bank of Canada, Canada's central bank. The Bank's major activities include financial and monetary analysis, securities analysis, and econometric modeling. These activities are largely enabled by and carried out through writing. The Bank could be characterized as "writing-intensive": Writing is one of the most pervasive and important value-adding activities in the organization.

The Bank of Canada is large, employing hundreds of analysts and economists; it is stable and mature, being more than fifty years old; and it is organized into a fairly rigid hierarchy. These characteristics help distinguish the Bank as a unique and partly self-regulating discourse community. Other aspects of the Bank that are relevant to its functioning as a discourse community include the importance of writing to the carrying out of virtually all of the Bank's business functions, and the high regard given good writing, especially by members of senior management; a high degree of caution and care in decision making; the research orientation and academic background of its economists and analysts; and the predominance of "document cycling" (Paradis, Dobrin, & Miller, 1985) as a vehicle for "massaging" texts and managing the documentation process.

The writing done by analysts and economists responds to the information needs of various senior decision makers in the organization. These information needs are not readily apparent to new employees, and senior decision makers are often many levels removed in the hierarchy from junior employees. New employees typically write analytic reports dealing with ad hoc issues, as well as regularly scheduled, more generic text such as "weekly notes", "updates," and "monthlies." Most of this writing involves research, analysis, and evaluation, and often in the case of the analytic reports, the development of a tightly reasoned argument.

The size and maturity of the Bank mean that newcomers accommodate and adapt themselves much more to the organization than in a small or "emergent" organization (Doheny-Farina, 1986). The strength and singularity of the Bank's culture are hinted at when employees speak of the "Bank culture," the "Bank point of view" on economic events, and the "Bank style" in writing. The Bank has its own style guide and in-house editors to review many of its most prestigious and important public documents.

METHOD

To maintain the required level of professional staff, the Bank regularly recruits newly graduated economists and financial analysts. This study involved investigating changes in the writing and writing-related knowledge of ten of these newly-arrived professional employees. The participants were the first ten economists and analysts to be hired following Bank approval to conduct this research. All of the participants were recent university graduates; several had masters' degrees. None of the participants, save one, perhaps, had ever written much outside of school. All of the participants, eight men and two women, were under the age of thirty.

Most of the data came from intensive two-hour interviews. The participants were interviewed twice. The first interview immediately followed the final draft of their first major paper (on average, about three months after joining the Bank) after they had revised on the basis of the final round of feedback. The second interview, ten to twenty months later, immediately followed the final draft of another major paper. Following the second interview, in-depth interviews were conducted with six managers (those who consented) of the participants. In addition to supervising the participants, these managers act as reviewers in the document-cycling process.

The interviews with the participants and the managers were document based. That is, the interviews tool place immediately following the production of a large document and focused largely on the concrete and discrete experience of writing the single document at hand. Because of the variation in document types and because of the effects of managerial intervention in the document cycling process, discourse analysis was not practicable. The first interview had participants respond to identical, open-ended questions and dealt with four areas: attitudes and beliefs regarding writing; knowledge of the social and organizational context of writing; the writing process; and the written product.

The second interview asked participants first to describe any possible global changes of significance in their writing, and then the four areas were probed both generally and with participant-specific prompts, such as quotations of salient comments made by the participant in the first interview. Participants were asked to refer to aspects of the document at hand to illustrate statements regarding possible change since the first interview. The interview concluded by asking participants about differences between writing at a university and the Bank, advice they would give an incoming employee about writing, and whether they thought their writing would continue to develop in the following year.

As preparation for their interviews, the managers were asked to review the first-stage and second-stage documents. The first part of this interview involved asking the manager about possible changes in the employee's writing, both generally and in the four areas investigated. The second part had the manager react to interesting or salient participant statements regarding change from the participant's second-stage interview. Immediately following the interview, the manager was asked to fill in a reaction sheet that posited possible writing-related changes in the employee. The manager checked changes that were seen as true and then ranked the changes in terms of significance.

In general, the methodology was naturalistic and qualitative in orientation and both emic and etic in data production. The largely retrospective data from the participants' reports were triangulated through reference to the documents (by the participant, manager, and researcher), through comparing the managers' general remarks with the participants', and through having the managers comment specifically on some of the participants' salient statements. The personal and collaborative nature of the writer-manager relationship meant that more "process-based" investigative methods would have been seen as intrusive.

FINDINGS

All participants appeared to acquire a wealth of writing-related knowledge over the ten to twenty month research time frame. It should be noted that the participants spent a good deal of time writing in this period. Seven of the participants reported spending 60% or more of their work time writing; one spent about 50%, one about 40%, and another about 35%. Many of the document types written over the research period were new to the participants.

The participants, managers, and researcher tended to have similar and corroborating perceptions regarding changes in the participants' writing and writing-related knowledge. The writers developed in the four areas examined, although many changes or developments did not fit neatly into a single area. Many of the findings show a complex, dynamic relationship between learning more about (and through) writing and learning more about the place of work, its employees, its culture and business functions, and the individual's job. [1]

Understanding of the social and organizational context. A critical aspect of the participants' development appeared to be learning a good deal about the social and organizational context in which they wrote: the business functions performed by their departments, the jobs performed by their readers (and thus their readers' needs for information), and the values and beliefs implicit in the organization's activities (more than one participant mentioned the "Bank view of the world").

Most participants said that increased knowledge of the social and organizational context had significant effects on their writing. "Understanding the power structure. . . knowing. . . my readers personally and how they fit into the workflow. . . helps me know how to convince my readers," said one participant. Another said, "You have to know what people know and don't know and that takes time." One participant who, according to his manager, had gained in several important ways as a writer over the course of his first year and a half, did not "put much value on knowing the organizational context" in the second interview. His manager, however, believed that he had acquired a good deal of knowledge in this area and thought this knowledge critical to his improved writing. The manager cited better understanding of the department's business functions, the participant's audience, and the participant's own job requirements as being the most important factors in helping the participant to write more useful papers. All managers interviewed confirmed the importance of this contextual knowledge for writers: "You want to understand what your reader is going to be doing with the information you're giving him in order to give it to him in a useful way," said one.

Another manager, in commenting on how useful it was for a writer to "know who's deciding what, and who needs what information" noted that in a large, hierarchic organization, this learning took time, typically more than a year. On the other hand, one participant noted that at the higher levels of a university, "you can go a long time without seeing your professor" while "here you see your boss every day."

In the first interview, most participants were confident in their knowledge of the audience and the underlying purpose of the document just written. Interestingly, ten to twenty months later, most participants looked back on these earlier certainties with disbelief. "I definitely wasn't clear [when I started working here] about the readership and how readers used the documents. . . Sometimes you're wrong," was one representative comment. All managers confirmed the participants' beliefs that they had developed a much better sense of audience and purpose over the research period; most managers also predicted that this knowledge would continue to grow.

Another significant, though largely unconscious, aspect of the writers' development was in how they learned to use and manipulate a social/organizational process—document cycling and complex feedback—in order to help them produce satisfactory documents. As they learned how to deal with feedback and documents cycling, which all participants found enormously frustrating at times, they were also learning more about their readers as individuals as well as their information needs, more about Bank discourse conventions, and more about the business functions enabled by or related to the documents they were writing. This increased acuity appeared to prompt consequent changes in the "macro" aspects of their writing processes, as suggested below.

The writing process. The participants reported more change in their writing process than the researcher had expected. Eight of the ten participants volunteered in the unguided, global part of the second interview that there had been major changes in their writing processes. These changes tended to be "high-level" or "macro" in nature, reflecting an accommodation to the larger dynamics of organizational life as well as an improved understanding of audiences, purposes, and writing-related business functions.

The following developments appeared to be the most common. In the second interview, the participants reported being more apt to:

- initiate ad hoc documents;
- clarify assigned papers orally to a greater degree when they were assigned and later while planning;

- be clearer before starting to draft about the "story" they would write, a story that economic or financial numbers would support;
- discuss writing with colleagues, especially in the early stages of writing;
- keep their readers in mind as they wrote ("On the last paper I went through so many drafting changes as a result of suggestions," said one participant, "that I began to think that there must be as easier way of making the paper acceptable to the reader earlier in the process. So, 'Would he [the manager] consider this acceptable' became my standard, my benchmark.");
- react to feedback effectively. Usually this meant better affective reaction as well as better instrumental strategies for using feedback. Virtually all of the participants felt less personally threatened and less depressed about feedback as time wore on: "A million red marks doesn't mean you aren't a good writer," said one participant. By the second interview, participants tended to take a more active role in the feedback sessions, understand more of the feedback, and revise on the basis of this understanding more effectively. Some participants had started asking colleagues for feedback. Participants said that "learning to use feedback" had been critical in making gains in writing; they also said that "feedback was the main [vehicle] for learning" about the Bank and its activities, readers' needs, and standards and expectations for documents.

Several participants obtained their own microcomputers during the study period. While the participants did not attribute significant changes in process or product to the use of word processing, more than one participant commented on increased speed and on not having to wait while a draft was tied up with a secretary.

Some participants reported an increase in efficiency and automatization in writing short, formulaic document types.

The written product. The participants' written products changed considerably over the research period to match more closely busy readers' narrow needs for information and analysis. Documents tended to be more focused, more visibly and hierarchically structured, more analytic, and less descriptive. Following are some representative comments from managers about changes in the participants' written products. The managers' interventions have a large effect on each draft; these comments therefore refer to changes in first or early drafts.

- "An improvement in making sure he's addressing the right audience."
- "Doesn't overburden the readers with details so much."
- "He [knows] the need to get a *story* out."
- "The structure and the purpose [are more] obvious [now]."
- "The content improved. . . After a year he knew. . . what information would help his readers."
- "An improvement in first drafts. . . [Better at] getting at the bottom line."

Attitudes and beliefs about writing. The participants' attitudes and beliefs about writing changed in various ways. Belief in the possibility of development was one important change. Few participants, when first interviewed, believed they would develop significantly as writers in their first year or two at the Bank. When asked about possible changes in their writing over the coming year, a few participants predicted no change ("Quite frankly, [I have] the same writing skills I've had since grade school," said one); several predicted possible change in rather narrow aspects of writing such as "better terminology" or "improved style"; a few were open to the idea that there would be broad change but had trouble making concrete the abstraction "something will change in relation to changing readerships."

By the second interview, ten to twenty months later, eight of the ten participants believed they had developed significantly as writers. In one revealing comment, a participant said: "Before I started. . . I would have definitely said that my writing wouldn't and needn't change all that much. But there have been significant changes and there need to be more changes. . . . I see there is a lot of learning to be done and a lot of improvement to be made. . . I guess I was being conceited." While few participants predicted they would develop as writers when they were new employees, in the second-stage interviews, most predicted further development.

Self-confidence in writing ability increased generally among the participants, although earlier self-confidence was seen later as unjustified, too "naïve" or "too high."[2] More than one manager made explicit reference to the relationship between attitude and writing development. One said, "Because he sees it [learning yet to be done and improvement yet to be made] he's making progress and he'll continue to make progress. It's the fellow who comes in saying, 'Hey, look. I did very well in English in university and I write poetry. You can't teach me: What are you, some bureaucrat?' Those guys don't do so well."

Participants' notions about the functions of writing broadened considerably. For many participants, this was reflected in a change from viewing the major function of writing as "to inform people about events and ideas" to the more dialogic sense of getting "reaction to ideas in hope of producing more research." Some participants developed a deeper understanding of the epistemic possibilities of writing as well. One participant said in her first interview that the role of writing was "to let other people know what you were doing and to get comments on your research." Eighteen months later, she added these three functions: "to answer questions" (analytic function); "to help people who come after us" (archival function); and "to clarify points of uncertainty. . . for ourselves" (exploratory or learning function). Significantly, most of the participants developed a powerful understanding over the research period that their writing needed to address the concrete information needs of their audience and not simply express what they, the writers, knew on a given topic.

Interestingly, many of the participants came to appreciate the "local" nature of writing in the Bank, and perhaps that "good" writing is not a fully generalizable notion: "If you don't know the culture, if you don't know the people you're working with, then you don't know your 'clients,'" said one participant. "It's a marketing strategy: if you don't know your readers, then your paper is not marketable. . . You have to write taking into account the environment in which you are working. This is definitely a change in the last year. . . You have to adapt to new environments and change your behavior."

How the participants learned. The developments outlined above appeared to be triggered by a variety of experiences, but two elements seemed critical: the new demands placed upon the participants as writers, and the feedback they received on their writing. The learning appeared to occur through various activities and media: through writing itself; through thinking about feedback, especially substantive, questioning feedback (from reviewers primarily, but also from colleagues, both in writing and orally); through reading, talking, listening, and observing; through development subject-matter expertise; and through being directed, admonished, questioned, and encouraged as an employee. All participants emphasized the importance of feedback for their development as writers. The document cycling process appeared to be a vehicle for building consensus and sharing and making knowledge, as well as for "massaging" texts. Most participants emphasized the importance of learning more about their readers, both personally and as consumers of specific information to perform specific business functions. Several managers stressed the importance of good listening and interpersonal skills to writing development. One manager said "she had a very good network of peers with whom she could talk and learn a lot that way. I think that's perhaps the major route through which people learn about this organization." Two managers mentioned "shyness," being "passive," and not being "challenging" in the feedback sessions as factors that would limit an employee's development of writing ability.

The participants' consciousness of the learning varied. A few participants appeared to develop much more than they were aware, perhaps because they had narrow, restricting notions of the nature and domain of writing. Some managers were "quite surprised" to hear that their employees had attributed a great deal of importance to the feedback they had received for their development as writers. One of these managers was "quite mystified" about how his employee had learned as much as he had. Other managers were more aware of their role in this regard.

Managers confirmed most of their employees' quoted comments on aspects of their learning. There were no major differences between participants' and managers' perceptions, save that the manager sometimes saw a less "dramatic" change than the writer. Interestingly, more than one manager seemed to think of their employees as still developing and predicted further rhetorical growth: "This is going to be the next stage of his development if he stays here," said one manager. "He sees himself as someone paid to answer questions. Well, you know, as he moves up, he's going to be paid to *ask* questions."

DISCUSSION

Overall, the writing-related changes were considerable, consequential, and a shock for some participants: "It's like going to China," said one. For most of the ten participants, the complex totality of the writing-related changes they experienced added up to a "sea change": a major shift in their understanding of what writing is and does in an organization, a revised understanding of the roles they saw for themselves as writing workers and as working writers, and often major changes in various aspects of the macro writing process.

The findings strongly support the idea that educated employees can develop writing ability on the job (Brown, 1988) as well as the notion that on-the-job experience is the major source of this learning (Paradis, Dobrin, & Bower, 1984), although the nature of this experience is complex and begs explanation. Interpreting development is always difficult—and usually needs explanation at some level—but perhaps it is useful to distinguish here between the developmental process and the developmental outcome.

The developmental process appears to have been instigated and informed by the new rhetorical demands of a new social-organizational context. If expertise in writing is "not categorical" but relatively domain-specific (Freedman, 1987; Nystrand, 1989) and cognitive strategies are less "portable" than previously assumed (e.g., Lave, 1985; Rogoff, 1984), then novel rhetorical demands are likely first to elicit inappropriate responses and then to spark new cognitive and social response strategies. In the Bank, the writing process is intertwined with a number of social processes, in particular, document cycling. It is through engagement in these social processes that newcomers to the Bank seem to acquire higher-level contextualizing knowledge, which in turn promotes gains in writing.

Over the research period, most participants acquired a wealth of knowledge that allowed them to function more effectively as writers. The most important knowledge gained, because of its catalytic nature, appeared to be contextualizing knowledge. Listed (roughly) from high-level, perhaps abstract and distant forms of context to lower-level, more immediate forms of context, the important elements of this knowledge included:

- aspects of the Bank as a discourse community that accomplishes specific (and at times unique) business functions through the specific discourse practices of a specific culture with its own distinct point of view;
- the information needs of senior, often remote readers;
- the business functions enabled through writing and the relationship of written texts to various larger institutional "value-adding" processes;
- the forms of argument and informal reasoning conventional to, and found persuasive in, the Bank;
- the participant's role as an employee and writer in the larger, ambient, business-function processes;
- the rationale for and the mechanics of the document cycling process;
- the participant's growing, job-specific professional knowledge.

As recent university graduates working in their first major jobs, the participants were writing, perhaps for the first time in their lives, to an audience with real needs-to-know and in order to make things happen. Instead of writing to display mastery of knowledge as they had done in school, they were writing, as they came to understand over time, to promote action and to inform decision making.

A key to the participants' development appeared to be a much greater awareness of, and sensitivity and adaptiveness to, the particular demands placed on writing by the Bank. Their growing awareness of these unique demands came as a surprise to some of the participants and appeared to be the precursor to writing more audience-sensitive text. In other words, the participants were not merely developing on some abstract continuum of "social decentration." Rather, some participants seemed to have come to an explicit understanding, perhaps for the first time in their lives, that audience counts. Indeed, some participants came to understand that meeting their readers' specific needs was the key determinant of success of written communication in the Bank. This growing realization appeared to precede later gains in audience sensitivity and adaptiveness.

Researchers have suggested that social cognitive ability—the ability to effectively represent one's social environment—may play an important role in the development of writing ability (e.g., Piche & Roen, 1987) and that "social cognition" may best be seen as a multidimensional ability that interacts in a complex manner with the composing process (Rubin, 1984b). These two notions cast light on the nature of a significant part of the participants' development: that of audience sensitivity and adaptiveness. Development occurred as the writers grew to understand some of the complexities and demands of the social structure in which they were writing.

Most of the participants contrasted their on-the-job writing with university writing by referring to the specificities of audience requirements in the Bank. It is tempting to speculate that in academic, a gradual, perhaps instrumentally motivated increase in "decentration" occurs, while initiation into an adult knowledge-working community might provoke a more sudden increase in social cognition, integratively as well as instrumentally motivated. Perhaps one needs to write to real, variegated audiences with real and divergent information needs to develop one's social cognition.

While social cognition and the demands of a new context account in part for the developmental process, a rhetorical perspective can help characterize the outcome of the participants' development. The participants were developing not only as writers, but also as members of a community they were still struggling to understand. In accepting and initiating writing assignments, submitting drafts at various stages of development to their reviewers, voluntarily seeking formal or informal feedback from colleagues, testing assumptions about their authority to make interpretive claims, responding in various fashions to feedback, and a plethora of other actions, the participants were assuming and adjusting writing-related roles for themselves and learning and conceptualizing the writing-related roles of others. The wideranging, role-related, and epistemic character of the writing-related changes, viewed as a totality, suggests that at least some participants experienced significant *rhetorical* development.

That is, over the research period, the participants enhanced their ability to engage the work community and carry out various tasks through discourse. As McCloskey (1985) and others have pointed out, writing in economics is highly rhetorical. In the Bank, most writing is analytic and argues a case. Writers pose and defend contestable ideas; they analyze and evaluate other people's ideas, all within an institutional forum. Senior managers have a pejorative term to describe merely descriptive writing: "elevator economics" (for example, housing starts are up; productivity is down). Young analysts and economists are often told that senior readers want analysis, evaluation, arguments, and "stories," but this must mystify the new employees at times. The making of persuasive stories in a community requires an understanding of the community's shared assumptions, beliefs, and values, as well as the forms of argument seen as legitimate in that community.

The participants in this study started to develop this knowledge by jumping into the fray, by joining the conversation, by making their own arguments. When their arguments failed to persuade, they knew so immediately, concretely and consequentially. When their arguments succeeded, it meant that they had succeeded in understanding the rhetorical dynamics of the Bank and the demands of their own rhetorical situation.

The growth of the economist/analyst as a writer can be viewed as the apprenticeship of a rhetor. The participants developed an effective, though largely tacit, understanding of the organization as a rhetorical domain. In acting rhetorically—in speaking, writing, talking, and listening—they were discovering what knowledge was socially significant and which forms of reasoned argument readers found persuasive. In so doing, the participants found a voice and conceptualized and assumed roles for themselves in a rhetorically bound, rhetorically functioning community. They joined in making and testing the community's knowledge and thus engaged more fully in the community's practices. In becoming rhetors, they became active participants in the community's business.

IMPLICATIONS OF THE STUDY

This study underscores the inadequacy of development theories based on the writing of children and adolescents in school settings to explain and interpret the development of educated adult writers in a work setting. If context is as important as it appears to be in determining the development of adult working writers, then an adequate *general* developmental theory may prove elusive. Developing domain-specific theory may be a more realistic, if more modest, goal.

Context would appear to be a critical element in any speculative model of rhetorical development of on-the-job writing ability. While theoreticians are starting to understand just how complex and multidimensional context can be (e.g., Piazza, 1987), it is useful to remember that writers and employees always start with the context at hand. In a large, hierarchic organization, "high-level context" (for example, the decision making of senior executives) is likely to be unknown and largely invisible to new employees. And yet in a large organization where knowledge is both the "raw material" and the goal of writing (cf. Gage, 1984), an understanding of this context will be critical to rhetorical growth. Development may occur as the employee engages and is engaged by a more immediate "mid-level context," the local organizational structure and forms of social interaction. Depending on the individual's ability to perceive, manipulate, and interpret that more immediate context, and depending on the supportive and inhibiting features of that context, the writer will over time gain a sense of the less immediate, though rhetorically important, high-level context. That is, developing writer interacts primarily with an immediate context that intermediates information about and from higher-level contexts.

For on-the-job writing trainers, knowing that important aspects of writing development may naturally follow an increased understanding of business functions, audience, and corporate culture will have implications for, among other things, the timing of training. In addition, if writers' managers are as important as they appear in developing new employees' writing, training these managers in the effective management of writing may be warranted.

For writing teachers in colleges and universities, one obvious implication of this research is that they need not expect students to "master" writing before they graduate. Writing development is likely to continue after students leave school, it would seem, if their writing situation demands the kind of changes that prompt rhetorical growth. For students in professional programs, being told the simple (but perhaps counterintuitive) fact that their writing will likely develop in important ways "even" after graduation may be the most important lesson of all.

This study examined development in newly hired, recently graduated employees only, and in just one workplace. Future research on the development of nonacademic writing ability might examine a variety of local, situationally specific development patterns. Research on highly educated adults shows particular promise and should be pursued.

Just as studies of the development of academic writing ability sometimes tell us more about the school system than the children (Bereiter, 1980), the study of the development of nonacademic writing ability is likely to tell us as much about the worlds of work as about writing and development. And it will always be difficult for the researcher to separate learning to write from learning the job. But if we view writing as a way of doing work, the study of these twin developmental processes will afford a strategic opportunity: the chance to further integrate research on development and nonacademic writing with rhetorical theory. Such an integrated approach is likely to generate additional elements of a theory for rhetorical development in context.

Notes

1. A summary of the findings of a naturalistic study is problematic in that the methodological and persuasive benefits of "thick description" are absent, and a rather "thin," decontextualized list of assertions is the apparent result. The findings presented here are the synthetic outcome of two earlier, "lower-level" rounds of classification and analysis. First, portraits of individual participants' writing-related changes were sketched. Second, thematically organized, generalizable changes were drawn from these portraits. The findings presented here, then, "hide their tracks," and represent only a few of the most important writing-related changes, those that suggest significant development of writing ability.

2. See Myers (1980) for a look at how "bounded rationality" and the "self-serving bias in the way we perceive ourselves" (p. 49) are powerful forces in shaping self-image.

DEVELOPING YOUR UNDERSTANDING

1. Referring to MacKinnon's findings, develop a set of goals he would suggest you set for yourself as an intern or newly employed writer. In addition, develop a plan that would help you reach these goals. Your plan should go beyond *what* you would want to accomplish to *how* you plan to accomplish it, in general.
2. Identify the two or three findings made by MacKinnon that most surprise and/or influence you. Explain why you find them surprising and/or influential.
3. Even though the subjects MacKinnon studied wrote a great deal (up to 60% of their work time), they could all be identified as "professionals who write" rather than "professional writers." Do you think MacKinnon's findings would have been different if his subjects had been professional writers? If so, explain why you think so and how they would be different. If not, explain why you believe his findings would be similar.

FOCUSING ON KEY TERMS AND CONCEPTS

Focus on the following terms and concepts while you read through this selection. Understanding these will not only increase your understanding of the selection that follows, but you will find that, because most of these terms or concepts are commonly used in professional writing and rhetoric, understanding them helps you get a better sense of the field itself.

1. facilitated performance

2. attenuated authentic participation
3. legitimate peripheral participation
4. scaffolding
5. social motive
6. writing as epistemic
7. writing as instrumental

Virtual Realities: Transitions from University to Workplace Writing

PATRICK DIAS, AVIVA FREEDMAN, PETER MEDWAY, ANTHONY PARÉ

In this chapter we consider contexts for learning that further exemplify the theoretical discussion provided in the previous chapter. We have argued that learning is a situated and contingent experience, and that school-based simulations of workplace writing fail to prepare students for professional writing because they cannot adequately replicate the local rhetorical complexity of workplace contexts. By this we do not mean to suggest that course-based case studies are a waste of time, but the goals of such learning opportunities need to be reexamined, especially as they relate to the development of professional writing ability. We believe that course-based case studies provide an essential introduction to the ways of thinking and knowing valued by disciplines and encouraged by the rhetorical practices of those disciplines. In order to learn about professional writing, however, that introduction must be followed by more extensive and integrated workplace experiences, such as work-study programs, internships, on-the-job training, and other forms of transition between school and work.

The learning continuum proposed in chapter 9 describes a range of experiences that provide ever greater distance from the deliberately fashioned educational contexts of the school and ever greater integration into the improvised and often spontaneous learning opportunities of the workplace. (Note: Although the movement from facilitated performance to legitimate peripheral participation does capture the professional development of many who pass through school and into the workplace, it is not a required sequence; some people learn their practice entirely within the workplace, without benefit of the careful scaffolding of facilitated performance.) In describing that transition, we have used the terms *facilitated performance* to refer to the conditions and contexts for teaching and learning in universities, *attenuated authentic participation* to point to the early stages of closely monitored and supervised learning in the workplace, and *legitimate peripheral participation* (Lave & Wenger, 1991) to describe the activities that lead to learning for newcomers who have achieved some degree of autonomy in their workplace involvement.

Source: Dias, Patrick, Aviva Freedman, Peter Medway, and Anthony Pare, "Virtual Realities: Transitions from University to Workplace Writing," In *Worlds Apart: Acting and Writing in Academic and Workplace Contexts.* Mahwah, NJ: Lawrence Erlbaum Associates, 1999, pp. 201–221.

As the learner moves out from the classroom toward professional practice, the moments and sites of learning become less clearly defined, and certain key features of learning and teaching change. There is a gradual increase in the authenticity of tasks; that is, their consequences and their influence on others and on activity escalates, and there is a parallel growth in their complexity or messiness: workplace tasks lack the exact moments of beginning and ending, the stated evaluation criteria, and the sharp divisions of labor that usually characterize school work. At the same time, there is a steady decrease in explicit attention to learning and in the degree of guidance or attenuation provided; students move from close supervision to autonomous practice as they make the transition from school to work. There are, too, subtle shifts in authority and expertise between learner and teacher; in school, the power and knowledge imbalance is marked, obvious, but newcomers to the workplace often bring certain kinds of expertise (technical, theoretical) that oldtimers who supervise them do not have. A similar shift occurs in terms of evaluation; in school contexts, the learner is evaluated by the teacher, but in the workplace the apprentice and supervisor are assessed together, because newcomer and oldtimer often work as a team. Underlying these changes is a shift from the epistemic function of school writing (the individual student's construction of knowledge) to the collective epistemic and instrumental goals of the workplace (the construction and application of institutional or community knowledge).

In this chapter we look at students learning to write in and for different contexts, some in school, some at work, and some in between. Although the writing tasks in the first two contexts attempted to simulate or suggest aspects of the rhetorical contexts of professional writing, neither could escape their ultimate location within an academic setting. And, it should be noted, neither was presented as an opportunity to learn how to write professionally. Writing in the school contexts was at the service of learning the discipline's values, beliefs, and ways of knowing. Learning occurred in the workplace as well, but there writing was at the service of the community. The first three settings described below mark points on a trajectory away from the school and into the workplace; the last, which we describe in greater detail, represents a final stage on that continuum. Our analysis of each stage demonstrates the extent to which learning is contextual and contingent.

LEARNING TO WRITE AT SCHOOL

As explained in the previous chapter, the theoretical frame that accounts for how students learned to write in the university disciplinary classes we observed is best captured by the term facilitated performance. Our argument is that this frame, based on Rogoff's (1990) notion of guided participation, accounts for how university students learn discipline-specific writing in much the same way as guided participation accounts for early child language acquisition or cognitive apprenticeship in middle-class homes.

The most salient commonality is that the guide in both cases, caregiver and teacher, is oriented entirely to the learner and to the learner's learning. In fact, the activity is being undertaken primarily for the sake of the learner, and although the conventions of neither writing nor language are explicitly taught, they are learned because of the carefully shaped context. There is a focused and centered concentration on the learner and the activity which is quite different from what we see in the instances of workplace-based learning described below (and that, as we noted in the previous chapter, is quite different from what Rogoff [1993] and Heath [1983] reported in non-middle-class-rearing).

Not only is the guide's attention focused on the learner, but the whole social context has been organized by the guide for the sake of the learner. Rogoff's caregiver organizes the storytime experience in much the same way that the instructors we observed orchestrate their courses (within certain temporal, spatial, organizational constraints): readings are set, lectures are delivered, seminars are or are not organized, working groups are or are not set up, assignments are specified—all geared toward enabling the learners to learn certain material.

Further, students did not, in the courses observed, learn to write new genres on the basis of explicit direction by their guides (the instructors and teaching assistants), except in the crudest terms with respect to format, length, and subject matter (Freedman, 1993). Nevertheless the writing was shaped and constrained from the first meeting of the course—that is, from its specification on the course outline and, more significantly, from the first words uttered by the instructor.

Our observations of these classes, and the students composing for them, revealed that learning new genres in the classroom came about as a result of carefully orchestrated processes of collaborative performance between course instructor and students: the students learned through doing, and specifically learned through performing with an attuned expert who structured the curriculum in such a way as to give to the students more and more difficult tasks. The instructor both specified the task and set that task in a rich discursive context. Both the collaborative performance and the orchestration of a richly evocative semiotic context enabled the acquisition and performance of the new genres.

At the beginning of a course in Financial Analysis, for example, the instructor assigned cases to be written up at home and then he himself modeled in class appropriate approaches to the data, identifying key issues and specifying possible recommendations for action. As they attempted to write up the cases themselves at home, the students were "extremely frustrated" because "you have to do a case before you have the tools to know how to do it." "It's like banging your head against a wall." However, the modeling in class, *especially in the context of the students' struggles to find meaning in the data themselves*, gradually enabled the students to make such intellectual moves themselves. At the beginning, "when he would tell us the real issue, we're like—where did that come from?" Then "When you're done and he takes it up in class, you finally know how to do it!" Modeling what the students would later do themselves, the instructor presented a number of cases at the beginning of the course. Like the mother with the storybook, the instructor showed the students first where to look and then what to say, picking out the relevant data from the information in the case:

> "What's the significance of 7 and 8 in the text? Did it add to your thinking about this case?"

"At what market share restriction would that growth strategy not work?"

"Assuming best case scenario, what will this company look like in 5 years?"

He constructed arguments, using the warrants of, and based on the values and ideology valorized in, the discipline. Drawing on the simulated purposes for the case, he pointed to the importance of looking at and presenting information in particular ways:

"As a consultant to the bank, is this a critical value to know?"

"In real life, you have to quantify this relationship between business risk and financial risk."

After students began presenting the cases themselves orally in class, he provided corrective feedback:

"Walk people through how you thought about the problem."

"Let people know what the agenda is and your role."

Gradually students were inducted into the ways of knowing, that is, the ways of construing and interpreting phenomena, valued in that discipline. That is, although they were learning a school genre that bore little resemblance to workplace genres, they were beginning to participate in the type of thinking encouraged by the rhetorical practices of their discipline.

We see in facilitated performance many of the elements specified by Wood, Bruner, and Ross (1976) as functions of the tutor in scaffolding (as quoted in Rogoff, 1990, pp. 93–94). The task is defined by the tutor; the tutor demonstrates an idealized version of the act to be performed; and the tutor indicates or dramatizes the crucial discrepancies of collaborative performance offer part of the answer to how the students learned to write the genres expected of them. In addition, the instructor set up a rich discursive context, through his lectures and through the readings, and it is through the mediation of these signs that the students were able to engage appropriately in the tasks set.

Wertsch (1991), drawing on Bakhtin, talks of the power of dialogism, and of echoing (or ventriloquating) of social languages and speech genres. The students that we observed responded—ventriloquistically—to the readings and the professor's discourse, as they worked through the tasks set for them. Initially, they picked up (and transformed in the context of their preexistent conversational patterns) the social language or register they had heard. Here are oral samples culled from the students' conversational interchanges as they worked on producing their case study:

Allison: I figured this is how we should structure it. . . First, how did they get there is the first thing.

Peter: So that's. . .

Allison: Business versus financial risk or operations versus debt. Whatever. . . Then. . . like we will get it from the bankers' perspective.

Peter: Yeah, that's pretty much like what I was thinking too.

Allison: So, right now I have their thing before '78. How do you want it, pre-'78 or post-'78? This is what I did. I went through all. . .

Peter: Internal comparison and stuff.

******

Allison: I guess the biggest thing is the debt-to-equity ratio. Notice that? X has way more equity. If you look at Y, their equity compared to their debts is nowhere near, it's not even in the ballpark.

Peter: Which company is it that took a whole bunch of short-term debt?

Then, in the writing of their papers, the conversational syntax, lexicon, and intonational contours of their conversation exchanges disappeared, as they reproduced the discipline-specific terms in the context of academic written English, achieving, thus, the written social register designated by their professor as that of a "financial analyst" (see Freedman et al., 1994). In the final draft of a case study, we find the following:

> Short-term debt restructuring is a necessity. The 60% ratio must be reduced to be more in line with past trends and with the competition. This will be achieved by extension of debt maturities, conversion of debt to equity, reduction of interest rates, as well as deferral of interest payments.

In other words, through the mediation of an appropriation of the social languages provided by the professor's discourse and the readings, the students created the new genres expected of them: ways of knowing became ways of saying. The epistemic social motive of school is met when the professor determines that the discipline's knowledge and ways of knowing are satisfactorily displayed in the completed text; then, a grade is assigned and the text reaches closure.

BETWEEN SCHOOL AND WORK

It is possible to distinguish between school-based attempts to introduce students to disciplinary ways of knowing, such as the case-study assignment described above, and those fuller immersions that occur when students balance their time and attention between the classroom and a workplace. Various forms of such an arrangement may allow students to make forays into work settings, receive visits from practitioners, handle professional documentation, and address actual workplace issues and problems, all in the context of a particular course. Evaluation in these situations, even when there is consultation with practitioners, is primarily an instructor's responsibility. Typically, students in this type of course prepare reports for the workplaces they visit, and representatives of the workplace may be invited to hear and respond to oral presentations of student research results. (Such an assignment is described in chap. 4.)

One of the most successful of this kind of learning experience that we have observed occurred in a 4th-year course in systems analysis. The course engaged students in writing that, in contrast to the simulation of a case study, succeeded in giving them a real experience of workplace discourse, though it was a highly scaffolded and protected introduction. However, as with the case study, there was no explicit writing instruction. However, as with the case study, there was no explicit writing instruction in the systems analysis course, and no stated intention to teach workplace writing.

The systems analysis course was organized as follows: every year that she teaches the course, the instructor finds six client-organizations who need to have their systems analyzed and redesigned; a total of 30 students are allowed to register, and students are selected by competition; the students are divided into groups of five. Criteria for nomination to specific groups include the following: each group is assigned someone with a background in accounting and another with a background in computing, as well as someone who has a car. A cardinal rule is that friends are to be separated, so that students gain experience working with people that they are not already familiar with or even fond of. Each group is assigned to a separate place of work, and must consequently solve the specific problem in that work environment. In order to do so, and in the course of doing so, they produce the following three kinds of documents:

1. a problem definition document (on the basis of a feasibility study);
2. systems specification document (on the basis of their systems analysis), which includes several alternatives as part of the structured specification;
3. a general design document.

In order to perform these tasks, students must go to the workplace to interview a range of practitioners there, the range determined by the needs and possibilities of that specific work environment. At the first interview, they are accompanied by the course instructor, who guides the interview when necessary and spends considerable time going over what was learned with the students involved.

In addition, each group reports back to the class as a whole three times, giving oral versions of their drafts of each of the three key documents. After these presentations, they receive feedback from the class, hand in their written document to the instructor, and then, at a separately scheduled interview, receive intensive feedback from her as to appropriate revisions. Sometimes further interviews with her are necessary. When she is satisfied, a written document is presented to the client, sometimes accompanied by an oral presentation, if the client so desires. The course is a half-course, the timeline is carefully spelled out at the beginning, and nothing is handed in late.

Clearly, this is school work with a difference, and that difference is the dual exploitation of both the academic and workplace settings. The workplace adds some critical features to the learning context. Perhaps most importantly, it provides authentic tasks. As the professor said about this course (Duxbury & Chinneck, 1993), "the process involves coping systematically with the complexities of the real world, including both human and technological limitations." Students have to "limit the area of study, abstract essential features subdivide a complex whole into parts of manageable size and model a real system to show the relationship among its components."" As opposed to traditional lectures in systems analysis and design, which give the students a tremendous amount of technical information, this course conveys the real-world people skills needed to be a good systems analyst. The communication skills, diplomacy and other human issues are not easily transmitted in a book or in the classroom.

Students gain experience in handling ambiguous situations and in developing solutions subject to real-world constraints. Working with actual problems also develops analytical skills and problem-solving abilities, gives an understanding of the importance of organizational influences on systems design and the impact that a user-requested change has on the analysis and design process.

More specifically, the instructor lists some of the strengths of the program as follows:

> students gain a "more perspective on the systems analysis process"; students "learn to work cooperatively in groups"; students learn to "critique each other"; the task "enables students to integrate the knowledge gained from other business courses."

In comparison to case-study writing, the writing in this course is situated quite differently along the continuum that is mapped in Table 1. Like case-study writing, this course is situated in an institutional context, to the extent that it has a course number, it counts as credit toward a degree, and it is constrained by the time limits of a single semester. But there are important ways in which it is different, ways that are revealed especially when we think in terms of the writing's double social motive, one served by the rhetorical exigences of school work, the other by the exigences of the workplace.

Our analysis of student interactions, composing sessions, and drafts, shows how the systems analysis course offered an interesting hybrid of features, a kind of bridge across the gap between school and work. The audience was both instructor and client. (The instructor was always the first reader and her views, as evaluator, were always important. At the same time, the students always referred to their clients and to their clients' needs and goals in their composing sessions.)

One textual indication of the hybrid nature of this transaction was that students kept having to adjust the amount of shared knowledge they provided: as they went from presenting their documents to class and teacher to preparing the document for their clients, they needed to revise the amount and kind of shared information. A recurrent feature of their revision was the removal of information that they could assume to be shared with their clients.

The social motive of the writing, like that of the workplace, was to recommend real-world action. (On the other hand, an incidental or possibly auxiliary purpose

Table 1. Changing Aspects of Writing in Three Contexts

	Case-Study	School/Workplace	Internship
1. audience:	instructor: colleagues	client; instructor	supervisor;
2. social motive:	epistemic	epistemic; instrumental	epistemic; instrumental
3. reader's concern:	writer's knowing (value to student)	writer's knowing; value to client	value to collective
4. knowledge:	shared	chared and new	new
5. reader's goal:	evaluation	evaluation; ensure text	ensure best text; apply knowledge
6. reader's comments:	justify grade	revision-oriented; response	revision-oriented; response
7. closure:	grade	grade; indeterminate	indeterminate

was always kept in mind: an epistemic goal for the writing was recognized by both the instructor and the students in their discussions.) The whole exercise was undertaken in order to learn how to perform in a real-world setting, with real-world constraints. Contrast this with the response, cited in chapter 9, of officers in a government agency who said with some degree of irritation that their novice employees could "learn on their own time." (It's not that learning does not occur; it's that no one sees this as an institutional goal.)

Perhaps the most revealing difference was the changed role of the teacher. Rather than teacher as evaluator, the instructor functioned in the role of supervisor: she collaborated with the students extensively. She accompanied them to the first interview, and discussed the interview with them afterward, carefully pointing to key features and eliciting appropriate understanding. For each piece of writing, she both evaluated (gave a low grade at the beginning) and wrote extensive suggestions for revision, insisting that they revise until they produced the best possible text.

This points to a larger kind of difference. This writing that the students were doing in the systems analysis course was different from typical student writing in that the instructor regarded it as reflecting on her (and the institution) in much the same way that a subordinate's writing in the workplace reflects on the manager. This stands in sharp contrast to most instructors' attitude toward their students' writing, which is considered to reflect on the students' rather than the teachers' competence.

A related point, explored more fully in chapter 4, is that instructors have a vested interest in a quality spread; it is indeed important that there be some A's (as confirmation that the appropriate teaching has taken place), but it is equally important that there be many B's and C's. Something that underlies all university activity, and that so easily becomes invisible to us in its normalcy, is the institution's function as a gatekeeper. This function is not incidental, something that we must contend with from time to time and can ignore until the actual time for reporting grades; it is pervasive and powerful, shaping the dynamics of the rhetorical exigence in ineluctable fashion.

The writing we observed in the systems analysis course describes a midway point in the gulf between academic and workplace settings. Though based in the university and shaped by some of that institution's epistemic motives, the writing was very much like that of the workplace—in the social roles taken on by writers and readers, in its sense of audience, in its textual features (e.g., shared information, surplus of corroborating detail), and most markedly in the responding and collaborative reading practices of its first reader, the instructor, who served as a guide to the realities of workplace rhetoric, much as a supervising practitioner might. Finally, to return to our comparisons in Table 1, we have observed that case-study writers feel closure when the grade is given, and the text cease to have importance (and often physical existence) at that point. For the systems analysis course, the writing was graded as well, but its existence was not entirely shaped nor ended by that grade: the documents led a continued and indeterminate existence in the clients' workplace. Like written documents we had observed in other workplace settings, their completion indicated their entry into a larger arena where their continued physical existence as documents along with their potential for material consequences could only be guess at.

The systems analysis course provides a useful perspective into workplace university differences. At the same time, it points as well to other kinds of programs (not so directly university-based) that might allow for easier and more easily scaffolded transitions to authentic workplace writing context: sheltered co-ops, work-study programs, and internships like those described below.

LEARNING TO WRITE AGAIN: ATTENUATED AUTHENTIC PARTICIPATION

Another step away from the university typically locates student in an actual workplace, and involves them in what we have called attenuated authentic participation. During this stage, students work alone, in pairs, or in small groups with a single supervisor. In some cases, interns are evaluated both by university and workplace supervisors, but for the most part the work they do and the texts they author belong to the workplace community. The interns that we observed learning the genres appropriate to the government agencies and social work settings to which they were assigned went through processes that, in some ways, were fundamentally similar to those entailed in the university settings. Common to both kinds of learning was the fact that learning took place as a result of collaboration, in the widest sense, or shared social engagement, as well as the fact that this learning took place through the mediation of sociocultural tools (primarily, but not solely, linguistic signs). There were important differences, however, and these differences are all the more significant for being tacit and implicit, complicating the transition into the workplace, and requiring that students learn in ways quite different from the ways they normally experience in classrooms.

The attenuated authentic participation that characterizes initial workplace experiences includes certain essential components: real but limited tasks, timely assistance, a grasp of rhetorical purposes, and a big picture or overview of the community. In addition, such participation, based as it is in the workplace rather than in the university, must offer the newcomer a sense of membership or belonging and the opportunity to play multiple roles in a variety of workplace relationships. When all these factors are in place, the newcomer is drawn toward mature practice.

We have noted that the participation of newcomers influences the workplace community in ways at once more obvious and more subtle than do course-based work projects, such as the systems analysis course described above. We believe this is a defining characteristic of situated learning in the workplace, marking it as essentially different from school-based learning experiences. A frequent form of influence can be found in the interaction between newcomers and their workplace supervisors (the oldtimers who usually volunteer to work with students). Eileen, one such oldtimer, said: "I still enjoy the feedback from the students and the sharing. I don't really see it as a one-way. . . I feel I get a lot back in return." Other comments from supervisors:

> I also find that [supervision] helps me to keep fresh in a way. Students obviously see things in a new kind of way and when you've been working in the field for 30 years you tend to see things in one way, so that I find that it keeps me thinking, stimulated. (Risa)

> I told her [a social work intern] that I felt when we had joint interviews that I learned from her because it was a different perspective. Questions that she may have thought of asking I just haven't. And also when you're in practice for so many years, sometimes you get quite stuck and you handle interviews the same way. And she was fresh and new and just out --just in school, actually—and the theory is so down pat. (Eliza)

Newcomers also influence the workplace community as a whole. Whereas the work done as part of course-based projects, including student-authored reports, may well be ignored or little valued in the workplace, the work of interns generally contributes to the community's efforts. Newcomers bring new ideas and theories, new approaches and practices, and these can alter the community. When newcomer participation results in a contribution—a meaningful contribution, one that affects and changes the community—newcomers experience a sense of belonging, a sense of membership. This last point is critical. Along with Lave (1991), we see a developing sense of membership as essential to effective workplace learning: "Developing an identity as a member of a community and becoming knowledgeably skillful are part of the same process, with the former motivating, shaping, and giving meaning to the latter, which it subsmes" (p. 65).

The following scene, reconstructed from field notes, is authentic and captures this process of joining the community, as well as many other significant features of workplace learning. Any extended analysis and commentary is in square brackets. The setting is a Canadian government office; Douglas is the learner and Richard is his mentor or supervisor.

> Douglas and Richard are observed as they respond to a sudden request to prepare a briefing note on the state of a particular set of negotiations for a new government minister. [Political events such as the appointments of new ministers often interrupt the anticipated flow of business in Canadian government offices. Mentors or supervisors must improvise, if they are to include the learners in the new tasks. Both must be agile.]
>
> Douglas and Richard are standing in front of a desk which has a pile of previous briefing notes and reports on these negotiations. Their task is to develop a new briefing document, summarizing succinctly what the new minister needs to know.
>
> The two discuss the potential content in global terms, brainstorming on a whiteboard, and then they sit down to write. Richard suggests that they work collaboratively. Douglas understands this to mean dividing up the task in two, with each taking responsibility for one half. [Presumably this reflects his notion of collaboration, based on what passed for collaboration as it was undertaken in university.]
>
> Richard corrects this misconception, explaining that he means that they will actually produce the whole text together: the two of them sitting together to generate and compose test, with one person assigned to do the actual inputting. There is some joking and jockeying about who will do the inputting, but Richard decides that Douglas's superior expertise in word-processing (he can use Windows) warrants his taking the seat in front of the computer. [It is not untypical in the workplace for novices to display superior expertise in relevant skills.]
>
> The two proceed to formulate and reformulate text together, with Richard taking the lead and providing feedback to each of Douglas' suggestions, but at the same time constantly eliciting suggestions and listening carefully to Douglas' comments about his own sugges-

tions. The two respond to each other conversationally in a series of half-sentences, which reveals the highly interactive nature of the interchange. Each half-sentence responds to and builds on the previous, so that the product becomes more and more jointly generated. [This kind of interactive generating and composing between a guide and learner is hard to imagine in a school context, even in a tutoring center. The co-participation often reaches a flow at which it is difficult to determine who is suggesting which words.]

Complicating this interactiveness further is the interaction with already extant texts. Each suggestion for the new text is based in large part on the briefing notes and reports that are already available in the documents in front of them, with the words and phrases being modified, echoed, reaccentuated, qualified. [These earlier text are cultural artifacts, which have been shaped by, and encode, the cultural practices and choices of the organization as it has evolved to that point. In Bakhtin's terms, the words in the new evolving documents are being echoed from the earlier ones, and reaccentuated in the light of the current "speech plan," so that the words become reinfused with slightly different meanings. This dialogism and mediation through cultural artifacts is true of university writing and hence learning as well, but without the complicating factor of the intersubjective activities of guide and learner.]

What we are seeing here is persons-in-activity-with-the-world as mediated through the technological tools (word-processing software) and the cultural artifacts available. The newcomer joins the network of distributed cognition. There is a hermeneutic grappling with notions, and it is sometimes difficult to discover where one thinker's processes ends and the other's begin, and where the new speech plan begins and the influence of older cultural artifacts recedes.

In order to begin feeling like a member of the community, especially in large institutions, social work student interns must quickly gain a picture of the entire collective endeavor and their own place in it. This picture includes the intricate and subtle geography of place and politics: the physical and organizational structure of the community. In the complex world of a large general hospital, for example, this geography includes all of the areas of specialized care, the laboratories, the wards, the hierarchy of the disciplines and individuals that inhabit the institution, and the community's network of complementary and conflicting interests. Eliza, a hospital-based social work supervisor, explains: "I find for the first. . . month there's an enormous amount of time spent really just orienting [students] to a hospital situation because that in itself is a complex world." She continues: "What I do is have them go to as many staff meetings and as many of structured kinds of things that are going on on the unit as possible at the beginning, both staff and patient things, so that they're sort of inundated with people and things happening."

The key to this orientation process is total immersion in the community's activities. Initial learning happens while newcomers are simply observing, or "sitting by Nellie," as it was called when apprentices observed veteran factory hands in the early days of industrialization (Clews & Magnuson, 1996). According to a social work newcomer, Natasha, "my first two weeks was a lot of observation: attending rounds, being introduced to people, and a few meetings." Newcomers are not told what to look for or what to learn; instead, "they're sort of inundated with people and things happening." The questions that arise belong to the students, and the supervisors answer them when the students need answers.

But early in their field placements, most social work students become more than passive observers at meetings or interviews: they respond to telephone calls, research available community resources, speak to medical staff. In short, they are included in community life, rather than studying it from afar. The supervisors we have interviewed spent little or no time preparing students before sending them on errands, giving them small tasks, bringing them to meetings or other events, and asking them to observe elements of practice. Indeed, most had nothing that could be called a deliberate curriculum, or even a set sequence of activities that constituted a learning path.

It is within the workplace community's public settings and in contacts with individuals from other professions in the hospital that the social work newcomers begin to feel like members of their own social work community and of the larger institutional community. They begin to learn the rituals and dynamics of community practice, the subtle tensions of institutional power and influence, and—most importantly for our purposes—they begin to learn about writing. That is, they begin to learn how, where, what, when, and why language occurs in the multiprofessional context of the institution. They participate in the heteroglossia of hospital life, the many specialized discourses about patients. They begin to learn what doctors, nurses, therapists, lab technicians, and psychologists care about, and how their own concerns are in concert or conflict with others. They begin to learn who can speak, when they can speak, and what they can say. They begin to enter the complex social world in which social work texts must operate, the world described in chapter 6.

When supervisors sense that this process has begun, they increase the newcomer's participation by turning cases over to the student. This case load is authentic, because newcomers work with actual people who have real problems, and attenuated, because the cases are not critical or overly difficult and supervisors step in to do or help with some aspects. As we noted in the previous chapter, learning is incidental and occurs as part of participation in the community of practice (COP), whose activities are oriented toward practical or material outcomes. Instruction is secondary to, or a byproduct of, institutional activity. In a sense, there is a reversal of the traditional order of instruction; rather than preceding performance or application, as is usual in school, teaching in the workplace often occurs during or even after he accomplishment of new tasks. As Eileen, a supervisor, noted: "After we do things, particularly do things together, [we] sit down and try to dissect things and I tell him, When I did that, this is what I was thinking." For such teaching to be successful, the oldtimers who act as field educators must carefully engineer and monitor tasks.

As social work interns move out of the initial stage of orientation and observation, and take on the beginnings of a case load, they must begin to produce the regular and occasional documents required by mature practice. In most agencies and institutions, this includes referral forms, assessments, progress reports, transfer and closing summaries, and all the usual rhetorical flora and fauna of institutional life: memos, letters, government forms, and so on. According to Colin, a supervisor: "Students who come here . . . should be expected to write all the different reports. It shouldn't be that they're following these cases but don't take on the paper work side of it and only do the clinical side. They should take on. . . all the different dimensions of what it means to be a social worker."

Newcomers receive very little advance assistance with this sort of task, although they receive as much feedback as they need, when they need it. Some scaffolding is provided, in the form of a format and some detail, but the newcomer is left to write the report, which must then be cosigned by the supervisor (following a collaborative revision process described below). Interviews, information searches, telephone calls, and other information gathering is done by the newcomer, closely monitored by the supervisor. Much of what is learned in this process is not addressed explicitly. Although we have seen workplace settings where some discourse practices are codified as guidelines, most rhetorical lore is implicit: for example, perhaps no one uses first-person pronouns in certain documents, but that interdiction is nowhere in print and cannot be traced to an authority. A major means of passing along this sort of rhetorical regularity is the practice of sending newcomers to the files to look at previous examples of required documentation, usually under the guidance of a supervisor. A supervisor in a group home for adolescents, Colin, describes the place of files in the abrupt transition from school to the "real world":

> [Students would] be assigned anywhere from three to five cases within the first week or two. And then they'd be told to go and look at the file and to pull the file and to look through and they would see the different kinds of reports that they're expected to do in there. . . We basically try and get the students to handle the normal responsibilities of the social worker. . . I guess the way I see it, you have to. In some ways, this is a kind of experience to the students of what it's like in the real world.

Reliance on existing texts realizes some of the key criteria for a successful transition, both for the individual and for the community: the newcomer gets to model the text produced by oldtimers, thus emulating mature practice, and the community gets to reproduce itself, while being revitalized by the unique contributions of each newcomer. Again, no explicit discipline-wide, or even agency-wide, policy on training appears o be in effect, and yet supervisor after supervisor does the same thing: sends the student newcomers to the files. Supervisors are careful to exploit the newcomer's early struggles with aspects of practice, because they offer what teachers often call "a teachable moment": a fortuitous opportunity to offer guidance. Eliza, a supervisor, explains: "Well, I think my basic philosophy is that you, as much as possible, try to allow the student[s] to go with it on their own, to have as much freedom—unless they run into, or you feel they're going to run into, real snags, at which point you intervene, or help, or support, or do for, or whatever." Again, opportunities to write appear to be apt moments for introducing newcomers to the subtleties of community life and practice, and a form of "document cycling" (Paradis, Dobrin, & Miller, 1985; Smart, 1993) procides plenty of opportunities, as this excerpt from an interview with Natasha, a student, indicates:

> I would do my assessments, [the supervisor] would look at them and then give me her feedback and suggestions and, uh, areas that might be reworked or what have you. So it was essentially like that: [the supervisor] sort of like corrected me, "here's it back, and let's talk about it and discuss."

In the workplace, as Natasha points out, the emphasis is different: "Something that we highlighted as a goal, in our initial contract, was to look at my writing and improve my writing skills. You know, not so much the grammar and the spelling, but the skill of writing in the field." Certainly, the rhetorical sophistication described by Glenn, a student newcomer, in the following interview excerpt seems tied closely to particular workplace circumstances:

> I began to learn how to be more concise, more focused in my writing. My supervisor at the [agency] tried to teach me how to say a lot but not very much, meaning, you know, there are certain innuendoes to what I was saying, um, while protecting certain things of confidentiality between myself and my client.

To "say a lot but not very much" demonstrates a rhetorical skill that cannot be taught outside a context and practice in which those measurements make sense. What constitutes an "innuendo"? For whom will the text "say a lot" without saying "very much," and what is too much? What is the worker-client relationship, and how would it be affected by breaking the standards of confidentiality, some of which are universally prescribed by law and others which are local and largely tacit? These are questions that have no answer outside of particular circumstances, and the generalities about confidentiality that are offered in a school setting are likely to mean little in the field, where each worker-client relationship is different, and where the secrets of a persons' life are defined by who reads reports and for what purpose. These lessons, the subtleties of culture, are learned in the centripetal pull of authentic attenuated participation, as the newcomer gradually transforms into an oldtimer.

To sum up: typically, early in the newcomer's workplace experience, learning takes place through engagement in active processes, guidance by mentors, and mediation through cultural tools. In this respect, the learning is like that of the university setting. What is different is the nature of the interactive co-participation and collaboration between mentor and learner; the improvisatory nature of the task; its authenticity and ecological validity in a larger context (the institution and indeed society as a whole); and the varied and shifting roles played by mentor and learner. Furthermore, there is no deliberate attention paid to the learner" learning; all attention is directed to the task at hand and its successful completion.

LATER STAGES: LEGITIMATE PERIPHERAL PARTICIPATION

In a final stage between school and full mature practice, students in the workplace begin to work autonomously. Even in their initial forays into the field as undergraduates, social work students are often expected to take on a version of mature practice. Colin, a supervisor we interviewed put it this way: "I'm a believer in trying to engage a student immediately in terms of doing and participating in clinical work. So as soon as the student comes in, literally the first day of placement, I have two cases on my desk waiting for her. . . I think that the most important thing to start off with is the case right away." When social work students reach the senior undergraduate or graduate level and enter their final field placements, or when they complete their programs and enter the profession, their experience even more closely resembles traditional apprenticeship in professional practice. Guidance is decreased, responsibility increased, and the newcomer blends into the community. This is the stage of legitimate peripheral participation:

> Once I began, my supervisor said, "Here's a case," gave me some background and said, "Go ahead." To me that was great. . . I wanted to learn to. . . have my own style and this was a great way for me to do it. It was a learning experience and I have achieved my goal. So, for me, that autonomy was good for me. (Kate)

As we note in the previous chapter, Hanks (1991) describes legitimate peripheral participation as "an interactive process in which the apprentice engages by simultaneously performing in several roles—status subordinate, learning practitioner, sole responsible agent in minor parts of the performance, aspiring expert, and so forth—each implying a different sort of responsibility, a different set of role relations, and a different interactive involvement" (p. 23). These factors were conspicuously missing from the following negative experience, a field placement that did not engage the newcomer Spiro in legitimate peripheral participation:

> In my experience at [agency], I was essentially isolated from the team. There was no team to go to. Even though I was a "member" [gestures to indicate quotation marks] of the team, I was not officially interacting with any of the team. . . I think that to feel you're on side you need to be with colleagues and to exchange clinical impressions, to learn from others. . . with their vast experience, I think, is essential.
>
> Q: Did you do any writing?
>
> A: [Yes, but] it was very confusing because it wasn't clear as to what exactly they wanted. So I would get back, you know, I would submit [the report] to my supervisor and my supervisor would look at it and most of the parts were okay. There were some parts that just needed total revision. Why I couldn't know this beforehand is beyond me. . . the expectations of my field work, my written work, was I would say, generally unclear. So I think the demand is left up to trial and error, I suppose. . . But it never really clicks in until the time comes when you're actually writing it yourself and you know the family or you have a good impression of where the family is at.

Lave and Wenger (1991) argue that learning is "an evolving form of membership" (p. 53); learning is thwarted when the learner is isolated, as was the student quoted above. He wrote a report and even received guidance from a supervisor, but the writing was out of context, he was not a member of a community and therefore had no role or responsibility, and the activity involved no authentic interaction. Compare this to another internship experience (excerpts from two interviews with the same student, Raymond):

> The first two weeks I was completely disoriented. The hospital was very intimidating to me, I was lost. I didn't know where to go and I didn't really feel grounded at all, so I felt lost. . . But then when I got my first case, I was able to feel a little more oriented and more like I had a role here.
>
> [By] participating in rounds. . . I'm understanding I guess the culture of it all. I'm feeling a part of it more than an outsider. . . The culture, you know, it's a big system, and there's a culture that goes on; like in rounds talking about patients, and the language that's used. . . And I guess I feel validated in a way I guess, you know, that what I'm doing, it's not just superficial, on a superficial level. . . now I just feel more involved and more part of it.

Like Lave and Wenger (1991), we see this movement toward membership as critical to the workplace learning experience. Knowledge is inseparable from a sense of identity and a sense of location within a group; and knowledge-making is always a collaborative activity. Engagement in the activities that produce the group's specialized knowledge leads to membership in the group. Spiro, a social work student we cited earlier, describes the collaboration:

> Q: How are you learning to do [assessment reports]?
>
> A: There are several ways. . . assessment workers do the same thing: gathering your information; gathering your data from the file material; doing interviews with people that have been involved with the child—whether it's a child care worker, a social worker, you get a different perspective on the problems; team feedback, because I present the case and get feedback on it, and based on the feedback we try to address the problems or the issues; and individuals—consultation with [fellow student], for example.

Note the changes that occur in this student's account of learning how to write a regular agency document: initially, he reports what others do; then he switches to the generic "you"; next he presents the case and gets feedback; and, finally, the team ("we") addresses the problem. Consider, too, the extensive interaction: with other texts, with child care and social workers, with team members, and with a fellow student. There is a sense here of membership, of working with others in partnership. Spiro elaborates on his field experience:

> We would have assessment team meetings, and there were various aspects within the assessment process that needed some clarification. Besides, I did the first couple of them with [fellow student]. . . We basically shared a lot of ideas and that was a very interesting process, that was. And we would pull our supervisor aside, too, and bounce off ideas, and other colleagues. So we had resources. We could access our team. We could access our supervisor. . . I was put, and [fellow student] was put, in a situation where we have to respond. . . not only to the family but. . . to the Director of Assessment—what the requirements, policies, and things of that nature need to reflect. We have to respond to the Director of Professional Services. . . I think it's very important to respond to what the social worker needs to know and what the child care worker needs to know. So, we're writing basically to a family of professionals that need to know where to go with this case.

Here, unmistakably, is a newcomer engaged in legitimate peripheral participation. He is a member of a team, operating as an equal (but still under supervision), and he is situated within the network of actions, interests, and individuals that constitute this COP. His tasks are no longer attenuated, and though reports are still cosigned at this late stage, they are rarely much revised by supervisors. The student's focus is on achieving a goal, on doing something with his text, on performing mature practice.

CONCLUSION

Our observations of school and workplace learning and writing have made us cautious about making generalizations. Clearly, learning is profoundly situated, as is writing: learning to write in particular contexts is indistinguishable from learning to participate in the full range of actions that constitute the activity in those contexts. As we argues in the previous chapter, students who move from university to the workplace and fail to recognize the differences in learning context and conditions risk missing the point. However, as the discussion in this chapter makes clear, there are ways to help student newcomers experience the practitioner's rhetorical life and, in so doing, learn something about workplace writing.

A condition central to the experiences of professional writing described in this chapter was their authenticity, or virtual rhetorical reality: the students in the two workplace settings had a real impact, they actually influenced action. Another important and closely related condition was the primacy of instrumentality. The students obviously learned something from their experience, but the learning was secondary to, or a by-product of, the instrumental purpose of their writing. Perhaps most importantly, the students' experiences were carefully monitored, scaffolded, and controlled by oldtimers who took advantage of specific moments of uncertainty or confusion to offer relevant advice. It is this careful balancing between actual practice and timely instruction that we feel characterizes successful transitions into workplace writing.

DEVELOPING YOUR UNDERSTANDING

1. Staring with Table 1 as the basis for a visual, and then adding rows or additional visuals you think more effectively capture the contents of the article, prepare a 10-minute oral presentation for your class summarizing the transition from school to workplace writing.
2. Dias et al., claim that internship experiences require you to learn in quite different ways than classroom settings do. Summarize these differences, and based on them, develop learning strategies interns can employ to maximize their internship learning experiences.
3. Internships can sometimes decay into filing/copying temp-jobs, but interns can keep that from happening by taking on an active role in structuring their internship experiences. Identify some of the essential components of an effective internship, as mentioned in the article, and develop a set of suggestions to fellow classmates that can help them actively structure these essential components into future internships.
4. After an extended example where Dias et al. describe a newcomer's successful entrance into a Canadian government office, they summarize the point of the example by saying:

   The newcomer joins the network of distributed cognition. There is a hermeneutic grappling with notions, and it is sometimes difficult to discover where one thinker's processes end and the other's begin, and where the new speech plan begins and the influence of older cultural artifacts recedes.

   Put this into your own words and explain why their summary of the example points to a significant aspect in the transition from school to workplace writing.

Chapter 8
Projects

1. The most obvious way you can "write yourself into the field" of professional writing and rhetoric is to apply for an internship or job. If you have not had a writing internship and are not yet ready to graduate, now is the time to get one. It is best if you can get such experience early, giving yourself time to have more than one internship before graduating. Not only do you get more experience, but you also get a better sense of what kind of writing work you may want to pursue. If you are ready for the market, internship experience or not, it is important that you carefully search for the kind of position for which you are ready and that will help you move toward your career goals. (Remember, though, that you probably won't start right off writing features for *National Geographic* or managing the technical publications team for the next "hot" software package.)

   Locate ten or more internships or jobs appropriate for you. You can use common online job search sites, as well as specialized professional writing organizational sites, like the Society of Technical Communicators and the American Medical Writers Association. Once you have located ten or more appropriate position openings, focus in on three. Do some research on the organizations. Find out what you can about the demands of the particular positions for which you will be applying. Design a resume and cover letter, and complete whatever application processes are required for each of the three positions. Your course instructor will help guide you through the various processes required to complete this project.

2. One of the ways you can "write yourself into the field" of professional writing and rhetoric is to create a portfolio. Traditionally, you would put together a printed collection of your work, designed to illustrate the breadth and depth of your professional knowledge. In our current e-culture, though, it makes sense to create paperless folios, or e-folios to complement print-based portfolios. Print-based portfolios cannot be widely distributed without great cost, and they cannot effectively present electronic media projects, such as Web sites, multimedia presentations, online help, and databases. Creating an e-folio helps you overcome these print-based portfolio limitations.

   In order to create an e-folio, take time to find Web sites that display some samples. As you conduct this initial research, notice that portfolios are more than mere collections of people's work; they are documents that define who someone is as a professional. Furthermore, since an e-folio is itself a professional writing document, your portfolio will become an example of your work. As a result, you should carefully plan, design, execute, and maintain yours.

   Create your own e-folio. In doing so, you will be faced will a great deal of the questions and issues that have been addressed in this textbook. For instance, when thinking about the technological context, you will need to determine if a Web site or CD is better suited to your purposes. You will need to carefully consider the multiple audiences who might use your e-folio: what will their needs and purposes be, and what visual design will best suit your content and context? If you are constructing a Web site, you should think about the kind of social space you wish to create: do you want to create spaces where your users can become authors, say through forms or e-mail links, and what kinds of links can you incorporate that produce a network advantageous to you and your readers? You will also need to think carefully through traditional portfolio questions, such as what you will include in your portfolio and how you will categorize the content.

3. Professional writing is a profession, and its health as a profession requires that it have advocates within both academic and workplace institutions. You can take on the role of an advocate in many ways, but one such way is by constructing new opportunities for professional writers.

   Individually or in teams of three or four, find a local client who has intensive writing demands but who does not currently have a writing internship program. You may already have a few clients in mind, perhaps from previous work experience, experiences friends have had, or from a nonprofit organization with which you work. You can also use your instructor, instructors who direct internships in any discipline, or a service learning director to help you locate a client.

   Once you have located a client, work with that client to develop a plan for initiating a writing internship. Your course instructor will help you get a sense of what such a plan requires in terms of what work you will need to do (process) and what possible shapes the plan might take (product).

INDEX

Note to the reader: All text selection titles and author names are included in this index. Page ranges that follow selection index entries show the complete location of the selection. Boldfaced page ranges cover those selections in which key terms are called out by the author.